Lecture Notes in Artificial Intelligence   8556

Subseries of Lecture Notes in Computer Science

Petra Perner (Ed.)

Machine Learning and Data Mining in Pattern Recognition

10th International Conference, MLDM 2014
St. Petersburg, Russia, July 21-24, 2014
Proceedings

 Springer

Volume Editor

Petra Perner
Institute of Computer Vision and Applied Computer Sciences, IBaI
Kohlenstrasse 2
04107 Leipzig, Germany
E-mail: pperner@ibai-institut.de

ISSN 0302-9743   e-ISSN 1611-3349
ISBN 978-3-319-08978-2   e-ISBN 978-3-319-08979-9
DOI 10.1007/978-3-319-08979-9
Springer Cham Heidelberg New York Dordrecht London

Library of Congress Control Number: 2014942573

LNCS Sublibrary: SL 7 – Artificial Intelligence

Typesetting: Camera-ready by author, data conversion by Scientific Publishing Services, Chennai, India

Printed on acid-free paper

Springer is part of Springer Science+Business Media (www.springer.com)

Preface

The 10th event of the International Conference on Machine Learning and Data Mining MLDM was held in St. Petersburg, Russia (www.mldm.de) running under the umbrella of the Worldcongress "The Frontiers in Intelligent Data and Signal Analysis, DSA2014" (www.worldcongressdsa.com).

For this edition the Program Committee received 128 submissions. After the peer-review process, we accepted 58 high-quality papers for oral presentation from these 40 are included in this proceeding book. The topics range from theoretical topics for classification, clustering, association rule and pattern mining to specific data mining methods for the different multimedia data types such as image mining, text mining, video mining and web mining. Extended versions of selected papers will appear in the *International Journal Transactions on Machine Learning and Data Mining* (www.ibai-publishing.org/journal/mldm).

A tutorial on Data Mining, a tutorial on Case-Based Reasoning, a tutorial on Intelligent Image Interpretation and Computer Vision in Medicine, Biotechnology, Chemistry & Food Industry, a tutorial on Standardization in Immunofluorescence, and a tutorial on Big Data and Text Analysis were held before the conference.

We are pleased to give out the best paper award for MLDM for the fourth time this year. There are four announcements mentioned at www.mldm.de. The final decision was made by the Best Paper Award Committee based on the presentation by the authors and the discussion with the auditorium. The ceremony took place at the end of the conference. This prize is sponsored by ibai solutions (www.ibai-solutions.de), one of the leading companies in data mining for marketing, Web mining and E-Commerce.

The conference was rounded up by an outlook of new challenging topics in machine learning and data mining before the Best Paper Award Ceremony.

We would like to thank all reviewers for their highly professional work and their effort in reviewing the papers. We also thank members of Institute of Applied Computer Sciences, Leipzig, Germany (www.ibai-institut.de) who handled the conference as secretariat. We appreciate the help and understanding of the editorial staff at Springer Verlag, and in particular Alfred Hofmann, who supported the publication of these proceedings in LNAI series.

Last, but not least, we wish to thank all the speakers and participants who contributed to the success of the conference. See you in 2015 in Hamburg for the next Worldcongress (www.worldcongressdsa.com) "The Frontiers in Intelligent Data and Signal Analysis, DSA2015" that combines under its roof the following three events: International Conferences Machine Learning and Data Mining MLDM, the Industrial Conference on Data Mining ICDM, and the International Conference on Mass Data Analysis of Signals and Images in Medicine, Biotechnology, Chemistry and Food Industry MDA.

July 2014   Petra Perner

International Conference on Machine Learning and Data Mining, MLDM 2014

Chair

Petra Perner   IBaI Leipzig, Germany

Committee

Agnar Aamodt	NTNU, Norway
Sergey V. Ablameyko	Belarus State University, Belarus
Jacky Baltes	University of Manitoba, Canada
Patrick Bouthemy	Inria-VISTA, France
Michelangelo Ceci	University of Bari, Italy
Xiaoqing Ding	Tsinghua University, China
Christoph F. Eick	University of Houston, USA
Ana Fred	Technical University of Lisbon, Portugal
Giorgio Giacinto	University of Cagliari, Italy
Patrick Gros	Inria-Texmex, France
Makato Haraguchi	Hokkaido University Sapporo, Japan
Robert Haralick	City University of New York, USA
Atsushi Imiya	Chiba University, Japan
Robert J. Hilderman	University of Regina, Canada
Abraham Kandel	University of South Florida, USA
Dimitrios A. Karras	Chalkis Institute of Technology, Greece
Adam Krzyzak	Concordia University, Canada
Tao Li	Florida International University, USA
Brian Lovell	University of Queensland, Australia
Pierre-Francois Marteau	Universite de Bretagne Sud, France
Mariofanna Milanova	University of Arkansas at Little Rock, USA
Jussi Parkinnen	Monash University, USA
Roman Palenichka	Universite du Quebec en Outaouais, Canada
Thang V. Pham	Intelligent Systems Lab Amsterdam (ISLA), The Netherlands
Maria da Graca Pimentel	Universidade de Sao Paulo, Brazil
Ioannis Pitas	Aristotle University of Thessaloniki, Greece
Petia Radeva	Universitat Autonoma de Barcelona, Spain
Michael Richter	University of Calgary, Canada
Gabriella Sanniti di Baja	CNR, Italy
Linda Shapiro	University of Washington, USA
Sameer Singh	Loughborough University, UK
Harald Steck	Bell Laboratories, USA

Tamas Sziranyi   MTA-SZTAKI, Hungary
Amit Thombre   Tech Mahindra, India
Francesco Tortorella   Universita' degli Studi di Cassino, Italy
Alexander Ulanov   HP Labs, Russia
Patrick Wang   Northeastern University, USA

Table of Contents

Classification and Learning

Clustering

Outlier Detection

Hierarchical Classification

Time Series and Sequential Pattern Mining

Signal, Image, and Video Mining

Medical Data Mining

Social Media Mining

Data Mining in Education

Data Mining in Marketing

Text and Document Mining

Neural Networks

Applications in Geography

Software Engineering

Aspects of Data Mining

Efficient Error Setting for Subspace Miners

Eran Shaham[1], David Sarne[1], and Boaz Ben-Moshe[2]

[1] Dept. of Computer Science, Bar-Ilan University, Ramat-Gan, 52900 Israel
{erans,sarned}@macs.biu.ac.il
[2] Dept. of Computer Science, Ariel University, Ariel, 44837 Israel
benmo@ariel.ac.il

Abstract. A typical mining problem is the extraction of patterns from subspaces of multidimensional data. Such patterns, known as a biclusters, comprise subsets of objects that behave similarly across subsets of attributes, and may overlap each other, i.e., objects/attributes may belong to several patterns, or to none. For many miners, a key input parameter is the maximum allowed error used which greatly affects the quality, quantity and coherency of the mined clusters. As the error is dataset dependent, setting it demands either domain knowledge or some trial-and-error. The paper presents a new method for automatically setting the error to the value that maximizes the number of clusters mined. This error value is strongly correlated to the value for which performance scores are maximized. The correlation is extensively evaluated using six datasets, two mining algorithms, seven prevailing performance measures, and compared with five prior literature methods, demonstrating a substantial improvement in the mining score.

Keywords: Biclustering, Subspace Mining, Error Setting.

1 Introduction

In recent years we have witnessed an ever increasing availability of information associated with people, the environment, machines and their interactions. In order to benefit from such huge amounts of data, efficient data mining and analysis tools are required. Such tools can facilitate uncovering hidden regulatory mechanisms governing the data objects. Take, for example, location-aware devices (e.g., smartphone, GPS, RFID) that leave behind electronic trails. Many commercial services (e.g., Google Latitude, Microsoft GeoLife, Facebook Places and others) collect the data in order to maintain location-based social networks (LBSN). The LBSN are later used for personal marketing purposes, which are based on finding similar patterns of behavior in other users.

A typical mining problem is the extraction of patterns from subspaces of multidimensional data. Such patterns would usually be noisy and overlap with each other, i.e., objects/attributes may belong to several patterns, or to none. A common way to represent a dataset is by a matrix of values, where the rows represent

P. Perner (Ed.): MLDM 2014, LNAI 8556, pp. 1–15, 2014.

objects, the columns represent attributes, and the data entries are the measurements of the objects over the attributes [12]. Early mining techniques used the key concept of mining patterns (clusters) formed by a subset of the objects over *all* attributes, or vice versa [12,4]. Seminal work by Cheng and Church [8] in the area of gene expression data, led to focusing on mining biclusters: a *subset* of objects that exhibit similar behavior across a *subset* of attributes, or vice versa. This problem, which for the most interesting cases was proved to be NP-complete [22], has attracted the attention of many researchers. This is mostly due to its extensive applicability, yielding various bicluster mining algorithms (miners) [12,16,20,24].

One key parameter that affects the ability to mine effective biclusters is the maximum allowed error. This parameter is commonly used in models of biclusters. Its primary goal is to deal with the inherent noise that characterizes real-life datasets (e.g., human or machine error), allowing some degree of error within the mined patterns. Taking a maximum error of zero will lead to biclusters with exact matching (mostly used in string matching), while taking a maximum error of 100% will encapsulate the entire dataset (i.e., contain all objects and attributes) which is futile in most cases. As such, this is the main parameter through which the user of a mining algorithm can control the quality of the mined clusters. Thus, correctly setting the maximum allowed error is of high importance. Since error is a feature of the dataset itself, the preferred value to be used when mining biclusters is dataset dependent [1,7], and testing the fit of a suggested value requires domain knowledge. The problem of setting the maximum allowed error is particularly challenging as in most cases the process is unsupervised. This means that the user is interested in finding the patterns in the data, but cannot a priori estimate whether a mined cluster is indeed the result of a regulatory mechanism in the data or a random effect.

Despite the great importance of this parameter, very little attention has been given to the way the error should be set for bicluster mining. Most prior work assumes the miner's user is capable of evaluating the mined clusters (e.g., has extensive domain knowledge or a way of validating the mined patterns) and thus is capable of suggesting trial-and-error methods for setting the error [15,34]. Other prior work suggests rules-of-thumb, such as setting the error value to 5% of the dataset range [19] or using the smallest value possible which still yields patterns [7]. Very few authors have suggested heuristic methods for setting the maximum allowed error [26,33]. Heuristic methods, as explained in detail in the next section, rely on an exogenous parameter thus posing a different problem of how to set the value of these new parameters.

In this paper we suggest an unsupervised method for automatically setting the maximum allowed error for bicluster miners. The method suggests using the *error value* which maximizes the number of the mined clusters. We evaluated our method extensively using six datasets of different domains, two state-of-the-art bicluster mining algorithms, and seven prevailing performance measures that evaluate the relevance of the mined biclusters. The results show that our method scores close to the maximum value for each of the above configurations.

In addition, we evaluated our method in comparison to five error-setting methods that have been proposed in prior literature. The results suggest that in the majority of cases, our method performs substantially better in terms of the performance measures, while in a small number of cases it results in a slight degradation. Overall, our method successfully releases the user from the task of setting the error parameter value, enabling effective mining even when the dataset properties are not fully known.

The remainder of the paper is structured as follows. The next section formally introduces and reviews the biclustering model, different approaches to mining biclusters (miners), the role of the error parameter value in the process, and prior approaches for setting the value of that parameter. Section 3 presents the main hypothesis and the reasoning regarding the correlation between the performance measure score and the number of clusters mined, as a function of the error value. Section 4 presents experimental design, including a detailed description of the datasets that were used, the main measures and their interpretation, and the mining algorithms with which biclusters were mined. Section 5 presents the results for validating the main hypothesis and their analysis. Subsection 5.1 presents the performance achieved for the various performance measures. The results of the comparison with prior methods for setting the maximum error value are given in Subsection 5.2. Finally, conclusions and directions for future research are outlined in Section 6. Related work is cited whenever applicable throughout the paper.

2   Models, Error-Setting Methods, and Performance Measures

Bicluster mining belongs to a class of problems where the goal is the extraction of patterns from subspaces of multidimensional data, where the rows represent objects, the columns represent attributes and the data entries are the measurements of the objects over the attributes [12]. Early mining techniques sought for a fully dimensional cluster: a subset of the objects over *all* attributes (or vice versa) exhibiting such a pattern [12]. However, when mining high-dimensional datasets, these methods become ineffective and inaccurate due to the "curse of dimensionality" [5].

Cheng and Church [8], in their seminal paper in the field of gene expression data, introduced a novel technique for mining clusters which focuses on mining **biclusters**: a *subset* of the objects over a *subset* of the attributes. Their approach was soon followed by many researchers (see surveys by [3,12,16]), applying a variety of mining techniques: greedy, projected clustering, exhaustive enumeration, spectral analysis, CTWC, bayesian networks, etc. The model they used, which is still the most common, is defined as follows.

Definition 1. *Let A be an $m \times n$ real number matrix of O objects ($|O| = m$) and T attributes ($|T|=n$). A bicluster of A, denoted (I,J), is a sub-matrix A_{IJ}, where I is a subset of the rows ($I \subseteq O$) and J is a subset of the columns ($J \subseteq T$).*

The desired bicluster is one which the sub-matrix elements exhibit some pre-defined pattern. As real world datasets are generally noisy, mining algorithms usually allow for some deviation from the "perfect pattern". For this purpose, the biclustering models are usually extended to include some similarity measure, which measures the deviation of the mined biclusters from the "perfect pattern", i.e., the "error" [6]. Commonly, the degree of deviation is constrained by a maximum allowed value, namely, the similarity measure score of the mined patterns would be lower than some pre-defined error $\delta \geq 0$. The choice of one similarity measure over the other would usually affect the nature of the mined clusters [31]. Next, we review several prevalent similarity measures found in literature.

2.1   Similarity Measures

Due to the "curse of dimensionality" [5], well-known distance functions (e.g., Euclidean distance, Manhattan distance, cosine distance and others) are commonly not adequate as similarity measures when mining multi-dimensional data [5,32]. To overcome this shortcoming, new similarity measures have been developed over the years. Among these, the two most commonly used are the *mean squared residue* and *pCluster*.

Mean Squared Residue. The *mean squared residue* measure was first used in the field of gene expression analysis by Cheng and Church [8] to assess the correlation between genes and conditions within a bicluster, and was soon adopted by many researchers [1,6]. The measure is defined as follows.

$$H(I, J) = \frac{1}{|I||J|} \sum_{i \in I, j \in J} (a_{ij} - a_{iJ} - a_{Ij} + a_{IJ})^2 \tag{1}$$

where a_{ij} is the element corresponding to the ith row and the jth columns of A and:

$$a_{iJ} = \frac{1}{|J|} \sum_{j \in J} a_{ij}, \; a_{Ij} = \frac{1}{|I|} \sum_{i \in I} a_{ij}, \; a_{IJ} = \frac{1}{|I||J|} \sum_{i \in I, j \in J} a_{ij},$$

are the row and column means and the mean of the submatrix (I,J), respectively. A sub-matrix A_{IJ} is called a δ-bicluster if the value of $H(I, J)$ is lower than some error value, i.e., $H(I, J) \leq \delta$ for an error $\delta \geq 0$.

The mean squared residue measure suffers from the following shortcomings: (i) the score of a bicluster can be smaller than the score of a sub-matrix of this bicluster. Therefore, a sub-cluster of a bicluster is not necessarily a bicluster [19,32,34]; (ii) the score is insensitive to outliers, i.e., biclusters may differ only by the few outliers they contain [19,32]; (iii) the score is biased toward flat biclusters with low row variance. This may result, in the domain of gene expression, in mining of biclusters which are less interesting as they contain genes with relatively unfluctuating expression levels [6]; and (iv) the relation of the score to a biological regulatory model is unclear [1,19]. These weaknesses have led in recent years to the emergence of an alternative measure known as *pCluster*.

pCluster. The *pCluster* similarity measure [23,32,34,36] is a score defined over a sub-matrix of 2×2. The concept is to find small sub-matrices, each of which is a δ-bicluster, to serve as building-blocks. Such building-blocks would then be assembled into a larger δ-bicluster. The definition of the measure is given as follows. Let $p, q \in I$ and $j, k \in J$, then:

$$pScore\left(\begin{bmatrix} A_{pj} & A_{pk} \\ A_{qj} & A_{qk} \end{bmatrix}\right) = |(A_{pj} - A_{qj}) - (A_{pk} - A_{qk})|. \tag{2}$$

The sub-matrix A_{IJ} forms a δ-bicluster if for any 2×2 sub-matrix X in A_{IJ}, we have $pScore(X) \leq \delta$ for some error $\delta \geq 0$. Alternatively, for every $p, q \in I$ and for every $j, k \in J$:

$$|(A_{pj} - A_{qj}) - (A_{pk} - A_{qk})| \leq \delta. \tag{3}$$

Intuitively, the meaning of the pScore measure is that the changes of values in the sub-matrix are limited by δ. It is worth noting that the pScore metric can be measured either horizontally or vertically as the inequality is equivalent:

$$|(A_{pj} - A_{qj}) - (A_{pk} - A_{qk})| \tag{4}$$

$$= |(A_{pj} - A_{pk}) - (A_{qj} - A_{qk})| \tag{5}$$

$$= \left(\begin{bmatrix} A_{pj} & A_{qj} \\ A_{pk} & A_{qk} \end{bmatrix}\right). \tag{6}$$

Over the years, several enhancements have been offered to the δ-bicluster definition. These extend the *pCluster* error model by detecting *larger* δ-bicluster building blocks. Guan et al. [10] use the following δ-bicluster definition. Let k be an arbitrary attribute ($k \in J$). The sub-matrix A_{IJ} is a δ-bicluster if for every $p, q \in I$:

$$\max_{j \in J}(A_{pj} - A_{pk}) - \min_{j \in J}(A_{qj} - A_{qk}) \leq \delta/2. \tag{7}$$

In a similar manner, a sub-matrix A_{IJ} is defined by Wang et al. [31] (see example in Fig. 1, based on [31]) to be a δ-bicluster if for every $p, q \in I$:

$$\forall j \in J, \ |(A_{pj} - A_{qj}) - (A_{pk} - A_{qk})| \leq \delta/2. \tag{8}$$

Fig. 1 depicts the reasoning behind the above inequality. Let $J = \{k, b, c, d, e, f\}$. The sub-matrix A_{IJ} would be a δ-bicluster if for every $p, q \in I$, the distance between p and q, with respect to dimension k, is $\leq \delta/2$. The distance between p and q on each column $j \in J$ would be $\leq \Delta \pm \delta/2$, where Δ is the distance between p and q on column $k \in J$.

Procopiuc et al. [26] and Califano et al. [7] use the following δ-bicluster definition. A sub-matrix A_{IJ} is a δ-bicluster if for every $p, q \in I$:

$$\forall j \in J, \ \max_{p \in I}(A_{pj}) - \min_{q \in I}(A_{qj}) \leq \delta/2. \tag{9}$$

Liu et al. [15] and Melkman et al. [19] define a sub-matrix A_{IJ} to be a δ-bicluster if for every $p, q \in I$:

$$\forall j \in J, \ |A_{pj} - A_{qj}| \leq \delta/2. \tag{10}$$

The above definitions of the *pCluster* error model are equivalent. A formal proof is available at supporting webpage [29].

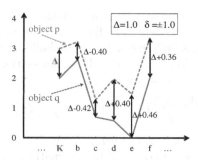

Fig. 1. An illustration of Wang et al. similarity measure [31]. For $\Delta=1$ and $\delta=1$: $\forall j \in J$, $|(A_{pj} - A_{qj}) - (A_{pk} - A_{qk})| \leq \delta/2$.

2.2 Existing Methods for Error Setting

One key parameter bicluster mining algorithms use, which has a critical effect over the results obtained is the *error*. The mining algorithm would supposedly report all biclusters which comply with the input error. Setting the error too low would usually yield flat (e.g., constant value) biclusters which are of an insignificant contribution. Setting the error too high might result in a single futile bicluster covering the entire dataset. The best error is usually dataset dependent [1,7], hence setting it correctly without prior domain knowledge is not trivial.

Many papers suggest a trial-and-error method for setting the error. This is usually done by testing a variety of error values and choosing the one which presents superior results [15,34]. As the mining of biclusters is usually an unsupervised process, such a way of evaluating the results is commonly unavailable. As a result, it is common to encounter different authors using different errors for the same dataset. For example, using a similar model of error, for the yeast dataset [8] containing 8,224 yeast genes (objects) and 17 conditions (attributes), Liu et al. [15] found that the best error to use is $\delta=0.1$, Yoon et al. [34] found an error of $\delta=1.0$ to be preferable, while Guan et al. [10] found the error of $\delta=10.0$ to be superior.

Melkman et al. [19] suggest a rule-of-thumb for setting the error to a value of 5% of the dataset range. Califano et al. [7] recommend setting the error to the smallest value possible which still yields patterns.

Procopiuc et al. [26] suggest the following technique. Let p^i be a data point and q^i its nearest neighbor ($p^i, q^i \in O$, $1 \leq i \leq m = |O|$). Let $\delta^i = \frac{1}{n} \sum_{j=1}^{n} |p^i_j - q^i_j|$ be the average distance between p^i and q^i. Then, choose the value of the error to be $\delta = C \cdot \frac{1}{m} \sum_{i=1}^{m} \delta^i$ for some small constant C. The drawbacks of this technique are: (i) the calculation of the average distance between the points p^i and q^i uses all attributes. The inclusion of noisy irrelevant attributes would result in an inaccurate measure; (ii) using all attributes is futile as the distance between any two multi-dimensional points tends to be the same [5]; (iii) the heuristic specifies the use of an unknown constant C, of which the user has no knowledge regarding its setting; and (iv) the constant C is arbitrary, i.e., forcing the user to specify the unknown value of C only transforms the problem of setting the unknown error δ into that of setting the unknown value C.

Yiu et al. [33] present the following technique. For each dimension j ($j \in T$), sort the values of A_{ij} ($i \in O$). Then, slide a window of size $\alpha|O|$, $0.0 \leq \alpha \leq 1.0$, along the values of the sorted set. For each position of the window, compute the value difference between the first and the last element of the window. Let δ^j be the average of the differences over all window positions. Then, set the error as the average over all δ^j, i.e., $\delta = \frac{1}{n}\sum_{j=1}^{n}\delta^j$. The disadvantage of the technique lies in its use of the unknown parameter α, which sets the size of the sliding window. In order for the user to set the unknown α, some rule-of-thumb or trial-and-error is needed.

2.3 Performance Measures

In order to evaluate the success of the bicluster algorithms, with the error set according to the different approaches, we use external performance measures. These measures characterize the quality, accuracy and relevance of the mined patterns (see survey in [9,11,21,27]). Among the main properties emphasized by these measures are: (i) object criterion – a mined cluster should only obtain the cluster's objects; (ii) attribute criterion – a mined cluster should only obtain the cluster's attributes; (iii) redundancy criterion – a cluster should only be identified and reported once (i.e., not as part of several sub-clusters); and (iv) identification criterion – all clusters should be identified.

To evaluate our method, we use the following seven popular performance measures: (i) F-score [30]; (ii) Entropy [37]; (iii) Purity [37]; (iv) Variation of Information – VI [18]; (v) Q_0 [9]; (vi) Misclassification Index – MI [35]; and (vii) Normalized Mutual Information – NMI [17]. For the computation of the above measures, we used the publicly available *ClusterEvaluator* package [27].

3   Main Hypothesis

Our main hypothesis is that *there is a strong correlation between the error values for which the number of clusters mined reaches its maximum and the error values for which the performance measure score reaches its maximum. In particular, the two reach their maximum for a similar error.*

The existence of the correlation enables the automatic setting of the error in a user-free, simple fashion manner, which does not require any specific domain knowledge nor any actual reasoning about the content of the mined clusters. In order to set the error, one should find the error in which the number of the mined clusters reaches its peak. To find that peak, one can start from a seed of zero error and adopt any of the various algorithms for finding a maximum [25], such as simplex method (also called 'amoeba' or the Nelder-Mead method) [25].[1]

---

[1] The simplex method is known for its simplicity, as it makes almost no assumptions regarding the function, and does not require any derivative evaluation. Furthermore, in our case of a single variable function, not only does this method converge, but it does so while using a relatively small number of function evaluations [14].

The reasoning behind the correlation is given within the scope of precision, recall and F-score.[2] When supplied with a maximum allowed error, mining algorithms would usually produce *many* biclusters, all associated with an error lower than the maximum allowed. The increase in the allowed error has two conflicting effects related to the number of clusters mined.

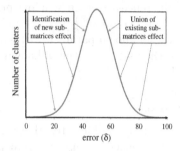

Fig. 2. Conflicting effects governing the number of mined clusters

The first is the identification of new sub-matrices associated with an error that was prohibited by lower maximum error (i.e., an increase in the number of clusters mined; see left side of Fig. 2). The second is the union of sub-matrices that were considered separate clusters for lower maximum error, but can now be considered as one cluster associated with an error below the maximum allowed error (i.e., a decrease in the number of clusters mined; see right side of Fig. 2). When the maximum allowed error is set to a low value, miners will mine a small number of trivial flat (e.g., constant value) biclusters. In this case, when the maximum allowed error is increased, the first effect dominates the second, as the maximum error is small enough to preclude union of clusters; however, any increase in the allowed error uncovers many new subsets of the "real" clusters as legitimate clusters. As the maximum allowed error keeps increasing, the second effect becomes more dominant, as the number of clusters that can be unified due to the increase in the allowed error grows exponentially. When the error reaches its maximum value, only one cluster is mined, i.e., the cluster which contains the entire dataset.

The effect of this behavior on the F-score is as follows. For low error values, where the number of generated clusters is relatively small, the F-score is low, mainly due to the recall measure – the mined clusters are indeed precise as the tight error precludes the inclusion of irrelevant elements. However, since these clusters are mostly subsets of the "real" clusters, the recall measure suffers. As the number of mined clusters increases, the recall measure improves. Nevertheless, once the effect of joining different sub-matrices due to the increase in the

---

[2] The typical use of this score is through its harmonic mean, which is the weighted harmonic mean of the precision and recall, and is defined as follows: $F_1 = 2 \cdot (precision \cdot recall)/(precision + recall)$. In terms of Type-I and type-II errors it is defined as: $F_1 = (2 \cdot true\ positives)/(2 \cdot true\ positives + false\ negatives + false\ positives)$.

allowed error becomes more apparent (i.e., when the number of clusters is on the decrease) the precision measure drops sharply, resulting in a total decrease in the F-score despite the marginal improvement in the recall. Therefore, when the number of mined clusters reaches its peak, it is likely that the mining using that error offers an effective tradeoff between precision and recall, in the sense that each of the mined clusters contains most elements of the "real" cluster it represents and does not include many elements which are either noise or belong to other "real" clusters.

4   Experimental Methodology

In the following paragraphs we describe the methodology of the experiments used to demonstrate the existence of the correlation hypothesized in the previous section.

4.1   Datasets

For our experiments, we used six published datasets, comprising a varied set of examples which differ in dimension, domain, number of classes, balanced vs. unbalanced classes (membership-wise), and natures (i.e., temporal vs. spatial). Each of the datasets contains labeled objects, which are used to evaluate (post factum) the mined clusters. The datasets are freely available (see the supporting webpage [29]).

The *Sonar* dataset [2] comprises 208 objects of sonar signals bounced off two object classes: metal cylinders (53%) and cylindrical rocks (47%). The dataset comprises 60 time readings as attributes. The *Mixed Bag* dataset [13] comprises the contour lines of 160 objects of nine shape classes: bone (12.5%), glass (12.5%), cup (12.5%), device (12.5%), fork (12.5%), pencil (12.5%), hand (7%), rabbit (7.5%), and tool (10.5%). The dataset comprises 1614 contour sampling as attributes. The *Waveform* dataset [2] comprises the synthetic generated waves of 5000 objects of three functions: function #1 (33.2%), function #2 (32.9%), and function #3 (33.9%). The dataset comprises 40 time readings as attributes. The *Arrhythmia* dataset [2] comprises the cardiac arrhythmias of 452 objects of 13 classes: normal (54%), ischemic (10%), anterior myocardial (3%), inferior myocardial (3%), tachycardia (3%), bradycardia (5.5%), PVC (1%), supraventricular (0.5%), left bundle (2%), right bundle (11%), ventricular (1%), fibrillation (1%), and others (5%). The dataset comprises 279 attributes, some categorial (e.g., Sex) and others temporal (e.g., time readings). The *Shapes* dataset [13] comprises the contour lines of shapes randomly selected from five datasets of objects: arrowhead, butterfly, fish, seashell, and shields (117 randomly selected objects from each of the five datasets). The dataset comprises 1024 contour sampling as attributes. The *Skylines* dataset [28] comprises skylines as viewed by various random observers on a topographic map. The dataset comprises 10 classes, each with 20 objects (observers), and 720 z-blocking angles as attributes.

4.2 Mining Algorithms

For the reasons considered in Section 2.1 we preferred the *pCluster* measure as the similarity measure to be used by the mining algorithms employed in our experiments. Consequently, our experiments used the following two widespread biclustering algorithms: *pCluster* [32], and *DOC* [26]. We chose these two biclustering algorithms as representatives of two different mining approaches.

The *pCluster* algorithm is deterministic in the sense that it discovers all qualified biclusters. The algorithm uses a depth-first search to simultaneously mine multiple biclusters. As the problem is NP-complete, the run-time of the algorithm may be exponential (although pruning techniques used in the early stages of the algorithm greatly reduce the overall complexity in practice). The *pCluster* algorithm executable was obtained from the author's website [32].

The *DOC* algorithm is probabilistic in the sense that it discovers approximations of the qualified biclusters. The algorithm uses a projected clustering approach. The use of the Monte-Carlo technique allows, with high probability, the mining of 2-approximated optimal biclusters. The algorithm has a polynomial time complexity. We have implemented the *DOC* algorithm ourselves based on [26].[3]

Both algorithms require, in addition to the maximum allowed error, a density criteria in the form of a minimum number of objects and a minimum number of attributes. A valid bicluster thus will need to have at least the specified number of rows and columns.

4.3 Experimental Design

The experiments were designed in the following way. Each of the mining algorithms (see Subsection 4.2) was run on each of the datasets (see Subsection 4.1) while setting the error to a variety of values. In order not to miss small biclusters (membership-wise), the minimum number of objects was set to two. To reduce the chance of mining artifacts, the minimum number of attributes was set to 5% of the total [19]. For the *DOC* algorithm [26], the number of iterations used for each error value was set to 10000. Due to the probabilistic nature of the *DOC* algorithm, and to reduce the effect of randomness on the results, we repeated the experiment with *DOC* and averaged the results of each error configuration for 10 trials.

For each of the above configurations, we recorded the mined biclusters, and computed the corresponding score according to each of the performance measures (see Subsection 2.3). Based on the common practice with mining algorithms [4], each bicluster was assigned to the class which is most frequent in the cluster (i.e., plurality, relative majority). In the cases where a cluster can be attributed to several classes (i.e., no plurality arises), it was randomly ascribed to one of the classes. For each class and for each performance measure, we choose the bicluster

---

[3] The implementation was of the *DOC* algorithm and not of the *FastDOC* algorithm as the probabilistic guarantee of mining the optimal biclusters was more important than the reduction in the run-time complexity.

which obtained the best score. For each performance measure, the score of a run (a tested error and a given dataset) was taken to be the average score over all classes.

5   Results

Next, we present the results of the experiments. Subsection 5.1 summarize the performance achieved by the proposed error-setting heuristic, presenting persistent high scores across various configurations. Subsection 5.2 presents a comparison of the results obtained with our method and those obtained by setting the error according to methods suggested in prior literature, demonstrating a substantial improvement in the mining score.

5.1   Heuristic Performance

Table 1 and Table 2 present the performance achieved by the proposed error-setting heuristic when using the DOC and $pCluster$ miners, respectively. Each result represents the ratio between the value obtained according to the specific measure when using the number-of-clusters-maximizing error and the maximum value that can be achieved according to that measure.

Table 1. Scores Obtained for DOC Miner

Dataset	MI	Entropy	VI	Purity	Q_0	F_1	NMI
Sonar	86%	100%	95%	85%	94%	99%	99%
Mixed Bag	100%	99%	100%	94%	95%	74%	72%
Waveform	94%	100%	100%	87%	97%	96%	100%
Arrhythmia	100%	77%	45%	69%	43%	94%	94%
Shapes	99%	100%	95%	97%	99%	95%	100%
Skylines	100%	93%	99%	100%	87%	86%	74%
average	96%	95%	89%	89%	86%	91%	90%

Tables 1 and 2 show that, other than four exceptions (the VI and Q_0 scores for the Arrhythmia dataset with the DOC miner and the $Purity$ and F_1 scores for the Skylines and Shapes dataset, respectively, with the $pCluster$ miner), setting the error to be equal to the number-of-clusters-maximizing error results in performance close to that achieved with the optimal error, for all the additional measures and with both miners. Furthermore, the tables show that even in those very few cases where a poor F-score was obtained when using the number-of-clusters-maximizing error (i.e., in the case of the Mixed Bag dataset with the DOC miner and Shapes dataset with the $pCluster$ miner) the performance results according to the other measures are very encouraging, indicating that

Table 2. Scores Obtained for *pCluster* Miner

Dataset	MI	Entropy	VI	Purity	Q_0	F_1	NMI
Sonar	100%	100%	100%	100%	96%	99%	100%
Mixed Bag	93%	99%	91%	98%	92%	99%	98%
Waveform	92%	100%	100%	95%	100%	100%	100%
Arrhythmia	100%	76%	100%	86%	89%	91%	100%
Shapes	93%	100%	85%	83%	96%	40%	84%
Skylines	90%	100%	82%	46%	98%	89%	71%
average	95%	96%	93%	85%	95%	86%	92%

the appropriate clusters were mined and that the low score according to the specific measure is probably a result of a unique inherent feature of the dataset.

Obviously, in order to get into the roots of how the number of mined clusters are correlated with any of these measures one needs to investigate how their different components are influenced, similarly to the way the effect over precision and recall was analyzed for the F-score in Section 3. While this is beyond the scope of the current paper, we stress that overall all these measures emphasize properties that are likely to be found within the optimal set of clusters and as such have strong correlation with the precision and recall. The fact that the same qualitatively positive results were obtained with all these additional measures substantially strengthen our confidence in the appropriateness of using the number-of-clusters-maximizing error as a default error value for biclustering miners.

5.2   Comparison with Alternative Methods

In order to compare the performance of our method with the ones used in literature we conducted the following experiment. We ran the two miners (*DOC* and *pCluster*; see Subsection 4.2) across six datasets (see Subsection 4.1). For each pair, we calculated the ratio between the F-score achieved when using our method and the one achieved when using any of the five alternative methods (see Subsection 2.2). For example, a ratio of 2 means that our method achieved twice the percentage of the maximum F-score than the alternative technique, while a ratio of 0.9 means that our method achieved 90% of the score achieved by the alternative technique.

To ease readability, we present only the results of the performance comparison between our method and the one of Yiu et al. [33], using the *DOC* miner, each of the six datasets, and various settings for the *unknown* $0.0 \le \alpha \le 1.0$ parameter (Table 3). The results for all other configurations (including the ones of the *pCluster* miner) are even better, and are given in details in the supporting webpage [29].

The important finding of Table 3 is that for most value settings, our method surpassed the alternative technique. Furthermore, in the cases in which it did

Table 3. Suggested Method vs. Yiu et al. Method

Dataset \ α	0.1	0.2	0.3	0.4	0.5	0.6	0.7	0.8	0.9	1.0
Sonar	2.00	1.14	1.01	1.00	1.01	1.02	1.04	1.04	1.84	1.08
Shapes	3.38	1.65	1.17	1.00	0.96	0.95	0.98	1.05	1.25	2.11
Arrhythmia	1.67	1.67	1.67	1.67	1.67	1.67	1.67	1.67	1.67	1.67
Waveform	9.64	5.27	1.76	1.15	1.00	0.96	0.96	0.98	1.02	1.42
Mixed Bag	1.12	0.80	0.76	0.74	0.76	0.79	0.88	1.05	1.65	2.30
Skylines	1.01	0.86	0.86	0.89	1.00	1.00	1.14	1.19	1.31	2.41

not, the achieved score was still high (88% on average) in comparison to the maximum score achieved by the alternative technique. The fact that a similar pattern was obtained for all other configurations further strengthens the results, and corroborates that the use of our method is preferable: not only is the user free from setting an *unknown* parameter, but also, in most cases, the achieved score is higher.

6   Discussion and Conclusions

The importance of mining clusters is unquestionable and has been thoroughly discussed and demonstrated in cited prior work. Similarly, the extensive biclustering literature includes many examples of the benefit of mining biclusters as opposed to traditional approaches. In order to identify a regulatory mechanism within a noisy dataset, a bicluster miner must rely on some degree of error. Setting the error parameter properly is crucial for the mining of coherent biclusters.

The paper presents a hypothesis regarding the correlation between the number of clusters and the errors for which performance scores are maximized. This hypothesis, which has important implications for setting the error parameter, is tested using extensive experimentation that spreads over six datasets, two state-of-the-art bicluster mining algorithms, and evaluated by seven prevailing performance measures. The encouraging results reported in Section 5 experimentally validate the correlation hypothesized. This enables setting the error parameter value automatically in an unsupervised, relatively simple manner. One needs only to follow the number of clusters mined for each error value and pick the one for which the number of mined clusters reaches its peak. As emphasized throughout the paper, this does not require any specific domain knowledge nor any actual reasoning about the content of the mined clusters. Furthermore, to find that peak, one can use various algorithms, as discussed in Section 3. The findings reported are of particular importance in light of the relatively poor performance of existing methods for setting the error.

The new method does not always result in the best error value; however, the performance achieved is very close to that of the performance-maximizing error. This is demonstrated with a set of measures that have been used in prior literature for evaluating the relevance of mined biclusters. Furthermore, the comparison with five existing error-setting methods reveals that, in the majority of cases, the new method results in substantially better performance. For a small number of cases it performs slightly worse than the competitor. These results were obtained for both algorithms used.

Finally, we note that the hypothesis tested in this paper may not hold when the mining is performed by an algorithm that uses an error model in which a sub-matrix of a bicluster is not necessarily a bicluster in itself (e.g., *mean squared residue*) and in cases of "incorrect" class labeling of the objects. The adaptation of the method for such cases is thus left for future research. Other future research that can benefit from the findings reported in this paper includes extensions to more complex biclustering models such as those that consider shifts (addition by a constant) and scales (multiplication by a constant), lag and fuzziness.

Acknowledgments. We are thankful to Haixun Wang for providing us with the source code of the *pCluster* algorithm [32]. Also, we are grateful to Andrew Rosenberg for the permission to use the *ClusterEvaluator* package [27], and for enlightening remarks.

References

1. Aguilar-Ruiz, J.S.: Shifting and scaling patterns from gene expression data. Bioinformatics 21(20), 3840–3845 (2005)
2. Bache, K., Lichman, M.: UCI Machine Learning Repository (2013)
3. Berkhin, P.: A survey of clustering data mining techniques. Grouping Multidimensional Data, pp. 25–71 (2006)
4. Berson, A., Smith, S., Thearling, K.: Building data mining applications for CRM. McGraw-Hill, New York (2000)
5. Beyer, K., Goldstein, J., Ramakrishnan, R., Shaft, U.: When is "nearest neighbor" meaningful? In: ICDT, pp. 217–235 (1999)
6. Bryan, K., Cunningham, P.: Bottom-up biclustering of expression data. In: CIBCB, pp. 1–8 (2006)
7. Califano, A., Stolovitzky, G., Tu, Y.: Analysis of gene expression microarrays for phenotype classification. In: ISMB, vol. 8, pp. 75–85 (2000)
8. Cheng, Y., Church, G.M.: Biclustering of expression data. In: ISMB, pp. 93–103 (2000)
9. Dom, B.E.: An information-theoretic external cluster-validity measure. In: UAI, pp. 137–145 (2002)
10. Guan, J., Gan, Y., Wang, H.: Discovering pattern-based subspace clusters by pattern tree. KBS 22(8), 569–579 (2009)
11. Günnemann, S., Färber, I., Müller, E., Assent, I., Seidl, T.: External evaluation measures for subspace clustering. In: CIKM, pp. 1363–1372 (2011)
12. Jiang, D., Tang, C., Zhang, A.: Cluster analysis for gene expression data: A survey. TKDE 16(11), 1370–1386 (2004)
13. Keogh, E., Wei, L., Xi, X., Lee, S.H., Vlachos, M.: LB_Keogh supports exact indexing of shapes under rotation invariance with arbitrary representations and distance measures. In: VLDB, pp. 882–893 (2006)

14. Lagarias, J., Reeds, J., Wright, M., Wright, P.: Convergence Properties of the Nelder–Mead Simplex Method in Low Dimensions. SIOPT 9(1), 112–147 (1998)
15. Liu, G., Sim, K., Li, J., Wong, L.: Efficient mining of distance-based subspace clusters. SADM 2(5-6), 427–444 (2009)
16. Madeira, S.C., Oliveira, A.L.: Biclustering algorithms for biological data analysis: a survey. TCBB 1(1), 24–45 (2004)
17. McDaid, A.F., Greene, D., Hurley, N.: Normalized mutual information to evaluate overlapping community finding algorithms. CoRR abs/1110.2515 (2011)
18. Meilă, M.: Comparing clusterings—an information based distance. J. Multivar. Anal. 98(5), 873–895 (2007)
19. Melkman, A.A., Shaham, E.: Sleeved CoClustering. In: KDD, pp. 635–640 (2004)
20. Moise, G., Zimek, A., Kroeger, P., Kriegel, H., Sander, J.: Subspace and projected clustering: experimental evaluation and analysis. KAIS 21(3), 299–326 (2009)
21. Patrikainen, A., Meila, M.: Comparing subspace clusterings. TKDE 18(7), 902–916 (2006)
22. Peeters, R.: The maximum edge biclique problem is NP-complete. DAM 131(3), 651–654 (2003)
23. Pei, J., Zhang, X., Cho, M., Wang, H., Yu, P.S.: Maple: A fast algorithm for maximal pattern-based clustering. In: ICDM, pp. 259–266 (2003)
24. Pio, G., Ceci, M., D'Elia, D., Loglisci, C., Malerba, D.: A Novel Biclustering Algorithm for the Discovery of Meaningful Biological Correlations between microRNAs and their Target Genes. BMC Bioinformatics 14(7), 1–25 (2013)
25. Press, W.H., Teukolsky, S.A., Vetterling, W.T., Flannery, B.P.: Numerical recipes in C: the art of scientific computing. Cambridge University Press (1992)
26. Procopiuc, C.M., Jones, M., Agarwal, P.K., Murali, T.: A Monte Carlo algorithm for fast projective clustering. In: SIGMOD, pp. 418–427 (2002)
27. Rosenberg, A., Hirschberg, J.: V-Measure: A Conditional Entropy-Based External Cluster Evaluation Measure. In: EMNLP-CoNLL, vol. 7, pp. 410–420 (2007)
28. Shaham, E., Sarne, D., Ben-Moshe, B.: Sleeved co-clustering of lagged data. KAIS 31(2), 251–279 (2012)
29. Supporting webpage (2013), http://tinyurl.com/Supporting-MLDM14
30. Van Rijsbergen, C.: Information retrieval, 2nd edn. Butterworths (1979)
31. Wang, H., Chu, F., Fan, W., Yu, P.S., Pei, J.: A fast algorithm for subspace clustering by pattern similarity. In: SSDBM, pp. 51–60 (2004)
32. Wang, H., Wang, W., Yang, J., Yu, P.S.: Clustering by pattern similarity in large data sets. In: SIGMOD, pp. 394–405 (2002)
33. Yiu, M.L., Mamoulis, N.: Iterative projected clustering by subspace mining. TKDE 17(2), 176–189 (2005)
34. Yoon, S., Nardini, C., Benini, L., De Micheli, G.: Enhanced pClustering and its applications to gene expression data. In: BIBE, pp. 275–282 (2004)
35. Zeng, Y., Tang, J., Garcia-Frias, J., Gao, G.R.: An adaptive meta-clustering approach: combining the information from different clustering results. In: CSB, pp. 276–287 (2002)
36. Zhao, L., Zaki, M.J.: TRICLUSTER: an effective algorithm for mining coherent clusters in 3D microarray data. In: SIGMOD, pp. 694–705 (2005)
37. Zhao, Y., Karypis, G.: Criterion functions for document clustering: Experiments and analysis. Machine Learning (2001)

Towards the Efficient Recovery of General Multi-Dimensional Bayesian Network Classifier

Shunkai Fu[1], Sein Minn[1], and Michel C. Desmarais[2]

[1] College of Computer Science and Technology, National Huaqiao University, Xiamen, China
{fusk,seinminn}@hqu.edu.cn
[2] Department of Computer Engineering, École Polytechnique de Montréal, Montréal, Canada
Michel.desmarais@polymtl.ca

Abstract. Multi-dimensional classification (MDC) aims at finding a function that assigns a vector of class values to a vector of observed features. Multi-dimensional Bayesian network classifier (MBNC) was devised for MDC in 2006, but with restricted structure. By removing the constraints, an undocumented model called general multi-dimensional Bayesian network classifier (GMBNC) is proposed in this article, along with an exact induction algorithm which is able to recover the GMBNC by local search, without having to learn the whole BN first. We prove its soundness, and conduct experimental studies to verify its effectiveness and efficiency. The larger is the problem, the more saving by IPC-GMBNC versus conventional approach (global structure learning by PC algorithm), e.g. given an example network with 200 nodes, around 99% saving is achieved.

Keywords: multi-dimensional classification, multi-dimensional Bayesian network classifier, Bayesian network, Markov blanket.

1 Introduction

Bayesian network (BN) (Fig. 1 (a)) classifiers (BNC) is actually a Bayesian network tailored for classification problem, and it contains a single class (variable) traditionally. However, in many applications, instances are connected to multiple classes meanwhile, e.g. we may have to classify an oesophageal tumour in terms of its depth of invation, its spread to lymph nodes, and whether or not is has given rise to haematogenous metastases [1]. Multi-dimensional Bayesian Network Classifiers (MBNCs) were proposed to deal with this kind of problem providing an accurate modeling of the probabilistic dependence relationships among selected features and class variables [2-5]. The *graphical structure* of MBNCs actually is composed of three different subgraphs: *class subgraph* representing the dependence relationships between class variables, *bridge subgraph* representing the dependence relationships between class and feature variables, and *feature subgraph* representing the dependence relationships between feature variables. Fig. 1 (b) is one such example. Although *parameter* is a critical part to MBNC, we focus on the recovery of its structure here considering that it is more challenging a task.

P. Perner (Ed.): MLDM 2014, LNAI 8556, pp. 16–30, 2014.

As noted in Fig. 1 (b), in the bridge subgraph of previously proposed MBNC, there exist only arcs like $C_i \to X_j$, but no $X_j \to C_i$. Here, we relax this constraint by allowing $X_j \to C_i$, as in general BN, and the corresponding classifier model therefore could be referred as General Multidimensional Bayesian Network Classifier(GMBNC) (see Fig. 1 (c)). We propose an algorithm named IPC-GMBNC to induce the structure of GMBNC, and it relies on *Iterative* local search of *Parents* and *Children*. We prove its soundness, and evaluate its accuracy and time efficiency with synthetic data generated from several known networks.

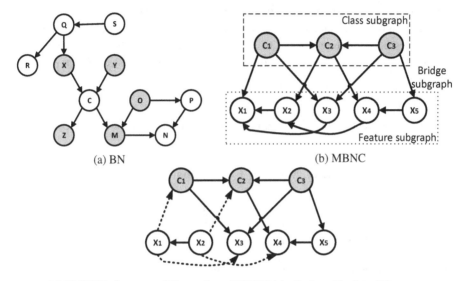

(a) BN

(b) MBNC

(c) GMBNC, those arcs different from MBNC(b) is displayed in dotted lines

Fig. 1. Example of BN (a) in which grey nodes are C's Markov blanket, MBNC (b) in which grey nodes are class variables and GMBNC (c) with the same class nodes as MBNC (b)

The remainder of this paper is organized as follows. Section 2 briefly presents basics of BN, BNC, MBNC and GMBNC. An exact algorithm to induce GMBNC is proposed in Section 3, with necessary proof. Section 4 is about experimental study, and related works are briefly reviewed in Section 5. Section 6 sums up the paper.

2   Theoretical Basis and Concepts

2.1   Notations

We use bold-faced Roman letters for sets of variables (e.g. $\mathbf{X}$), non-bold-faced Roman letters for singleton variables (e.g. X) and lower-case Roman letters for values of the variables (e.g. $\mathbf{X} = \mathbf{x}$, $X = x$). For a distribution P, we use $I_P(\mathbf{X}, \mathbf{Y} \mid \mathbf{Z})$ to denote the fact that in P, $\mathbf{X}$ is independent of $\mathbf{Y}$ given conditioning set $\mathbf{Z}$. When the conditioning set is empty, we use $I_P(\mathbf{X}, \mathbf{Y})$ instead.

To simplify notation, we omit the standard set notation when considering a singleton variable in any position. For example, we use $I_P(X,Y|\mathbf{Z})$ instead of $I_P(\{X\},\{Y\}|\mathbf{Z})$. Similarly, we use $I_G(\mathbf{X},\mathbf{Y}|\mathbf{Z})$ to denote the conditional independence relation encoded in a graph model G.

2.2   Bayesian Network and Bayesian Network Classifier

A BN over a set of discrete random variables $\mathbf{U}=\{x_1,x_2,...,x_n\}$, $n\geq 1$, is a pair $B=(\mathbf{G},\Theta)$, and $\mathbf{G}=(\mathbf{V},\mathbf{A})$ is a directed acyclic graph (DAG) in which vertices $\mathbf{V}$ corresponds to variables in $\mathbf{U}$ and arcs $\mathbf{A}$ represents direct probabilistic dependencies between the vertices, Θ is a set of conditonal probability distributions such that $\Theta_{x_i|\mathbf{pa}(x_i)}=p(x_i|\mathbf{pa}(x_i))$ defines the conditonal probability of each possible value x_i given a set value $\mathbf{pa}(x_i)$ of $\mathbf{pa}(X_i)$, where $\mathbf{pa}(X_i)$ denotes the set of parents of X_i in G.

A BN B represents a joint probability distribution over $\mathbf{V}$ factorized according to structure G:

$$p(X_1,X_2,...,X_n)=\prod_{i=1}^{n}p(X_i|\mathbf{pa}(X_i)) \tag{1}$$

Definition 1 (Faithfulness). A BN B and a joint distribution P are faithful to one another, if and only if every conditional independence encoded by the graph of B is also presented in P, i.e. $I_P(X,Y|\mathbf{Z})=I_G(X,Y|\mathbf{Z})$

Definition 2 (Markov Blanket). The Markov blanket of X, denoted as $\mathbf{MB}(X)$, is a minimal set of variables with the following property: $I_P(X,Y|\mathbf{MB}(X))$ holds for every variable $Y\in\mathbf{U}\setminus\mathbf{MB}(X)\setminus\{X\}$.

Theorem 1. Given a BN structure G, $X\in\mathbf{V}$ and faithfulness assumption, (1) the Markov blanket of X is unique and it is the union of its parents, children and spouses[6]; (2) $I_G(X,Y|\mathbf{MB}(X))$ holds as well.

Definition 3 (Bayesian Network Classifier). A Bayesian network classifier about a class X, denoted as $\mathbf{BNC}(X)$, is the DAG over $\mathbf{MB}(X)\cup\{X\}$ (actually a BN as well) [7].

Given X, the cardinality of its Markov blanket $|\mathbf{MB}(X)|$ normally is much smaller than $|\mathbf{V}|$, hence the $\mathbf{BNC}(X)$ is only a very small sub-graph of the whole G.

Theorem 2. Two DAGs are equivalent if and only if they have the same underlying undirected graph and the same v-structures (i.e. the same set of uncoupled head-to-head converging like $X\rightarrow Y\leftarrow Z$)[6].

2.3   Multi-Dimensional Bayesian Network

Definition 4 (Multi-Dimensional Bayesian Network Classifier, MBNC). An MBNC [8] is a BN $B = (\mathbf{G}, \Theta)$ where the structure $G = (\mathbf{V}, \mathbf{A})$ has a restricted topology. The set of n vertices $\mathbf{V}$ is partitioned into two subsets: class variables $\mathbf{V}_C = \{C_1, ..., C_m\}$, $m \geq 1$, and feature variables $\mathbf{V}_X = \{X_1, ..., X_k\}$, $k \geq 1$ ($m + k = n$). The set of arcs $\mathbf{A}$ is partitioned into three subsets $\mathbf{A}_C$, $\mathbf{A}_X$ and $\mathbf{A}_{CX}$ such that:

— $\mathbf{A}_C \subseteq \mathbf{V}_C \times \mathbf{V}_C$ is composed of the arcs between the class variables having a subgraph $G_C = (\mathbf{V}_C, \mathbf{A}_C)$ - *class subgraph* – of G induced by $\mathbf{V}_C$.
— $\mathbf{A}_X \subseteq \mathbf{V}_X \times \mathbf{V}_X$ is composed of the arcs between the feature variables having a subgraph $G_X = (\mathbf{V}_X, \mathbf{A}_X)$ - *feature subgraph* – of G induced by $\mathbf{V}_X$.
— $\mathbf{A}_{CX} \subseteq \mathbf{V}_C \times \mathbf{V}_X$ is composed of the arcs from the class variables to the feature variables having a subgraph $G_{CX} = (\mathbf{V}, \mathbf{A}_{CX})$ - *bridge subgraph* – of G induced by $\mathbf{V}$ [2]. Note: there exist only $C_i \rightarrow X_j$ in $\mathbf{A}_{CX}$.

Given a vector of observation $\mathbf{x} = \{x_1, ..., x_m\}$ and 0-1 loss function, classifiation with an MBNC is equivalent to solving the most probable explanation (MPE) problem:

$$c^* = (c_1^*, ..., c_m^*) = \arg\max_{c_1, ..., c_m} p(C_1 = c_1, ..., C_m = c_m \mid \mathbf{x}) \tag{2}$$

Example 1. An example of an MBNC structure is shown in 错误！未找到引用源。(b). $\mathbf{V}_C$ contains four class variables, $\mathbf{V}_X$ includes seven features. Using (1), we have

$$\max_{C_1, C_2, C_3} p(C_1 = c_1, C_2 = c_2, C_3 = c_3 \mid \mathbf{x}) = \max_{C_1, C_2, C_3} p(c_1) p(c_2 \mid c_1, c_3) p(c_3) \bullet$$
$$p(x_1 \mid c_1, x_2, x_3) p(x_2 \mid c_2, x_4) p(x_3 \mid c_1, c_3) p(x_4 \mid c_2, x_5) p(x_5 \mid c_3)$$

2.4   General Multi-dimensional Bayesian Network

To provide more general modeling of multi-dimensional classification problems, we remove the constraint of bi-partitie, i.e. allowing any to any dependencies, resulting in a so-called general MBNC.

Definition 5 (GMBNC). An GMBNC is a BN $B = \langle \mathbf{G}, \Theta \rangle$ where the structure $G = \langle \mathbf{V}, \mathbf{A} \rangle$ is a general DAG. The set of vertices is partitioned into subsets, $\mathbf{V} = \bigcup_{i=1}^m (\mathbf{MB}(C_i) \cup \{C_i\})$. The set of arcs is partitioned into subsets corresponding to $\mathbf{MB}(C_i) \cup \{C_i\}$, such that $\mathbf{A}_i \subseteq (\{C_i\} \cup \mathbf{MB}(C_i)) \times (\{C_i\} \cup \mathbf{MB}(C_i))$ is composed of arcs between C_i and its markov blanket $\mathbf{MB}(C_i)$ having a subgraph $G_{C_i} = \langle \{C_i\} \cup \mathbf{MB}(C_i), \mathbf{A}_i \rangle$. (see Fig. 1(c) for an example)

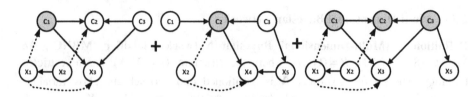

Fig. 2. The local DAG components as composed in the GMBNC shown in Fig. 1(c)

Hence GMBNC actually is composed of a series of local DAGs, and each such local DAG is a BNC describing general dependencies among $\{C_i\} \cup \mathbf{MB}(C_i)$. There is no limit on the parents of C_i, so it is not bi-partitie graph any more and it provides more general modeling ability as compared to MBNC. Fig. 2 demonstrates the three local DAGs as contained in the example GMBNC of Fig. 1(c). Existing algorithms for the induction of MBNC cannot work here, so we propose one in the next section.

3   An Exact Algorithm to Induce GMBNC

3.1   Algorithm Specification

In the past decade, many efficient Markov blanket learning methods have been proposed such as GS, IAMB and its variants, PCMB and IPC-MB (see [9-11] and the references therein). In this paper, we extend IPC-MB to the context of GMBNCs since it demonstrates with the best tradeoff among time and data efficiency, also with some topology information induced [9]. Algorithm *IPC-GMBNC* (Table 1) consists of a sequential search of the BNC about each $C \in \mathbf{V}_C$ by *IPC-BNC* (Table 2), and *IPC-BNC* primarily relies on iterative search of parents and children via the recognition of false positives by *RemoveFalseOnes* (Table 3). Connectivity ($\mathbf{A}_{Ind}$) or dis-

Table 1. Algorithm IPC-GMBNC – Induce GMBNC

```
IPC-GMBNC ( V_C , D :dataset, ε :significance value)
    A_Ind = ∅ ;
    A_Del = ∅ ;
    for ( ∀C ∈ V_C )do
        IPC-BNC ( C,D,ε,A_Ind ,A_Del ) ;
    end for
    Specify the directionality of non-oriented arcs in A_Ind
    using the orientational rules proposed in [12];
    return A_Ind ;
```

Table 2. Algorithm IPC-BNC - Induce the Bayesian network classifier

IPC-BNC(C :class variable being studied, **D** , ε ,
   $\mathbf{A}_{Ind}$:arcs known to exist in the target GMBNC,
   $\mathbf{A}_{Del}$:arcs known not to exist in the target GMBNC)
   $\mathbf{V}_{Studied} = \{X \mid (C-X) \in \mathbf{A}_{Del}\}$;
   $\mathbf{A}_{Can} = \{(C-X) \mid \forall X \in \mathbf{V} \setminus \mathbf{V}_{Studied} \setminus \{C\}\}$;
   RemoveFalseOnes ($C, \mathbf{A}_{Can}, \mathbf{A}_{Ind}, \mathbf{A}_{Del}, \mathbf{D}, \varepsilon$) ;
   $\mathbf{V}_{CanPC(C)} = \{X \mid (C-X) \in \mathbf{A}_{Can}\}$;
   $\mathbf{V}_{Studied} = \mathbf{V}_{Studied} \cup \{C\}$;
   for ($\forall X \in \mathbf{V}_{CanPC(C)}$) do
      $\mathbf{A}_{Can} = \mathbf{A}_{Can} \cup \{(X-Y) \mid \forall Y \in \mathbf{V} \setminus \mathbf{V}_{Studied}\}$;
      RemoveFalseOnes ($X, \mathbf{A}_{Can}, \mathbf{A}_{Ind}, \mathbf{A}_{Del}, \mathbf{D}, \varepsilon$) ;
      $\mathbf{V}_{Studied} = \mathbf{V}_{Studied} \cup \{X\}$
   end for
   $\mathbf{V}_{PC(C)} = \{X \mid (C-X) \in \mathbf{A}_{Can}\}$;
   $\mathbf{V}_{Spouse(C)} = \varnothing$;
   for ($\forall X \in \mathbf{V}_{PC(C)}$) do
      $\mathbf{V}_{CanPC(C)} = \{Y \mid (X-Y) \in \mathbf{A}_{Can}\}$;
      for ($\forall Y \in \mathbf{V}_{CanPC(X)}$ and $Y \notin \mathbf{V}_{Studied}$) do
         if ($I_D(C,Y \mid \mathbf{V}_{Sepset(C,Y)} \cup \{X\}) < \varepsilon$) then
            Specify $C \rightarrow X \leftarrow Y$ in $\mathbf{A}_{Can}$;
            $\mathbf{V}_{Sepset(C,Y)} = \mathbf{V}_{Sepset(C,Y)} \cup \{Y\}$;
            RemoveFalseOnes ($Y, \mathbf{A}_{Can}, \mathbf{A}_{Ind}, \mathbf{A}_{Del}, \mathbf{D}, \varepsilon$) ;
         end if
      end for
   end for
   Remove arcs connecting $\mathbf{V} \setminus \mathbf{V}_{PC(C)} \setminus \mathbf{V}_{Spouse(C)}$ to from $\mathbf{A}_{Can}$;
   $\mathbf{A}_{Ind} = \mathbf{A}_{Ind} \cup \mathbf{A}_{Can}$;

connectivity ($\mathbf{A}_{Del}$) information inferred in an earlier call of *IPC-BNC* can be referred in a late *IPC-BNC* to avoid redundant study, which helps to save great computing resource considering there is intersection part between BNCs (see Fig. 2).

Some important notations used in the specification are exaplained as below:

- $\mathbf{V}$ and $\mathbf{V}_C$: All variables (nodes) and all class variables;
- $\mathbf{V}_{Studied}$: Nodes which are studied;
- $\mathbf{V}_{CanPC(X)}$: Candidate parents and children of X ;

- $\mathbf{V}_{PC(X)}$: Induced parents and children of X ;
- $\mathbf{A}_{Ind}$. Arcs determined to existing in the target GMBNC;
- $\mathbf{A}_{Del}$. Arcs determined to NOT existing in the target GMBNC;
- $\mathbf{A}_{Can}$. Candidate arcs of BNC, initiated with all potential connectivity to the target, and it contains only true arcs belonging to the target GMBNC by the end of *IPC-BNC*.
- $I_D(X,Y\mid \mathbf{Z})$. Some statistical testing about the conditional probability of X and Y given $\mathbf{Z}$ based on sample set $\mathbf{D}$. For example, it can be known χ^2 test.

Besides, our algorithm requires assumptions including (1) faithfulness, (2) no hidden variable(s), (3) no missing values and (4) discrete observations.

Table 3. RemoveFalseOnes - Remove false connections between C and its neighbors

```
RemoveFalsePC ( C , A_Can , A_Ind , A_Del , D , ε )
```
$\mathbf{V}_{NotPC(C)} = \varnothing$;   // To contain non-PC variables
$css = 0$;   // Cut set size
$\mathbf{V}_{CanPC(C)} = \{X \mid (C-X)\in \mathbf{A}_{Can}\}$;
```
while ( V_CanPC(C) > css ) do
    for ( ∀X ∈ V_CanPC(C) and (X-C)∉ A_Del ) do
        for ( ∀S ⊆ V_CanPC(X) \{X} and |S|= css ) do
            if ( I_D(C,X|S) ≥ 1-ε ) then
```
$\qquad\qquad\qquad\mathbf{A}_{Can} = \mathbf{A}_{Can}\setminus\{(C-X)\}$;
$\qquad\qquad\qquad\mathbf{A}_{Del} = \mathbf{A}_{Del}\cup\{(C-X)\}$;
$\qquad\qquad\qquad\mathbf{V}_{NotPC(C)} = \mathbf{V}_{NotPC(C)}\cup\{X\}$;
$\qquad\qquad\qquad\mathbf{V}_{Sepset(C,X)} = \mathbf{S}$;
```
                break;
            end if
        end for
    end for
```
$\quad\mathbf{V}_{CanPC(C)} = \mathbf{V}_{CanPC(C)}\setminus\mathbf{V}_{NotPC(C)}$;
$\quad css = css+1$;
```
end while
```

3.2   Soudness of IPC-BNC

In this section, we will prove that IPC-BNC will induce the correct parents, children and spouses of C , so as their connections.

Definition 6 (Markov Condition). In a BN B, given the value of parents, X is conditional independent with all its non-descendants excluding its parents.

Definition 7 (Descendant). Y is a descendant of X, if there exists a directed path from X to Y, but there exists NO directed path from Y to X. The set of descendants of X is denoted with $\mathbf{V}_{Des(X)}$.

Definition 8 (Non-Descendant). Given $\mathbf{V}$, those other than the descendants are known as non-descendants of X, denoted as $\mathbf{V}_{NonDes(X)} (= \mathbf{V} \setminus \mathbf{V}_{Des(X)})$.

Theorem 3. If a BN B is faithful to a joint probability distribution P, then for each pair of nodes X and Y in B, X and Y are adjacent in B iff. $I_D(X,Y \mid \mathbf{Z})$ is not satisfied for all $\mathbf{Z}$ such that X and $Y \notin \mathbf{Z}$.

Lemma 1. Parents and children of C will NOT be recognized as false positives in *RemoveFalseOnes(C)*.

Proof. $\mathbf{V}_{CanPC}$ contains all candidate parents and children initially. If some parent or child $X \in \mathbf{V}_{CanPC}$ is added to $\mathbf{V}_{NotPC}$, which means it fails some $I_D(X,Y \mid \mathbf{S})$ (at the line 7). It contradicts with Theorem 3, hence all parents and children of C will still in $\mathbf{V}_{CanPC}$ by the end of *RemoveFalseOnes*. ∎

Lemma 2. All false positives (i.e., false parents and children) could be recognized in *RemoveFalseOnes*, with the exception of some descendant(s) of C.

Proof. Initially, $\mathbf{V}_{CanPC}$ contains P/Cs, descendants and non-descendants of C. (1) Since all P/Cs will always in $\mathbf{V}_{CanPC}$ all alone accordign to Lemma 1, for each $X \in \mathbf{V}_{NonDes}$, it will fail some test $I_D(C,X \mid \mathbf{S})$ where $\mathbf{S} \subseteq \mathbf{V}_{PC(C)}$ given the Markov condition. (2) Descendant may fail to be removed, and Fig. 3 is one such example. When $css = 0$, P and R are successfully removed from $\mathbf{V}_{CanPC}$. Then, when $css = 1$, S fail to be removed from $\mathbf{V}_{CanPC}$ due to the missing of P and R. When $css = 2$, the loop has to terminate in *RemoveFalseOnes* due that $\mid \mathbf{V}_{CanPC} \mid \geq css$. ∎

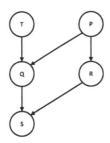

Fig. 3. Example in which descendant may fail to be recognized by *RemoveFalsePC*

Theorem 4. IPC-BNC could induce the correct parents and children of C (by the Line 11).

Proof. (1) All parents and children of C are contained in $\mathbf{V}_{CanPC(C)}$ at line 5 in *IPC-BNC* (Lemma 1). (2) For $\forall X \in \mathbf{V}_{CanPC(C)}$, *RemoveFalseOnes*(X) will be called. If X is a true parent or child of C, it is easy to know that $C \in V_{CanPC(X)}$ by the end of *RemoveFalseOnes*(X). (3) If X is a false positive in $V_{CanPC(C)}$, $X \in \mathbf{V}_{Des(C)}$ and $C \in \mathbf{V}_{NonDes(X)}$ (Lemma 2 and Definition 8). Then it is trivial to infer that C won't appear in $V_{CanPC(X)}$ by the end of *RemoveFalseOnes*(X) (Lemma 2). Hence, all false positives in $\mathbf{V}_{CanPC(C)}$ will be correctly recognized, resulting in a correct parent-child set of C at line 11 in *IPC-BNC*, i.e. $\mathbf{V}_{PC(C)} = \{X \mid (C - X) \in \mathbf{A}_{Can}\}$. ∎

Theorem 5. The spouse of C can be correctly induced in *IPC-BNC*.

Proof. The proof can be found in [10]. ∎

Theorem 6. *IPC-BNC* could induce the skeleton of local DAG about C (BNC), and some v-structure(s) $C \rightarrow X \leftarrow Y$ where Y is C's spouse and X is their child.

Proof. Since we are able to induce the correct parents/children (Theorem 4) and spouses (Theorem 5) of C in *IPC-BNC*, it is trivial to know that the corresponding skeleton is contained in $\mathbf{A}_{Can}$. Upon the recognition of spouses (at line 17), the important v-structures then are induced as well. ∎

3.3   Soudness of IPC-GMBNC

Theorem 7. *IPC-GMBNC* will output the correct general multi-dimensional Bayesian network classifier given $\mathbf{V}_C$.

Proof. (1) From Theorem 6, it is known that *IPC-BNC* enables us to induce the correct skeleton structure of BNC(C_i), given $C_i \in \mathbf{V}_C$. (2) This induced structure information is stored in $\mathbf{A}_{Ind}$, in an incremental manner. Hence, after the call of IPC-BNC on each $C_i \in \mathbf{V}_C$, $\mathbf{A}_{Ind}$ will contain the skeleton of GMBNC plus some v-structure information. (4) Orientation rules have been studied widely in earlier study of Bayesian network [12, 13], so ignored here. They are directly applied to $\mathbf{A}_{Ind}$ to determine the directions of non-oriented arcs. ∎

4   Experimental Study

4.1   Data Sets

We randomly generate samples from two real networks, *Asia* and *Alarm*, and two synthetic networks randomly generated by Weka Bayesian Network toolkit, named

Test100 and *Test200* respectively. Basic description about each network can be found in Table 4. For each network, 10 folders are prepared for different data set size (1k/2k/3k/4k/5k).

Table 4. Basic information about three networks

Networks	# Nodes	# Arcs
Asia	8	8
Alarm	37	46
Test100	100	130
Test200	200	260

4.2   Approaches

Given each network, we randomly select dozens of combinations of 2 and 3 nodes as class vectors (indicated by $|V_C|=2$ and $|V_C|=3$ in Table 5 and Table 6) and then apply IPC-GMBNC to recover the corresponding GMBNC. We also use PC algorithm [14] to induce the global BN (with same orientation rules as IPC-GMBNC), and then extract the GMBNCs regarding the corresponding class vectors. Structures induced by both approaches are compared with the underlying true models (in this study, we only care about the critical structure information – skeleton and v-structures, based on Theorem 2) respectively to get the absolute accuracy performance (F-measure). We use χ^2 as the conditional independence test, and the threshold value (ε) for significance is set as 0.05.

We do not compare our work with existing algorithms for recovering MBNC considering the fundamental difference between MBNC and GMBNC structures. We may leave the prediction performance evaluation in future studies.

4.3   Evaluation Metrics

We measure the accuracy by the F-measure (the harmonic mean of precision and recall), and the time efficiency by the number of CI tests totally conducted during the learning by PC and IPC-GMBNC, considering that this is independent of the underlying computing machine.

4.4   Results and Discussions

Accuracy performance is shown in Table 5, and it is noticed that, (1) the accuracy increases for both IPC-GMBNC and PC algorithms when more samples are available; (2) IPC-GMBNC, although making decision only by local search, even outperforms PC; (3) Worse performance is observed by both algorithms in complex problems, i.e. $|V_C|=3$.

Table 5. The accuracy performance of   IPC-GMBNC vs. PC algorithm, in term of F-measure

Networks	# Instances	$\mid V_C \mid = 2$		$\mid V_C \mid = 3$	
		IPC	PC	IPC	PC
Asia	1k	.65 ± .08	.42 ± .05	.63 ± .08	.43 ± .05
	2k	.74 ± .05	.48 ± .03	.74 ± .04	.50 ± .03
	3k	.71 ± .06	.48 ± .03	.72 ± .06	.50 ± .03
	4k	.73 ± .06	.49 ± .02	.73 ± .06	.50 ± .02
	5k	.77 ± .05	.50 ± .04	.77 ± .05	.51 ± .04
Alarm	1k	.83 ± .03	.79 ± .03	.81 ± .03	.77 ± .04
	2k	.90 ± .02	.87 ± .03	.88 ± .03	.86 ± .03
	3k	.91 ± .03	.90 ± .02	.89 ± .04	.88 ± .03
	4k	.92 ± .03	.92 ± .04	.90 ± .04	.90 ± .04
	5k	.95 ± .01	.90 ± .05	.94 ± .01	.89 ± .05
Test100	1k	.62 ± .02	.54 ± .02	.54 ± .02	.48 ± .01
	2k	.68 ± .02	.58 ± .01	.61 ± .02	.53 ± .01
	3k	.71 ± .01	.60 ± .01	.65 ± .01	.55 ± .01
	4k	.71 ± .01	.61 ± .01	.65 ± .01	.56 ± .01
	5k	.73 ± .01	.61 ± .01	.68 ± .01	.58 ± .01
Test200	1k	.50 ± .02	.40 ± .02	.44 ± .02	.36 ± .02
	2k	.58 ± .02	.45 ± .01	.53 ± .02	.41 ± .01
	3k	.62 ± .01	.47 ± .01	.56 ± .01	.43 ± .01
	4k	.65 ± .01	.49 ± .01	.58 ± .01	.45 ± .01
	5k	.65 ± .02	.50 ± .01	.60 ± .01	.46 ± .01

Table 6. The time efficiency of IPC-GMBNC vs. PC algorithm, in term of # CI tests

Networks	# Instances	$\mid V_C \mid = 2$		$\mid V_C \mid = 3$	
		IPC	PC	IPC	PC
Asia	1k	101 ± 17	375 ± 47	136 ± 22	375 ± 47
	2k	99 ± 13	376 ± 46	132 ± 17	376 ± 46
	3k	107 ± 14	410 ± 50	143 ± 18	410 ± 50
	4k	117 ± 11	423 ± 33	148 ± 13	423 ± 33
	5k	119 ± 18	448 ± 49	157 ± 20	448 ± 49
Alarm	1k	1034 ± 66	3150 ± 185	1530 ± 103	3150 ± 185
	2k	1143 ± 51	3507 ± 139	1733 ± 65	3507 ± 139
	3k	1205 ± 49	3718 ± 107	1860 ± 80	3718 ± 107
	4k	1294 ± 57	3907 ± 127	1998 ± 77	3907 ± 127
	5k	1374 ± 55	4084 ± 123	2123 ± 62	4084 ± 123
Test100	1k	745 ± 46	8188 ± 114	1099 ± 63	8188 ± 114
	2k	918 ± 39	8884 ± 81	1359 ± 47	8884 ± 81
	3k	1020 ± 17	9349 ± 152	1506 ± 36	9349 ± 152
	4k	1061 ± 35	9543 ± 236	1580 ± 50	9543 ± 236
	5k	1130 ± 52	9802 ± 498	1684 ± 65	9802 ± 498
Test200	1k	2841 ± 183	307436 ± 114	4373 ± 300	307436 ± 114
	2k	3536 ± 232	328346 ± 578	5552 ± 310	328346 ± 578
	3k	3805 ± 143	338706 ± 514	5956 ± 224	338706 ± 514
	4k	4176 ± 234	349626 ± 583	6491 ± 322	349626 ± 583
	5k	4253 ± 247	356146 ± 1034	6682 ± 443	356146 ± 1034

Time efficiency performance is shown in Table 6, and we notice that, (1) When there are more samples, more CI tests are conducted. This is because there are more 'trustable' CI tests available in large data sets, which allows the search to go further to induce more information. Consequently, higher induction accuracy is achieved as we observe in Table 5; (2) More computing resource is required for higher dimensional problems by IPC-GMBNC, e.g. 30% more for 3-Class-Variable problems than 2-Class-Variable in Alarm network; (3) Local learning strategy used by IPC-GMBNC is much more efficient than global search by PC, e.g. only 1% of the total number of CI tests necessary to PC algorithm given Test200; (4) The larger is the network, the more significant saving achieved by IPC-GMBNC. Hence, we can conclude that IPC-GMBNC is much more scalable than PC.

5   Motivation and Related Works

Many application domains include classification problems where an instance has to be assigned to a most likely combination of classes. Since the number of class variables in a Bayesian classifier is restricted to one, such classification problems cannot be modelled straightforwardly. One option is to construct a compound class variable that models all possible combinations of classes. The class variable may then easily end up with an inhibitvely large number of value. Moreover, the structure of the classification problem is not reflected in the model. Another option is to develop multiple classifiers, one for eah of the original classes. Obvious, it fails to model the possible interaction effects among the classes. Furthermore, even if the classifiers indicate multiple classes, the implied combination of classes may not be the most likely explanation of the observed features [8].

An MBNC includes one or more class variables and one or more feature variables. It models the relationships between the variables by directed acyclic graphs (DAGs) over the class variables and over the feature variables separately, and further connects the two sets of variables by means of a bi-partite directed graph.

Because learning a general Bayesian network is intractable in the most general setting (actually it is an NP-hard problem, [15]), constraints are imposed on the target structure to shrink the potential search space. For instance, the so-called fully tree-augmented multi-dimensional classifiers require both class and feature subgraphs ,i.e. $\mathcal{G}_C$ and $\mathcal{G}_X$, to be directed trees. They use a wrapper approach to do feature selection, i.e. the bridge subgraph $\mathcal{G}_{CX}$, and then apply Chow and Liu's algorithm [16] to learn the $\mathcal{G}_C$ and $\mathcal{G}_X$. De Waal and van der Gaag later present a theoretical approach for learning <polytree>−<polytree> MBNCs in [17]. Class and feature subgraphs are separately generated using Rebane and Pearl's algorithm [18]; however, the induction of the bridge subgraph was not specified. Zaragoza et al. [19] also propose a two-step method to learn <polytree>−<polytree> MBNCs.

Rodriguez and Lozano [20] use a multi-objective evolutionary approach to learn <DAG>−<DAG> MBNCs. Each permitted MBNC structure is coded as an individual with three substrings, one per subgraph. Based on different classification measures, joint and marginal, they define the objective functions as k-fold cross-validated estimators of each class classification error. The aim is to find non-dominated structures according to the objective functions.

Bielza et al. [2] propose three MBNC learning algorithms: pure filter (guided by any filter algorithm based on a fixed ordering among the variables), pure wrapper (guided by the classification accuracy) and a hybrid algorithm (a combination of pure filter and pure wrapper). None of these three algorithms places any constraints on the subgraph structures of the generated MBNCs.

In [3], Borchani et al. propose the CB-MBNC learning algorithm for class-bridge decomposable MBNCs[2], instead of general MBNCs, based on a greedy forward selection wrapper approach. Class or feature subgraphs can have any type of structure.

More recently, Zaragoza et al. Present a two-step method in [5]. In the first phase, a tree-based Bayesian network that represents the dependency relations between the class variables is learned. In the second phase, several chain classifiers are built using selective naive Bayes models, such that the order of the class variables in the chain is consistent with the class subgraph. At the end, the results of the different generated orders are combined in a final ensemble model.

All of these works require some constraints on the subgraphs, so that the computing complexity could be reduced to an affordable level. For the first time, Borchani et al. propose a Markov blanket based approach[21, 22], with no additional restrict on the structure. Their method uses HITON algorithm [23] which originally proposed for the inducton of Markov blanket. However, according to [24], HITON algorithm could not always produce correct outcomes. Therefore, those published works relying on HITON are NOT theoretically sound.

6   Conclusion

In this paper, we introduce a modeling technique called general multi-dimensional Bayesian network classifier which differs from existing multi-dimensional Bayesian network classifier by having no constraints on the underlying topology. An exact algorithm is proposed to recover the GMBNC by local search only. Although the work of [21, 22] is similar by relying on the induction of Markov blanket, they are devised for MBNC and there is obvious doubt on its soudness. Experimental studies by real and synthetic networks indicate that great savings may be achieved due to the local search strategy as compared to global search. Future work would be on the further improvement on the time efficiency, especially via inference given known conditional independence relations, which may be an effective way to avoid high-dimensional CI tests.

Acknowledgement. This research is supported by National Natural Science Foundation of China (Grant No: 61305085, 61300139 and 61102163), Science and Technology Foundation of Xiamen (Grant No. 3505Z20133027), Huaqiao University Grant (Grant No: 11Y0274 and 12HJY18) and the Fundamental Research Funds for the Central Universities (Grant No: 11J0263).

References

1. van der Gaag, L.C., et al.: Probabilities for a probabilistic network: a case study in oesophageal cancer. Artificial Intelligence in Medicine 25(2), 123–148 (2002)
2. Bielza, C., Li, G., Larranga, P.: Multi-dimensional classification with Bayesian networks. International Journal of Approximate Reasoning 52(6), 705–727 (2011)
3. Borchani, H., Bielza, C., Larranaga, P.: Learning CB-decomposable multi-dimensional Bayesian network classifiers. In: The 5th European Workshop on Probabilistic Graphical Models (PGM 2010) (2010)
4. Borchani, H., Bielza, C., Larranaga, P.: Learning multi-dimensional Bayeisan network classifiers using Markov blankets: A case study in the prediction of HIV Protease inhibitors. In: AIME Workshop on Probabilistic Problem Solving in Biomedicine, Bled, Slovenia (2010)
5. Zaragoza, J.H., et al.: Bayesian chain classifiers for multidimensional classification. In: IJCAI (2011)
6. Verma, T., Peal, J.: Equivalence and Synthesis of Causal Models. In: The 6th Annual Conference on Uncertainty in Artificial Intelligence (UAI). Elsevier Science (1990)
7. Cheng, J., Greiner, R.: Comparing Bayesian Network Classifiers. In: The 5th Conference on Uncertainty in Artificial Intelligence (UAI) (1999)
8. van der Gaag, L.C., de Waal, P.R.: Multi-dimensional Bayesian network classifiers. In: Third European Workshop on Probabilistic Graphical Models, Prague (2006)
9. Fu, S., Desmarais, M.C.: Tradeoff analysis of different markov blanket local learning approaches. In: Washio, T., Suzuki, E., Ting, K.M., Inokuchi, A. (eds.) PAKDD 2008. LNCS (LNAI), vol. 5012, pp. 562–571. Springer, Heidelberg (2008)
10. Fu, S., Desmarais, M.C.: Fast Markov blanket discovery algorithm via local learning within single pass. In: Bergler, S. (ed.) Canadian AI. LNCS (LNAI), vol. 5032, pp. 96–107. Springer, Heidelberg (2008)
11. Fu, S., Desmarais, M.C.: Markov blanket based feature selection: A review of past decade. In: World Congress on Engineering (2010)
12. Spirtes, P., Glymour, C., Scheines, R.: Causation, Prediction and Search, 2nd edn. The MIT Press (2001)
13. Pearl, J.: Causality: Models, Reasoning, and Inference, 2nd edn. Cambridge University Press (2009)
14. Spirtes, P., Glymour, C., Scheines, R.: Causation, Prediction, and Search. A Bradford Book (2001)
15. Chickering, D.M., Geiger, D., Heckerman, D.: Learning Bayesian Network is NP-Hard, p. 22. Microsoft (1994)
16. Chow, C.K., Liu, C.N.: Approximating discrete probability distributions with dependence trees. IEEE Transactions on Information Theory 14(3), 462–467 (1968)
17. de Waal, P.R., van der Gaag, L.C.: Inference and learning in multi-dimensional Bayesian network classifiers. In: Mellouli, K. (ed.) ECSQARU 2007. LNCS (LNAI), vol. 4724, pp. 501–511. Springer, Heidelberg (2007)
18. Rebane, G., Pearl, J.: The Recovery of Causal Poly-Trees from statistical Data. In: Third Conference on Uncertainty in Artificial Intelligence (UAI) (1987)
19. Zaragoza, J.C., Sucar, E., Morales, E.: A Two-Step Method to Learn Multidimensional Bayesian Network Classifiers Based on Mutual Information Measures. In: Twenty-Fourth International FLAIRS Conference. AAAI Press, Florida (2011)

20. Rodriguez, J.D., Lozano, J.A.: Multi-Objective Learning of Multi-Dimensional Bayesian Classifiers. In: Eighth International Conference on Hybrid Intelligent Systems, Barcelona (2008)
21. Borchani, H., et al.: Markov blanket-based approach for learning multi-dimensional Bayesian network classifiers: An application to predict the European Quality of Life-5 Dimensions (EQ-5D) from the 39-item Parkinson's Disease Questionnaire (PDQ-39). Journal of Biomedical Informatics 45(6), 1175–1184 (2012)
22. Borchani, H., et al.: Predicting human immunodeficiency virus inhibitors using multi-dimensional Bayeisan network classifiers. Artificial Intelligence in Medicine 57(3), 219–229 (2013)
23. Aliferis, C., Tsamardinos, I., Statnikov, A.: HITON, A Novel Markov Blanket Algorithm for Optimal Variable Selection. In: Annual Symposium on American Medical Informatics Association (AMIA) (2003)
24. Pena, J.M., et al.: Towards scalable and data efficient learning of Markov boundaries. International Journal of Approximate Reasoning 45(2), 211–232 (2007)

A Cost-Sensitive Based Approach for Improving Associative Classification on Imbalanced Datasets

Kitsana Waiyamai and Phoonperm Suwannarattaphoom

Data Analysis and Knowledge Discovery Laboratory (DAKDL), Computer Engineering
Department, Engineering Faculty, Kasetsart University, Bangkok, Thailand
{g521455020,kitsana.w}@ku.ac.th

Abstract. Associative classification is one of rule-based classifiers that has
been applied in many real-world applications. Associative classifier is easily
interpretable in terms of classification rules. However, there is room for
improvement when associative classification applied for imbalanced
classification task. Existing associative classification algorithms can be limited
in their performance on highly imbalanced datasets in which the class of
interest is the minority class. Our objective is to improve the accuracy of the
associative classifier on highly imbalanced datasets. In this paper, an effective
cost-sensitive rule ranking method, named (SSCR Statistically Significant Cost-
sensitive Rules), is proposed to estimate risks of a rule in classifying unseen
data. Risk of a statistically significant association rule is estimated based on its
classification cost induced from the training data. SSCR combines statistically
significant association rules with cost-sensitive learning to build an associative
classifier. Experimental results show that SSCR achieves best performance in
terms of true positive rate and recall on real-world imbalanced datasets,
compared with CBA and C4.5.

Keywords: Associative classification, Imbalanced class datasets, Cost-sensitive
learning.

1 Introduction

In general, the problem of class imbalance happens when there is a difference in class
distribution between minority and majority classes. The minority class is highly
skewed with very small proportion compared to the majority class. *Absolute rarity*
and *relative rarity* are among the active class imbalance problems [1]. *Absolute rarity*
happens when the number of examples associated with the minority class is small in
an absolute sense, where *relative rarity* happens when objects are not rare in an
absolute sense, but rare with respect to their relation with the other objects. Large
number of solutions to these problems has been previously proposed. For absolute
rarity, over-sampling can be used to address this problem; however this technique
leads to the over-fitting situation [15,16]. For relative rarity, appropriate evaluation
metrics, cost-sensitive learning, boosting and sampling are among the solutions that
have been investigated [12,13,14]. With inappropriate rule evaluation metrics, many
potentially interesting rules of the minority class are eliminated and the classifier

P. Perner (Ed.): MLDM 2014, LNAI 8556, pp. 31–42, 2014.
© Springer International Publishing Switzerland 2014

performance will decrease. Effective rule evaluation metrics to estimate risk of classification rules in classifying unseen data on imbalanced datasets are needed.

Many efficient associative classification algorithms have been proposed in the literature [2-4]. Based-on the support-confidence paradigm, these algorithms are biased toward the majority class on imbalanced datasets. They yield good accuracy for the majority class, however do not perform well for the minority class. Indeed, large number of potentially interesting rules of the minority class is eliminated by the minimum support threshold. In [5], the authors introduced a Class Correlation Ratio (CCR) measurement to evaluate how a rule is correlated with the class it predicts.

They combine statistically significant association rules with CCR to build an associative classifier. Although rule's CCR provides correlation with the class it predicts, higher CCR means the rule is positively correlated with the class it predicts than the other classes. However, performance in minority class is not guaranteed since false negative error is neglected. Rules with high CCR may contain more false negative errors than rules with lower CCR.

In this paper, we propose a cost-sensitive based approach for improving associative classification on imbalanced datasets. Statistically Significant Cost-sensitive Rules (SSCR) are generated to build an associative classifier. An effective cost-sensitive rule ranking method, named SSCR, is proposed to estimate risk of a rule in classifying unseen data. Risk of a statistically significant association rule is estimated based on its classification cost induced from the training data. Rules with lowest risk are preferred to rules with high risk because these rules will have more chances to predict the true classes. SSCR combines statistically significant association rules with cost-sensitive learning to build an associative classifier. Experimental results show that SSCR achieves best performance in terms of true positive rate and recall on real-world imbalanced dataset where minority class is highly skewed, compared with CBA [2] and C4.5 [9].

The rest of this paper is organized as follows. Section 2 describes the background knowledge related to associative classification, statistically significant association rules and cost-sensitive learning. Section 3 describes our SSCR technique. Section 4 compares the experimental results with related methods. Conclusion is given in Section 5.

2   Background Knowledge

This section gives background knowledge about associative classification, statistically significant rules selection and cost-sensitive learning.

2.1   Associative Classification

Associative classification is a data mining technique that combines the association rule mining (ARM) and classification together by using association rules in the form of class association rules (CAR) to classify new instances.

Association rule mining is to discover rules from a training dataset of transactions $T = \{t_1, t_2, ..., t_n\}$ where n is the number of transactions. Let $I = \{i_1, i_2, ..., i_d\}$ be the set of all items where d is the number of items. Each transaction $t_i \subseteq I$ represents an itemset.

An association rule R is an implication expression of the form $X \rightarrow Y$, where X and Y are disjointed itemsets. In general, the strength of an association rule can be measured in terms of support $sup(X \rightarrow Y)$ and confidence $conf(X \rightarrow Y)$ of rules. The support of rule R is the percentage of the instances in transactions T satisfying the Equation 1. The confidence of R rule is the percentage of transactions where Y itemset appears together with X itemset in all the transactions as shown in Equation 2.

$$sup(X \rightarrow Y) = \frac{\sigma(X \cup Y)}{n} \tag{1}$$

$$conf(X \rightarrow Y) = \frac{\sigma(X \cup Y)}{\sigma(X)} \tag{2}$$

Class association rule is an association rule and an implication of the form of $X \rightarrow Y$ which Y contains with itemsets of classes label only. Let $Y \in \{c_1, c_2, ..., c_m\}$ and c be the class and m is the number of all classes.

Associative classifier (AC) consists of three main processes which are rule generator, rule selection, and classification. In rule generator process, association rules in the form of CARs are discovered by using association rule generation algorithms such as Apriori [6] or its variation. In rule selection process, pruning and ranking techniques are performed to select small subset of CARs that yields highest accuracy. In the classification process, unseen instances are classified using CARs from the previous step.

2.2 Statistically Significant Rules

Webb [7] introduced the concept of statistically significant rules. In a traditional rule generation process, the output contains both productive rules and unproductive rules. Given an association rule of the form $X \rightarrow Y$, and $Z \in X$ if $P(Y|X) = P(Y|X\backslash\{Z\})$. Given $X\backslash\{Z\}$, Y and Z are conditionally independent. Therefore, unproductive and redundant rules should be eliminated.

To eliminate unproductive rules, Fisher's Exact Test (FET) is used to test the null hypothesis of instances with rules $X \rightarrow Y$ and $X\backslash\{Z\} \rightarrow Y$ that are conditionally independent. This test compares $X \rightarrow Y$ with each of immediate generalization $X\backslash\{Z\} \rightarrow Y$, where $Z \in X$. The probability (p_{value}) is calculated to test immediate generalization shown in Equation 3. If data is separated into two groups independently, the first group is $\{X \subseteq t_i \mid 1 \leq i \leq n\}$ and second group is $\{X\backslash\{Z\} \subseteq t_i \mid 1 \leq i \leq n\}$. Each group is separated into two classes Y and $\neg Y$ with null hypothesis At $H_0: P_1 = P_2$, that H_0 is an opportunity of both groups divided dataset into classes Y and $\neg Y$ as the same. For the alternative hypothesis $H_1: P_1 \neq P_2$, that H_1 is an opportunity of both groups to divide dataset into classes Y and $\neg Y$ differently.

To test the statistical hypothesis, H_0 will be rejected if p_{value} less than the critical value (α). In general $\alpha = 0.05$ is used, if testing $p_{value} > \alpha$, the null hypothesis H_0 will be accepted. This indicates that the rule $X \rightarrow Y$ is an unproductive rule. Inversely, if testing $p_{value} \leq \alpha$ the null hypothesis H_0 will be rejected. This indicates that the rule $X \rightarrow Y$ is a productive rule and it cannot be eliminated. Thus, it follows that the statistical hypothesis testing is able to discover all the significant association rules.

The process of mining significant rules is a time-consuming process. In order to reduce the number of rules, we consider only positively correlated rules using the interest factor measurement [10]. Although, its limitation on imbalanced datasets where the interested class is the majority class. However, the interest factor measure can be used to eliminate negatively correlated rules reduce the number of rules to be tested using the FET.

2.3   Cost-Sensitive Learning

Cost-sensitive classification considers the varying costs of different misclassification types. Cost-sensitive learning may be used to enhance the performance of rare class detection on imbalanced class distribution [10]. It uses a cost matrix to encode the penalty of classifying instances from one class to another class, as shown in Table 1.

Table 1. Cost matrix penalty

		Predited Class	
		Class = +	Class = −
Actual Class	Class = +	-1	100
	Class = −	1	0

Let $C(i,j)$ denote the cost of prediction an instance from class i to class j. When this notation $C(+,-)$ is the cost of misclassifying a positive (rare class) instances as the negative (prevalent class) instance and $C(-,+)$ is the cost of the contrary case. Cost-sensitive classifier uses the cost matrix to evaluate the model for lowest total cost. In highly imbalanced datasets, cost of positive class is much more than the one of negative class. To cover instances of positive class, cost-sensitive learning reduces number of false negative errors while increasing number of false positive errors.

The total cost of a model can be evaluated using Equation 4.

$$C(M) = TP \times C(+,+) + FP \times C(-,+) + FN \times C(+,-) + TN \times C(-,-) \qquad (4)$$

Where TP, FP, FN and TN correspond to true positive, false positive, false negative and true negative, respectively.

In this paper, cost-sensitive learning is used to estimate risk of a significant association rule to classify unseen data. Rules with lowest risk should be preferred to rules with high risk because these rules will have more chances to predict true actual classes. Risk of a significant association rule is measured based on its classification cost induced from the training data.

3   SSCR: Statistically Significant Cost-Sensitive Rules for Associative Classification

SSCR is composed of four components:

- Rule generation step generates all the association rules from the training data. Explanation of this step is given in Section 3.1.
- Rule pruning step determines all the potentially interesting rules with statistically significant test. Explanation of this step is given in Section 3.2.

- Rule ranking step determines risk of rules and assigns each rule a cost to be used for classification. Explanation of this step is given in Section 3.3.
- Classification step classifies an unseen instance by using the selected cost-sensitive association rules. Explanation of this step is given in Section 3.4.

3.1 Rules Generation Step

In this step, class association rules (CARs) are generated based-on Weka's Apriori implementation [8]. Minimum support and minimum confidence thresholds are set to zero in order to obtain the complete set of class association rules.

3.2 Rules Pruning Step

In this step, two pruning strategies are performed. The first pruning strategy is to keep only positive correlated rules with $corr(X \rightarrow Y) > 1$ and prune negative correlated rules. The correlation is calculated by using Equation 5. The idea behind is to reduce number of FET statistical tests in the next step.

$$corr(X \rightarrow Y) = \frac{P(X \cup Y)}{P(X) \times P(Y)} \tag{5}$$

The second pruning strategy is to keep only statistically significant rules with statistical hypothesis testing using FET. Table 2, is a contingency table for Fisher Exact Test (FET). For testing, we compare $X \rightarrow Y$ against each of immediate generalization $X \backslash \{Z\} \rightarrow Y$ where $Z \in X$. We calculate the probability p_{value} of every member of the Z itemset through Equation 6. Let $a = \sigma(X \rightarrow Y)$, $b = \sigma(X \backslash \{Z\} \rightarrow Y) - \sigma(X \rightarrow Y)$, $c = \sigma(X \rightarrow \neg Y)$ and $d = \sigma(X \backslash \{Z\} \rightarrow \neg Y) - \sigma(X \rightarrow \neg Y)$.

$$p_{value} = min_{Z \in X}(p[a, b; c, d]) \tag{6}$$

Table 2. A 2-way contingency table for rule $X \rightarrow Y$. Represent with the notation $[a, b; c, d]$

	X	$\neg X$	
Y	$a = \sigma(X \cup Y)$	$b = \sigma(\neg X \cup Y)$	$a + b$
$\neg Y$	$c = \sigma(X \cup \neg Y)$	$d = \sigma(\neg X \cup \neg Y)$	$c + d$
	$a + c$	$b + d$	$n = a + b + c + d$

From Equation 6, we represent rule $X \rightarrow Y$ in the form of contingency table that is shown in Table 3 and 4. For $\{X \rightarrow Y; |X| > 1\}$, the p_{value} is calculated using Table 3. For $\{X \rightarrow Y; |X| = 1\}$, the p_{value} is calculated using Table 4. For rule pruning, we apply general critical value of $\alpha = 0.05$ and consider only rules that are statistically significant when $p_{value} < \alpha$. Then, these rules will be selected. Otherwise, $p_{value} \geq \alpha$, these rules will be pruned.

Table 3. The contingency table $[a, b; c, d]$ for significance test of rule $X \rightarrow Y$ when $|X| > 1$

	$i: X \subset t_i$	$i: X \backslash \{Z\} \subset t_i \wedge Z \notin t_i$	$i: X \backslash \{Z\} \subset t_i$
$i: Y \in t_i$	$a = \sigma(X \rightarrow Y)$	$b = \sigma(X \backslash \{Z\} \rightarrow Y) - \sigma(X \rightarrow Y)$	$a + b = \sigma(X \backslash \{Z\} \rightarrow Y)$
$i: \neg Y \in t_i$	$c = \sigma(X \rightarrow \neg Y)$	$d = \sigma(X \backslash \{Z\} \rightarrow \neg Y) - \sigma(X \rightarrow \neg Y)$	$c + d = \sigma(X \backslash \{Z\} \rightarrow \neg Y)$
	$a + c = \sigma(X)$	$b + d = \sigma(X \backslash \{Z\}) - \sigma(X)$	$a + b + c + d = \sigma(X - \{Z\})$

Table 4. The contingency table $[a, b; c, d]$ for significance test of rule $X \rightarrow Y$ when $|X| = 1$

	$i: X \subset t_i$	$i: X \backslash \{Z\} \subset t_i \wedge Z \notin t_i$	$i: X \backslash \{Z\} \subset t_i$
$i: Y \in t_i$	$a = \sigma(X \rightarrow Y)$	$b = \sigma(Y) - \sigma(X \rightarrow Y)$	$a + b = \sigma(Y)$
$i: \neg Y \in t_i$	$c = \sigma(X \rightarrow \neg Y)$	$d = \sigma(\neg Y) - \sigma(X \rightarrow \neg Y)$	$c + d = \sigma(\neg Y)$
	$a + c = \sigma(X)$	$b + d = \sigma(X)$	$a + b + c + d = \sigma(X)$

In the experiments, the two pruning strategies are considered. The first strategy considers pruning of rules $\{X \rightarrow Y; |X| = 1\}$, and the second strategy without pruning rules $\{X \rightarrow Y; |X| = 1\}$. Result of this step is the set of productive interesting rules that are positively correlated and statistically significant. Details of the pruning algorithm are given in Figure 1.

1. Let T denote the set of training instances
2. Let R denote the set of rules found
3. Let α denote the significance level
4. Let m denote the pruning method
5. $R_p = \textbf{pruneRules}(T, R, \alpha, m)$
6. $R_p = \emptyset$
7. **for each** $r \in R$ **do**
8. $\quad r.corr = \frac{P(X \cup Y)}{P(X) \times P(Y)}$
9. $\quad$ **if** $(r.corr > 1)$ **then**
10. $\quad$ **if** $(|r.X| > 1)$ **then**
11. $$p_{value} = \left\{ r.X \backslash \{z\}.p_{value} \,\middle|\, \min_{z \in r.X} FET \begin{pmatrix} \sigma(r.X \rightarrow r.Y), \\ \sigma(r.X \backslash z \rightarrow r.Y) - \sigma(r.X \rightarrow r.Y); \\ \sigma(r.X \rightarrow \neg r.Y), \\ \sigma(r.X \backslash z \rightarrow \neg r.Y) - \sigma(r.X \rightarrow \neg r.Y) \end{pmatrix} \right\}$$
12. $\quad\quad$ **if** $(p_{value} < \alpha)$ **then**
13. $\quad\quad\quad R_p = R_p \cup r$
14. $\quad\quad$ **end if**
15. $\quad$ **else if** (m is prune $|r.X| == 1$) **then**
16. $$p_{value} = \left\{ r.X \backslash \{z\}.p_{value} \,\middle|\, \min_{z \in r.X} FET \begin{pmatrix} \sigma(r.X \rightarrow r.Y), \\ \sigma(r.Y) - \sigma(r.X \rightarrow r.Y); \\ \sigma(r.X \rightarrow \neg r.Y), \\ \sigma(\neg r.Y) - \sigma(r.X \rightarrow \neg r.Y) \end{pmatrix} \right\}$$
17. $\quad\quad$ **if** $(p_{value} < \alpha)$ **then**
18. $\quad\quad\quad R_p = R_p \cup r$
19. $\quad\quad$ **end if**
20. $\quad$ **else**
21. $\quad\quad R_p = R_p \cup r$
22. $\quad$ **end if**
23. $\quad$ **end if**
24. $\quad$ **end if**
25. **end for**
26. **return** R_p

Fig. 1. Pruning algorithm

3.3   Rule Ranking

To reduce the chance of misclassification, rules generated from the previous step can be ranked according to their degree of interestingness. We propose two ranking methods which are cost-based ranking and combination of rule size and cost-based ranking. The first ranking method is based on the concept of rule risk. Rules with lowest risk should be preferred to rules with high risk because these rules will have more chances to predict true actual classes. Risk of a significant association rule is measured based on its classification cost induced from the training data. Cost of a rule can be obtained by the following Equation 7.

$$C(X \rightarrow Y) = TP \times C(+,+) + FP \times C(-,+)$$
$$+FN \times C(+,-) + TN \times C(-,-)$$
(7)

Figure 2 shows the algorithm that determines cost of each rule, and assigns its cost. Notice that, we use the cost matrix in Table 1 in our experiments.

The second ranking method is based on the combination of rule size and cost-based ranking. Rules are ordered in descending order of rule' size $|X|$. If they have same size, then rules will be ordered in ascending order of their cost $C(X \rightarrow Y)$.

1.	Let R denote the set of rules found
2.	Let T denote the set of training instances
3.	Let $cost$ denote the cost matrix encodes
4.	Let pc denote the positive class
5.	**riskEstimate**$(T, R, cost, pc)$
6.	**for each** $r \in R$ **do**
7.	$tp, fp, fn, tn = 0$
8.	**for each** $t \in T$ **do**
9.	$tp += |\{t|r.X \subseteq t.X \wedge r.Y \in t.Y \wedge t.Y \in pc\}|$
10.	$tn += |\{t|r.X \subseteq t.X \wedge r.Y \in t.Y \wedge t.Y \notin pc\}|$
11.	$fp += |\{t|r.X \subseteq t.X \wedge r.Y \notin t.Y \wedge t.Y \notin pc\}|$
12.	$fn += |\{t|r.X \subseteq t.X \wedge r.Y \notin t.Y \wedge t.Y \in pc\}|$
13.	**end for**
14.	$r.cost = tp \times cost.C(+,+) + fp \times cost.C(-,+)$
	$+ fn \times cost.C(+,-) + tn \times cost.C(-,-)$
15.	**end for**

Fig. 2. Risk estimate algorithm

3.4   Classification

Once all the class association rules have been ranked, they together constitute a classifier which can be used to predict unseen instance. Figure 3 shows how SSCR algorithm performs rule selection to predict unseen instance. SSCR makes a decision by selecting rules with lowest cost of misclassification among all the matching rules.

If all the matching rules have equal cost and predict the same class, then SSCR will assign that class to the unseen instance.

If all the matching rules have equal cost but they predict different classes, then SSCR assign the class with majority vote to the unseen instance. In case of non majority class, SSCR will proceed to an alternative rule matching with higher cost to classify unseen instance according to the method described above.

If SSCR cannot choose any classes from the previous steps to predict unseen instance, SSCR will assign the class with majority vote from all the matching rules. In case of non majority class, SSCR will assign the default class (positive class) to the unseen instance. In the case of no matching rules with the unseen instance, SSCR will assign the default class to the unseen instance.

1.	Let R denote the set of rules found
2.	Let t denote an instance to classify
3.	Let pc denote the positive class
4.	Let c denote the predicted class
5.	**c=classifyInstance**(R, t, pc)
6.	$c_d = pc$ //default positive class
7.	$R_m = \{r \lvert r.X \subseteq t \wedge r \in R\}$
8.	$C_m = \{r.Y \lvert r \in R_m\}$
9.	**for each** $c \in C_m$ **do**
10.	$c.count += \lvert\{r \lvert r.Y == c\}\rvert$
11.	**end for**
12.	$gc = \{c \lvert \max_{c \in C_m} c.count\}$
13.	**if** ($\lvert gc \rvert > 1$) **then**
14.	$gc = \emptyset$
15.	**end if**
16.	$c_p = \emptyset$
17.	**repeat**
18.	$R_g = \{r \lvert \min_{r \in R_m} r.cost\}$
19.	**if** ($\lvert R_g \rvert == 1$) **then**
20.	$c_p = \{c \lvert c = r.Y \wedge r \in R_g\}$
21.	**else**
22.	$C_g = \{r.Y \lvert r \in R_g\}$
23.	**for each** $c \in C_g$ **do**
24.	$c.count += \lvert\{r \lvert r.Y == c\}\rvert$
25.	**end for**
26.	$lc = \{c \lvert \max_{c \in C_g} c.count\}$
27.	**if** ($\lvert lc \rvert > 1$) **then**
28.	$R_m = R_m - R_g$
29.	**else**
30.	$c_p = lc$
31.	**end if**
32.	**end if**
33.	**until** $c_p \neq \emptyset \vee R_m = \emptyset$
34.	**if** ($c_p = \emptyset \wedge gc \neq \emptyset$) **then**
35.	$c_p = gc$
36.	**end if**
37.	**if** ($c_p = \emptyset$) **then**
38.	$c_p = c_d$
39.	**end if**
40.	**return** c_p

Fig. 3. SSCR classifier algorithm

4   Experimental Results

In this section, we present experimental results of SSCR method. We select four datasets from the UCI machine learning repository [11] which are breast-cancer, yeast3, yeast6, and abalone19. All the four datasets have the imbalanced ratio (IR) varied from low to high. Table 5 summaries the properties of the selected datasets. For each dataset, we give number of instances (#Ins), number of attributes (#Atts), number of instanced in positive class (#Pcs), number of instanced in negative class (#Ncs) and number of classes (#Cls).

Table 5. Imbalanced datasets information

Datasets	#Ins	#Atts	#Pcs	#Ncs	#Cls
Breast-Cancer	286	10	85	201	2
Yeast3	1484	9	163	1321	2
Yeast6	1484	9	35	1449	2
Abalone19	4174	9	32	4142	2

Algorithm	Threshold	Measure	Ranking	Breast Cancer	Yeast3	Yeast6	Abalone19
CBA	$minSup = 1\%$	TP Rate		36.5%	23.3%	0.0%	0.0%
	$minCon = 25\%$	FN Rate		63.5%	76.7%	100.0%	100.0%
C4.5	$conf = 100\%$	TP Rate		37.6%	69.9%	42.9%	0.0%
		FN Rate		62.4%	30.1%	57.1%	100.0%
SSCR	$\alpha = 0.1$ $C = [-1, 100; 1,0]$ Prune R1	TP Rate	c	85.9%	90.8%	80.0%	50.0%
			rc	72.9%	90.8%	80.0%	56.3%
		FN Rate	c	14.1%	9.2%	20.0%	50.0%
			rc	27.1%	9.2%	20.0%	43.8%
	$\alpha = 0.1$ $C = [-1, 100; 1,0]$	TP Rate	c	85.9%	90.8%	80.0%	50.0%
			rc	72.9%	90.8%	80.0%	56.3%
		FN Rate	c	14.1%	9.2%	20.0%	50.0%
			rc	27.1%	9.2%	20.0%	43.8%
	$\alpha = 0.05$ $C = [-1, 100; 1,0]$ Prune R1	TP Rate	c	82.4%	91.4%	82.9%	62.5%
			rc	71.8%	**92.6%**	82.9%	65.6%
		FN Rate	c	17.6%	8.6%	17.1%	37.5%
			rc	28.2%	*7.4%*	17.1%	34.4%
	$\alpha = 0.05$ $C = [-1, 100; 1,0]$	TP Rate	c	84.7%	91.4%	82.9%	56.3%
			rc	71.8%	**92.6%**	82.9%	65.6%
		FN Rate	c	15.3%	8.6%	17.1%	43.8%
			rc	28.2%	*7.4%*	17.1%	34.4%
	$\alpha = 0.01$ $C = [-1, 100; 1,0]$ Prune R1	TP Rate	c	72.9%	**92.6%**	**85.7%**	68.8%
			rc	61.2%	**92.6%**	**85.7%**	**75.0%**
		FN Rate	c	27.1%	*7.4%*	*14.3%*	31.3%
			rc	38.8%	*7.4%*	*14.3%*	*25.0%*
	$\alpha = 0.01$ $C = [-1, 100; 1,0]$	TP Rate	c	**88.2%**	92.0%	82.9%	59.4%
			rc	61.2%	**92.6%**	**85.7%**	68.8%
		FN Rate	c	*11.8%*	8.0%	17.1%	40.6%
			rc	38.8%	*7.4%*	*14.3%*	31.3%

Fig. 4. TPR, FNR of minority class on imbalanced datasets

To test the performance of SSCR classifier on imbalanced datasets, we compare the experimental results with well-known classifiers that are CBA and decision tree classifier C4.5 using WEKA software. CBA requires the minimum support and confidence thresholds. We set $minSup = 1\%$ and $minConf = 25\%$ for experiments to obtain sufficient number of association rules. C4.5 requires the confidence factor for leaf node pruning, so we set $conf = 100\%$ to have all leaf nodes pure. SSCR requires significant level for statistical testing and cost matrix penalty as parameters. We set three significant levels $\alpha = \{0.1, 0.05, 0.01\}$, cost matrix $C = [-1, 100; 1,0]$, similar to Table 1. We use 10-fold cross validation to measure the performance of all the classifiers. Figure 4 shows the true positive rate (TPR) and the false negative rate (FNR) of the minority class.

Compared with CBA and C4.5, empirical analysis from the experimental results demonstrates that SSCR provides higher true positive rate (TPR) and lower false negative rate (FNR). For highly imbalanced dataset such as abalone19, SSCR also yields better result compared with CBA and C4.5 which predict toward the majority class with false negative rate (FNR) equal to 100%. Further improvement of the results can be obtained by using high quality of the statistically significant association rules. We obtained higher TPR when the significant level is set to $\alpha = 0.01$.

We can evaluate SSCR based on its ranking methods, $c = C(r)$ and $rc = |r.X|, C(r)$. Experimental results show that rc gives the better results with pruning or without pruning of rule size one. The best result can be obtained when rule size one

Algorithm	Threshold	Measure	Ranking	Breast Cancer	Yeast3	Yeast6	Abalone19
CBA	$minSup = 1\%$ $minConf = 25\%$	precision		**47.7%**	**92.7%**	0.0%	0.0%
		recall		36.5%	23.3%	0.0%	0.0%
		f-measure		41.3%	37.3%	0.0%	0.0%
		ROC		59.8%	61.5%	50.0%	50.0%
C4.5	$conf = 100\%$	precision		44.4%	67.5%	**55.6%**	0.0%
		recall		37.6%	69.9%	42.9%	0.0%
		f-measure		40.8%	**68.7%**	**48.4%**	0.0%
		ROC		58.1%	**88.9%**	72.8%	69.0%
SSCR	$\alpha = 0.1$ $C = [-1, 100; 1,0]$ Prune R1	precision	rc	35.8%	36.7%	13.1%	2.3%
		recall	rc	**72.9%**	90.8%	80.0%	56.3%
		f-measure	rc	48.1%	52.3%	22.5%	4.4%
		ROC	rc	58.9%	85.7%	83.6%	68.9%
	$\alpha = 0.05$ $C = [-1, 100; 1,0]$ Prune R1	precision	rc	37.7%	36.7%	12.6%	**2.4%**
		recall	rc	71.8%	**92.6%**	82.9%	65.6%
		f-measure	rc	**49.4%**	52.5%	21.8%	**4.6%**
		ROC	rc	60.8%	86.4%	**84.5%**	72.4%
	$\alpha = 0.01$ $C = [-1, 100; 1,0]$ Prune R1	precision	rc	40.9%	38.7%	9.5%	2.2%
		recall	rc	61.2%	**92.6%**	**85.7%**	**75.0%**
		f-measure	rc	49.1%	54.6%	17.1%	4.2%
		ROC	rc	**61.9%**	87.3%	83.0%	**74.5%**

Fig. 5. Precision, recall, f-measure , ROC of minority class on imbalanced datasets

pruning is performed. This can be explained from the observation that positive class occurs often rarely with minority class, thus rules size one have a strong relationship with majority classes.

Figure 5 shows experimental results of CBA, C4.5 and SSCR in terms of precision, recall, f-measure and ROC. For SSCR, we choose the rule ranking method rc only to compare that gives the best average results. CBA and C4.5 give better results in terms of precision. We can explain that both classifiers can handle the majority class better than the minority class with higher TNR and FNR. However, SSCR is able to handle the minority class better with higher FPR compared to CBA and C4.5. Notice that error in predicting actual positive class as a negative class is of higher cost.

C4.5 gives the best f-measure which is a harmonic mean of both precision and recall. However, on highly imbalanced dataset such Abalone19, SSCR yields higher accuracy compared to CBA and C4.5. Further, SSCR gives a better result in terms of ROC. Figure 6 shows performance of the three algorithms in terms of accuracy. For all the experimental datasets, SSCR obtains the best accuracy.

Algorithm	Threshold	Measure	Ranking	Breast Cancer	Yeast3	Yeast6	Abalone19
CBA	$minSup = 1\%$, $minConf = 25\%$	Accuracy		59.8%	61.6%	50.0%	50.0%
C4.5	$conf = 100\%$	Accuracy		58.9%	82.9%	71.1%	50.0%
SSCR	$\alpha = 0.1$, Prune R1	Accuracy	rc	58.9%	85.8%	83.6%	68.9%
	$\alpha = 0.05$, Prune R1	Accuracy	rc	60.8%	86.4%	**84.5%**	72.4%
	$\alpha = 0.01$, Prune R1	Accuracy	rc	**62.0%**	**87.3%**	83.0%	**74.5%**

Fig. 6. Accuracy of minority class on imbalanced datasets

5 Conclusion

In this paper we present a technique, SSCR for improving associative classification on imbalanced datasets. SSCR combines statistically significant association rules with cost-sensitive learning to build a associative classifier. We show that statistically significant association rules are efficient on highly imbalanced datasets. Cost sensitive learning is used to estimate risk of a significant association rule to classify unseen data. Rules with lowest risk are preferred to rules with high risk because these rules will have more chances to predict true actual classes.

SSCR yields good accuracy, higher true positive rate (TPR), lower false negative rate (FNR) and higher recall to handle imbalanced classes. SSCR has lower precision and f-measure because it loses some association rules of the majority class, thus there is room for improvement. Optimization of rule pruning, rule ranking and cost matrix should be further investigated.

References

1. Weiss, G.M.: Mining with Rarity: A Unifying Framework. Sigkdd Explorations 6(1), 7–19 (2004)
2. Liu, B., Hsu, W., Ma, Y.: Integrating Classification and Association Rule Mining. In: KDD, pp. 80–86 (1998)
3. Li, W., Han, J., Pei, J.: CMAR: Accurate and Efficient Classification Based on Multiple Class-Association Rules. In: ICDM, pp. 369–376 (2001)
4. Yin, X., Han, J.: CPAR: Classification based on Predictive Association Rules. In: SDM (2003)
5. Verhein, F., Chawla, S.: Using Significant, Positively Associated and Relatively Class Correlated Rules for Associative Classification of Imbalanced Datasets. In: ICDM, pp. 679–684 (2007)
6. Agrawal, R., Srikant, R.: Fast Algorithms for Mining Association Rules in Large Databases. In: VLDB, pp. 487–499 (1994)
7. Webb, G.I.: Discovering significant rules. In: KDD, pp. 434–443 (2006)
8. Hall, M., Frank, E., Holmes, G., Pfahringer, B., Reutemann, P., Witten, I.H.: The WEKA data mining software: an update. SIGKDD Explorations, 10–18 (2009)
9. Quinlan, J.R.: C4.5: Programs for Machine Learning
10. Tan, P., Steinbach, M., Kumar, V.: Introduction to Data Mining
11. Asuncion, A., Newman, D.J.: UCI Machine Learning Repository (2007)
12. Chai, X., Deng, L., Yang, Q., Ling, C.X.: Test-Cost Sensitive Naïve Bayesian Classification. In: Proceedings of the Fourth IEEE International Conference on Data Mining. IEEE Computer Society Press, Brighton (2004)
13. Domingos, P.: MetaCost: A general method for making classifiers cost-sensitive. In: Proceedings of the Fifth International Conference on Knowledge Discovery and Data Mining, pp. 155–164. ACM Press (1999)
14. Sheng, V.S., Ling, C.X.: Thresholding for Making Classifiers Cost-sensitive. In: Proceedings of the 21st National Conference on Artificial Intelligence, Boston, Massachusetts, July 16-20, pp. 476–481 (2006)
15. Japkowicz, N.: The Class Imbalance Problem: Significance and Strategies. In: Proceedings of the 2000 International Conference on Artificial Intelligence (IC-AI 2000): Special Track on Inductive Learning, Las Vegas, Nevada (2000)
16. Solberg, A., Solberg, R.: A Large-Scale Evaluation of Features for Automatic Detection of Oil Spills in ERS SAR Images. In: International Geoscience and Remote Sensing Symposium, Lincoln, NE, pp. 1484–1486 (1996)

Multiple Regression Method
Based on Unexpandable and Irreducible Convex
Combinations

Oleg Senko[*] and Alexander Dokukin

CC RAS, 40 Vavilova str., Moscow, Russia
senkoov@mail.ru, dalex@ccas.ru
http://www.ccas.ru

Abstract. A new multiple regression method based on optimal convex combinations of simple univariate regressions is discussed, where the simple regressions are searched with an ordinary least squares technique. Convex combination is considered optimal if it correlates with the response variable in the best way. It is shown that the developed approach is equivalent to a least squares technique variant regularized by constraints on signs of regression parameters. A method of optimal convex combination search is discussed that is based on unexpandable and irreducible ensembles. The developed method is compared with elastic networks at simulated data.

Keywords: regression, convex combinations, regularization.

1 Introduction

The standard multiple regression task is considered. The response variable Y is predicted by variables $X_1, \ldots, X_n$ with the help of linear regression function $\beta_0 + \sum_{i=1}^{n} \beta_i X_i$. It is assumed that the vector $\boldsymbol{\beta} = (\beta_0, \ldots, \beta_n)$ is chosen from $\Re^{n+1}$ by observations $(y_j, x_{1j}, \ldots, x_{nj})$, where $j \in \{1, \ldots, m\}$. Ridge regression, lasso [10] and elastic net [12] are now popular methods for searching regression coefficients in task with highdimensional data. These methods are based on solving optimization task of the type

$$\min_{\beta \in \Re^{n+1}} \left\{ \sum_{j=1}^{m} \left(y_j - \beta_0 - \sum_{i=1}^{n} \beta_i x_{ij} \right)^2 + \gamma P(\boldsymbol{\beta}) \right\},$$

where $\gamma \geq 0$ and $P(\boldsymbol{\beta}) = P(\beta_0, \ldots, \beta_n)$ is penalty function. At that the lasso penalty is $\sum_{i=1}^{n} |\beta_i|$, the ridge penalty is $\sum_{i=1}^{n} \beta_i^2$, the elastic net penalty is

$$(1 - \alpha) \sum_{i=1}^{n} \beta_i^2 + \alpha \sum_{i=1}^{n} |\beta_i|, \tag{1}$$

---

[*] The publication is supported by Russian Foundation for Basic Research grant No. 14-07-00819.

P. Perner (Ed.): MLDM 2014, LNAI 8556, pp. 43–57, 2014.
© Springer International Publishing Switzerland 2014

where parameter $\alpha \in [0,1]$ characterizes compromise between ridge and lasso penalty.

Let's note that another regularization way exists that is based on solving optimization task

$$\min_{\beta \in \Re^{n+1}} \left\{ \frac{1}{m} \sum_{j=1}^{m} \left(y_j - \beta_0 - \sum_{i=1}^{n} \beta_i x_{ij} \right)^2 \right\} \qquad (2)$$

$$C_1(\beta_0, \ldots, \beta_n) \geq 0,$$

$$\ldots$$

$$C_k(\beta_0, \ldots, \beta_n) \geq 0,$$

where $C_1, \ldots, C_k$ are some functions of regression parameters. This way of regularization is used for example in lasso variant described in [3]. We consider constraints based on demand that sign of regression coefficient for variable X_i is equal to sign of Pearson correlation coefficient between Y and X_i. This coefficient will be denoted as $\rho(Y, X_i)$. It will be shown that optimization task of the type (2) with constraints corresponding to the mentioned demand may be reduced to search of optimal convex combination of simple regressions that are built by each of variables $X_1, \ldots, X_n$.

Convex combinations are widely used in pattern recognition. The bagging and boosting techniques [1,5] may be mentioned as an example, as well as methods based on collective solutions by sets of regularities [11]. Convex correction is used in regression tasks also. Thus, neural networks ensembles based on optimal balance between individual forecasting ability of predictors and divergence between them are discussed in [2]. In works [8,7] a method was discussed that was based on searching optimal convex combinations of simple one-dimensional linear regression functions $(R_1, \ldots, R_n)$ calculating Y from variables $(X_1, \ldots, X_n)$ correspondingly. At that the minimal forecasting error is the target. Experiments with simulated data demonstrated rather high correlation between optimal convex combination and Y. Experiments also showed that variances of convex combinations usually are small. This fact may be related to structure of convex combinations variances. Let $\widetilde{Z} = \{Z_1, \ldots, Z_l\}$ be some set of random variables defined at probability space Ω and $P(\widetilde{Z}, \mathbf{c}) = \sum_{i=1}^{l} c_i Z_i$ is arbitrary convex combination of variables from $\widetilde{Z}$, where $\mathbf{c}$ belongs to standard simplex. Then decomposition

$$P(\widetilde{Z}, \mathbf{c}) = \sum_{i=1}^{l} c_i V(Z_i) - \frac{1}{2} \sum_{i=1}^{l} \sum_{j=1}^{l} c_i c_j \varrho(Z_i, Z_j) \qquad (3)$$

is true for variance of $\widetilde{Z}$, where

$$\varrho(Z_i, Z_j) = E_\Omega (Z_i - E_\Omega Z_i - Z_j + E_\Omega Z_j)^2$$

and $V(Z_i)$ is the variance of Z_i. The decomposition was discussed in ([7]). It follows from (3) that $V(\widetilde{Z}) \leq \sum_{i=1}^{l} c_i V(Z_i)$. So an additional linear transformation is necessary to improve forecasting ability. It is natural to search such a transformation as a simple linear regression calculating Y from $P(\widetilde{R}, \mathbf{c})$. But forecasting error of simple regression $\beta_0 + \beta_1 X$ monotonically depends on $\rho(Y, X)$. So, the search of convex combination with the best forecasting ability is naturally reduced to a search of convex combination that in the best way correlates with the response [9].

2   Optimal Convex Combination as a Way of Regularization

Let's consider a method of optimal convex combination that consists of three stages. At the first stage a set of n simple linear regression models for predicting response Y from each of covariates $X_1, \ldots, X_n$ is built by OLS technique:

$$\widetilde{R} = \{R_1 = \beta_0^{u1} + \beta_1^{u1} X_1, \ldots$$

$$\ldots, R_n = \beta_0^{un} + \beta_1^{un} X_n\}. \tag{4}$$

At the second stage such convex combination of regression models $R_1, \ldots, R_n$ is searched that correlation coefficient between this combination and response Y is maximal. In other words we search such $\mathbf{c}^b = (c_1^b, \ldots, c_n^b)$ in a standard simplex that for arbitrary $\mathbf{c} = (c_1, \ldots, c_n)$ from a standard simplex inequality $\rho[Y, P(\widetilde{R}, \mathbf{c}^b)] \geq \rho[Y, P(\widetilde{R}, \mathbf{c})]$ is true. At the third stage a simple linear regression function $\beta_0^c + \beta_1^c P(\widetilde{R}, \mathbf{c}^b)$ calculating response Y from convex combination $P(\widetilde{R}, \mathbf{c}^b)$ is searched from observations with the help of OLS. The described method will be further referred to as Convex Regression (CR) and the regression function $\beta_0^c + \beta_1^c (\sum_{i=1}^{n} c_i^b \beta_0^{ui}) + \beta_1^c (\sum_{i=1}^{n} c_i^b \beta_1^{ui} X_i)$ will be called a CR solution.

Let $I_p = \{i | i = 1, \ldots, n, \rho(Y, X_i) \neq 0\}$, $I_0 = \{1, \ldots, n\} \setminus I_p$. It is evident that coefficient β_i^{u1} is equal 0 if $\rho(Y, X_i) = 0$. Let some CR solution be built by convex combination $\sum_{i=1}^{n} c_i R_i$ where there is such $i \in I_0$ that $c_i > 0$. Then identical convex combination exists where $c_i > 0$ only when $i \in I_p$. Let's try to show that CR is equivalent to the following variant of LS technique with constraints of the type (2). A vector of regression coefficients is a solution of the following optimization task:

$$\min_{\beta \in \Re^{n+1}} \left\{ \frac{1}{m} \sum_{j=1}^{m} \left(y_j - \beta_0 - \sum_{i=1}^{n} \beta_i x_{ij} \right)^2 \right\} \tag{5}$$

$$\beta_i \rho[Y, X_i] \geq 0, i \in I_p, \beta_i = 0, i \in I_0.$$

Two statements are true.

Statement 1. *Let $P(\widetilde{R}_p, \mathbf{c})$ be some convex combination of regression functions from $\widetilde{R}_p = \{R_i | i \in I_p\}$ and β_0^c, β_1^c are coefficients of linear OLS regression calculating response Y from the convex combination $P(\widetilde{R}_p, \mathbf{c})$. Then the constraints from the task (5) are satisfied for coefficients $(\beta_1, \ldots, \beta_n)$ of a CR solution.*

Indeed, we must show that $sign(\beta_i) = sign[\rho(Y, X_i)]$. For each linear regression $R_i = \beta_0^{ui} + \beta_1^{ui} X_i$ from $\widetilde{R}_p$ $sign(\beta_1^{ui}) = sign[\rho(Y, X_i)]$. So, $\rho(Y, R_i) > 0$ for each R_i from $\widetilde{R}_p$. Correlation coefficient $\rho[Y, P(\widetilde{R}_p, \mathbf{c})] > 0$ because $P(\widetilde{R}_p, \mathbf{c})$ in any way is sum of functions that are positively correlated with Y. So, $\beta_1^c \geq 0$. It follows from equality $\beta_i = c_i \beta_1^c \beta_1^{ui}$ that $sign(\beta_i) = sign(\beta_i^u)$ or $\beta_i = 0$ when $i \in I_p$ and always $\beta_i = 0$ when $i \in I_0$. Q.E.D.

Statement 2. *For any regression coefficients β_0 and β_i, $(i \in I_p)$ satisfying constraints of the task (5) there are such points $c_1, \ldots, c_n$ in a standard simplex and numbers β_0^c and β_1^c that the following equality is true*

$$\beta_0 + \sum_{i \in I_p} \beta_i X_i = \beta_0^c + \beta_1^c \sum_{i \in I_p} c_i R_i . \tag{6}$$

Indeed, let $\hat{c}_i = |\beta_i|/|\beta_1^{ui}|$ when $i \in I_p$. Equality (6) is true when $c_i = \hat{c}_i / \sum_{j \in I_p} \hat{c}_j$ and coefficients $\beta_1^c = \sum_{j \in I_p} \hat{c}_j$, $\beta_0^c = \beta_0 - \sum_{j \in I_p} c_i \beta_0^{ui}$.
The following theorem follows from statements (1) and (2).

Theorem 1. *Any solution of the task (5) is a CR solution. Any CR solution is also a solution of the task (5).*

Let vector $(\beta_0^*, \ldots, \beta_n^*)$ be a solution of the optimization task (5). Let's note that according to the Statement (2) coefficients $(\beta_0^*, \ldots, \beta_n^*)$ can be received from regressions $R_1, \ldots, R_n$ by use of corresponding convex combination $P(\widetilde{R}, \mathbf{c}^*)$ and linear transformation $\beta_0^{c*} + \beta_1^{c*} P(\widetilde{R}, \mathbf{c}^*)$. The latter must be OLS regression calculating Y from $P(\widetilde{R}, \mathbf{c}^*)$. Let's suppose that $\beta_0^{c*}, \beta_1^{c*}$ are not coefficients of OLS regression. Then squared error for $(\beta_0^*, \ldots, \beta_n^*)$ is less than squared error for coefficients $(\beta_0^{**}, \ldots, \beta_n^{**})$ that are received from convex combination $P(\widetilde{R}, \mathbf{c}^*)$ with the help of OLS regression. Besides according the statement (1) constraints (5) are satisfied for $(\beta_0^{**}, \ldots, \beta_n^{**})$. So vector $(\beta_0^*, \ldots, \beta_n^*)$ is not a solution of task (5).

Let's suppose that $\rho[Y, P(\widetilde{R}_p, \mathbf{c}^*)]$ is not maximal. In other words there is such a point $\mathbf{c}' = (c_1', \ldots, c_n')$ in a standard simplex that for $P(\widetilde{R}, \mathbf{c}')$

$$\rho[Y, P(\widetilde{R}, \mathbf{c}^*)] < \rho[Y, P(\widetilde{R}, \mathbf{c}')]. \tag{7}$$

Suppose that $\beta_0^{c'}, \beta_1^{c'}$ are coefficient of OLS regression calculating Y from $P(\widetilde{R}, \mathbf{c}')$. Let's $(\beta_0', \ldots, \beta_n')$ are such coefficients that the equation

$$\beta_0' + \sum_{i \in I_p} \beta_i' X_i = \beta_0^{c'} + \beta_1^{c'} P(\widetilde{R}, \mathbf{c}') \tag{8}$$

is true. It follows from the statement (1) that constraints of task (5) are satisfied for coefficients $(\beta_0', \ldots, \beta_n')$. But it is known that squared error of OLS regression calculating Y from some variable Z monotonically depends on $\rho(Y, Z)$.

So squared error for $(\beta_0', \ldots, \beta_n')$ is less than squared error for $(\beta_0^*, \ldots, \beta_n^*)$. Thus, there is a contradiction and any solution of task (5) is also CR solution. The first statement of theorem is proved.

Let $(\beta_0^*, \ldots, \beta_n^*)$ be a CR solution. In other words there is such a convex combination $P(\widetilde{R}, \mathbf{c}^*)$ that the equality $\rho[Y, P(\widetilde{R}_p, \mathbf{c}^*)] \geq \rho[Y, P(\widetilde{R}_p, \mathbf{c})]$ is true for any convex combination $P(\widetilde{R}_p, \mathbf{c})$ and equation

$$\beta_0^* + \sum_{i \in I_p} \beta_i^* X_i = \beta_0^{c*} + \beta_1^{c*} P(\widetilde{R}, \mathbf{c}^*) \tag{9}$$

is true for coefficients of OLS regression calculating Y from $P(\widetilde{R}_p, \mathbf{c}^*)$.

Let's suppose that $(\beta_0^*, \ldots, \beta_n^*)$ is not a solution of the task (5). Then there are such coefficients $(\beta_0', \ldots, \beta_n')$ that squared error of Y prediction for function $\beta_0^* + \sum_{i=1}^n \beta_i^* X_i$ is greater than squared error for function $\beta_0' + \sum_{i=1}^n \beta_i' X_i$. It follows from the statement (2) that there is such convex combination $P(\widetilde{R}_p, \mathbf{c}')$ and numbers $\beta_0^{c'}$ and $\beta_1^{c'}$ that equation

$$\beta_0' + \sum_{i \in I_p} \beta_i' X_i = \beta_0^{c'} + \beta_1^{c'} P(\widetilde{R}_p, \mathbf{c}') \tag{10}$$

is true.

Let $\beta_0^{c''}$ and $\beta_1^{c''}$ be coefficients of OLS regression calculating Y from $P(\widetilde{R}, \mathbf{c}')$. It is evident that squared error of Y prediction for linear function $\beta_0^{c''} + \beta_1^{c''} P(\widetilde{R}, \mathbf{c}')$ is less than squared error for $\beta_0^* + \sum_{i=1}^n \beta_i^* X_i$. But then equation the $\rho[Y, \widetilde{R}, \mathbf{c}')] > \rho(Y, \widetilde{R}, \mathbf{c}^*)]$ holds. So, there is a contradiction and $(\beta_0^*, \ldots, \beta_n^*)$ is a solution of the task (5). Thus, the second statement is true and the theorem is proved.

3 Optimal Convex Combination Search

3.1 Optimization Task

Our goal is a convex combination $P(\widetilde{R}, \mathbf{c}^b)$ with maximal Standard Pearson correlation coefficient between Y and $P(\widetilde{R}, \mathbf{c}^b)$. Pearson correlation coefficient for arbitrary $P(\widetilde{R}, \mathbf{c})$ is defined as the ratio:

$$\rho(Y, P(\widetilde{R}, \mathbf{c})] = \frac{cov\left(Y, P(\widetilde{R}, \mathbf{c})\right)}{\sqrt{V(Y)V[P(\widetilde{R}, \mathbf{c})]}} \quad ,$$

where $cov(Y, P) = E_\Omega\{(Y - E_\Omega Y)[P - E_\Omega P)\}$.

The following well known equalities are true for any linear OLS regression R:

$$E_\Omega R = E_\Omega Y, cov(Y, R) = V(R) \tag{11}$$

Let's note that $cov[Y, P(\widetilde{R}, \mathbf{c})] = \sum_{i=1}^n c_i cov(Y, R_i)$. But R_i is a simple regression that is built with the help of OLS method. So taking into account (3), (11) we receive that

$$\rho[Y, P(\tilde{R}, \mathbf{c})] = \sum_{i=1}^{n} c_i V(R_i) \left[\sqrt{V(Y)} \right]^{-1} \times$$

$$\times \left[\sqrt{\sum_{i=1}^{n} c_i V(R_i) - \frac{1}{2} \sum_{i=1}^{n} \sum_{j=1}^{n} c_i c_j \varrho(R_i, R_j)} \right]^{-1},$$

where $\varrho(R_i, R_j) = E_\Omega(R_i - R_j)^2$. The discussed further optimization procedure is based on irreducible ensemble concept.

3.2   Irreducible Ensemble

Let $\tilde{r}$ be some set of regression functions predicting Y (predictors). Small letter r is used in this section to distinguish between initial set of predictors $\tilde{R}$ and ensemble $\tilde{r}$ that consists of predictors from $\tilde{R}$. Set $\tilde{r}$ is called irreducible ensemble if convex combination of predictors from $\tilde{r}$ with the best prognostic ability includes all predictors with nonnegative coefficients. In other words all predictors are necessary to achieve the best prognostic ability. Let's give a strict definition of ensembles irreducibility.

Definition 1. *Sets* $\overline{D}_l$, D_l *from* $\Re^l$ *are defined as*

$$\overline{D}_l = \left\{ \mathbf{c} \mid \sum_{i=1}^{l} c_i = 1, c_i \geq 0, i = 1, \dots, l \right\},$$

$$D_l = \left\{ \mathbf{c} \mid \sum_{i=1}^{l} c_i = 1, c_i > 0, i = 1, \dots, l \right\}.$$

Definition 2. *A set of predictors* $r_1, \dots, r_l$ *is called irreducible ensemble relative to Pearson correlation coefficient if a)* $l = 1$; *b) there is such a vector* $\mathbf{c}^* \in D_l$, *that for any* $\mathbf{c}' \in \overline{D}_l \backslash D_l$, $\rho[Y, P(\tilde{r}, \mathbf{c}^*)] > \rho[Y, P(\tilde{r}, \mathbf{c}')]$.

Definition 3. *An irreducible ensemble* $\tilde{r}$ *consisting of* l *predictors will be called unexpandable irreducible ensemble (UIE) if there is no such an irreducible ensemble* $\tilde{r}'$ *with the number of predictors equal* $l + 1$ *that each predictor from* $\tilde{r}$ *belongs to the* $\tilde{r}'$.

Let,s suppose that ρ_b is maximal value of $\rho[Y, P(\tilde{r}, \mathbf{c})]$ when $\mathbf{c} \in \overline{D}_n$.

Theorem 2. *There is such UIE* $\tilde{r}^b \subseteq \tilde{R}$ *that* $\rho[Y, P(\tilde{r}^b, \mathbf{c}^b)] = \rho_b$, *where* $\mathbf{c}^b = (c_1^b, \dots, c_{|\tilde{r}^b|}^b)$ - *coefficients of optimal convex combination for UIE* $\tilde{r}^b$.

Let's suppose that $\rho[Y, P(\tilde{R}, \mathbf{c}^b)] = \rho_b$ for some $\mathbf{c}^b \in \overline{D}_n$. Let's show that there is such irreducible ensemble $\tilde{r}^i$ that $\rho[Y, P(\tilde{r}^i, \mathbf{c}^b)] = \rho_b$. Let's $\tilde{R}$ be not irreducible ensemble. So by definition (2) vector $\mathbf{c}^b \in \overline{D}_n \backslash D_n$. However it means that there is such ensemble $\tilde{r}^1 \subseteq \tilde{R}$ that $\tilde{r}^1 < n$ and $\rho[Y, P(\tilde{r}^1, \mathbf{c}^b)] = \rho_b$. Let $\tilde{r}^1$ be not irreducible ensemble. So by definition (2) vector $\mathbf{c}^b \in \overline{D}_{n-1} \backslash D_{n-1}$.

However it means that there is such ensemble $\tilde{r}^2 \subseteq \tilde{r}^2$ that $\tilde{r}^2 < (n-1)$. We may continue this process until irreducible ensemble is achieved.

Let's suppose that there is such irreducible ensemble $\tilde{r}^b \subseteq \tilde{R}_p$ that equality $\rho[Y, P(\tilde{r}^b, \mathbf{c}^b)] = \rho_b$ is true. Let's show that $\tilde{r}^b$ is unexpandable. Let's suppose that $\tilde{r}^b$ is not unexpandable. Then such set $\tilde{r}' \in \tilde{R}_p$ must exist that $\tilde{r}^b \subset \tilde{r}'$, $|\tilde{r}'| = |\tilde{r}^b| + 1$ and $\rho[Y, P(\tilde{r}', c^b)] > \rho[Y, P(\tilde{r}^b, \mathbf{c}^b)]$. So, $\rho[Y, P(\tilde{r}', c^b)] > \rho_b$ and we come to contradiction and $\tilde{r}^b$ is not irreducible. This process may be continued until irreducible ensemble is achieved. Q.E.D.

So to find convex combination $P(\tilde{R}, \mathbf{c}^b)$ with maximal $\rho[Y, P(\tilde{R}, \mathbf{c}^b)]$ it is sufficient to find optimal convex combination for each UIE from $\tilde{R}$ and to select UIE $\tilde{r}^b_{irr}$ with maximal $\rho[Y, P(\tilde{r}^b_{irr}, \mathbf{c}^b)]$. So it is necessary to have effective methods allowing to evaluate irreducibility and to calculate optimal convex combination. Rather easy procedures exist when $l = 2$.

A set of points from $\mathfrak{R}^l$ simultaneously satisfying constraints: $\sum_{i=1}^{l} c_i = 1$, $\sum_{i=1}^{l} c_i V(r_i) = \Theta$ and $c_i \geq 0$, $i = 1, \ldots, l$ will be further referred to as $\mathbf{W}(\Theta)$. In case ensemble $\tilde{r} = \{r_1, r_2\}$ with $V(r_1) > V(r_2)$ set $\mathbf{W}(\Theta)$ contains single point $\mathbf{c} = (c_1, c_2) : c_1 = [\Theta - V(r_1)]/[V(r_1) - V(r_2)]$, $c_2 = [V(r_2) - \Theta]/[V(r_1) - V(r_2)]$. Let note that $\mathbf{c} \in D_2$ only when $\Theta \in (V(r_2), V(r_1))$. Let's consider Pearson correlation coefficient as function of Θ. Then $\rho[Y, P(\tilde{r}, \Theta)] = \dfrac{\Theta}{\sqrt{V(Y)}} \times$

$[\sqrt{\Theta - \varrho(r_1, r_2)\dfrac{[\Theta - V(r_2)][V(r_1) - \Theta]}{[V(r_1 - V(r_2)]^2}}]]^{-1}$. It is rather easily to show that stationarity point of $\rho[Y, P(\tilde{r}, \Theta)]$ is unique and may be calculated as

$$\Theta^* = \frac{-2\varrho(r_1, r_2)V(r_1)V(r_2)}{[V(r_1) - V(r_2)]^2 - \varrho(r_1, r_2)[V(r_1) + V(r_2)]}.$$

Extremum of $\rho[Y, P(\tilde{r}, \Theta)]$ is achieved inside interval $(V(r_1), V(r_2))$ only when $\Theta^* \in (V(r_1), V(r_2))$. Otherwise extremum is achieved at $\overline{D}_2 \backslash D_2$. So ensemble $\tilde{r}$ is irreducible only when $\Theta^* \in (V(r_1), V(r_2))$ and $\rho[Y, P(\tilde{r}, \Theta^*)] > \rho\{Y, P[\tilde{r}, V(r_1)]\}$ or $\rho[Y, P(\tilde{r}, \Theta^*)] > \rho\{Y, P[\tilde{r}, V(r_2)]\}$.

Discussion for cases when $l > 2$ is based on following theorem.

Theorem 3. *A necessary condition of irreducibility of predictors set $r_1, \ldots, r_l$ relative to Pearson correlation coefficient is an existence of such real positive Θ that quadratic functional*

$$\mathbf{Q}_f(\mathbf{c}) = \sum_{i=1}^{l} \sum_{j=1}^{l} c_i c_j \varrho(r_i, r_j)$$

achieves its strict maximum on $\mathbf{W}(\Theta)$ at a point $\mathbf{c}^* = (c_1^*, \ldots, c_l^*)$ that satisfies conditions $c_i^* > 0$, $i = 1, 2, \ldots, l$.

Let's suppose that a set $r_1, \ldots, r_l$ is irreducible ensemble relative to Pearson correlation coefficient. According to the classical Weierstrass theorem about attainability of the extreme values by continuous functions on compact sets

$\rho[Y, P(\widetilde{r}, \mathbf{c})]$ attains its maximum on set $\overline{D}_1$ at a point $\mathbf{c}^* \in D_1$. Let $\Theta^* = \sum_{i=1}^{l} c_i^* V(r_i)$. Since $\rho[Y, (\widetilde{r}, \mathbf{c})]$ is monotonic function of $\mathbf{Q}_f(\mathbf{c})$ on $\mathbf{W}(\Theta^*)$ the functional $\mathbf{Q}_f(\mathbf{c})$ attains its maximum on $\mathbf{W}(\Theta^*)$ at the point $\mathbf{c}^* \in D_1$.

Lets suppose also that $\mathbf{c}_n$ is some point in $\mathbf{W}(\Theta^*)$ that don't coincide with $\mathbf{c}^*$ and $\mathbf{v} = \mathbf{c}_n - \mathbf{c}^*$. Let $S(\mathbf{c}^*, \mathbf{v})$ be a straight line that is defined by the direction $\mathbf{v}$ and the point $\mathbf{c}^*$. Let $Cut\,[S(\mathbf{c}^*, \mathbf{v}), \mathbf{c}^*, \mathbf{c}_b]$ is a cut of $S(\mathbf{c}^*, \mathbf{v})$ with boundary points $\mathbf{c}^*$ and $\mathbf{c}_b \in \overline{D}_L \backslash D_1 \bigcap \mathbf{W}(\Theta^*)$. It follows from the stationarity at $\mathbf{c}^*$ that the first derivative along direction $\mathbf{v}$ is equal 0 at $\mathbf{c}^*$. Inequality $\mathbf{Q}_f(\mathbf{c}^*) > \mathbf{Q}_f(\mathbf{c}_b)$ is correct due to irreducibility condition. Second directional derivative of $\mathbf{Q}_f(\mathbf{c})$ for an arbitrary direction is constant in $\Re^l$. So, first directional derivative along $\mathbf{v}$ is positive in every point of interval of $S(\mathbf{c}^*, \mathbf{v})$ between points $\mathbf{c}^*$ and $\mathbf{c}_b$. Consequently, $\mathbf{Q}_f(\mathbf{c}^*) > \mathbf{Q}_f(\mathbf{c}')$ for every point $\mathbf{c}'$ from $Cut[S(\mathbf{c}^*\mathbf{v}), \mathbf{c}^*, \mathbf{c}_b]$ different from $\mathbf{c}^*$. Thus, the point $\mathbf{c}^*$ is the strict maximum point on $\mathbf{W}(\Theta^*)$. Q.E.D.

Let's consider necessary conditions for a point $\mathbf{c}^* = (c_1^*, \dots, c_l^*)$ to be a maximum point of function $\sum_{i=1}^{l} \sum_{j=1}^{l} c_i c_j \varrho(r_i, r_j)$ at set $\mathbf{W}(\Theta^*)$.

One of the necessary conditions is stationarity of Lagrange functional

$$L = \sum_{i=1}^{l} \sum_{j=1}^{l} c_i c_j \varrho(r_i, r_j) +$$

$$+ \lambda \left(\sum_{i=1}^{l} c_i V(r_i) - \Theta \right) + \mu \left(\sum_{i=1}^{l} c_i - 1 \right),$$

at point $\mathbf{c}^* = (c_1^*, \dots, c_l^*)$. In other words partial derivative of Lagrange functional

$$\frac{\partial L}{\partial c_k} \bigg|_{\mathbf{c}^*} = \sum_{i=1}^{l} c_i^* \varrho(r_i, r_k) + \lambda V(r_k) + \mu = 0 , \qquad (12)$$

$$\frac{\partial L}{\partial \lambda} \bigg|_{\mathbf{c}^*} = \sum_{i=1}^{l} c_i^* V(r_i) - \Theta = 0, \quad \frac{\partial L}{\partial \mu} \bigg|_{\mathbf{c}^*} = \sum_{i=1}^{l} c_i^* - 1 = 0 .$$

Suppose that matrix $\mathbf{P}^{-1}$ inverse to $\mathbf{P}$ exists. Then it is possible to find a function calculating $\mathbf{c}^*$ from Θ and $\varrho(r_i, r_j)$, $i, j \in \{1, \dots, l\}$. Moving from (12) to a vectorial form we get

$$\mathbf{c}^* \mathbf{P} + \lambda \mathbf{V} + \mu \mathbf{I} = \mathbf{O} , \qquad (13)$$

$$\mathbf{c}^* \mathbf{V} = \Theta , \qquad (14)$$

$$\mathbf{c}^* \mathbf{I} = 1 , \qquad (15)$$

where $\mathbf{P} = \|\varrho(r_i, r_j)\|_{l \times l}$, $\mathbf{V} = \|V(r_i)\|_{l \times 1}$, $\mathbf{I} = \|1\|_{l \times 1}$, $\mathbf{O} = \|0\|_{l \times 1}$.

Let's multiply (13) by $\mathbf{P}^{-1}$. Taking into account that $\mathbf{c}^*\mathbf{P} = \mathbf{P}(\mathbf{c}^*)^T$ we get $(\mathbf{c}^*)^T + \lambda\mathbf{P}^{-1}\mathbf{V} + \mu\mathbf{P}^{-1}\mathbf{I} = \mathbf{O}$, that can be further transformed with the use of (14) and (15) to $\Theta + \lambda\mathbf{V}^T\mathbf{P}^{-1}\mathbf{V} + \mu\mathbf{V}^T\mathbf{P}^{-1}\mathbf{I} = 0$, and $1 + \lambda I^T\mathbf{P}^{-1}\mathbf{V} + \mu I^T\mathbf{P}^{-1}I = 0$ accordingly. Lets denote $\alpha = \mathbf{V}^T\mathbf{P}^{-1}\mathbf{V}$, $\beta = I^T\mathbf{P}^{-1}\mathbf{V} = \mathbf{V}^T\mathbf{P}^{-1}\mathbf{I}$, $\gamma = \mathbf{I}^T\mathbf{P}^{-1}\mathbf{I}$ for short. The received equation system gets the following form $\Theta + \lambda\alpha + \mu\beta = 0$, $1 + \lambda\beta + \mu\gamma = 0$. From these equations the dependence between $\mathbf{c}^*$ and Θ can be derived $\lambda = \frac{\beta - \Theta\gamma}{\alpha\gamma - \beta^2}$, $\mu = \frac{\alpha - \Theta\beta}{\beta^2 - \alpha\gamma}$, whence

$$(\mathbf{c}^*)^T + \frac{\beta - \Theta\gamma}{\alpha\gamma - \beta^2}\mathbf{P}^{-1}\mathbf{V} + \frac{\alpha - \Theta\beta}{\beta^2 - \alpha\gamma}\mathbf{P}^{-1}\mathbf{I} = \mathbf{O} .$$ (16)

Moving back to scalar form we get

$$c_k^* = \frac{\Theta\gamma - \beta}{\alpha\gamma - \beta^2}\sum_{i=1}^{l}\varrho_{ki}^- V(r_i) + \frac{\Theta\beta - \alpha}{\beta^2 - \alpha\gamma}\sum_{i=1}^{l}\rho_{ki}^-,$$ (17)

where ϱ_{ij}^- is an element of the $\mathbf{P}^{-1}$ matrix, k=1,...,l. Thus, for cases when inverse matrix $\mathbf{P}^{-1}$ exists the necessary conditions of irreducibility from Theorem (3) are satisfied when the following inequalities

$$\frac{\Theta\gamma - \beta}{\alpha\gamma - \beta^2}\sum_{i=1}^{l}\varrho_{ki}^- V(r_i) + \frac{\Theta\beta - \alpha}{\beta^2 - \alpha\gamma}\sum_{i=1}^{l}\varrho_{ki}^- > 0,$$ (18)

are simultaneously correct $k = 1, \ldots, l$.

Let Θ_{min} is the minimal and Θ_{max} is the maximal value of Θ for which one of inequalities (18) becomes an equality.

Let's find a function that describes the dependency of $\rho[Y, P(\widetilde{r}, \mathbf{c}^*)]$ on Θ inside $[\Theta_{min}, \Theta_{max}]$. Let

$$\Phi_k^v = \sum_{i=1}^{l} V(r_i)\varrho_{ki}, \ \Psi_k = \sum_{i=1}^{l}\varrho_{ki} ,$$

$$\Gamma_k^1 = \frac{\gamma\Phi_k^v - \beta\Psi_k}{\alpha\gamma - \beta^2}, \ \Gamma_k^0 = \frac{\gamma\Phi_k^v + \beta\Psi_k}{\alpha\gamma - \beta^2} ,$$

then

$$c_k^* = \Gamma_k^0 + \Gamma_k^1\Theta.$$

Thus, $\mathbf{Q}_f = B_0 + B_1\Theta + B_2\Theta^2$, where

$$B_0 = \sum_{i=1}^{l}\sum_{j=1}^{l}\Gamma_i^0\Gamma_j^0\varrho_{ij} ,$$

$$B_1 = \sum_{i=1}^{l}\sum_{j=1}^{l}(\Gamma_i^0\Gamma_j^1 + \Gamma_i^1\Gamma_j^0)\varrho_{ij} ,$$

$$B_2 = \sum_{i=1}^{l} \sum_{j=1}^{l} \Gamma_i^1 \Gamma_j^1 \varrho_{ij}.$$

It is easy to show that $\rho[Y, P(\widetilde{r}, \mathbf{c}^*)] = \frac{\kappa(\Theta)}{\sqrt{V(Y)}}$, where

$$\kappa(\Theta) = \frac{\Theta}{\sqrt{(1 - B_1)\Theta - B_2\Theta^2 - B_0}}.$$

Let's prove the theorem that allows to receive easily evaluated necessary conditions of irreducibility in case when matrix $\mathbf{P}\|\varrho(r_i, r_j)\|_{l \times l}$ is not singular.

Theorem 4. *Simultaneous correctness of inequalities* $\Theta_{min} < \frac{2B_0}{(1-B_1)} < \Theta_{max}$, $\kappa(\frac{2B_0}{(1-B_1)}) > \kappa(\Theta_{min})$ *and* $\kappa(\frac{2B_0}{(1-B_1)}) > \kappa(\Theta_{max})$ *is a necessary condition of irreducibility of the predictors set* $(r_1, \dots, r_l)$ *when* $\mathbf{P}\|\varrho(r_i, r_j)\|_{l \times l}$ *is not singular.*

According to the theorem (3) predictors set $(r_1, \dots, r_l)$ may be irreducible only when such Θ exists that strict maximum of $\mathbf{Q}_f$ on $\mathbf{W}(\Theta)$ is achieved at a point from D_1. It follows from Θ_{min} and Θ_{max} definitions that stationarity point of $\mathbf{Q}_f$ at $\mathbf{W}(\Theta)$ belongs to D_1 if and only if $\Theta \in (\Theta_{min}, \Theta_{max})$. So strict maximum of $\mathbf{Q}_f$ on $\mathbf{W}(\Theta)$ may be achieved at a point from D_1 if $\Theta \in (\Theta_{min}, \Theta_{max})$. First derivative of $\kappa(\Theta)$ is equal $\frac{(1-B_1)\Theta - 2*B_0}{2[(1-B_1)\Theta - B_2\Theta^2 - B_0]^{3/2}}$ and is equal 0 at the single point $\frac{2B_0}{(1-B_1)}$. So, maximum of $\kappa(\Theta)$ is achieved inside interval $(\Theta_{min}, \Theta_{max})$ when $\Theta_{min} < \frac{2B_0}{(1-B_1)} < \Theta_{max}$, $\kappa(\frac{2B_0}{(1-B_1)}) > \kappa(\Theta_{min})$ and $\kappa(\frac{2B_0}{(1-B_1)}) > \kappa(\Theta_{max})$. Let's suppose that at least one of these conditions is violated. Then evidently for any $\Theta \in (\Theta_{min}, \Theta_{max})$ the two possibilities exist: $\kappa(\Theta) < \kappa(\Theta_{min})$, or $\kappa(\Theta) < \kappa(\Theta_{max})$. But stationarity points $\mathbf{c}_m^*$ corresponding to Θ_{min} or Θ_{max} belongs to $\overline{D}_l \backslash D_l$. Thus, for any stationary point $\mathbf{c}'$ corresponding $\Theta \in (\Theta_{min}, \Theta_{max})$ the inequality $\rho[Y, P(\widetilde{r}, \mathbf{c}')] < \rho[Y, P(\widetilde{r}, \mathbf{c}_m^*)]$ holds. So violation of at least one condition of $\Theta_{min} < \frac{2B_0}{(1-B_1)} < \Theta_{max}$, $\kappa\left(\frac{2B_0}{(1-B_1)}\right) > \kappa(\Theta_{min})$, $\kappa\left(\frac{2B_0}{(1-B_1)}\right) > \kappa(\Theta_{min})$ leads also to a violation of irreducibility. Q.E.D.

Statement 3. *For irreducible* $\widetilde{r}$ *coefficient* $\rho[Y, P(\widetilde{r}, \mathbf{c})]$ *achieves maximum at* $\overline{D}_l$ *in point* $\mathbf{c}^*$ *that is calculated by (17) when* $\Theta = \frac{2B_0}{(1-B_1)}$.

It follows from theorem (3) and stationarity conditions that in case of irreducibility maximum of $\rho[Y, P(\widetilde{r}, \mathbf{c})]$ at $\overline{D}_l$ is achieved in point $\mathbf{c}^*$ that is calculated by (17) with use of some Θ^* from $(\Theta_{min}, \Theta_{max})$. Besides it follows from maximality condition $\rho[Y, P(\widetilde{r}, \mathbf{c}^*)] > \rho[Y, P(\widetilde{r}, \mathbf{c}')]$ for any $\mathbf{c}'$ calculated by (17) with use of some $\Theta' \neq \Theta^*$ from $(\Theta_{min}, \Theta_{max})$. According theorem (4) for any irreducible ensemble $\frac{2B_0}{(1-B_1)} \in (\Theta_{min}, \Theta_{max})$ and the only Θ^* that satisfies maximality condition is $\Theta^* = \frac{2B_0}{(1-B_1)}$. So the only possible point of maximum is $\mathbf{c}^*$ that is calculated by (17) when $\Theta = \frac{2B_0}{(1-B_1)}$. Q.E.D.

Theorem 5. *Let suppose that matrix of mutual distances between predictors from set $\widetilde{r} = \{r_1, \ldots, r_l\}$ is not singular. Let $\mathbf{c}^*$ be calculated by (17) and Pearson correlation coefficient $\rho[Y, P(\widetilde{r}, \mathbf{c}^*)]$ at $\Theta = \frac{2B_0}{(1-B_1)}$ be greater than $\rho[Y, P[\widetilde{r}_{irr}(\mathbf{c})]$, where $\widetilde{r}_{irr}$ is any irreducible ensemble that is subset of $\widetilde{r}$. Then set $\widetilde{r}$ is irreducible.*

It is sufficient to prove that $\rho[Y, P(\widetilde{r}, \mathbf{c}^*)] > \rho[Y, P(\widetilde{r}', \mathbf{c})]$ for any $\widetilde{r}' \subset \widetilde{r}$. Let's suppose that $\widetilde{r}'$ is irreducible. Then inequality is true according theorem condition. Let's suppose that $\widetilde{r}'$ is not irreducible and $\rho_b(\widetilde{r}')$ is maximal value of Pearson correlation coefficient between Y and convex combination of predictors from $\widetilde{r}'$. But it follows from theorem 4 that there is such irreducible ensemble $\widetilde{r}'_{irr} \subset \widetilde{r}'$ that maximal value of Pearson correlation coefficient between Y and convex combination of predictors from $\widetilde{r}'_{irr}$ is equal $\rho_b(\widetilde{r}')$. So inequality is correct. Q.E.D.

Theorems (4) and (5) may be used to evaluate irreducibility of some ensemble $\widetilde{r}$ with number of predictors equal l in case when matrix of mutual distances between predictors in ensemble is not singular. It is supposed that all irreducible ensembles with number of predictors less than l are known.

Procedure for irreducibility testing. At first it is necessary to calculate Θ_{min}. and Θ_{max}. In case $\frac{2B_0}{1-B_1} \in (\Theta_{min}, \Theta_{max})$ inequality $\kappa\left(\frac{2B_0}{(1-B_1)}\right) > \kappa_b$ is evaluated, where κ_b is maximal value of κ ($\rho[Y, P(\widetilde{r}_{irr}, \mathbf{c})] \times \sqrt{V[P(\widetilde{r}_{irr}, \mathbf{c})]}$) among all irreducible ensembles $\widetilde{r}_{irr} \subset \widetilde{r}$.

Let some ensemble $\widetilde{r}$ be tested by discussed procedure.

Theorem 6. *In case $\widetilde{r}$ is evaluated as irreducible it is irreducible indeed. In case $\widetilde{r}$ is irreducible it is evaluated as irreducible also.*

The first statement is true because sufficient condition of irreducibility from theorem 5 is correct for $\widetilde{r}$. Let's prove second statement. According statement (3) discussed procedure calculates maximal value of $\rho[Y, P(\widetilde{r}, \mathbf{c})]$ that is $\rho(\widetilde{r})_{max}$. Due to irreducibility $\rho(\widetilde{r})_{max} > \rho[Y, P(\widetilde{r}', \mathbf{c})]$ for any $\widetilde{r}' \subset \widetilde{r}$ including of course any irreducible ensemble. So for irreducible ensemble $\kappa\left(\frac{2B_0}{1-B_1}\right)$ will be always greater than κ_b. So $\widetilde{r}$ will be evaluated by discussed procedure as irreducible. Q.E.D.

In experiments discussed below irreducibility testing is based on evaluating of inequality $\kappa\left(\frac{2B_0}{\beta_1}\right) > \tau_{irr}\kappa_b$, where $\tau_{irr} \in [0, 1]$ is chosen by user parameter that allows to select irreducible ensembles of sufficient quality.

4 Regression Models Based on Sets of Unexpandable Irreducible Ensembles

At the first stage we construct by OLS technique regressions set (4).

Variances of relevant single predictors V and distances ρ between these predictors, that are used to calculate if some ensemble is irreducible, are evaluated from training data by standard formulae $V(R_i) = \frac{1}{m}\sum_{j=1}^{m}[R_i(j) - \hat{R}_i]^2$,

$\rho(R_{i1}, R_{i2}) = \frac{1}{m} \sum_{j=1}^{m} [R_{i1}(j) - R_{i2}(j)]^2$, where $R_i(j)$ is prognosis calculated by R_i for observation j in training set; m is training set size, $\hat{R}_i = \sum_{j=1}^{m} R_i(j)$.

Experiments showed that such type of estimates leads to the selection of too many variable and conseqently to decrease of the prognostic ability. However, effectiveness may be systematically improved by using additional penalty multiplier for ρ equal $\frac{1}{1+\frac{5}{m}}$. This effect demands mathematical explanation.

Initially, set of all UIE ensembles $\widetilde{\mathbf{R}}_{UIE}$ is empty.

Step 1. Set $\widetilde{\mathbf{R}}_2$ of all irreducible ensembles that consist of two predictors is searched using rules described in the previous section. In case some single uni-dimensional regression R_i does not belong to any ensemble from $\widetilde{\mathbf{R}}_2$ and $\rho(Y, X_i) \neq 0$ it is declared UIE and is added to $\widetilde{\mathbf{R}}_{UIE}$. In case set $\widetilde{\mathbf{R}}_2$ is empty the searching procedure is stopped.

Step k. Let $\widetilde{\mathbf{R}}_{k+1}$ be set of all irreducible ensembles that are received by adding to each ensemble from $\widetilde{\mathbf{R}}_k$ new additional predictor. Then irreducibility of such widened ensemble is tested with the help of procedures for cases with $l > 2$. In case some irreducible ensemble $\tilde{r}$ from $\widetilde{\mathbf{R}}_k$ does not belong to any ensemble from $\widetilde{\mathbf{R}}_{k+1}$ it is declared UIE and is added to $\widetilde{\mathbf{R}}_{UIE}$. In case set $\widetilde{\mathbf{R}}_{k+1}$ is empty the searching procedure is stopped.

Final CR solution may be built by best UIE $\tilde{r}_{uie}^{max}$ satisfying inequality $\rho[Y, P(\tilde{r}_{uie}^{max}, \mathbf{c}^b)] \geq \rho[Y, P(\tilde{r}_{uie}, \mathbf{c}^b)]$ for any $\tilde{r}_{uie} \in \widetilde{\mathbf{R}}_{UIE}$.

However, usually for some UIE from $\widetilde{\mathbf{R}}_{UIE}$ correlation coefficients between Y and corresponding optimal convex combinations are close to $\rho[Y, P(\tilde{r}_{uie}^{max}, \mathbf{c}^b)]$. Due to overfitting effect the forecasting ability of CR solutions based on such UIE may be even better than forecasting ability of $P(\tilde{r}_{uie}^{max}, \mathbf{c}^b)$. It is natural to use all such UIE in final CR solution. So final CR solution was built by heuristic convex combination with the help of working set of UI ensembles. Working set $\widetilde{\mathbf{R}}_{uie}$ is formed with the help of threshold parameter $\tau \in [0, 1]$. Some UIE $\tilde{r}_{uie}$ from $\widetilde{\mathbf{R}}_{uie}$ belongs to a working set if $\rho[Y, P(\tilde{r}_{uie}, \mathbf{c}^b)] > \tau \times \rho[Y, P(\tilde{r}_{uie}^{max}, \mathbf{c}^b)]$. Let $\widetilde{R}^w$ is set of all predictors that belong to ensembles from $\widetilde{\mathbf{R}}_{uie}^w$. An optimal convex combination $P(\widetilde{R}^w, \mathbf{c}^{ws})$ that is used further to calculate the final forecasting function in the second method is evaluated by $\widetilde{\mathbf{R}}_{uie}^w(\tau)$ as weighted sum:

$$P(\widetilde{R}^w, \mathbf{c}^{ws}) = \sum_{\tilde{r}_{uie} \in \widetilde{\mathbf{R}}_{uie}^w} P(\tilde{r}_{uie}) W(\tilde{r}_{uie}), \qquad (19)$$

where $W(\tilde{r}_{uie}) = 1 - \rho[Y, P(\tilde{r}_{uie}, \mathbf{c}^b)]^2$ is heuristic weighting coefficient that is searched from observations with the help of OLS. Thus, the final CR solution is received.

4.1 Simulations

Convex regression was compared with elastic net regression in a series of experiments with simulated data. Matlab version of glmnet implementation [4] was used. The Matlab function was downloaded from http://www.stanford.edu/

~hastie/glmnet_matlab/ website. For each experiment 100 pairs of data sets were generated. The first dataset in each pair was used for estimation of regression model and the second (control) one was used to evaluate forecasting ability of the calculated model. An average value of correlation coefficients of 100 control data sets was calculated to evaluate performance of the regression method in the experiment as a whole.

In all studies latent variables $U_1, \ldots, U_g$ were used to calculate simultaneously response Y and variables $X_1, \ldots, X_n$. Vector levels of variables U are independently distributed multivariate normal with mean 0 and standard deviation 1. A value of the forecasted variable Y for an observation j is generated by formula $y_j = \sum_{k=1}^{g} u_{jk}\psi_k + e_y^j$ where u_{jk} is the value of the latent variable U_k, ψ_k is a parameter of the experiment scenario, e_j^y is value of a random error term distributed $N(0, d_y)$ at observation j. At that 85% of cases were generated with $d_y = d$, 15% of cases were generated with $d_y = 2d$. Set of X-variables consists of variables relevant to Y and "noisy" ones irrelevant to Y. For each experiment $\{1, \ldots, n\}$ was randomly split to subsets I_{ir} corresponding to irrelevant X-variables and subset I_{rel} corresponding to relevant ones. Let $e_{x_i}^j (i = 1, \ldots, n)$ are values of mutually independent error terms distributed $N(0, d)$ for observation j. For $i \in I_{rel}$ value of variable X_i at observation j (x_{ij}) were generated by real vector $\boldsymbol{\phi^i} = (\phi_1^i, \ldots, \phi_g^i)$: $x_{ij} = \sum_{k=1}^{g} u_{jk}\phi_k^i + e_{x_i}^j$ where $e_{x_i}^j$ is error term. At that parameters $\phi_1^i, \ldots, \phi_g^i$ were randomly taken from the $[-0.5, 1.5]$ cut. For $i \in I_{irr}$ x_{ij} was equal $e_{x_i}^j$.

Table 1. Parameters of simulations

No.	m	n	g	ν	d
1	40	550	6	0.02	0.3
2	40	550	6	0.05	0.3
3	40	550	4	0.05	0.3
4	30	550	4	0.05	0.3
5	20	550	4	0.05	0.3
6	40	550	6	0.02	0.5
7	40	550	6	0.02	0.2
8	40	550	6	0.05	0.2
9	40	550	6	0.05	0.5
10	30	550	12	0.05	0.3
11	30	550	6	0.05	0.3
12	40	550	4	0.02	0.3
13	50	1200	6	0.02	0.3
14	20	1000	6	0.02	0.3

Thus, varied parameters of experiments were the initial number of X-variables n, the size of data sets m, the number of latent variables g, fraction of relevant variables ν, and the variance of the noisy component d. The specific values of the parameters for the experiments are given in Table (1), where the first column identifies the experiment number.

Optimal level of threshold τ corresponding to the best forecasting ability was searched at τ_{irr} levels equal 0.99 and 0.95. In first and second steps thresholds 0.995 and 0.99 were used. In step i the threshold $0.99 - (i-2) * 0.01$ was tested. The enumeration stopped when the forecasting ability begin to decrease. Also in each experiment the best value of open elastic net parameter α from set $\{0.1 * i | i = 0, \ldots, 10\}$ was selected, see reference (1) at page 1.

Table 2. Simulation results

No.	CR		glmnet			
	τ	ρ_{CR}^b	α^b	ρ_{glm}^b	ρ_{glm}^0	ρ_{glm}^1
1	0.97	0.802	0.4	0.758	0.713	0.685
2	0.99	0.89	0.2	0.924	0.919	0.759
3	0.94	0.893	0.2	0.883	0.866	0.631
4	0.98	0.819	0.2	0.803	0.79	0.651
5	0.95	0.617	0.4	0.568	0.537	0.539
6	0.93	0.659	1	0.628	0.538	0.628
7	0.97	0.87	0.3	0.85	0.828	0.758
8	0.99	0.906	0	0.957	0.957	0.781
9	0.97	0.841	0.3	0.85	0.826	0.823
10	0.65	0.725	0.8	0.703	0.647	0.698
11	0.96	0.843	0.2	0.866	0.853	0.746
12	0.6	0.667	0.5	0.655	0.55	0.617
13	0.98	0.831	0.3	0.807	0.793	0.707
14	0.98	0.938	0.4	0.944	0.934	0.934

Experiment results are represented in the Table (2). For each experiment the following characteristics are given: ρ_{CR}^b is the best value of average correlation coefficient between Y and the CR solution found; τ_b is the threshold τ corresponding ρ_{CR}^b; ρ_{glm}^b is the best value of average correlation coefficient between Y and the glmnet solution found; α^b is the parameter α corresponding ρ_{glm}^b; ρ_{glm}^0 is the average correlation coefficient for the ridge regression penalty $(\alpha = 0)$; ρ_{glm}^1 is the average correlation coefficient for the lasso penalty $(\alpha = 1)$.

Let note that optimal prognostic ability was achieved at $\tau_{irr}^b = 0.95$ for experiments 1, 3, 5, 6, 10, 12 and at $\tau_{irr}^b = 0.99$ for experiments 2, 4, 7, 8, 13, 14. It is seen from the table that forecasting effectiveness of CR is rather close to forecasting ability of glmnet algorithm. Maximal difference between ρ_{CR}^b and ρ_{glm}^b is 0.051 at experiment 8. CR performance was better for experiments 1, 3-7, 10, 12, 13 where forecasting ability of both methods was relatively low: ρ_{CR}^b and ρ_{glm}^b were less than 0.9. At that glmnet was winner in experiments 2, 8, 14 where at least ρ_{glm}^b was greater 0.9.

5   Conclusion

The represented results can be summed as follows. A new regression method was developed that is based on optimal convex combinations and is referred to as Convex regression (CR). It was shown that CR is equivalent to a least squares variant with constraints on signs of regression coefficients. A method of optimal convex combination search was developed that is based on search of unexpandable and irreducible ensembles of simple linear regressions. Experiments with high-dimensional simulated data demonstrated that efficiency of CR appeared to be better than that of glmnet in tasks with higher level of noise and/or small size of training data sets.

The main difference of the described results and that of [8,9] is a new version of optimization algorithm that doesn't require positive definiteness tests. Moreover, the present article demonstrates the relation between convex procedures based regression and regularization and gives strict derivation of every fact skipped previously.

References

1. Breiman, L.: Random forests – random features. Statistics department, University of California, Berkley (1999)
2. Brown, G., Wyatt, J.L., Tino, P.: Diversity in Regression Ensembles. Journal of Machine Learning 6, 1621–1650 (2005)
3. Efron, B., Hastie, T., Jonnstone, I., Tibshirani, R.: Least Angle Regression. Annals of Statistics 32(2), 407–499 (2004)
4. Friedman, J.H., Hastie, T., Tibshirani, R.: Regularization Paths for Generalized Linear Models via Coordinate Descent. Journal of Statistical Software 33(1) (2010)
5. Kuncheva, L.I.: Combining Pattern Classifiers. Methods and Algorithms. Wiley Interscience, New Jersey (2004)
6. Senko, O.V.: The Use of Collective Method for Improvement of Regression Modeling Stability. InterStat. Statistics on the Internet (2004),
   http://statjournals.net
7. Senko, O.V.: An Optimal Ensemble of Predictors in Convex Correcting Procedures. Pattern Recognition and Image Analysis 19(3), 465–468 (2009)
8. Senko, O.V., Dokukin, A.A.: Optimal Forecasting Based on Convex Correcting Procedures. In: New Trends in Classification and Data Mining, pp. 62–72. ITHEA, Sofia (2010)
9. Senko, O., Dokukin, A.: On Some Properties Of Regression Models Based On Correlation Maximization Of Convex Combinations. International Journal "Information Models and Analyses" 1(1), 112–122 (2012)
10. Tibshirani, R.: Regression shrinkage and selection via the lasso. J. Roy. Stat. Soc. 58, 267–288 (1996)
11. Zhuravlev, Y.I., Kuznetsova, A.V., Ryazanov, V.V., Senko, O.V., Botvin, M.A.: The Use of Pattern Recognition Methods in Tasks of Biomedical Diagnostics and Forecasting. Pattern Recognition and Image Analysis 58(2), 195–200 (2008)
12. Zou, H., Hastie, T.: Regularization and variable selection via the elastic net. J. Roy. Stat. Soc. 67(2), 301–320 (2005)

A Novel Approach for Identifying Banded Patterns in Zero-One Data Using Column and Row Banding Scores

Fatimah Binta Abdullahi, Frans Coenen, and Russell Martin

The Department of Computer Science, The University of Liverpool, Ashton Street,
Liverpool, L69 3BX, United Kingdom
{f.b.abdullahi,coenen,russell.martin}@liverpool.ac.uk

Abstract. Zero-one data is frequently encountered in the field of data mining. A banded pattern in zero-one data is one where the attributes (columns) and records (rows) are organized in such a way that the "ones" are arranged along the leading diagonal. The significance is that rearranging zero-one data so as to feature bandedness enhances the operation of some data mining algorithms that work with zero-one data. The fact that a dataset features banding may also be of interest in its own right with respect to various application domains. In this paper an effective banding algorithm is presented designed to reveal banding in 2D data by rearranging the ordering of columns and rows. The challenge is the large number of potential row and column permutations. To address this issue a column and row scoring mechanism is proposed that allows columns and rows to be ordered so as to reveal bandedness without the need to consider large numbers of permutations. This mechanism has been incorporated into the Banded Pattern Mining (BPM) algorithm proposed in this paper. The operation of BPM is fully discussed. A Complete evaluation of the BPM algorithm is also presented clearly indicating the advantages offered by BPM with respect to a number of competitor algorithms in the context of a collection of UCI Datasets.

Keywords: Banded Patterns, Zero-One data, Data Mining and Pattern Mining.

1 Introduction

Many data sets take the form of a $n \times m$ two dimensional, binary ("zero-one"), matrix whereby a one indicates the presence of some attribute and a zero its absence with respect to a particular record. More formally if we consider a data set of attributes $A = \{a_1, a_2, \ldots, a_m\}$ with the value set $\{0, 1\}$, and a set of records $R = \{r_1, r_2, \ldots, r_n\}$ such that each record is some subset of A, then we have a binary valued, zero-one, data set. Many application domains exist where zero-one data is typically found, examples include: the field of information retrieval [5], bioinformatics and computational biology (genes, probe mappings) [3],[18] and paleontology (sites and species occurrences) [4], [10].

P. Perner (Ed.): MLDM 2014, LNAI 8556, pp. 58–72, 2014.
© Springer International Publishing Switzerland 2014

Of note with respect to this paper is the field of data mining, especially frequent pattern mining [1,2,8] where it is necessary to process large collections of zero-one data stored in the form of a set of feature vectors (drawn from a vector space model of the data). Although outside the scope of this paper, banding will also have benefits with respect to non-binary data. For example in the context of co-clustering algorithms where the objective is to simultaneously cluster the rows and columns in 2D data set (see for example the work presented in [12] and [16].

A zero-one matrix is fully banded if both the columns and rows can be presented in such a manner that the "ones" are arranged along the leading diagonal as illustrated in Figure 1. In practice data can typically not be perfectly banded, but in many cases some form of banding can be achieved.

 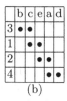

(a)   (b)

Fig. 1. Banding Example: (a) raw data (b) rearrangement of columns and rows to reveal a banding

While the concept of banded matrices has its origins in numerical analysis, it has been studied within the data mining community. The benefits of banding may be summarized as follows:

1. The fact that a zero-one data set can be banded is interesting in its own right as it is indicative of the existence of a relationship between the columns and rows, for example between records and certain attribute values (assuming rows equate to records and columns to attribute values).
2. Working with banded data is seen as preferable from a computational point of view as the computational cost involved in performing certain operations falls significantly for banded matrices, often leading to significant savings in terms of processing time [9]. One example where this is the case is the use of affinity matrices in spectral clustering [15], another example is where adjacency matrices are used in the context of graph mining [13]. A third example is in the context of co-clustering algorithms [12] and [16].
3. Related to 2 above, working with banded patterns requires less storage. When a matrix is banded only the non-zero diagonal need to be stored. Thus, for the banded storage schemes the amount of memory required to store the matrix is directly proportional to the bandwidth. Therefore, finding a row-column ordering that minimizes the bandwidth is significant with respect to reducing storage space and consequently results in algorithmic speed up [17].

The main issue with the identification of banding in data is the large number of permutations that need to be considered. In this paper we present an approach whereby bandedness in data can be revealed using a scoring mechanism thus avoiding the need to consider permutations. This concept has been built into the Banded Pattern Mining (BPM) algorithm; the central contribution of this paper. This differs Minimum Banding Augmentation (MBA) algorithm proposed in [14], see above, which allowed for the discovery of banded structures in binary matrices by assuming a fixed column permutation.

The rest of this paper is organized as follows. Section 2 disuss related work. A formalism for the banded pattern problem is then presented in Section 3. This is followed, in Section 4, with an overview of the proposed scoring mechanism. Section 5 presents the BPM Algorithm whilst Section 6 provides a worked example illustrating the proposed approach. The evaluation of the BPM algorithms is then reported in Section 7; and finally, in Section 8, some conclusions are presented.

2   Related Work

The property of bandedness with respect to data analysis was first studied by Gemma *et al.* [11]. Where the minimum banding problem was addressed by computing "how far" a 2-D data matrix (data set) was away from being banded. The authors in [11] define the banding problem as: given a binary matrix M, find the minimum number of 0s entries that needs to be modified into 1s entries and the minimum number of 1s entries that needs to be modified into 0s entries so that M becomes fully banded. In [11], the authors use the principle of assuming "a fixed column permutation" on a given Matrix M. The basic idea is to solve optimally the consecutive one property on the permuted matrix M and then resolve Sperner conflicts between each row of the permuted matrix M, by going through all the extra rows and making them consecutives. While it can be argued that the fixed column permutation assumption is not a very realistic assumption with respect to many real world situations, heuristical methods were proposed in [11] to determine a suitable fixed column permutation.

The current state of the art algorithm is the MBA algorithm [14] which also adopts the fixed column permutation assumption. The MBA algorithm focuses on minimizing the distance of non-zero entries from the main diagonal of the matrix by reordering the original matrix. The MBA algorithm operates by "flipping zero entries (0s) into one entries (1s) and vice versa to identify a banding.

Another Strategy for transposing a zero-one matrix is the Barycentric (BC) algorithm that was previously used to draw graphs [20] to seriate paleontological data [7], and more recently used to reorder binary matrices [19]. In essence, the Barycentric algorithm finds permutations for both rows and columns such that 1s are as close to each other as possible. It is based on the Barycentric measure, which is the average position of 1s in a row/column. The algorithm first computes the barycenter for all rows, then orders the rows from smallest to largest, transposes the matrix accordingly and then iterates again until convergence is reached.

Given the above the MBA and BC algorithms are the two exemplar banding algorithms with which the operation of the proposed BPM algorithm was compared and evaluated as discussed later in this paper (see Section 7).

3   Problem Definition

A binary matrix is a matrix with entries $\{0,1\}$. Let A be an $n \times m$ binary matrix comprising rows $\{1, 2, \ldots, n\}$ and columns $\{1, 2, \ldots, m\}$. We indicate a particular row i using row_i $(1 \leq i \leq n)$, a particular column j using col_j $(1 \leq j \leq m)$ and a particular element using a_{ij}. We denote the set of '1' valued elements associated with row i using R_i, $R_i = \{r | \forall a_{ij} \in row_i, a_{ij} = 1\}$. Similarly the set of '1' valued elements associated with column j using C_j, $C_j = \{c | \forall a_{ij} \in col_j, a_{ij} = 1\}$. (It may ease understanding to note that C_j is sometimes referred to as a Transaction ID list or a TID list, a concept well established in transaction rule mining [1,2]). A zero-one matrix A can be "perfectly banded" if there exist a permutation of columns $\{1, 2, \ldots, m\}$ and rows $\{1, 2, \ldots, n\}$ such that: (i) for every element in C_j the 1 values occur consecutively at row indexes $\{i_k, i_{k+1}, i_{k+2}, \ldots\}$ and the "starting index" for C_j is less than or equal to the starting index for C_{j+1}; and (ii) for every element in R_i the 1 values occur consecutively at column indexes $\{j_k, j_{k+1}, j_{k+2}, \ldots\}$ and the starting index for R_i is less than or equal to the starting index for R_{i+1}.

4   The Banding Score Mechanisms

The discovery of the presence of banding in zero-one matrices requires the rearrangement of the columns and rows in the matrix so as to "reveal" a banding (or at least an approximate banding if no perfect banding exists). As noted above the discovery of banded patterns in zero-one data requires some mechanism for automatically determining whether one potential configuration features a "better" banding (according to the above definition) than some other configuration. The idea presented in this paper is to use the concept of "banding scores" that reflect the relative position of each "one" value both horizontally and vertically. Consider the perfect banding given in Figure 2(a) where we have three records and an attribute set $\{a, b, c\}$, each record features one of the attributes. We wish to assign some kind of banding score to this configuration. We achieve this by considering two types of banding score: (i) *column scores* (BS_{col}) and (ii) *row scores* (BS_{row}). Both are determined in a similar manner thus the following discussion will be focused on column score calculation.

With reference to Figure 2(a), which features a perfect banding, we wish to assign a greater weight to locations corresponding to the first column than locations associated with the second column, and greater weight to locations associated with the second column than the third column (etc). In real datasets a perfect banding may not exist. However, we are interested in revealing the arrangement that is as close to a perfect banding as possible. Thus for each column i we can calculate a Column Score CS as follows (we assume that column and row numberings start from 1):

$$CS = \sum_{k=1}^{k=|C_i|} m - r_k + 1$$

Fig. 2. Example zero-one data configurations: (a) perfect banding, and (b) an alternative banding

where m is the number of rows, C_i is the TID list for column i, and r_k is the row index at location k in C_i. Thus the column scores for the configuration in Figure 2(a) will be:

$$CS_A = 3-1+1 = 3 \qquad CS_B = 3-2+1 = 2 \qquad CS_C = 3-3+1 = 1$$

However, we would prefer it if our columns scores were normalized. Thus:

$$CS' = \frac{\sum_{k=1}^{k=|C_i|} m - r_k + 1}{\sum_{k=1}^{k=|C_i|} m - k + 1}$$

The normalised column scores will then be:

$$CS'_A = \frac{3-1+1}{3} = \frac{3}{3} = 1 \qquad CS'_B = \frac{3-2+1}{3} = \frac{2}{3} = 0.67$$

$$CS'_C = \frac{3-3+1}{3} = \frac{1}{3} = 0.33$$

We could now simply sum the individual column scores to obtain an overall column banding score. However, this would mean that the banding score for the configuration presented in Figure 2(a) would be equal to that for the configuration presented in Figure 2(b) (which features an entirely different kind of banding). Thus we need to weight the columns as well. Thus the final column banding score, BS_{col}, should be calculated as follows:

$$BS_{col} = \sum_{k=1}^{k=m} CS'_k(m - k + 1)$$

As a result the configuration in Figure 2(a) will have a score of:

$$BS_{col} = 1 \times (3 - 1 + 1) + 0.67 \times (3 - 2 + 1) + 0.33 \times (3 - 3 + 1)$$

$$1 \times 3 + 0.67 \times 2 + 0.33 \times 1 = 4.67$$

The maximum normalized column score is 1, thus the maximum column banding score (BS_{col}), calculated as described above, will be 6. Consequently the normalized banding score should be calculated as follows:

$$BS'_{col} = \frac{\sum_{k=1}^{k=m} CS'_k (m - k + 1)}{m(m + 1)/2}$$

which means that the normalized column banding score for the configuration in Figure 2(a) is 0.78 while that for the configuration in Figure 2(b) is 0.56; thus the desired effect.

Considering the data set given in Figure 2 this can be configured in 6 ways. The different configurations are illustrated in Figure 3 together with their associated normalized column banding scores (BS'_{col}). From Figure 3, it can clearly be seen that the column banding score values serve to differentiate between the different configurations.

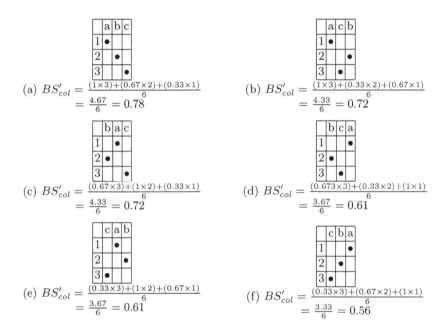

Fig. 3. Permutations for $A = \{a,b,c\}$

With respect to row banding scores (BS_{row}) we can formulate a similar argument. We can calculate individual Row Scores (RS) as follows:

$$RS = \sum_{k=1}^{k=|R_j|} n - c_k + 1$$

where n is the number of columns, R_j is the "column ID list" for row j, and c_k is the column index at location k in R_j. The normalized row scoes (RS') are then calculated as follows:

$$RS' = \frac{\sum_{k=1}^{k=|R_j|} n - c_k + 1}{\sum_{k=1}^{k=|R_j|} n - k + 1}$$

Thus the normalized row scores for the configuration presented in in Figure 2(a) will be:

$$RS'_a = \frac{3 - 1 + 1}{3} = \frac{3}{3} = 1 \qquad RS'_b = \frac{3 - 2 + 1}{3} = \frac{2}{3} = 0.67$$

$$RS'_c = \frac{3 - 3 + 1}{3} = \frac{1}{3} = 0.33$$

In this case the row and column scores are identical, this is because of the idealized banding featured in the configuration given in Figure 2(a).

The overall row banding score (BS_{row}) will then be given by:

$$BS_{row} = \sum_{k=1}^{k=n} RS'_k(n - k + 1)$$

As a result the configuration in Figure 2(a) will have a score of:

$$BS_{row} = 1 \times (3 - 1 + 1) + 0.67 \times (3 - 2 + 1) + 0.33 \times (3 - 3 + 1)$$

$$1 \times 3 + 0.67 \times 2 + 0.33 \times 1 = 4.67$$

normalizing this:

$$BS'_{row} = \frac{\sum_{k=1}^{k=n} RS'_k(n - k + 1)}{n(n + 1)/2}$$

The overall normalized banding Score (BS) is then calculated as follows:

$$BS = \frac{BS'_{col} + BS'_{row}}{2}$$

The overall banding score for the configuration given in Figure 2(a) will then be 0.78 while that for the configuration given in Figure 2(b) will be 0.56.

5   Banded Pattern Mining Algorithm

From the above it was noted that to identify the "best" banding we need to maximize the banding score (BS). In this section the Banded Pattern Mining (BPM) algorithm is presented. The BPM algorithm was designed to identify a column and row configuration that serves to maximize BS. The algorithm proceeds in an iterative manner, on each iteration it sequentially rearranges the columns and rows according to their individual column and row scores (CS and RS). The process is continued until a maximum value for BS is reached. The BPM algorithm is presented in 1. The input is a zero-one matrix measuring $m \times n$ (line1).The algorithm proceeds in an iterative manner until BS is maximized or no changes have been made; initially BS is set to 0 (line 5). On each iteration the columns in A are first rearranged in descending order of BS'_{col} to produce a new matrix A' (lines 8-15), and then the rows in A' are rearranged in descending order of BS'_{row} to produce a new matrix A'' (lines 17-24). The normalized column and row banding scores are calculated (lines 16 and 25 respectively), and then a new banding score $newBS$ is determined (line 26). If $newBS$ is greater than the previously recorded normalized banding score we continue, if not we exit with matrix A as processed on the previous iteration. Thus if no changes (line 33) are made we also exit.

A disadvantage of the banding score calculation mechanism, as described above, might be that sparse data sets will generate very low banding scores as the normalization is conducted assuming a column and row score of one for every column and row. In some cases this might make it difficult to conduct comparisons depending on the precision of the numeric types used when implementing the above.

6   Worked Example

To illustrate the operation of the BPM algorithm, as described in the foregoing section, a worked example is presented in this section using the 5×5 input matrix shown in Figure 4. We commence, on the first iteration, by calculating the normalized column scores: $CS'_A = 0.7500$, $CS'_B = 0.7500$, $CS'_C = 0.8333$, $CS'_D = 0.5000$ and $CS'_E = 0.9167$. Using this set of scores the columns are rearranged to produce the matrix A' as shown in Figure 5. The normalized column banding score is now $BS'_{col} = 0.7444$. Next we calculate the normalized row scores: $RS'_1 = 1.0000$, $RS'_2 = 0.9167$, $RS'_3 = 0.7500$, $RS'_4 = 0.7500$ and $RS'_5 = 0.5000$. Using this set of scores the rows are rearranged to produce the matrix A'' as shown in Figure 6. The normalized row banding score is now $BS'_{row} = 0.8111$. The normalized banding score after this first iteration is then:

$$BS = \frac{BS'_{col} + BS'_{row}}{2} = \frac{0.7444 + 0.8111}{2} = 0.7776$$

On the second iteration we repeat the process. The normalized column banding scores are now: $CS'_E = 1.0000$, $CS'_C = 0.9167$, $CS'_A = 0.6667$, $CS'_B = 0.6667$ and $CS'_D = 0.5000$; with this set of scores the columns remain unchanged.The overall normalized column banding score is $BS'_{col} = 0.8334$ (previously this was 0.7444). The normalized row banding scores are now: $RS'_1 = 1.0000$, $RS'_2 = 0.9167$, $RS'_4 = 0.6667$, $RS'_5 = 0.6667$ and $RS'_5 = 0.5000$. Again, with this set of scores the rows remain unchanged, thus produce the same matrix A''

Algorithm 1. The BPM Algorithm

1: **Input** A, a zero-one matrix measuring $n \times m$
2: **Output** the matrix A rearranged so that the columns and rows serve to maximize BS
3: Column ordering = $\{1 \dots m\}$
4: Row ordering = $\{1 \dots n\}$
5: $BS \leftarrow 0$
6: **loop**
7:     $change \leftarrow false$
8:     CS' = empty set of column scores for columns 1 to m
9:     **for all** $j \in \{1 \dots m\}$ **do**
10:         $CS'_j \leftarrow$ Normalized column score for column j
11:     **end for**
12:     $A' \leftarrow$ matrix A rearranged with columns in descending ordered as per CS'
13:     **if** $(A' \neq A)$ **then**
14:         $change \leftarrow true$
15:     **end if**
16:     $BS'_{col} \leftarrow$ Normalized column banding score
17:     RS' = empty set of row scores for row 1 to n
18:     **for all** $i \in \{1 \dots n\}$ **do**
19:         $RS'_i \leftarrow$ Normalized row score for row i
20:     **end for**
21:     $A'' \leftarrow$ matrix A' rearranged with rows in descending ordered as per RS'
22:     **if** $(A'' \neq A')$ **then**
23:         $change \leftarrow true$
24:     **end if**
25:     $BS'_{row} \leftarrow$ Normalized row banding score
26:     $newBS' \leftarrow \frac{BS'_{col} + BS'_{row}}{2}$ {Normalized Banding Score}
27:     **if** $(newBS > BS)$ **then**
28:         $BS \leftarrow newBS$
29:         $A \leftarrow A''$
30:     **else**
31:         Exit with A
32:     **end if**
33:     **if** $(\neg change)$ **then**
34:         Exit with A
35:     **end if**
36: **end loop**

Fig. 4. Raw data

Fig. 5. Raw data with columns rearranged

Fig. 6. Raw data with rows rearranged

as was shown in Figure 6. The overall normalized row banding score is now $BS'_{row} = 0.8334$ (was 0.8111). The normalized banding score after this second iteration is now:

$$BS = \frac{BS'_{col} + BS'_{row}}{2} = \frac{0.8334 + 0.8334}{2} = 0.8334$$

On the previous iteration it was 0.7776, however no changes have been made on the second iteration so the algorithm terminates.

7   Evaluation

To evaluate the BPM algorithm, its operation was compared with the established MBA and BC algorithms, two exemplar algorithms illustrative of alternative approaches to identifying banding in zero-one data as described on Section 2. For the evaluation, eight data sets taken from the UCI machine learning data repository [6] were used, which featured nineteen columns (attributes) or more. The first set of experiments, reported in sub-section 7.1 below, considered the efficiency of the BPM algorithm in comparison with the MBA and BC algorithms. The second set of experiments (Section 7.2) considered the effectiveness of the BPM algorithm, again in comparison with the MBA and BC algorithms. The third set of experiments, reported in sub-section 7.3 below, considered the effectiveness of banding with respect to a frequent pattern mining scenario.

7.1   Efficiency

To determine the efficiency of the proposed BPM algorithm with respect to the MBA and BC algorithms, and with respect to the selected data sets, we recorded the number of iterations and run time required to maximize the banding score BS in each case. The data sets were normalized and discretized using the LUCS-KDD ARM DN Software[1] to produce the desired zero-one data sets (continuous values were ranged using a maximum of five ranges). Table 1 shows the results obtained. The table presents the run-time and the final BS value obtained in each case. The table also records the number of columns (after discretization)

---

[1] http://www.csc.liv.ac.uk/~/frans/KDD/Software/LUCS_KDD_DN_ARM

the number of rows and the density for each data set. From the table it can be observed that there is a clear correlation between the number of rows in the dataset and run time. For example the largest dataset, Annealing, required considerably less processing time to identify a banding using BPM than when using either MBA and BC.

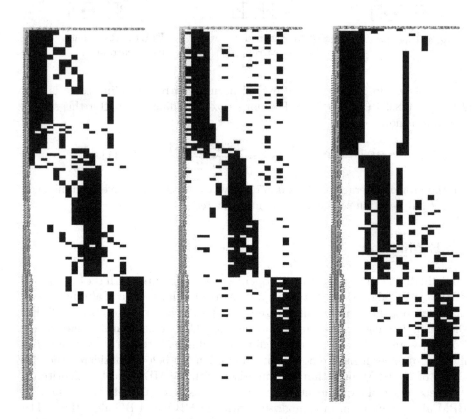

Fig. 7. Banding resulting from BPM algorithm as applied to Iris dataset (BS = 0.8404)

Fig. 8. Banding resulting from MBA algorithm as applied to Iris dataset (BS = 0.7806)

Fig. 9. Banding resulting from BC as applied to Iris dataset (BS = 0.7796)

7.2   Effectiveness with Respect to Banding Score

Note that, unsurprisingly, it was not possible to identify a perfect banding with respect to any of the UCI data sets. However, in terms of banding score, Table 1 clearly shows that the proposed BPM algorithm outperforms the previously proposed MBA and BC algorithms (best scores highlighted using bold font). Figures 7, 8, 9 and 10, 11, 12 show the bandings obtained using the Iris and Wine data sets using the BPM, MBA and BC algorithms respectively. Inspection of these Figures indicates that a banding can be identified in all cases. However,

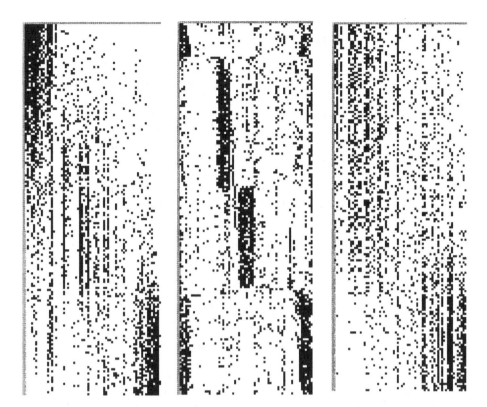

Fig. 10. Banding resulting from BPM algorithm
as applied to Wine
dataset (BS = 0.7993)

Fig. 11. Banding resulting from MBA algorithm
as applied to Wine
dataset (BS = 0.7123)

Fig. 12. Banding resulting from BC algorithm as
applied to Wine dataset
(BS = 0.7123)

from inspection of the figures it is suggested that the bandings produced using
the proposed BPM algorithm are better. For example considering the banding
produced when the MBA algorithm is applied to the Wine dataset (Figure 11)
the resulting banding includes "1"s in the top-right and bottom-left corners while
the BPM algorithm does not (it features a smaller bandwidth). When the BC
algorithm is applied to the Wine dataset (Figure 12) the banding is less dense
than in the case of the BPM algorithm. Similar observations can be made with
respect to the Iris data set.

7.3   Effectiveness with Respect to Frequent Pattern Mining

As already noted earlier in this paper banding has application with respect to
increasing the efficiency of algorithms that use zero-one matrices or tabular information stored in the form of 2D data storage structures. One example is
algorithms that use $N \times N$ affinity matrices, such as spectral clustering algorithms [15], to identify communities in networks (where N is the number of

Table 1. Efficiency Experimental Results (best results presented in bold font), BS = Banding Score, RT = Run time (secs.)

Datasets	# Rec s	# Cols	BPM BS	MBA BS	BC BS	BPM RT	MBA RT	BC RT
annealing	898	73	**0.8026**	0.7305	0.7374	**0.150**	0.260	0.840
heart	303	52	**0.8062**	0.7785	0.7224	**0.050**	0.160	0.170
horsecolic	368	85	**0.8152**	0.6992	0.7425	**0.070**	0.200	0.250
lympography	148	59	**0.8365**	0.7439	0.7711	**0.030**	0.140	0.110
wine	178	68	**0.7993**	0.7123	0.7021	**0.040**	0.150	0.110
hepatitis	155	56	**0.8393**	0.7403	0.7545	**0.050**	0.150	0.090
iris	150	19	**0.8404**	0.8205	0.7516	**0.020**	0.080	0.060
zoo	101	42	**0.8634**	0.7806	0.7796	**0.020**	0.100	0.050

Table 2. FIM runtime with and without banding with BPM ($\sigma = 2\%$), best results highlighted in bold font

Datasets	#Rows	#Cols	Banding Time(s)	FIM time (s) with Banding	Total	FIM time (s) without Banding
adult	48842	97	346.740	**2.274**	349.014	**5.827**
anneal	898	73	0.150	**0.736**	**0.086**	2.889
chessKRvk	28056	58	95.370	**0.082**	95.452	**0.171**
heart	303	52	0.050	**0.294**	**0.344**	0.387
hepatitis	155	56	0.030	**0.055**	**0.085**	22.416
horseColic	368	85	0.070	**0.899**	**0.969**	1.242
letRecog	20000	106	42.420	**3.004**	45.424	**6.763**
lympography	148	59	0.030	**7.997**	**8.022**	12.658
mushroom	8124	90	14.400	**874.104**	**888.504**	1232.740
penDigits	10992	89	21.940	**2.107**	24.047	**2.725**
waveForm	5000	101	3.030	**119.220**	**122.250**	174.864
wine	178	68	0.010	**0.155**	**0.165**	0.169

network nods). Another example is Frequent Itemset Mining (FIM) algorithms [1,2] where it is necessary to process large collections of zero-one data stored in the form of a set of feature vectors (drawn from a vector space model of the data). To test the effectiveness of banding with respect to such algorithms a FIM algorithm was applied to the banded data sets produced using the proposed BPM and the established MBA algorithm as a result of the experiments reported in Sub-section 7.1 above. More specifically the Total From Partial (TFP) algorithm [8] was used, but any alternative FIM algorithm would equally well have sufficed. The results are presented in Tables 2 and 3 (a FIM support threshold, σ, of 2% was used). From the tables it can be seen that FIM is always much more efficient when using both BPM and MBA banded data than when using non banded data if we do not include the time to conduct the banding. If we include the banding time, in 8 out of the 12 cases for BPM and 4 out of the 12 cases for MBA it is still more efficient.

Table 3. FIM runtime with and without banding with MBA ($\sigma = 2\%$)

Datasets	#Rows	#Cols	Banding Time(s)	FIM time (s) with Banding	Total	FIM time (s) without Banding
adult	48842	97	370.942	**10.525**	381.467	**5.827**
anneal	898	73	0.260	**1.733**	1.993	2.889
chessKRvk	28056	58	97.767	**0.075**	97.842	**0.171**
heart	303	52	0.160	**0.461**	0.621	**0.387**
hepatitis	155	56	0.150	**19.104**	19.254	22.416
horseColic	368	85	0.200	**2.134**	2.334	**1.242**
letRecog	20000	106	43.480	**6.221**	49.701	**6.763**
lympography	148	59	0.140	**11.187**	11.329	12.658
mushroom	8124	90	16.304	**1595.949**	1612.253	**1232.740**
penDigits	10992	89	22.002	**2.731**	24.733	**2.725**
waveForm	5000	101	3.135	**125.624**	**128.759**	174.864
wine	178	68	0.150	**0.211**	0.361	**0.169**

8 Conclusions

In this paper the authors have described an approach to identifying bandings in zero-one data using the concept of banding scores. The idea is to iteratively rearrange the columns and rows in a given zero-one matrix, according to normalized column (CS') and row (RS') scores until an overall banding score BS is maximized or no more changes can be made. These ideas have been incorporated into the Banded Pattern Mining (BPM) algorithm which was also described and illustrated. The BPM algorithm was evaluated by comparing its operation with the established MBA and BC banding algorithms using eight data sets taken from the UCI machine learning repository. The reported evaluation established that the proposed BPM approach could identify banding in zero-one data in a more effective manner (in terms of Banding score and run time) than in the case of the MBA and BC comparator algorithms. The reported evaluation also confirmed that, at least in the context of FIM, efficiency gains can be realized using the banding concept. For future work the authors intend to extend their research on banding to address firstly banding in the context of 3D volumetric data, and then in the context of N dimensional data. The authors have been greatly encouraged by the results produced so far as presented in this paper.

References

1. Agrawal, R., Imielinski, T., Swami, A.: Mining association rules between sets of items in large databases. In: SIGMOD 1993, pp. 207–216 (1993)
2. Agrawal, R., Srikant, R.: Fast algorithms for mining association rules in large databases. In: Proceedings 20th International Conference on Very Large Data Bases (VLDB 1994), pp. 487–499 (1994)

3. Alizadeh, F., Karp, R.M., Newberg, L.A., Weisser, D.K.: Physical mapping of chromosomes: A combinatorial problem in molecular biology. Algorithmica 13, 52–76 (1995)
4. Atkins, J., Boman, E., Hendrickson, B.: Spectral algorithm for seriation and the consecutive ones problem. SIAM J. Comput. 28, 297–310 (1999)
5. Baeza-Yates, R., RibeiroNeto, B.: Modern Information Retrieval. Addison-Wesley (1999)
6. Blake, C.L., Merz, C.J.: Uci repository of machine learning databases (1998), http://www.ics.uci.edu/$\sim$mlearn/MLRepository.htm
7. Brower, J.C., Kile, K.M.: Seriation of an original data matrix as applied to paleoecology. Lethaia 21, 79–93 (1988)
8. Coenen, F., Goulbourne, G., Leng, P.: Computing association rules using partial totals. In: Siebes, A., De Raedt, L. (eds.) PKDD 2001. LNCS (LNAI), vol. 2168, pp. 54–66. Springer, Heidelberg (2001)
9. Cuthill, A.E., McKee, J.: Reducing bandwidth of sparse symmentric matrices. In: Proceedings of the 1969 29th ACM national Conference, pp. 157–172 (1969)
10. Fortelius, M., Puolamaki, M.F.K., Mannila, H.: Seriation in paleontological data using markov chain monte method. PLoS Computational Biology, 2 (2006)
11. Garriga, G.C., Junttila, E., Mannila, H.: Banded structures in binary matrices. Knowledge Discovery and Information System 28, 197–226 (2011)
12. Pio, G., Ceci, M., D'Elia, D., Loglisci, C., Malerba, D.: A novel biclustering algorithm for the discovery of meaningful boiological correlaions between micrornas and their target genes. BMC Bioiinformatics 14, 8 (2013)
13. Inokuchi, A., Washio, T., Motoda, H.: An apriori-based algorithm for mining frequent substructures from graph data. In: Zighed, D.A., Komorowski, J., Żytkow, J.M. (eds.) PKDD 2000. LNCS (LNAI), vol. 1910, pp. 13–23. Springer, Heidelberg (2000)
14. Junttila, E.: Pattern in Permuted Binary Matrices. Ph.D. thesis (2011)
15. Luxburg, U.V.: A tutorial on spectral clustering. Statistical Computation 17, 395–416 (2007)
16. Deodhar, M., Gupta, G., Ghosh, J., Cho, H., Dhillon, I.: A scalable framework for discovering coherent co-clusters in noisy data. In: Proceedings of 26th International Conference on Machine Learning (ICML), Montreal Canada, p. 31 (2009)
17. Mueller, C.: Sparse matrix reordering algorithms for cluster identification. Machune Learning in Bioinformatics (2004)
18. Myllykangas, S., Himberg, J., Bohling, T., Nagy, B., Hollman, J., Knuutila, S.: DNA copy number amplification profiling of human neoplasms. Oncogene, 7324–7332 (2006)
19. Mäkinen, E., Siirtola, H.: The barycenter heuristic and the reorderable matrix. Informatica 29, 357–363 (2005)
20. Sugiyama, K., Tagawa, S., Toda, M.: Methods for visual understanding of hierarchical system structures. IEEE Transactions on Systems, Man and Cybernetics 11, 109–125 (1981)

Fast Pattern Recognition and Deep Learning Using Multi-Rooted Binary Decision Diagrams

Dmitry Bugaychenko and Dmitry Zubarevich

Saint-Petersburg State University
{DmitryBugaychenko,Zubarevich.Dmitry}@gmail.com

Abstract. Binary decision diagrams (BDD) is a compact and efficient representation of Boolean functions with extensions available for sets and finite-valued functions. The key feature of the BDD is an ability to employ internal structure (not necessary known upfront) of an object being modelled in order to provide a compact in-memory representation. In this paper we propose application of the BDD for machine learning as a tool for fast general pattern recognition. Multiple BDDs are used to capture a sets of training samples (patterns) and to estimate the similarity of a given test sample with the memorized training sets. Then, having multiple similarity estimates further analysis is done using additional layer of BDDs or common machine learning techniques. We describe training algorithms for BDDs (supervised, unsupervised and combined), an approach for constructing multi-layered networks combining BDDs with traditional artificial neurons and present experimental results for handwritten digits recognition on the MNIST dataset.

Keywords: pattern recognition, binary decision diagrams.

1 Introduction

Binary Decision Diagrams (BDD) were initially proposed to model Boolean functions, but they didn't attracted much attention due to lack of unified algorithms for dealing with them. The second birth was brought to the BDD by Bryant [1], who proposed a canonical ordered reduced form of the BDD capable to exploit internal regularity of the represented function and developed a set of unified algorithms for all common logical operations. Multiple extensions were proposed to deal with finite-valued functions and other kinds of objects, many of those extension and classical BDD are implemented as an open source packages. BDD based algorithms has been presented in many areas [2], including the symbolic model checking [3], tests generations [4], similarity joins [5], graph analysis [6].

The ability of BDD to construct an efficient representation of a large set of objects makes them worth considering for application in machine learning and data mining. Given a set of objects described by a vector of properties BDD can be constructed to store all the property vectors and to evaluate similarity of a new objects with the stored set. Having multiple BDDs representing different sets of objects multiple similarity estimates can be evaluated and used for detailed

P. Perner (Ed.): MLDM 2014, LNAI 8556, pp. 73–77, 2014.

analysis either by an additional layer of BDDs or in combination with traditional machine learning techniques (ANN, SVM, regression analysis and etc.). One of the BDD's modifications — Zero-suppressed BDD (ZDD) — was applied to data mining as well to represent item sets in-memory [7,8]. The main difference of this work from the these works is that multi-rooted version of BDD (or ZDD) is used to represent pattern similarity function instead of encoding pattern set itself. This allows for faster similarity evaluation in runtime at the expense of larger memory requirements and training time.

The goal of this work is to present algorithms for constructing BDDs representations of pattern similarity functions and for grouping objects into pattern. The paper is structured as follows. In section 2 we provide preliminary information about BDD and their modifications. Section 3 describes the main contribution — algorithms for constructing pattern similarity functions and interpreting their results. Experimental results for MNIST [9] database are presented in section 4 and section 5 concludes the work.

2   Binary Decision Diagrams

Formally, an ordered binary decision diagram is an oriented rooted acyclic graph with vertex set $V = N \cup T$, where $N \cap T = \emptyset$. The vertices from the set N are *nonterminals*; for every such vertex $v \in N$, an *order index*$(v) \in \{1, \dots, n\}$ and precisely two child vertices $low(v), high(v) \in V$ are defined. The vertices from the set T are *terminals* and have no child vertices. For such vertices $v \in T$, value $value(v) \in \{0, 1\}$ is defined. The following *ordering condition* holds: for any nonterminal $v \in N$ and its child vertex v', either $v' \in T$ or $index(v) < index(v')$. Each vertex $v \in V$ represents the function $f_v : \{0,1\}^n \to \{0,1\}$ of n variables:

- If $v \in T$, then $f_v(x_1, \dots, x_n) \doteq value(v)$;
- If $v \in N$, and $index(v) = i$ then $f_v(x_1, \dots, x_n) \doteq f_{high(v)}(x_1, \dots, x_n)$ iff $x_i = 1$ and $f_{low(v)}(x_1, \dots, x_n)$ otherwise.

In practice *reduced* decision diagrams are used. In a reduced decision diagram there can be no different vertices each representing the same function. An alternative set of reduction rules were proposed at [10] optimized for the case when most values of the function are zeros. The resulting decision diagram is called Zero-suppressed binary decision diagram (ZDD) and it is very efficient for representation of the relatively small sets and the key benefits of ZDD is that its size never exceeds the size of explicit representation of the objects.

In order to model a function taking arguments from a finite set S *binary encoding bin* $: S \to \{0,1\}^n$ is required. Similarly, in order to model a function returning values from a finite set S *binary decoding bin*$^{-1} : \{0,1\}^n \to S$ is required. With both encoding and decoding in place function $f : S \to S$ can be modelled with a set of Boolean functions $\{f_i : \{0,1\}^n \to \{0,1\}\}_1^n$ as follows: $f(x) = bin^{-1}(f_1(bin(x)), \dots, f_n(bin(x)))$. Each of the functions f_i can be represented with a decision diagram v_i (either BDD or ZDD). The decision diagrams $\{v_i\}_1^n$ can be seen as a single graph with multiple entry points which we call *multi-rooted binary decision diagram* (MRBDD).

3   Pattern Recognition with BDD

Given a finite non empty domain set S, pattern training sequence $P \in S^+$, similarity measure $sim : S \times S \to \mathbb{R}$ and aggregate function $agg : \mathbb{R} \times \mathbb{R} \to \mathbb{R}$ we define a pattern similarity function $f_{P,sim,agg} : S \to \mathbb{R}$ as follows

$$f_P(s) \doteq \begin{cases} agg(sim(s_0, s), f_{\{s_1, \dots s_n\}}(s)), \text{iff } P = \{s_0, s_1, \dots s_n\} \\ sim(s_0, s), \text{iff } P = \{s_0\} \end{cases} . \quad (1)$$

The pattern similarity function f_P represented using MRBDD constructed recursively using functions composition as described at [3] or using simple arithmetic operations if aggregate allows it (for example, sum). For long pattern sequences construction can be speeded up by splitting sequence into parts and then aggregating result for sub-sequences.

Depending on the concrete tasks different similarity measures and aggregates might be used, but the efficiency of BDD representation should be kept in mind. MRBDD are good at representing additive functions, most of bitwise functions and conditional functions, thus for similarity functions good choices are, for example, Hamming distance, L1 norm, Chebyshev distance, discrete metric. For aggregate function efficient candidates are, for example, sum, minimum, maximum, count of samples above threshold. Using simple arithmetic post-processing (eg. divide result by a constant) such aggregates as average or percentage of samples above threshold can be constructed.

There are multiple approaches for constructing pattern training sequences. The simplest is used in *supervised* learning when the patterns are pre-defined. In case if patterns are not known upfront *unsupervised* learning might be used in a form of *competitive learning* or *clustering*. In the competitive learning single training sequence is presented to the system with multiple pattern similarities function and each sample is assigned to the function with highest current similarity value. Alternatively, pre-processing can be applied to the initial sequence using clustering algorithms (eg. K-means or affinity propagation) and then pattern similarities trained for all identified clusters. Supervised and unsupervised schemas might be combined assuming that in the pre-defined patterns there might be group of objects with different structure. In the combined schema multiple similarity functions constructed for a single training sequence can be treated independently or aggregated.

After training a family of pattern similarity functions $\{f_{P_i} : S \to \mathbb{R}\}_{i=1}^N$ can be used to assign new object $s \in S$ to the most probable pattern $s \simeq P_j$ as $j = argmax_{i=1}^N f_{P_i}(s)$ or to construct a pattern probability distribution $p(s \simeq P_j) = f_{P_j}(s) / \sum_{i=1}^N (f_{P_i}(s))$ (depending on the actual similarity function range additional normalization might be needed to handle negative values and zeros). Furthermore, this family of functions can be used as a mapping $S \to \mathbb{R}^N$ converting original representation to n-dimensional vector space and the new representation can be used as a source of input to train second-order pattern similarity functions or as an input to other machine learning techniques.

Another approach to combination of supervised/unsupervised learning and layering is *conflict detection*. In case when after training an exemplar of pattern

$j \; s_j \in P_j$ is assigned to pattern P_k ($k = argmax_{i=1}^{N} f_{P_i}(s_j)$) additional pattern patterns and similarities for objects belonging to P_j but classified (or having high probability of being classified) as P_k ($P_{j \to k}$) and vice versa (belonging to P_k but classified as P_j — $P_{k \to j}$). In case if both functions f_{P_j} and f_{P_k} report high similarity values for s then functions $f_{P_{j \to k}}$ and $f_{P_{k \to j}}$ are used instead to make the decision. New similarity functions might be seen as a new layer in the decision making network which either transmits the signal from the underlying layer (if only one of the nodes active there) or replaces with the new signal. It is also possible to apply more advance machine learning techniques to make j-or-k decision based on all four signals. Conflict detection schema might be applied iteratively taking two most confused classes first, replacing classifiers for them with conflict resolution layer and choosing two most confused classes again.

4   Experiments

The experimental evaluation of BDD was done on the MNIST handwritten digits database [9] using BddFunctions [11] package for MRBDD implementation and Weka [12] for other machine learning techniques. Classifier was trained on the whole MNIST training set and then classification accuracy is measured on the entire MNIST test dataset. Normalized Hamming distance was used as a similarity measure with average and minimum as aggregate functions. Perceptron and SVM classifiers were used in combination with MRBDD and as a baseline. Result are shown in the table 1. As we can see MRBDD pattern matching is at least ten times faster comparing to brute force approach and the accuracy is comparable with the baselines. Introduction of conflict resolution layers and hybridization increase accuracy above the baselines. As expected, ZDD-based MRBDD outperform BDD-based in terms of both size and time.

Table 1. Experimental results on MNIST database

Classifier	Accuracy	Classification time			Number of nodes		Training time	
		BDD	ZDD	Brute force	BDD	ZDD	BDD	ZDD
Max similarity:								
Average	60.66%	2m	2m	30m	28M	1,1M	88m	10m
Min	91.31%	2m	2m	30m	28M	1,1M	12h	14m
Conflict detection:								
Average	93.64%	3m	3m	1h	30M	1,6M	6h	91m
Combining with:								
SVM	95.75%	3m	3m	1h	30M	1,6M	6h	139m
Perceptron	94.37%	3m	3m	1h	30M	1,6M	6h	101m
Baseline:								
Perceptron	91.95%	3m					38m	
SVM	92.34%	3m					10m	

5 Conclusion and Further Work

Results presented in the paper show that pattern similarity evaluation using MRBDD has a potential as a machine learning technique. It can be used both as a classification tool and as a feature extraction tool and thus combined with existing tools for classification and feature extraction. Usage of the pattern similarity functions as a base for classification provides a more coherent base for explanation of the classification which is important in some cases. Capabilities of MRBDD are worth investigating: new similarity and aggregate functions for different tasks, new techniques for arranging samples into pattern sequences and interaction with other methods are some of the most interesting goals for future researches.

References

1. Bryant, R.E.: Symbolic boolean manipulation with ordered binary-decision diagrams. ACM Computing Surveys 24(3), 293–318 (1992)
2. Minato, S.I.: Techniques of bdd/zdd: Brief history and recent activity. IEICE Transactions on Information and Systems 96(7), 1419–1429 (2013)
3. Bugaychenko, D.: On application of multi-rooted binary decision diagrams to probabilistic model checking. In: Kuncak, V., Rybalchenko, A. (eds.) VMCAI 2012. LNCS, vol. 7148, pp. 104–118. Springer, Heidelberg (2012)
4. Segall, I., Tzoref-Brill, R., Farchi, E.: Using binary decision diagrams for combinatorial test design. In: Proceedings of the 2011 International Symposium on Software Testing and Analysis, ISSTA 2011, pp. 254–264. ACM, New York (2011)
5. Shirai, Y., Takashima, H., Tsuruma, K., Oyama, S.: Similarity joins on item set collections using zero-suppressed binary decision diagrams. In: Meng, W., Feng, L., Bressan, S., Winiwarter, W., Song, W. (eds.) DASFAA 2013, Part I. LNCS, vol. 7825, pp. 56–70. Springer, Heidelberg (2013)
6. Toda, T.: Hypergraph transversal computation with binary decision diagrams. In: Bonifaci, V., Demetrescu, C., Marchetti-Spaccamela, A. (eds.) SEA 2013. LNCS, vol. 7933, pp. 91–102. Springer, Heidelberg (2013)
7. Minato, S.I.: Data mining using binary decision diagrams. Synthesis Lectures on Digital Circuits and Systems, 1097 (2010)
8. Shirai, Y., Tsuruma, K., Sakurai, Y., Oyama, S., Minato, S.-i.: Incremental set recommendation based on class differences. In: Tan, P.-N., Chawla, S., Ho, C.K., Bailey, J. (eds.) PAKDD 2012, Part I. LNCS, vol. 7301, pp. 183–194. Springer, Heidelberg (2012)
9. LeCun, Y., Cortes, C., Burges, C.J.: The MNIST database of handwritten digits, http://yann.lecun.com/exdb/mnist/
10. Minato, S.I.: Zero-suppressed bdds for set manipulation in combinatorial problems. In: 30th Conference on Design Automation, pp. 272–277. IEEE (1993)
11. Bugaychenko, D.: BddFunctions: Multi-rooted binary decision diagrams package, http://code.google.com/p/bddfunctions
12. Bouckaert, R.R., Frank, E., Hall, M., Kirkby, R., Reutemann, P., Seewald, A., Scuse, D.: Weka manual for version 3-7-8 (2013)

ACCD: Associative Classification over Concept-Drifting Data Streams

Kitsana Waiyamai, Thanapat Kangkachit, Bordin Saengthongloun,
and Thanawin Rakthanmanon

Department of Computer Engineering,
Kasetsart University, Bangkok, Thailand
{fengknw,g5185041,g5314550113,thanawin.r}@ku.ac.th

Abstract. Associative classification has shown good result over many classification techniques on static datasets. However, little work has been done on associative classification over data streams. Different from data in traditional static databases, data streams typically arrive continuously and unboundedly with occasionally changing data distribution known as concept drift. In this paper, we propose a new Associative Classification over Concept-Drifting data streams, ACCD. ACCD is able to accurately detect concept drift in data streams and reduce its effect by using an ensemble of classifiers. A mechanism for statistical accuracy bounds estimation is used for supporting concept-drift detection. With this mechanism, accuracy recovering time is decreased and a situation where ensemble of classifiers drops accuracy is avoided. Compared to AC-DS (first technique on associative classification algorithm over data streams), AUEH (Accuracy updated ensemble with Hoeffding tree) and VFDT(Very Fast Decision Trees) on 4 real-world data stream datasets, ACCD exhibits the best performance in terms of accuracy.

Keywords: concept drift, changing environments, data streams classification, associative classification, multiple class-sssociation rules.

1 Introduction

Unlike classification from static databases, classification from data streams poses many new challenges such as ability to produce results in real time, process data in a single pass, and maintain accuracy for on-line classification, while using limited memory. Associative classification is a classification technique that predicts class from association rules. Compared to other classification techniques, associative classification is very accurate and easily explainable. However, almost of existing associative classification algorithms [1,2,3] have been designed to work only on static datasets. Classifying a data stream with an associative classifier is a newly explored area of research [4], and poses many challenges such as how to generate and maintain frequent itemsets in concept drifting data streams and when to update class association rules.

Concept drift is one of the most crucial issues in data streams classification. It is a situation where class distribution changes as a time passes, thus result in

P. Perner (Ed.): MLDM 2014, LNAI 8556, pp. 78–90, 2014.

distribution mismatch problem [5,6]. Accuracy of a classifier can decrease when concept drift occurs because it does not fit to the new data. An efficient stream classification method should be able to detect occurrence of concept drift. For rule-based classifiers, techniques have proposed to detect and reduce effect of concept drift. In [7], the authors introduced Hoeffding trees for learning on-line from high-volume data streams. VFDT (*Very Fast Decision Tree*) uses a Hoeffding tree as the classification model for data streams. This incremental tree can evolve according to the current data in the data streams. AUEH (*Accuracy updated ensemble with Hoeffding tree*) [8] uses also Hoeffding trees as its classification model. Several Hoeffding trees can be generated from different data blocks, then AUEH combines the result using weighted ensemble from these trees.

Almost existing associative classification algorithms [1,2,3] are developed to work only on static data and, thus, cannot be applied directly over data streams. To the best of our knowledge, AC-DS [9] is the first and unique associative classification algorithm developed to work on data streams. AC-DS is based on the estimation of support threshold and a landmark window model [4]. However, AC-DS has two main drawbacks. First, it uses single rule for prediction of unseen data. There is a bias on general rules, with single rule for predicting a new data stream. This is not appropriate for streams that are changing from time to time. Second, it does not support concept-drifting data streams. Although AC-DS solves this problem by fading old data concepts, new data concept still not supported. Thus, accuracy recovering time is increased from concept change. To the best of our knowledge, no existing associative classification technique over concept-drifting data streams has been reported.

In this paper, we propose ACCD (Associative Classification over Concept-Drifting Data Streams) that is a technique for classification over concept-drifting data streams. ACCD is able to accurately detect concept drift and reduce concept drift effect by using an ensemble of classifiers. An mechanism for statistical accuracy bounds estimation is used for concept-drift detection. For each data concept, an ensemble of associative classifiers is generated, and reinitialized when concept drift is detected. With this mechanism, accuracy recovering time is decreased and situation when ensemble of classifiers drops accuracy is avoided. To avoid bias on general rules (not appropriate for streams that change from time to time), ACCD classifies new record using multiple rules. An interval estimated Hoeffding-bound is used as a gain to approximate best number of rules, k. We compared our associative classification technique, ACCD, with AC-DS, VFDT and AUEH. Three real stream datasets and one generated dataset are used to evaluate their performance. Compared to other methods, experimental results show that ACCD gives the best performance in terms of accuracy.

2 Preliminaries and Related Works

In this section, preliminaries and related works are given. Assume that a transaction data streams $D_s = \{B_0, B_1, \ldots, B_N\}$ is an infinite sequence of ordered blocks. Each block $B_i = \{T_1, T_2, \ldots, T_k\}$ is set of transactions. $|B_i|$ is number of transaction in each block. Current total number of transactions in data streams is given as $DL = |B_0| + |B_1| + |B_2| + \ldots + |B_n|$.

Let $I = i_1, i_2, \ldots, i_k$ is a set of singleton itemsets. An itemset subset of I that contains k items is called an k-itemset. Let $C = c_1, c_2, \ldots, c_l$ is a set of classes.

2.1   Associative Classification over Data Streams

Associative classification over data streams (AC-DS) was first proposed by [9]. Like the traditional associative classification, AC-DS is composed of two main steps: rule generation and classifier building. Unfortunately, data streams can not be processed in a single pass due to huge volumes of unbounded data which continuously arrive during progression of time. This is a reason why AC-DS performs in an incremental manner. The detail of each steps is further described as follows.

In rule generation step, AC-DS starts by performing the frequent-itemsets extraction process. To estimate the support count of each itemset, many strategies, [4] [9] and [10], use a landmark window model which partitions data streams into many chunks with respected to their arrival time. Support of an $itemset_i$ can be estimated at different times and can be defined as follows.

$$sup(itemset_i) + minSup \times (t_f - 1) \times DL < minSup \times t \times DL \qquad (1)$$

where $minSup$ is the minimum support threshold specified by user, t is current time, t_f is the first time timestamp. DL is data length. This support estimation constraint is also used as a decay gain for pruning frequent itemsets. That is, support of a frequent itemset is decreased by the value of minimum support each time there is an update.

After retrieving all frequent itemsets, all class association rules (CARs) are generated. CAR is defined as $X \to c$ where $X \subseteq I$, $c \in C$ and $X \cap c = \varnothing$. X is called an itemset and c is a class.

In the classifier building step, a classifier is constructed based on the generated CARs from previous step. Only a subset of CARs is selected based on some heuristic orders such as confidence and coverage. Then, rules are ranked based on precedence order. Assume R_i is rule. Given $R_1 \subset R_2$. R_2. R_2 is generated if $Measurement(R_2) > Measurement(R_1)$. To predict class of unseen data, the resulted class is derived from the highest-ranked rule that satisfies such data.

2.2   Classification of Concept-Drifting Data Streams

Concept drift [5,6] is one of the most crucial issues in data streams classification. It is a situation where class distribution changes as a time passes. If adjustment of a decision boundary of an incremental learner is not appropriate, it will result in a performance degradation [5] of the classifiers. There are many works that have been proposed to solve concept-drift problem. We focus here on the ones that are rule-based classifiers which can be categorised into single classifier approach and ensemble of classifiers approach. With single classifier approach, concept drift is detected then different solutions are proposed to response the situation such as on-line classifier reinitialization. This mechanism helps classifier learning new data concept. Ensemble of classifiers have several advantages over single classifier methods: they can adapt to change quickly by pruning under-performing parts of the ensemble, and they therefore usually generate more accurate concept descriptions [11]. Compared to single classifier, ensemble of classifiers combine many models to reduce bias toward class distribution of the training dataset. This result in better classifier performance.

AWE [5] uses ensemble of classifiers for mining concept-drifting data streams. The ensemble of classifiers is made from sequential static data block size then ensemble will weight by its accuracy that it is difference between mean square predict random (MSE_r) and mean square error (MSE_i) on current data block. Classifier with error equal or larger than MSE_r is discarded. If number of classifiers is larger than the user-specified threshold then the ensemble is replaced with the one with lowest weight. Weighted ensemble then prediction using majority vote with weighted classier. Formal proof and experimental results show that accuracy of AWE is greater than using single classifier.

AUE [8] is an accuracy updated ensemble for concept drifting data streams. To solve rare case that is mute all of classifier in AWE, AUE propose new weight ensemble function and allow ensemble be updated by training data block. In order to avoid MSE_i equal zero, new weight function allow add term offset that is small constant value. The ensemble can be updated by current training data block, if it enough accurate. Enough accuracy is classifier accuracy than $MSEr$ on current data block. AUE select top weighted classifiers use to prediction. If number of classifiers more than user specific threshold, replace the lowest weight classifier with new one.

3   ACCD Algorithm

Although several algorithms have been recently proposed for data stream classification, concept drifting is a really challenge. In this paper, a novel associative classification framework to handle concept drift is proposed and demonstrated by a concrete algorithm named ACCD. Intuitively, ACCD is an on-line classifier which has ability to evolve itself to support the change of the data over the time. On the other hand, if the concept of the data is immediately drifted, ACCD copes with either of these two strategies: using learned classifiers to classify this concept, or creating a new on-line classifier from this new concept.

In our framework, data streams will be represented as a series of data blocks and, at any time step, one data block would be considered. However, without loss of generality and to reduce the complexity of the algorithm, we split each data block to training part and testing part. The training part is used to create or update the classifiers and the testing part is only used for measuring the final accuracy. Precisely, ACCD is composed of two types of classifiers. Firstly, at any time, ACCD maintains an on-line classifier by creating a new one or updating the current one with respect to how the new data (i.e., training part) similar to the previous data (i.e., accuracy of the current on-line classifier). Secondly, ACCD maintains a list of batch classifiers and uses them as an ensemble classifier whose weights depend on how accurate they perform on the training part of the current data block. Fig 1 gives the overview of ACCD framework.

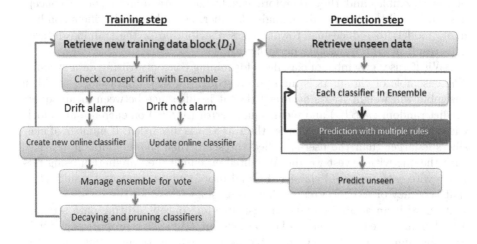

Fig. 1. Overview ACCD algorithm

3.1   Training Steps of ACCD

Table 1 shows the main process of ACCD algorithm to learn a classification model from the training data. The key ideas of these steps is to detect a concept drift, take a response (line 1 to 6), manage the ensemble classification (line 8-13), and decay the weights of classifiers over time (line 14).

Concept-Drifting Detection and Response. First of all, after the training part of the current data block has been retrieved, our algorithm detects whether the concept of the current data drifts from the previous one. If the drift has been detected, the algorithm will stop the current on-line classifier, store it as a batch classifier. Then, a new on-line classifier is created from scratch based on the current data. For normal situation or when the data is slightly changed, the algorithm will update the current on-line classifier according to the new coming data.

Intuitively, if the concept of the new data is drifted or dramatically changed, the accuracy when applying the previous classifier will be dropped significantly. Therefore, ACCD learns this by comparing the maximum accuracy of current time with the estimated statistical lower bound of maximum accuracy [6] at last time.

To create a new classifier, there are three steps to perform: find the frequent itemsets, generate class associaton rules, and build the classifier. To update the current classifier, support and confidence of all the itemsets and rules will be

Table 1. ACCD algorithm

Function ACCD(D_i, BC, oc)
Parameter D_i : training part of the current data block BC : set of batch classifiers oc : current online classifier
1   **if** $(ConceptDrift(oc, D_i))$
2       $BC \leftarrow BC \cup \{oc\}$
3       $oc \leftarrow CreateNewClassifier(D_i)$
4   **else**
5       $UpdateCurrentClassifier(oc, D_i)$
6   **endif**
8   **for each** batch classifier bc in BC
9   **if** $Accuracy(bc, D_i) \geq Accuracy(oc, D_i)$
10          $UpdateRuleWeight(bc, D_i)$
11          $AddToEnsemble(bc, Accuracy(bc, D_i))$
12   **endif**
13   **endfor**
14   $DecayAndPrune(BC)$

updated as some new rules may be introduced. Notice also that not only classifier but also some itemsets may evolve over time. Hence, the support of itemsets are faded by time too. Moreover, to avoid generate duplicate association rules, only longer rules which achieve higher confidence are allowed to be generated.

Managing Weights of Classifiers. At any step, there are only one on-line classifiers and several batch classifiers in ACCD. The on-line classifier is basically updated with respect to the new coming data. However, when an on-line classifier is obsoleted or not suitable for the new data any more, it will be stored in a set of batch classifiers. Then, a new on-line classifier will be created. After that, ACCD will combine the results from all the classifiers together using an ensemble technique [5]. Based on AdaBoost, the weight of each classifier is defined as a logarithm function of the classifier's accuracy, as shown in Equation 2:

$$W_c(t) = \begin{cases} log(\frac{acc_c(t)}{1-acc_c(t)}) & if(acc_c(t) \geq Est(acc(t-1))) \\ 0 & Otherwise \end{cases} \quad (2)$$

where $W_c(t)$ is a weight for majority vote of classifier c at time t, and $acc_c(t)$ is the accuracy of the classifier. Notice that if the accuracy of a given classifier c is less than the estimated lower bound accuracy of all the classifiers from previous time step $Est(acc(t-1))$, that classifier c will not involve in the vote of this time step but will contribute later. Moreover, a classifier is *active*, if it has a positive weight.

$$Est(acc(t)) = \frac{|D_i|}{|D_i|+Z_{\frac{\alpha}{2}}^2} \left(acc_t + \frac{Z_{\frac{\alpha}{2}}^2}{2|D_i|} - Z_{\frac{\alpha}{2}} \left(\sqrt{\frac{acc_t(1-acc_t)}{|D_i|} + \frac{Z_{\frac{\alpha}{2}}^2}{4|D_i|^2}} \right) \right) \quad (3)$$

Equation 3: is estimated lower bound function [6]. $|D_i|$ is number of training data block at time t. $Z_{\frac{\alpha}{2}}^2$ is z score (Standard normal distribution).

Decaying and Pruning Classifiers. Because all batch classifiers cannot be stored forever as the line classifier, there is a mechanism to decide which one to be kept and which to be removed. This step is composed of few intuitive ideas. Firstly, weight of classifiers depends on the number of times they are active. More precisely, if they are active, they will receive one point of weight; if not, they receive nothing. Secondly, weights of all the classifiers are exponentially decayed by time. Thirdly, if the number of batch classifiers is larger than a user-defined threshold, the classifier which has the smallest weight will be removed. This threshold is not sensitive at all because after time passing, weight of many classifiers will be dropped to its lowest value.

Managing Rule Weights in Classifiers. Because the online classifier is up-to-dated, all rules will be used in classification. Contrary, not all the rules in batch classifiers are useful becuase all rules have been created in the past. For batch classifiers, ACCD will use only the rules that can explain the current data (i.e., trainning part) correctly. The rule's weight w_r at time t is defined as the accuracy of this rule when applied to the current data, as in Equation 4

$$w_r(t) = \frac{\sum correct_r(t)}{\sum cover_t(t)} \qquad (4)$$

where $correct_t$ is number of training rules that predict correct answer. $cover_t$ is number of data that is covered by rule. High accuracy of rule means fact.

Moreover, as in Equation 5, a rule is defined as an *active* rule if its weight (i.e., accuracy) is not less than a threshold.

$$w_r(t) \geq threshold_{activerule} \qquad (5)$$

where $threshold_{activerule}$ is describe as following. ACCD calculates mean of all rules' weights in each classifier and then select the maximum mean. The threshold is the estimated statistical maximum mean upper bound as Equation 3: [6], however minus operator before square root is changed to add operator. Please note that, ACCD will not consider to use any rule which is not suitable to explain the concept of the current data block, and only active rules in active classifiers will be used in the majority vote.

3.2   Prediction Steps of ACCD

As mention in the previous section in Fig. 1, this prediction process consists of three steps which are rule selection, number of rules estimation, and computing the score for each class label as shown in Table 2. After the training process in the previous section, all active classifiers are stored and will be used as an ensemble classifier.

ACCD uses multiple rules from each classifier to avoid the bias of single rule prediction. Precisely, only active rules in active classes are used to predict the final result. In addition, to reduce the bias that some classes may have more rules than others, ACCD borrows techniques from [10] to select the equal number of rules from each classes, and choose the class that has the maximum score (i.e., summation of all rules' weight).

Multiple rules prediction use majority vote to predict the best class which has maximum summation confidence. An interval estimated Hoeffding-bound [7] is used as a gain to separate the best class from other classes to determine top k number of rules. The idea is to find significant difference between summation confidence of the best class and the second best class. The best class will be selected as the answer only if the difference of predictive accuracy is significantly large (i.e., greater than the Hoeffding-bound gain).

After specifying all active rules in all active classifiers, the ensemble technique by weighted voting is used as score for each class label, and the label which has the maximum score will be return as the final answer.

Table 2. ACCD prediction steps

Function Prediction(x, EC)	
Input	x : unseen data, EC : set of ensemble classifiers
1	**for each** classifier c in EC
2	$ActiveRuleSelection(c)$
3	$ApproxNumberOfRules(c,x)$
4	**for each** class lable l
5	$UpdateScore(l,c,x)$
6	**end for**
7	$label_{final} \leftarrow MaxLabel(l)$

4   Experiment Results

To assess performance of ACCD, its performance is compared with other three algorithms which are *VFDT* [7], *AUEH* [8], and *AC-DS* [9]. ACCD and AC-DS are implemented based on LUCS-KDD libraries [12]. Meanwhile, AUEH and VFDT are implemented based on Massive Online Learning(MOA) [13] by using the same block size as ACCD and AC-DS. Baseline classifier is an implementation of ACCD using single classifier, and without concept-drifting detection.

The performance is evaluated in terms of both accumulative accuracy and current block accuracy. The accumulative accuracy at time t is an accuracy from the beginning of data streams till time t. The current block accuracy at time t is the accuracy from current measured at time t. To evaluate performance of ACCD, we use both synthetic and real world data streams e.g. Airline, Cover-Type, Electricity and KDD1999 .

The parameter settings are as follows: minimum support = 1%, minimum confidence = 50%, minimum length of itemset = 4 and maximum number of classifiers = 15. For each dataset, training and testing data are select 50% both. To limit quantity of memory usage, the maximum number of frequent itemsets is fixed.

Synthetic Data Streams. Based on Agrawal (AgrawalGenerator), 10 data concepts are generated, and each one contains 10000 instances. Each data block has 2000 instances. Thus, a data concept is changed for every 5 blocks. Fig.2 illustrates the predictive performance of ACCD for current block accuracy. We denote that ACCD is much more faster than the other algorithms for its accuracy recovering time, i.e. one data block faster for almost of the data concepts. This is due to the ability of ACCD to detect concept-drift, and use the new on-line classifier to predict unseen data. ACCD also yields the best accuracy (88.29% in average) for all the streams progressions. This can be explained by the fact that ACCD has the best accuracy recovery mechanism after concept-drift is detected. Its old-concept classifiers together with its on-line classifier can reduce errors when unseen data need explanation from the old concepts. Further, ACCD is able to avoid risk rules in classifiers and uses majority vote to predict unseen data.

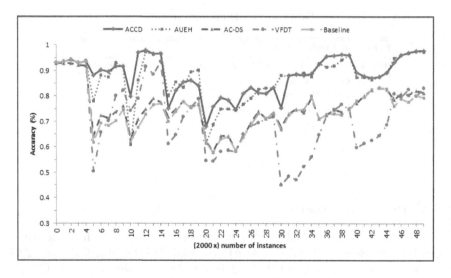

Fig. 2. Current accuracy comparison with different algorithms on synthetic data

Real World Data Streams. Real world data can contain both concept-drifting (i.e. sudden change) and gradual-change streams data. Here, CoverType is used as a benchmark dataset for evaluating classification performance on concept-drifting data streams. Fig.3 and 4 show performance of ACCD in terms of ac-cumulative and current block accuracy, respectively, on CoverType dataset. It can be clearly seen that accumulative accuracy of ACCD is better than the oth-ers in most of the times. Notice that, VFDT and baseline classifier offer higher accumulative accuracy in the first period of data streams. However, their per-formance decrease from 30th to 110th block, respectively, after facing several concept changes.

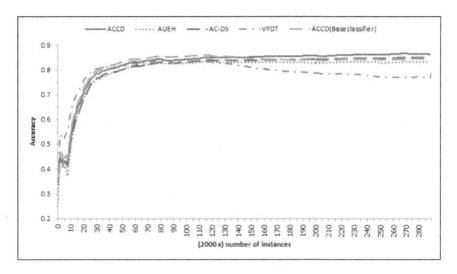

Fig. 3. Accumulative accuracy comparison with different algorithms on CoverType dataset

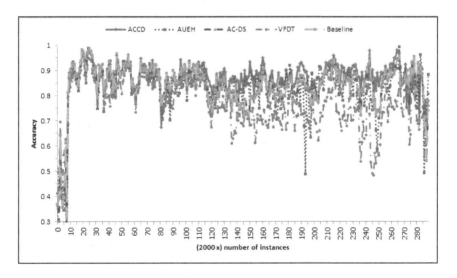

Fig. 4. Current accuracy comparison with different algorithms on CoverType dataset

4.1   Gradual Change Data Streams

We have previously seen that ACCD has a very good performance over concept-drifting data streams. Here, we evaluate its performance on gradual change data streams. Table 3. shows predictive performance of ACCD compared with the other methods on gradual change data streams. Compared to AUEH (which also produces good performance on concept-drifting data streams), ACCD still yields significantly higher accuracy in all the datasets. Compared to AC-DS, VFDT and the baseline classifier which do not support concept-drifting data streams, its performance is considered equal. Thus, ACCD is the associative classification technique that is able to generate very good performance for both concept-drifting and gradual change data streams.

Table 3. Average accuracy comparison on gradual change data streams

Datasets	ACCD	AUEH	AC-DS	VFDT	Baseline
Airline	0.6660	0.6246	0.6647	0.6503	0.6646
Electricity	0.7972	0.7626	0.7924	0.7775	0.7970
KDD1999	0.9811	0.9458	0.9826	0.9849	0.9883
AVG	0.8147	0.7777	0.8132	0.8042	0.8166

4.2   Effect of Parameter Settings

Appropriate maximum length of rules and active rule threshold can affect the performance of ACCD. Length of rule corresponds to the number of itemsets in its antecedence. active rule threshold is used in the process of rule selection for building a classifier. Table 4 shows the average accuracy of ACCD when varying the length of rule from 1 to 7. When length of rules is equal to 4, ACCD produces its best accuracy. When length of rules is more than 4, the accuracy of ACCD remains un-changed for all the datasets. This is a reason why we set maximum length of rules equal to 4.

Instead of using a fixed active rule threshold, ACCD uses the estimation of the upper bound of its maximum mean. Thus, this threshold can be changed during the streams progressions. Experiments are conducted on several datasets with

Table 4. Average accuracy with varying maximum length of rules

Datasets	length=1	length=2	length=3	length=4	length=5	length=6	length=7
CoverType	0.7445	0.8192	0.8503	0.8529	0.8529	0.8529	0.8529
Airline	0.6537	0.6572	0.6577	0.6578	0.6576	0.6576	0.6576
Electricity	0.7633	0.7403	0.7453	0.7453	0.7435	0.7435	0.7435
KDD1999	0.9673	0.9668	0.9648	0.9648	0.9648	0.9648	0.9648
Agrawal	0.7906	0.8420	0.8666	0.8689	0.8689	0.8689	0.8689

Table 5. Average accuracy with varying active rule thresholds

Datasets	0.1	0.2	0.3	0.4	0.5	0.6	0.7	0.8	0.9	1.0
CoverType	0.8527	0.8521	0.8525	0.8538	0.8550	0.8558	0.8581	0.8605	0.8664	0.8684
Airline	0.6578	0.6578	0.6580	0.6589	0.6609	0.6641	0.6671	0.6681	0.6657	0.6652
Electricity	0.7448	0.7464	0.7496	0.7597	0.7642	0.7730	0.7913	0.7988	0.7909	0.7742
KDD1999	0.9650	0.9643	0.9643	0.9643	0.9644	0.9644	0.9692	0.9678	0.9806	0.9813
Agrawal	0.8682	0.8682	0.8683	0.8689	0.8700	0.8744	0.8792	0.8910	0.8911	0.8909

varying active rule threshold from 0.1 to 1.0. Experimental results are shown in Table 5. With active rule threshold equal to 0.9, ACCD obtains its best accuracy for all the datasets. These results are comparable to what we obtained in table 3.

5 Conclusion and Future Work

In this work, we present ACCD, a new technique for associative classification over concept-drifting data streams. To determine when to update class association rules, ACCD uses estimation of statistical accuracy bounds for concept-drift detection. An ensemble composed of on-line and batch associative classifiers is used to detect concept-drift in data streams. While on-line classifier is updated for every step in a data block, only weight of class association rules is updated in the batch classifier. Once concept drift detected, ACCD adds current on-line classifier into the ensemble system, then a new on-line classifier from the new data block in created. To avoid inaccurate ensemble and risky rules in the batch classifier, a majority vote mechanism is used to predict unseen data.

We compare our method ACCD with AUEH (Accuracy updated ensemble with Hoeffding tree) and VFDT (very fast decision tree) on 3 real-world UCI data stream datasets and 1 generated data stream dataset. Experimental results show that ACCD exhibits a better performance in terms of accuracy on the 4 datasets.

As a future work, we plan to further improve performance of ACCD in several aspect such: dimension projection to support high-dimensional data streams, adaptive windows to support on-demand classification.

References

1. Li, W., Han, J., Pei, J.: Cmar: accurate and efficient classification based on multiple class-association rules. In: Proceedings IEEE International Conference on Data Mining, ICDM 2001, pp. 369–376 (2001)
2. Yin, X., Han, J.: Cpar: Classification based on predictive association rules. In: Proc. SIAM Int. Conf. on Data Mining, SDM 2003, pp. 331–335 (2003)
3. Phan-Luong, V., Messouci, R.: Building classifiers with association rules based on small key itemsets. In: 2nd International Conference on Digital Information Management, ICDIM 2007, pp. 200–205 (2007)
4. Jiang, N., Gruenwald, L.: Research issues in data stream association rule mining. In: SIGMOD Record (2006)
5. Haixun, W., Fan, W., Yu, P., Han, J.: Mining concept-drifting data streams using ensemble classifiers. In: Proceedings of the Ninth ACM SIGKDD International Conference on Knowledge Discovery and Data Mining, KDD 2003, pp. 226–235. ACM, New York (2003)
6. Nishida, K., Yamauchi, K.: Adaptive classifiers-ensemble system for tracking concept drift. In: International Conference on Machine Learning and Cybernetics, vol. 6, pp. 3607–3612 (2007)
7. Domingos, P., Hulten, G.: Mining high-speed data streams. In: Proceedings of the Sixth ACM SIGKDD International Conference on Knowledge Discovery and Data Mining, KDD 2000, pp. 71–80. ACM, New York (2000)

8. Brzeziński, D., Stefanowski, J.: Accuracy updated ensemble for data streams with concept drift. In: Corchado, E., Kurzyński, M., Woźniak, M. (eds.) HAIS 2011, Part II. LNCS, vol. 6679, pp. 155–163. Springer, Heidelberg (2011)
9. Su, L., Liu, H., Song, Z.: A new classification algorithm for data stream. I. J. Modern Education and Computer Science, 32–39 (2011)
10. Saengthongloun, B., Kangkachit, T., Rakthanmanon, T., Waiyamai, K.: Ac-stream: Associative classification over data streams using multiple class association rules. In: The 10th International Joint Conference on Computer Science and Software Engineering (JCSSE 2013), Khonkaen, Thailand (May 2013)
11. Bifet, A., Holmes, G., Kirkby, R., Pfahringer, B.: Data Stream Mining A Practical Approach. The University of WEIKATO (2011)
12. Coenen, F., Yamauchi, K.: The lucs-kdd software library
13. Bifet, A., Holmes, G., Kirkby, R., Pfahringer, B.: Massive online analysis (moa)

Integrating Weight with Ensemble to Handle Changes in Class Distribution

Nachai Limsetto and Kitsana Waiyamai

Data Analysis and Knowledge Discovery Lab (DAKDL),
Department of computer engineering Faculty of Engineering Kasetsart University BangKhen
campus Bangkok, 10900, Thailand
khim1204@gmail.com, fengknw@ku.ac.th

Abstract. Concept drift can be considered as a distribution mismatch problem where class distribution changes as a time passes. This problem is commonly found in classification task of data mining. Among the proposed solutions, the cost-based Class Distribution Estimation (CDE) shows the best performance in coping with difference in class distribution between train and test datasets. However there is still some problem, as CDE lost its performance when there is too much change in class distribution. In this paper, CDE-weight is proposed to reduce the impact of high change in class distribution. The idea is to use many models suitable with many class distributions along with dynamic weighting method that adjusts weight of each model according to its class distribution. Experimented results indicate that CDE-Weight methods are able to reduce the impact of misestimating and improve the classifier performance when train and test data are different.

Keywords: Concept drift, Classification, Cost-sensitive learning, Quantification, Class distribution estimation, Ensemble method.

1 Introduction

Situation where concepts of data are non-static and changeable is common problem in real-world classification application. The problem is also known as *concept drift* [15] and needed to handle properly otherwise it will significantly lower classifier performance. The reason for this is that as a time passed, data is bound to changes causing concept of data to be different from when the classifier was built. This makes previously built model to become obsolete and unsuitable for the current data due to the change in decision boundary.

One of the ways to compensate for changes in class distribution is probability-based expectation maximization (EM) method [8, 14] which drastically improves classification performance compared to traditional classifiers. EM uses probability to estimate change in class distribution and adjusts the original classifier's posterior probability outputs. However, limited number of classifiers EM can be applied poses utility problem to this method, and making it hard for EM to cope with various situations that can arise since no classification method is suitable for all situations.

P. Perner (Ed.): MLDM 2014, LNAI 8556, pp. 91–106, 2014.

Another approach for this problem is robust classifier [2, 10], while this method has its advantage when there is no way to obtain information about future changes in class distribution. The fact that robust classifier doesn't use any information from test data's class distribution which can be reliably estimated, diminish the efficiency of this method. Sampling [5, 15] can also be used to compensate for the difference in class distributions. While these methods have their own advantages, decreasing of majority class members usually has negative effect to classification accuracy and increasing minority class members consumes a lot of time. Therefore, these methods are not suitable for solving the concept drift problem

Among cost-based methods, CDE (Class Distribution Estimation) [19] has been proposed to solve concept drift problem. CDE estimates change in class distribution then use it to adjust cost of cost sensitive classifier to make classifier better suited for current data. When compared with the other proposed solutions [1, 8, 14], CDE is able to cope with more situations while gaining significant improvement in term of accuracy and f-measure. However, CDE lost its performance when face high change in class distribution between train and test datasets. CDE-EM [9], Class Distribution Estimation-Ensemble method, offers a decent solution to concept drift problem by reducing the impact of high change in class distribution via using many models (instead of one model) suitable with many class distributions. Experimental results on several UCI datasets show that CDE-EM produces significantly less error in class distribution estimation than CDE.   However, accurately estimating the class distribution of unseen data proves to be quite challenging, even with all the improvements done by CDE and CDE-EM. The error in the class distribution estimation process is about 20% impacting the performance of the classification.

In this research work, our idea is to improve CDE-EM with weight to make it better at estimating the class distribution of unseen data. The problem of the previous methods is that all ensemble models are given the same weight. This hinders the class distribution estimation performance; weight of each model should be accounted accordingly to reflect the true properties of datasets. We propose CDE-weight method to estimate class distribution with dynamic weighting algorithm that adjusts weight of each model according to its class distribution. To evaluate its performance, CDE-weight is compared with its predecessors in terms of accuracy and robustness. Experimental results show that CDE-Weights yields better performance when class distribution of train data is very different from test data due to the improvement in class distribution estimation process. The remainder of this paper is structured as follows. Section 2 describes important basic knowledge and previous works. Section 3 presents details of our CDE-weight methods. Experiment methodology and experimented results are presented in section 4. Section 5 summarizes our conclusions then referents are present in section 6.

2   Related Works

In this section we describe Class Distribution Estimation-EM methods (CDE-EM) which we intend to improve, including some of its basic knowledge.

2.1   Cost Sensitive Classifier (CSC)

CSC [6] is a wrapper based algorithm working over other classifiers. Normally, a classifier model is usually built with each class having equal incorrectly classify cost. However, CSC will adjust these costs to be different, resulting in better performance in terms of accuracy and f-measure if the costs are properly adjusted.

2.2   Class Distribution

Class distribution is the percentage of examples in a class varying from 0% to 100%. Class distribution will be different depending on its number of examples. In case of binary classification, class distribution can be represented with positive rate (pr) and negative rate (nr). Positive rate is a percentage of positive examples in the dataset and negative rate is a percentage of negative examples in the dataset. Notice that the difference in class distribution between training and testing sets will be resulted in performance degradation of a classifier.

2.3   Positive-to-Negative Ratio (pr/nr)

pr/nr is a ratio of positive examples to negative examples in a dataset. In case of binary classification, (pr/nr) can be calculated using equation 1.

$$(pr/nr) = \frac{\text{Number of positive class examples}}{\text{Number of negative class examples}} \tag{1}$$

2.4   Distribution Mismatch Ratio (DMR)

DMR is ratio of distribution mismatch between two datasets. It represents the different of class distribution between the two datasets and its direction. DMR can be calculated using (pr/nr) of each dataset and equation 2.

$$DMR = \frac{(pr/nr)'}{(pr/nr)} \tag{2}$$

This ratio can also be used in conjunction with CSC to adjust the incorrectly classify cost in each class. For example: If the train data has 800 positive examples and 200 negative examples (pr = 8/10, nr=2/10) and the test data has 200 positive and 800 negative examples (pr'=2/10, nr'=8/10). The ratio of positive to negative examples changes between the original and the new distribution can be indicated using DMR. As for this example, DMR = 4:¼ or, equivalently, 1:16. Hence, based on the positive rate to negative rate ratio, the positive example are 16 times more prevalent in the train distribution than in the test distribution. This means if we train model with this train data and expect this model to work well in new distribution of test data, then the cost of negative example miss-classification should be 16 times more than the cost of positive example miss-classification to compensate for lack of negative example information.

2.5   CDE-EM

CDE-EM [9] the improved version of CDE [19] which work on the basis that if reliable estimation of unseen data class distribution is obtainable, then the cost of CSC can
be adjusted to make classifier, built from train data, able to account for the differences
in the class distribution between the two datasets. For CDE-EM, this estimated class
distribution is acquired, from using the ensemble method which mitigates the impact
of severe drifts in class distribution.   In this paper we show results from two CDE-
EM methods which is CDE-AVG-N and CDE-AVGBM-M-N

In CDE-AVG-N, multiple models are built with different class distributions to increase the diversity. The average estimated class distribution is then obtained from all
these models. Their objective is to use many models built with different costs obtained from many class distributions to reduce biasing toward train data. The number
of ensemble models being built is indicated by the variable N. While this method
performance is quite acceptable, the accumulated error from the models built with
class distribution too different from the test data hinders the performance of classification. The CDE-AVGBM-M-N is an improved version of CDE-AVG. It selects the
pivot model (best model) likely to best represent the unseen data and uses it in conjunction with its neighbor models to estimate the class distribution of unseen data.
The variable M indicates how many neighbor models are used to find the average
value. Its objective is to reduce the impact of error in pivot model selection. By testing CDE-AVGBM method in various situations, we found that some problems still
remain, mainly in the pivot model (best model) selection process which is prone to
error when train and test data are too different resulting in lower classification performance, even with the help from neighbor models.

The errors of the selection process are shown in figure 1 and 2 which represent
overestimation and underestimation found in the CDE-AVGBM's process of pivot
model (Best Model) selection. The dark line is the correct class distribution while the
grey one is the estimated class distribution used to select neighbor models from CDE-
AVGBM. The bright tab indicates the class distribution of the train data. Note that in
these experiments the train data has a fixed class distribution while test data class
distribution is varying from positive rate of 0.5 to 0.95 with 0.1 increments.

In best case, the grey line has to be the same as the dark one. But as shown in these
figures, there are still some errors. These errors occur in pattern, biasing toward class
distribution of train data. If the train data positive rate is lower than the test data, the
estimated class distribution using to select the pivot model will likely to be underes-
timated (as shown by the grey line is under the dark line in figure 2). The same could
be said when the train data positive rate is higher, which will likely result in an overestimation (figure 1). These errors grow as the different in train and test data increase.
The reason for this behavior is mainly due to lack of data and classifier's biasing.
These problems lead to lower classification performance which will be handled in this
paper.

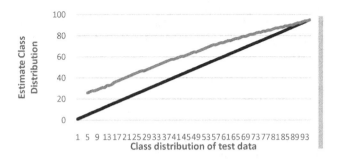

Fig. 1. Estimate Class Distribution from Spambase training dataset with positive rate 0.95

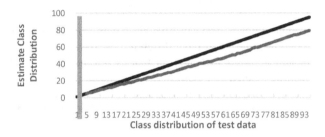

Fig. 2. Estimate Class Distribution from Spambase training dataset with positive rate 0.05

3 Our Method

In this section, we describe our proposed CDE-Weight methods and their related knowledge. Its objective is to reduce impact of inaccurate estimation by combining CDE-AVGBM with weight graph, thus improve performance of the classification process.

3.1 Preprocessing

In this section, we describe our algorithm before the weight is applied. Our CDE-Weight is an improved version of CDE-AVGBM. Aside from variable N (that indicate the total number of ensemble classifier models) and variable M (indicating the number of models used in actual class distribution estimation process), this method has two additional variables, which are T and σ. The T variable is used to decide whether the class distribution of train data is too distinct from test data or not, the different weight functions are used depending on this factor. As for the σ, the shape of weight function is determined by its value.

This method starts by take value N then the models interval is calculated using equation 3. This interval indicates the distance of class distribution between each ensemble model.

$$\text{interval} = \frac{100}{N} \qquad (3)$$

After that we calculate (pr_i/nr_i) to build classifier model i by using equation 4, where (pr_i/nr_i) is a class distribution of dataset that model i expected to be able to accurately estimating (It can also view as virtual class distribution of train data used to build classifier$_i$).

$$(pr_i/nr_i) = \frac{i \times interval}{100 - (i \times interval)} \quad \text{i varying from 1 to N} - 1 \tag{4}$$

Cost calculation for cost sensitive classifier is performed by using equation 5. The (pr/nr) represents the class distribution of train data.

$$DMR_i = \frac{(pr_i/nr_i)}{(pr/nr)} \tag{5}$$

Then, CDE-Weight builds N cost sensitive classifiers from the train data using the obtained cost. Each classifier$_i$ will have different cost based on DMR_i . These classifiers will be labeled according to (pr_i/nr_i) that used to build it. The classifier built from (pr_i/nr_i) will be labeled as i classifier, which mean these classifiers are sorted based on the virtual class distribution it's represent. These cost sensitive classifiers are used on new data (test data) to obtain estimate class distributions and (pr_i/nr_i) of test data from each model.

After obtain (pr_i/nr_i) from every ensemble models, CDE-Weight selects classifier that has (pr_i/nr_i) nearest to the estimated (pr/nr) as much as possible. The selected model will serve as pivot model (S) while CD_S is equal to the Class Distribution model S is suitable for.

The pseudo code of CDE-Weight: Preprocessing is shown in Algorithm 1. Since we assume that we have no idea about how class distribution of train data should be, line 4 through line 9 is to get information about class distribution of test data from various perspectives. In line 10, the pivot model which likely to able to accurately estimate class distribution of test data, is selected. This is done by comparing each ensemble model estimated positive rate with virtual class distribution using to build that model. In line 12, the models in Selected List will be used later to estimate the class distribution of test data for the final cost adjustment. Loop in line 13 – 17 is to add neighbor models that are responsible of higher class distribution than the pivot model to the Selected List. And, the loop in line 18 – 22 is to add the lower one.

3.2   Case I: Weight When Train and Test Data Are Similar

If the different between (pr/nr) and (pr_i/nr_i) of the selected model don't exceed the T value, then our algorithm decides that the class distribution of train and test data are not too different. Thus M models having the most similar (pr_i/nr_i) to the selected model are chosen and labeled as neighbor models. These models are used to estimate the final class distribution of unseen data or AVG(pr/nr) which is calculated in equation 6 while S is a label of pivot model

$$AVG(pr/nr) = \frac{\sum_{S-M/2}^{S+M/2}(pr/nr)_i}{M} \tag{6}$$

Algorithm 1. CDE-Weight: Preprocessing

1.   **Input** (M, N, T, σ)
2.   CD_Train = Class distribution of train data
3.   Calculate the interval between ensemble models
4.   **While** i is less than or equal to N
5.   Build ensemble model i suitable with class distribution i from train data
6.   Use model i to classify and obtain estimated class distribution from test data
7.   Calculate the estimated (pr/nr) from model i
8.   Add model i to the Model List
9.   Add one to i
10.   Select a classifier and Label it as pivot model (S)
11.   CD_S = Class distribution that model S is suitable for
12.   Add pivot model to the Selected List
13.   **While** j is less than or equal to (M-1)/2
14.   Select model i that has the lowest i value but still higher than S
15.   Add model i to the right side of the Selected List
16.   Remove model i from the Model List
17.   Add one to j
18.   **While** k is less than or equal to (M-1)/2
19.   Select model i that has the highest t i value but still lower than S
20.   Add model i to the left side of the Selected List
21.   Remove model i from the Model List
22.   Add one to k

Algorithm 2. CDE-Weight: Case I

23.   Sum(pr/nr) = 0
24.   SumWeight = 0
25.   **If** | CD_S – CD_Train | ≤ T
26.   Sum(pr/nr)   = (Sum Class Distribution of all models in the Selected List)
27.   SumWeight = M
28.   DMR = Sum(pr/nr)/ SumWeight
29.   Adjust the cost of cost sensitive classifier according to DMR
30.   Build new classifier from train data

Now, CDE-Weight calculates DMR again from AVG(pr/nr) . It substitutes (pr_i/nr_i) in equation 5 by AVG(pr/nr). Then it builds new cost sensitive classifier from the train data using this cost. Note that in this case, which the T value is not exceeded, all models using in the estimation are given the same weight (1). However, if that is not the case the weight of each model will be different as described in the next section. Following is the pseudo code of CDE-Weight: Case I.

3.3   Case II: Weight When Train and Test Data Are Too Different

In case that the different between (pr/nr) and (pr_i/nr_i) of the selected model exceeds the T value; the weight of each models will be adjusted. There are two ways the T value can be exceeded.

First is the situation where (pr_i/nr_i) is more than (pr/nr) indicating that the estimate class distribution is higher than the train data. This situation would result in underestimation of test data class distribution which will have a negative effect on the pivot model selection process. To compensate for this problem our CDE-Weight gives higher weight to the neighbor models labeled higher than pivot model. As for the lower ones, they are given less weight the further they are from the pivot model. The σ value indicates the rate of weight reduction in the lower part while the higher ones are given the same weight. The weight calculation is shown in equation 7 and graph shape is shown in Figure 4.

$$Weight_i = f(i, \sigma^2) = \begin{cases} if\ (i < S) \gg \quad f(i, \sigma^2) = \dfrac{1}{\sigma\sqrt{2\pi}}\ e^{-\frac{1}{2}\left(\frac{i - CD\_S}{\sigma}\right)^2} \\ \\ if\ (i \geq S) \gg\ f(BM, \sigma^2) = \dfrac{1}{\sigma\sqrt{2\pi}} \end{cases} \tag{7}$$

While i is the class distribution of ensemble models ranging from model S-M/2 to S+M/2 (in other words the neighbor models of S).

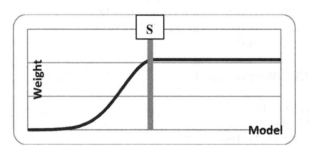

Fig. 3. Shape of weight graph

Second is the situation where is less than (pr/nr) indicating that the estimate class distribution is lower than the train data. In this case, the estimating process is likely to result in overestimation, thus the weight calculation and graph shape are the reversal of equation 7 and Figure 4.

Then we calculate AVG(pr/nr)$_{Weight}$ using equation 8.

$$AVG(pr/nr)_{Weight} = \frac{\sum_i (Weight_i \times (pr_i/nr_i))}{\sum_i Weight_i} \tag{8}$$

DMR for adjusting cost sensitive classifier is calculated with equation 9.

$$DMR = \frac{AVG(pr/nr)_{Weight}}{(pr/nr)} \tag{9}$$

This cost is then applied to cost sensitive classifier to build the new classifier from the train data. This classifier will able to efficiently handle the new class distribution of unseen data.

Following is the pseudo code of CDE-Weight: Case II

Algorithm 3. CDE-Weight: Case II

31. Sum(pr/nr) = 0
32. SumWeight = 0
33. **If** (CD_S - CD_Train) > T
34.   **While** there is still models in the right side of S in the Selected List
35.   ES_A = Estimated class distribution of test data from model A
36.   WeightA $= \frac{1}{\sigma\sqrt{2\pi}}$
37.   Sum(pr/nr) = Sum(pr/nr) + (WeightA×ES_A)
38.   SumWeight = SumWeight + WeightA
39.   Remove model A from the Selected List
40.   **While** there is still models in the left side of S in the Selected List
41.   b = Class distribution using to build current model B
42.   ES_B = Estimated class distribution of test data from model B
43.   WeightB $= \frac{1}{\sigma\sqrt{2\pi}} \, e^{-\frac{1}{2}\left(\frac{b-S}{\sigma}\right)^2}$
44.   Sum(pr/nr) = Sum(pr/nr) + (WeightB×ES_B)
45.   SumWeight = SumWeight + WeightB
46.   Remove model B from the Selected List
47. **Else if** (CD_Train - CD_S) > T
48.   **While** there is still models in the right side of S in the Selected List
49.   c = Class distribution using to build current model C
50.   ES_C = Estimated class distribution of test data from model C
51.   WeightC $= \frac{1}{\sigma\sqrt{2\pi}} \, e^{-\frac{1}{2}\left(\frac{c-S}{\sigma}\right)^2}$
52.   Sum(pr/nr) = Sum(pr/nr) + (WeightC×ES_C)
53.   SumWeight = SumWeight + WeightC
54.   Remove model C from the Selected List
55.   **While** there is still models in the left side of S in the Selected List
56.   ES_D = Estimated class distribution of test data from model D
57.   WeightD $= \frac{1}{\sigma\sqrt{2\pi}}$
58.   Sum(pr/nr) = Sum(pr/nr) + (WeightD×ES_D)
59.   SumWeight = SumWeight + WeightD
60.   Remove model D from the Selected List
61. DMR = Sum(pr/nr)/ SumWeight
62. Adjust the cost of cost sensitive classifier according to DMR
63. Build new classifier from train data

Line 33 determines whether test data fall into the situation in Figure 2 or not. If it falls into that situation, the following loop in line 34 through 39 is used to calculate weights of models on the right side of pivot model (i.e. the neighbor models with higher class distribution). These models are likely to give higher estimations than the one from the left side which is exactly what we need since the error in this case is underestimating; therefore models on this side are given the higher weight. Note that the weight on this side is static, since even though the prediction of the underestimating situation occurrence is possible, we still unable to determine the range of it.

On the contrary, in line 40 – 46, models on the left side are given less and less weight the farther they are from the pivot model, as we know that there will be an underestimation and that model building with lower class distribution is likely to give a lower class distribution estimation too.

4   Experimental Results

In order to evaluate the effectiveness of our proposed method, CDE-Weight is compared with CDE-Iterated [19], CDE-AVG [9] and CDE-AVGBM [9]. Note that the CDE-Iterated is the base method of CDE-AVG and is here for comparison. In this research, the evaluation is performed in terms of accuracy and robustness in various class distributions. This demonstrates the performance of CDE-Weight.

Experimental setups, consisting three UCI data sets, are described below while a description of the three datasets is given in Table 1.

Table 1. Description of 3 UCI Data Sets

Dataset	Size	% of Positive class (positive rate)	Partition size
Adult	32,561	24.08%	8000
Magic Gamma	19,020	64.84%	7000
Spambase	4,601	60.6%	1900

All the datasets used here contain only two classes. The original dataset size, the percentage of the examples that belong to the positive class and the number of examples in each partition are given. As described earlier, the investigated problem is required two datasets: the train data and test data.

Experiment 1 and 2 are performed to measure the effectiveness of our method in case that class distribution of train data is very different from test data. To accomplish this, the train data is partitioned from original dataset by 5% and 95% of positive class, respectively.

The performance in general situation is examined in experiment 3 in which the class distribution of train data is not adjusted and train data is exactly the same as original data. The test data generated in the same way have varying positive rate between 5% and 95% with 1% incremental rate. The size of each partition is limited, in order to generate multiple class distributions without any duplication. The maximum possible partition sizes that don't cause duplication problem are then used which are specified in the last column of Table1.

All the experiments in this paper utilize the J48 [17] implementation of C4.5 [12] from WEKA released 3.6.3. Changes in class distribution are compensated using WEKA's cost-sensitive learning capabilities with example reweighting. Next are our experimental results and their analysis.

For notations used, the number that follows the method name *Iterated* indicates the number of iterations used in CDE-iterated. For CDE-AVGBM and CDE-Weight, M is the number of neighbor models and N is the number of ensemble models used in the experiments. Note that for visualization reason, the results shown in the following tables are only a part of its (11 of 91).

4.1   Experimental Result I-II: Train and Test Data Are Different

Following is the detailed accuracy result when the class distributions of train and test data are different. The Magic Gamma data set with modified positive rate to 5% is shown in Figure 4 and Table 2. The result for Spambase with positive rate 95% is displayed in Figure 5 and Table 3.

As shown below, the results demonstrate that our method offer significant improvement when class distributions of train and test data are different. This is due to the decreasing impact of error in pivot model (best model) selection.

Table 2. Experiment 1: Accuracy Performance on Magic Gamma dataset with pr 5 %

	CDE Iterated		CDE-EM methods		Our Method	
	Iterated 1	Iterated 2	CDE-AVG-100	CDE--AVGBM-61-100	Weight-99-100 σ-5 T-0.2	Weight-99-100 σ-5 T-0.3
0.05	95.44%	94.60%	98.54%	96.11%	97.97%	97.97%
0.1	90.90%	93.61%	95.67%	92.31%	95.50%	95.50%
0.2	84.53%	84.53%	89.31%	87.67%	89.07%	89.07%
0.3	77.44%	78.30%	83.40%	83.59%	83.14%	83.14%
0.4	76.03%	77.79%	78.06%	77.79%	78.09%	78.09%
0.5	69.29%	71.31%	72.77%	70.94%	70.70%	70.70%
0.6	65.49%	65.63%	67.80%	65.63%	69.37%	67.80%
0.7	58.31%	59.46%	64.20%	60.63%	64.81%	66.49%
0.8	54.83%	53.00%	60.56%	56.51%	59.99%	60.43%
0.9	47.99%	52.59%	55.31%	54.27%	63.37%	55.37%
0.95	44.03%	49.93%	61.73%	56.26%	69.13%	68.19%
AVG	69.26%	70.57%	74.44%	71.72%	75.21%	74.62%

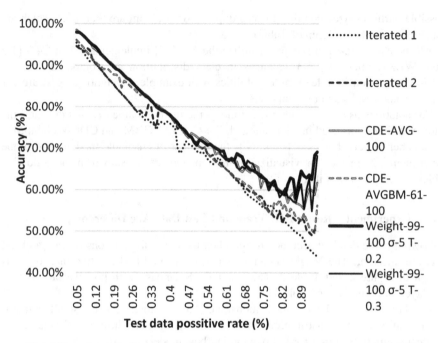

Fig. 4. Experiment 1: Accuracy Performance on Magic Gamma dataset with pr 5 %

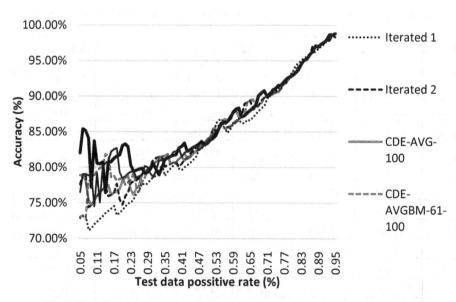

Fig. 5. Experiment 2: Accuracy Performance on Spambase dataset with pr 95 %

Table 3. Experiment 2: Accuracy Performance on Spambase dataset with pr 95 %

	CDE Iterated		CDE-EM methods		Our Method	
	Iterated 1	Iterated 2	CDE-AVG-100	CDE-AVGBM-61-100	Weight-99-100 σ-5 T-0.2	Weight-99-100 σ-5 T-0.3
0.05	72.84%	77.58%	78.89%	73.00%	82.00%	76.47%
0.1	72.00%	75.32%	75.32%	78.84%	83.68%	74.47%
0.2	74.00%	74.74%	78.11%	78.47%	83.05%	78.47%
0.3	78.26%	80.16%	80.16%	80.16%	79.79%	79.16%
0.4	79.84%	80.11%	81.21%	81.21%	82.74%	82.74%
0.5	83.37%	83.79%	83.95%	83.79%	83.37%	83.37%
0.6	86.26%	86.37%	88.00%	86.53%	88.21%	87.95%
0.7	88.63%	89.79%	90.79%	89.79%	90.79%	90.79%
0.8	93.37%	92.84%	92.95%	93.32%	92.95%	92.95%
0.9	97.00%	97.26%	97.37%	97.37%	97.37%	97.37%
0.95	98.68%	98.32%	98.84%	98.74%	98.84%	98.84%
AVG	83.88%	85.05%	85.24%	85.37%	86.34%	85.80%

4.2 Experimental Result III: Train and Test Data Are Similar

The experiment this section is to demonstrate the performance of CDE-weight and other methods in the situation where train and test data are similar. The experiment in figure 6 and table 4 is perform on none modification adult dataset (using its natural class distribution).

The results show that CDE-Weight performance is lower when class distribution of train and test data is similar. This is due to the process of weight shifting between the different weight graphs, the weight graph when T is not exceeded and when is exceeded, which is too rough, resulting in lower classification performance when its start shifting. This transition problem should be alleviated with smoother weight function with we intend to do in a future.

This experiment also shows that by increasing the value T, the effect of weight shifting can be significantly lower in trade of accuracy gain when class distribution of train and test are different.

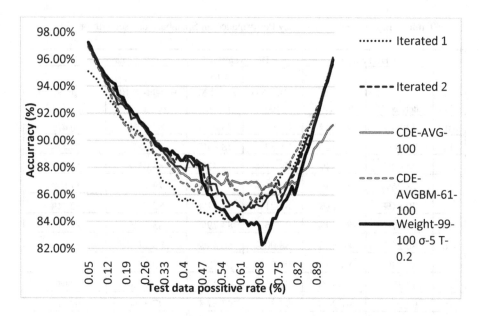

Fig. 6. Experiment 3: Accuracy Performance on Adult dataset

Table 4. Experiment 3: Accuracy Performance on Adult dataset

	CDE Iterated		CDE-EM methods		Our Method	
	Iterated 1	Iterated 2	CDE-AVG-100	CDE-AVGBM-61-100	Weight-99-100 σ-5 T-0.2	Weight-99-100 σ-5 T-0.3
0.05	95.10%	97.20%	97.10%	96.93%	97.24%	97.10%
0.1	93.81%	95.11%	95.10%	94.88%	95.26%	95.06%
0.2	91.06%	92.49%	92.28%	91%	92%	92%
0.3	88.83%	89.83%	89.95%	89.01%	89.64%	89.84%
0.4	85.54%	88.58%	87.55%	86.84%	88.53%	87.93%
0.5	84.39%	86.64%	87.45%	87.43%	85.97%	86.78%
0.6	84.44%	85.18%	86.91%	86.34%	84.49%	85.38%
0.7	85.99%	86%	86.54%	86.53%	82.49%	85.46%
0.8	87.89%	87.30%	87.03%	88.70%	86.64%	87.58%
0.9	92.74%	92.36%	89.93%	92.70%	92.35%	92.54%
0.95	96.20%	96%	91.18%	95.68%	96.01%	96.18%
AVG	88.49%	89.44%	89.18%	89.37%	88.90%	89.36%

5   Conclusion

In this paper, we present a method for handling the situation where class distribution after a classifier is built is not stable. We demonstrate that a previous ensemble method still have some error in model selection, especially when there is a different between train and test data, and improves it with weight function to mitigate the misestimating impact in pivot model selection. The results indicate that our method performs significantly better when class distribution of train and test (unseen) data are different.

References

1. Alaiz-Rodriguez, R., Guerrero-Curieses, A., Cid-Sueiro, J.: Minimax.: Regret Classifier for Imprecise Class Distributions. Journal of Machine Learning Research 8 (2007)
2. Alaiz-Rodríguez, R., Japkowicz, N.: Assessing the Impact of Changing Environments on Classifier Performance. In: Bergler, S. (ed.) Canadian AI. LNCS (LNAI), vol. 5032, pp. 13–24. Springer, Heidelberg (2008)
3. Asuncion, A., Newman, D.J.: UCI Machine Learning Repository (2007), http://www.ics.uci.edu/~mlearn/MLReposito-ry.html
4. Chapelle, O., Scholkopf, B., Zien, A.: Semi-Supervised Learning. MIT Press, Cambridge (2006)
5. Chawla, N.V., Bowyer, K.W., Kegelmeyer, W.P.: SMOTE: Synthetic Minority Over-sampling Technique. Journal of Artificial Intelligence Research 16 (2002)
6. Elkan, C.: The foundations of cost-sensitive learning. In: The Proceedings of the 17th International Joint Conference on Artificial Intelligence, pp. 973–978 (2001)
7. Forman, G.: Quantifying counts and costs via classification. Data Mining Knowledge Discovery 17, 2 (2008)
8. Latinne, P., Saerens, M., Decaestecker, C.: Adjusting the Outputs of a Classifier to New a Priori Probabilities May Significantly Improve Classification Accuracy: Evidence from a multi-class problem in remote sensing. In: The Proceedings of the 18th International Conference on Machine Learning, pp. 298–305 (2001)
9. Limsetto, N., Waiyamai, K.: Handling Concept Drift via Ensemble and Class Distribution Estimation Technique. In: Tang, J., King, I., Chen, L., Wang, J. (eds.) ADMA 2011, Part II. LNCS, vol. 7121, pp. 13–26. Springer, Heidelberg (2011)
10. Provost, F., Fawcett, T.: Robust Classification for Imprecise Environments. Machine Learning 42, 3 (2001)
11. Provost, F., Fawcett, T., Kohavi, R.: The Case against Accuracy Estimation for Comparing Induction Algorithms. In: The Proceedings of the 15th International Conference on Machine Learning, pp. 445–453 (1998)
12. Quinlan, R.J.: C4.5: Programs for Machine Learning. Morgan Kaufmann (1993)
13. Rokach, L.: Ensemble-based classifiers. Artificial Intelligence Review 33 (2010)
14. Saerens, M., Latinne, P., Decaestecker, C.: Adjusting the outputs of a classifier to new a priori probabilities: A simple procedure. Neural Computing 14 (2002)
15. Tsymbal, A.: The problem of concept drift: Definitions and related work. Computer Science Department, Trinity College Dublin (2004)

16. Weiss, G.M.: Mining with rarity: a unifying framework. SIGKDD Explorations Newsletter 6, 1 (2004)
17. Weiss, G.M., Provost, F.: Learning when training data are costly: The effect of class distribution on tree induction. Journal of Artificial Intelligence Research 19 (2003)
18. Witten, I.H., Frank, E.: Data Mining: Practical Machine Learning Tools and Techniques, 2nd edn. Morgan Kaufmann (2005)
19. Xue, J.C., Weiss, G.M.: Quantification and semi-supervised classification methods for handling changes in class distribution. In: Proceedings of the 15th ACM SIGKDD International Conference on Knowledge Discovery and Data Mining, Paris, France, pp. 897–906. ACM (2009)
20. Zadrozny, B., Elkan, C.: Learning and making decisions when costs and probabilities are both unknown. In: The Proceedings of the 7th International Conference on Knowledge Discovery and Data Mining, pp. 204–213 (2001)

A Robust One-Class Classification Model with Interval-Valued Data Based on Belief Functions and Minimax Strategy

Lev V. Utkin[1], Yulia A. Zhuk[1], and Anatoly I. Chekh[2]

[1] Department of Control, Automation and System Analysis,
Saint-Petersburg State Forest Technical University, Russia
[2] Department of Computer Science,
Saint-Petersburg State Electrotechnical University, Russia

Abstract. A robust model for solving the one-class classification problem by interval-valued training data is proposed in the paper. It is based on using Dempster-Shafer theory for getting the lower and upper risk measures. The minimax or pessimistic strategy is applied to state an optimization problem in the framework of the modified support vector machine (SVM). The algorithm for computing optimal parameters of the classification function stems from extreme points of the convex polytope produced by the interval-valued elements of a training set. It is shown that the hard non-linear optimization problem is reduced to a finite number of standard SVMs.

Keywords: novelty detection, support vector machine, kernel, interval-valued data, minimax strategy, quadratic programming, Dempster-Shafer theory.

1 Introduction

A special important problem of the statistical machine learning is the classification problem which can be regarded as a task of classifying some objects into classes (group) in accordance with their properties or features. However, for many real-world problems, the task is not to classify but to detect novel or abnormal instances [1–5] which can be carried out by using the so-called novelty detection models or the one-class classification (OCC) models. Comprehensive and interesting reviews of OOC approaches are provided by Markou and Singh [6], by Bartkowiak [7], by Khan and Madden [8], by Hodge and Austin [9]. A typical feature of OCC models is that only unlabeled samples are available. One of the most common ways to define anomalies is by saying that anomalies are not concentrated [10].

A lot of OOC models use kernel-based methods and the support vector machine (SVM). The corresponding models are called OCC SVMs. There are two main approaches for constructing the OCC SVM. The first approach is proposed by Tax and Duin [11, 12]. This is one of the well-known OOC models, which can

P. Perner (Ed.): MLDM 2014, LNAI 8556, pp. 107–118, 2014.

be regarded as an unsupervised learning problem. Tax and Duin [11, 12] solve the OCC problem by distinguishing the positive class from all other possible data points. They find a hyper-sphere around the positive class data that contain almost all points in the data set with the minimum radius. This approach is also called the Support Vector Data Description. By adapting the kernel function, this approach becomes more flexible than just a sphere in the input space. Markou and Singh [6] pointed out that a drawback of the approach is that it often requires a large data set.

An alternative way to geometrically enclose a fraction of the training data is via a hyperplane and its relationship to the origin proposed by Scholkopf et al. [4, 5]. Under this approach, a hyperplane is used to separate the training data from the origin with maximal margin, i.e., the objective is to separate off the region containing the data points from the surface region containing no data. Here the authors consider the trade-off between the number of errors made on the training set (number of target objects rejected) and the margin separation between the training points and the origin. This is achieved by constructing a hyperplane which is maximally distant from the origin with all data points lying on the opposite side from the origin. The data is mapped into the feature space corresponding to the kernel and is separated from the origin with maximum margin.

There are other interesting OCC models [2, 13, 14]), which can be used in many applications. In particular, Campbell and Bennett [2] propose a linear programming approach to the OCC problem, which is much simpler than the standard quadratic programming in the OCC SVM. The authors of [14] propose the so-called single-class minimax probability machines which offer robust OCC with distribution-free worst case bounds on the probability that a pattern will fall inside the normal region. Bicego and Figueiredo [13] study a weighted OCC model. According to their approach, every data point has a weight indicating the importance assigned to each point of the training set. A question of getting the corresponding weights for every point is a main drawback of the approach.

A general approach for constructing the imprecise robust OCC models is proposed by Utkin [15], Utkin and Zhuk [16]. The main idea of the approach is to replace the empirical probability distribution which is exploited in the standard SVM [17, Chapter 10.] by a set of probability distributions produced by some imprecise inference models [18]. The robustness of the proposed models is achieved by considering sets of probability distributions of data points produced by a number of imprecise probability models. Another imprecise OCC model incorporating imprecise prior statistical information is studied by Utkin and Zhuk [19].

At the same time, we have to point out that the above OCC models use a training set consisting of precise or point-valued data. However, training patterns in many real applications can be obtained only in the interval form. Interval-valued data may result from imperfection of measurement tools or imprecision of expert information. There may also be some missing data when some features of an example are not observed [20]. Many methods in machine learning have been presented for dealing with interval-valued data [21–23]. In some methods, interval-valued observations are replaced by precise values based on some additional assumptions, for example by taking middle points of the intervals [24]. Another part of methods use the standard interval analysis for constructing the classification and regression models [25, 26].

In this paper, we propose quite different approach which uses the framework of belief functions or Dempster-Shafer theory in order to process the interval-valued observations. The minimax or pessimistic strategy is applied to the OOC problem for getting the separating function in the OOC SVM. It is shown that the OCC problem with interval-valued training data can be reduced to a finite set of standard quadratic programming problems. One of the main ideas applied to the proposed model is to change the Gaussian kernel in the OOC SVM by a kernel of a special form which leads to simplification of optimization problems for computing parameters of the separating function.

The paper is organized as follows. Section 2 presents the standard OCC problem proposed by Scholkopf et al. [4, 5]. Some elements of Dempster-Shafer theory are considered in Section 3. Moreover, the OCC problem is stated in the framework of Dempster-Shafer theory. The corresponding method for solving the stated OCC problem by interval-valued data is proposed in Section 4. In Section 5, concluding remarks are made.

2   One-Class Classification by Precise Data

Suppose we have unlabeled training data $\mathbf{x}_1, ..., \mathbf{x}_n \subset \mathcal{X}$, where n is the number of observations, $\mathcal{X}$ is some set, for instance, it is a compact subset of $\mathbb{R}^m$. Let $\mathcal{D} = \{\mathbf{x}_i\}_{i=1}^n$ be drawn i.i.d. from a distribution on $\mathcal{X}$. The sample space $\mathcal{D}$ is finite and discrete. According to papers [4, 5], a well-known novelty detection or OCC model aims to construct a function f which takes the value $+1$ in a "small" region capturing most of the data points and -1 elsewhere. It can be done by mapping the data into the feature space corresponding to a kernel and by separating them from the origin with maximum margin.

Let ϕ be a feature map $\mathcal{X} \to G$ such that the data points are mapped into an alternative higher-dimensional feature space G. In other words, this is a map into an inner product space G such that the inner product in the image of ϕ can be computed by evaluating some simple kernel $K(\mathbf{x}, \mathbf{y}) = (\phi(\mathbf{x}), \phi(\mathbf{y}))$, such as the Gaussian kernel

$$K(\mathbf{x}, \mathbf{y}) = \exp\left(- \|\mathbf{x} - \mathbf{y}\|^2 / \sigma^2\right).$$

σ is the kernel parameter determining the geometrical structure of the mapped samples in the kernel space. It is pointed out by [27] that the problem of selecting a proper parameter σ is very important in classification. When a very small σ is used ($\sigma \to 0$), $K(\mathbf{x}, \mathbf{y}) \to 0$ for all $\mathbf{x} \neq \mathbf{y}$ and all mapped samples tend to be orthogonal to each other, despite their class labels. In this case, both between-class and within-class variations are very large. On the other hand, when a very large σ is chosen ($\sigma^2 \to \infty$), $K(\mathbf{x}, \mathbf{y}) \to 1$ and all mapped samples converge to a single point. This obviously is not desired in a classification task. Therefore, a too large or too small σ will not result in more separable samples in G.

It is shown [1] that the data points lie on the surface of a hypersphere in feature space since $\phi(\mathbf{x}) \cdot \phi(\mathbf{x}) = K(\mathbf{x}, \mathbf{x}) = 1$ (translation invariant kernels). Now we have to find a hyperplane

$$f(\mathbf{x}, \mathbf{w}, \rho) = \langle \mathbf{w}, \phi(\mathbf{x}_i) \rangle - \rho = 0$$

that separates the data from the origin with maximal margin, i.e., we want ρ to be as large as possible so that the volume of the halfspace $\langle \mathbf{w}, \phi(\mathbf{x}_i) \rangle \geq \rho$ is minimized. Let us introduce the parameter $\nu \in [0; 1]$ which is analogous to ν used for the ν-SVM [3, 10]. Roughly speaking, it denotes the fraction of input data for which $\langle \mathbf{w}, \phi(\mathbf{x}_i) \rangle \leq \rho$. To separate the data set from the origin, we solve the following quadratic program:

$$\min_{w, \xi, \rho} \frac{1}{2} \|\mathbf{w}\|^2 + \frac{1}{\nu n} \sum_{i=1}^{n} \xi_i - \rho, \tag{1}$$

subject to

$$\langle \mathbf{w}, \phi(\mathbf{x}_i) \rangle \geq \rho - \xi_i, \ \xi_i \geq 0, \ i = 1, ..., n. \tag{2}$$

Slack variables ξ_i are used to allow points to violate margin constraints.

Since nonzero slack variables ξ_i are penalized in the objective function, we can expect that if $\mathbf{w}$ and ρ solve this problem, then the decision function $\text{sgn}(f(\mathbf{x}, \mathbf{w}, \rho))$ will be positive for most examples $\mathbf{x}_i$ contained in the training set, while the SV type regularization term $\|\mathbf{w}\|$ will still be small. The actual trade-off between these two goals is controlled by ν.

Using multipliers $\alpha_i, \beta_i \geq 0$, we introduce a Lagrangian

$$L(\mathbf{w}, \xi, \rho, \alpha, \beta) = \frac{1}{2} \|\mathbf{w}\|^2 + \frac{1}{\nu n} \sum_{i=1}^{n} \xi_i - \rho$$
$$- \sum_{i=1}^{n} \alpha_i \left(\langle \mathbf{w}, \phi(\mathbf{x}_i) \rangle - \rho + \xi_i \right) - \sum_{i=1}^{n} \beta_i \xi_i.$$

It is shown that the dual problem is of the form:

$$\max_{\alpha} \left(-\frac{1}{2} \sum_{i=1}^{n} \sum_{j=1}^{n} \alpha_i \alpha_j K(\mathbf{x}_i, \mathbf{x}_j) \right), \tag{3}$$

subject to

$$0 \leq \alpha_i \leq \frac{1}{\nu n}, \ \sum_{i=1}^{n} \alpha_i = 1. \tag{4}$$

The value of ρ can be obtained as

$$\rho = (\langle \mathbf{w}, \phi(\mathbf{x}_j) \rangle) = \sum_{i=1}^{n} \alpha_i K(\mathbf{x}_i, \mathbf{x}_j).$$

After substituting the obtained solution into the expression for the decision function f, we get

$$f(\mathbf{x}, \mathbf{w}, \rho) = \text{sgn} \left(\sum_{i=1}^{n} \alpha_i K(\mathbf{x}_i, \mathbf{x}) - \rho \right).$$

The above optimization problems can be obtained by rewriting the OCC problem in a general form of minimizing the expected risk [17, Section 1.2.]

$$R(\mathbf{w},\rho) = \int_{\mathbb{R}^m} l(\mathbf{x}) \mathrm{d}F_0(\mathbf{x}). \tag{5}$$

Here the loss function $l(\mathbf{x})$ can be represented as the so-called hinge loss function:

$$l(\mathbf{x}) = \max \{0, \rho - \langle \mathbf{w}, \phi(\mathbf{x}) \rangle\} - \rho \nu.$$

The standard SVM technique is to assume that F_0 is the empirical (non-parametric) cumulative probability distribution function whose use leads to the empirical expected risk

$$R_{\text{emp}}(\mathbf{w},\rho) = \frac{1}{n} \sum_{i=1}^{n} l(\mathbf{w}, \phi(\mathbf{x}_i)). \tag{6}$$

If we substitute the hinge loss function into (6) and add the standard Tikhonov regularization term $\frac{1}{2} \|\mathbf{w}\|^2$ (this is the most popular penalty or smoothness term) [28] to the objective function (6), then we get the same optimization problem (1)-(2) proposed by [4, 5].

The assumption of the empirical probability distribution means that every point $\mathbf{x}_i$ has the probability $p_i = 1/n$. This is a rather strong assumption when the number of points is not large. Its validity might give rise to doubt in this case.

We apply the expected risk definition in order to study interval-valued training data.

3 Interval-Valued Training Data and Belief Functions

We consider classification problems where the input variables (patterns) $\mathbf{x}$ may be interval-valued. Suppose that we have a training set $(\mathbf{A}_i)$, $i = 1, ..., n$. Here $\mathbf{A}_i \subset \mathbb{R}^m$ is the Cartesian product of m intervals $[\underline{a}_i^{(k)}, \overline{a}_i^{(k)}]$, $k = 1, ..., m$, which again are not restricted so could even include intervals $(-\infty, \infty)$. In other words, every feature of every observation or training example is interval-valued. We aim to construct a function f which takes the value $+1$ in a "small" region capturing most of the interval-valued examples and -1 elsewhere.

There are several ways in which one could deal with such an interval-valued data set. In this paper, we consider the expected risk by interval-valued data in the framework of belief functions or Dempster-Shafer theory. Below, we give some basic definitions of the framework of belief functions.

Let $\mathcal{X}$ be a universal set under interest, usually referred to in evidence theory as the frame of discernment. Suppose n observations were made of an element $u \in \mathcal{X}$, each of which resulted in an imprecise (non-specific) measurement given by a set $\mathbf{A}$ of values. Let c_i denote the number of occurrences of the set $\mathbf{A}_i \subseteq U$, and $Po(\mathcal{X})$ the set of all subsets of $\mathcal{X}$ (power set of $\mathcal{X}$). A frequency function m, called basic probability assignment (BPA), can be defined such that [29, 30]:

$$m : Po(\mathcal{X}) \rightarrow [0,1], \quad m(\varnothing) = 1, \quad \sum_{\mathbf{A} \in Po(\mathcal{X})} m(\mathbf{A}) = 1.$$

According to [29], this function can be obtained as follows:

$$m(\mathbf{A}_i) = c_i/n.$$

According to [30], the belief $Bel(A)$ and plausibility $Pl(\mathbf{A})$ measures of an event $\mathbf{A} \subseteq \mathcal{X}$ can be defined as

$$\mathrm{Bel}(\mathbf{A}) = \sum_{\mathbf{A}_i : \mathbf{A}_i \subseteq \mathbf{A}} m(\mathbf{A}_i), \quad \mathrm{Pl}(\mathbf{A}) = \sum_{\mathbf{A}_i : \mathbf{A}_i \cap \mathbf{A} \neq \varnothing} m(\mathbf{A}_i).$$

As pointed out in [31], a belief function can formally be defined as a function satisfying axioms which can be viewed as a weakening of the Kolmogorov axioms that characterize probability functions. Therefore, it seems reasonable to understand a belief function as a generalized probability function [29] and the belief $Bel(\mathbf{A})$ and plausibility $Pl(\mathbf{A})$ measures can be regarded as lower and upper bounds for the probability of $\mathbf{A}$, i.e., $\mathrm{Bel}(\mathbf{A}) \leq \mathrm{Pr}(\mathbf{A}) \leq \mathrm{Pl}(\mathbf{A})$. This implies that for a function $l(\mathbf{x})$, we can define the lower $\underline{R}$ and upper $\overline{R}$ expectations of the function $l(\mathbf{x})$ in the framework of belief functions as follows [32, 33]:

$$\underline{R} = \sum_{i=1}^{n} m(\mathbf{A}_i) \inf_{\mathbf{x}_i \in \mathbf{A}_i} l(\mathbf{x}_i), \quad \overline{R} = \sum_{i=1}^{n} m(\mathbf{A}_i) \sup_{\mathbf{x}_i \in \mathbf{A}_i} l(\mathbf{x}_i). \tag{7}$$

In fact, the above bounds for the expected risk coincide with (5) under condition that we have a set of probability distribution functions delimited by some upper and lower bounds which are called p-boxes [34] sometimes and represent the epistemic uncertainty about the distribution function of a random variable. The set of distributions is produced due to available interval-valued data. However, it is difficult to determine the lower and upper probability distributions in the explicit form by using a set of intervals (see an example of the corresponding algorithm in [35]). The definition (7) provides a more simple way for determining the bounds for the expected risk.

It follows from the above that we have the interval $[\underline{R}, \overline{R}]$ of the expected risk measure instead of its precise value. In order to use this interval in solving the classification problem, we have to determine a strategy of decision making which selects one point within this interval for searching optimal classification parameters $\mathbf{w}$ and ρ.

One of the well-known and popular ways for dealing with the interval of the expected risk is to use the minimax (pessimistic) strategy. According to the minimax strategy, we select a probability distribution from the set of distributions such that the expected risk R achieves its maximum for fixed values of parameters. It should be noted that the "optimal" probability distributions may be different for different values of parameters. If to return to the interval $[\underline{R}, \overline{R}]$, then the minimax strategy assumes the largest risk, i.e., the upper bound $\overline{R}$. The minimax strategy can be explained in a simple way. We do not know a precise probability distribution and every distribution from their predefined set can be selected. Therefore, we should take the "worst" distribution providing the largest value of the expected risk. The minimax criterion can be interpreted as an insurance against the worst case because it aims at minimizing the expected loss in the least favorable case [36, Subsection 2.4.2.]. This criterion of decision making can be regarded as the well-known Γ-minimax [37, 38].

Finally, we can write the optimization problem for computing the optimal classification parameters $\mathbf{w}$ and ρ as follows:

$$R(\mathbf{w}_{\mathrm{opt}}, \rho_{\mathrm{opt}}) = \min_{\mathbf{w}, \rho} \overline{R}(\mathbf{w}_{\mathrm{opt}}, \rho_{\mathrm{opt}}) = \min_{\mathbf{w}, \rho} \left(\sum_{i=1}^{n} m(\mathbf{A}_i) \sup_{\mathbf{x}_i \in \mathbf{A}_i} l(\mathbf{x}_i) \right).$$

By using the assumption accepted in the empirical expected risk, we can conclude that $m(\mathbf{A}_i) = 1/n$ for all $i = 1, ..., n$. Hence, by substituting the hinge loss function into the above objective function, we get

$$R(\mathbf{w}_{\mathrm{opt}}, \rho_{\mathrm{opt}}) = \min_{\mathbf{w}, \rho} \left(\frac{1}{n} \sum_{i=1}^{n} \sup_{\mathbf{x}_i \in \mathbf{A}_i} (\max\{0, \rho - \langle \mathbf{w}, \phi(\mathbf{x}_i) \rangle\} - \rho\nu) \right).$$

This problem is equivalent to the following problem:

$$R(\mathbf{w}_{\mathrm{opt}}, \rho_{\mathrm{opt}}) = \min_{\mathbf{w}, \rho} \left(-\rho + \frac{1}{n\nu} \sum_{i=1}^{n} \sup_{\mathbf{x}_i \in \mathbf{A}_i} \max\{0, \rho - \langle \mathbf{w}, \phi(\mathbf{x}_i) \rangle\} \right). \tag{8}$$

4 Optimal Classification Parameters by Interval-Valued Data

Note that problem (8) can be rewritten as

$$R(\mathbf{w}_{\mathrm{opt}}, \rho_{\mathrm{opt}}) = \sup_{\mathbf{x}_i \in \mathbf{A}_i, i=1,...,n} \min_{\mathbf{w}, \rho} \left(-\rho + \frac{1}{n\nu} \sum_{i=1}^{n} \max\{0, \rho - \langle \mathbf{w}, \phi(\mathbf{x}_i) \rangle\} \right).$$

In other words, we can fix some precise values $\mathbf{x}_1, ..., \mathbf{x}_n$ from the corresponding intervals $\mathbf{A}_1, ..., \mathbf{A}_n$ and compute the optimal parameters $\mathbf{w}$ and ρ for the fixed precise values. The same can be done for other precise values $\mathbf{x}_1, ..., \mathbf{x}_n$ from the intervals. By enumerating hypothetically all precise values $\mathbf{x}_1, ..., \mathbf{x}_n$ from the intervals, we get sets of parameters $\mathbf{w}$ and ρ. The parameters providing the largest value of the expected risk are optimal. Of course, it is practically impossible to enumerate the precise values of intervals except for some special cases.

Suppose that we have fixed precise values $\mathbf{x}_1, ..., \mathbf{x}_n$ from $\mathbf{A}_1, ..., \mathbf{A}_n$, respectively. After adding the standard Tikhonov regularization term $\frac{1}{2} \|\mathbf{w}\|^2$ to the objective function (8) and introducing the optimization variables

$$\xi_i = \max \left\{ 0, \rho - \langle \mathbf{w}, \phi(\mathbf{x}_i) \rangle \right\},$$

we obtain the standard primal form (1)-(2) and the corresponding dual form (3)-(4). However, we should remind that this problem is based on selecting fixed precise values $\mathbf{x}_1, ..., \mathbf{x}_n$. Therefore, the optimal parameters $\mathbf{w}$ and ρ can be found from the problem:

$$\sup_{\mathbf{x}_i \in \mathbf{A}_i, i=1,...,n} \max_{\alpha} \left(-\frac{1}{2} \sum_{i=1}^{n} \sum_{j=1}^{n} \alpha_i \alpha_j K(\mathbf{x}_i, \mathbf{x}_j) \right).$$

Here Lagrange multipliers α_i, $i = 1, ..., n$, depend on the selected values $\mathbf{x}_1, ..., \mathbf{x}_n$.

It would seem that we can minimize the kernels in order to maximize the objective function by maximizing the distances $\|\mathbf{x}_i - \mathbf{x}_j\|^2$. However, another very important difficulty here is that the values selected in intervals are connected each other. By selecting points $\mathbf{x}_1$, $\mathbf{x}_2$, $\mathbf{x}_3$ in order to minimize or to maximize the distances between $\mathbf{x}_1$ and $\mathbf{x}_2$, $\mathbf{x}_1$ and $\mathbf{x}_3$, we cannot change the distance between $\mathbf{x}$ and $\mathbf{x}_3$. This implies that it is necessary to consider all distances together.

In order to overcome the above difficulties, it is proposed to introduce a new kernel function

$$T(\mathbf{x}, \mathbf{y}) = \max\{0, 1 - \|\mathbf{x} - \mathbf{y}\|^1 / \sigma^2\}.$$

It can be regarded as an approximation of the Gaussian kernel. The introduced kernel is bounded by 0 and 1. Its largest value is 1 takes place when $\mathbf{x} = \mathbf{y}$. A main peculiarity of the function is that it is linear. This peculiarity allows us to solve the above optimization problem.

Now we fix the values of α and write the dual optimization problem with the introduced kernel T:

$$\inf_{\mathbf{x}_i \in \mathbf{A}_i, i=1,...,n} \left(\frac{1}{2} \sum_{i=1}^{n} \sum_{j=1}^{n} \alpha_i \alpha_j G_{ij} \right), \tag{9}$$

subject to (4) and

$$G_{ij} \geq 1 - \sum_{k=1}^{m} \left| x_i^{(k)} - x_j^{(k)} \right| / \sigma^2, \ G_{ij} \geq 0, \ i = 1, ..., n, \ j = 1, ..., n,$$

$$\underline{a}_i^{(k)} \leq x_i^{(k)} \leq \bar{a}_i^{(k)}, \ k = 1, ..., m, \ i = 1, ..., n. \tag{10}$$

Here $x_i^{(j)}$ is the value of the j-th feature of the i-th example; G_{ij} is a new variable such that $G_{ij} = T(\mathbf{x}_i, \mathbf{x}_j)$. In order to minimize G_{ij}, we have to maximize $\left| x_i^{(k)} - x_j^{(k)} \right|$. Let us introduce the variables $u_{ij}^{(k)} \leq 0$ and $v_{ij}^{(k)} \leq 0$ such that $x_i^{(k)} - x_j^{(k)} = u_{ij}^{(k)} - v_{ij}^{(k)}$. Then the solution has either $u_{ij}^{(k)}$ or $v_{ij}^{(k)}$ equal to 0, depending on the sign of $x_i^{(k)} - x_j^{(k)}$, so

$$\left| x_i^{(k)} - x_j^{(k)} \right| = -u_{ij}^{(k)} - v_{ij}^{(k)}.$$

Then the constraints for G_{ij} can be rewritten as

$$G_{ij} \geq 1 + \sum_{k=1}^{m} \left(u_{ij}^{(k)} + v_{ij}^{(k)} \right) / \sigma^2, \ G_{ij} \geq 0, \tag{11}$$

$$u_{ij}^{(k)} - v_{ij}^{(k)} = x_i^{(k)} - x_j^{(k)}, \ u_{ij}^{(k)} \leq 0, \ v_{ij}^{(k)} \leq 0, \ k = 1, ..., m. \tag{12}$$

Finally, we have got the linear programming problem having objective function (9) and constraints (10)-(12). It should be note that the values of α are fixed yet. According to some general results from linear programming theory, an optimal solution to the above problem is achieved at extreme points or vertices of the polytope produced by the above constraints. Since these constraints do not depend on α, then the vertices can be found by means of the well known methods. Only vertices with different $\mathbf{x}_i$, $i = 1, ..., n$, are important for us. Let us denote the corresponding vertices as X_k, $k = 1, ..., t$, where t is the total number of vertices with different values of $\mathbf{x}_i$, $i = 1, ..., n$.

In order to get the optimal parameters $\mathbf{w}_{\mathrm{opt}}$ and ρ_{opt}, we solve t quadratic optimization problems (3)-(4) by substituting the values $\mathbf{x}_1, ..., \mathbf{x}_n$ from X_k into every problem. The largest value of the objective function or $R(\mathbf{w}, \rho)$ corresponds to the optimal values $\mathbf{x}_1, ..., \mathbf{x}_n$ and to the optimal parameters $\mathbf{w}_{\mathrm{opt}}$ and ρ_{opt}.

Let us consider in detail all variables in the problem with constraints (10), (11), (12) in order to find extreme points. If we have M variables, then every extreme point can be derived by considering the system of M equalities among all constraints. For simplicity, we assume that $x_i^{(k)} \neq x_j^{(k)}$ for all $i \neq j$. It is obvious from (12) that either $u_{ij}^{(k)} = 0$ and $v_{ij}^{(k)} < 0$ or $u_{ij}^{(k)} < 0$ and $v_{ij}^{(k)} = 0$. If we take into account the equality $u_{ij}^{(k)} - v_{ij}^{(k)} = x_i^{(k)} - x_j^{(k)}$, then we have one equality for every variable $u_{ij}^{(k)}$ and $v_{ij}^{(k)}$. The same can be said about G_{ij}. Indeed, if $x_i^{(k)} \neq x_j^{(k)}$, then exactly one of two inequalities in (11) becomes to be the equality. This implies that in fact there are $n \cdot m$ bounded variables $x_i^{(k)}$ and $2n \cdot m$ constraints for these variables. Every $x_i^{(k)}$ can take one of two its bounds $\underline{a}_i^{(k)}$ or $\overline{a}_i^{(k)}$. This implies that exactly mn constraints with $x_i^{(k)}$ will be equalities.

So, we have to enumerate all combinations of vertices of hyper-rectangles $\mathbf{A}_i$, $i = 1, ..., n$, for assigning the values to $\mathbf{x}_i$, $i = 1, ..., n$. A vector $\mathbf{x}_1, ..., \mathbf{x}_n$ can be regarded as the extreme point, if constraints (11), (12) will be valid after its substituting into (11), (12).

The above reasoning give us a simple algorithm for computing extreme points X_k, $k = 1, ..., t$. We do not need to solve many optimization problems by using the well-know technique. We just take different vertices of hyper-rectangles $\mathbf{A}_i$ and check whether these vertices satisfy (11)-(12).

Finally, we get a set of standard quadratic programming problems with different objective functions (9) depending on the set of possible values G_{ij}, $i = 1, ..., n$, $j = 1, ..., n$, but coinciding with (3), and with the same constraints (4).

So, we have ruduced the OCC problem to the set of standard "precise" OCC problems by selecting suitable points in intervals corresponding to the pessimistic strategy. The optimal solution corresponds to the largest value of objective function (3) or to the smallest value of objective function (9).

The function $f(\mathbf{x}, \mathbf{w}, \rho)$ can be rewritten in terms of Lagrange multipliers as

$$f(\mathbf{x}, \mathbf{w}, \rho) = \sum_{i=1}^{n} \alpha_i T(\mathbf{x}_i, \mathbf{x}) - \rho.$$

Hence, we find the optimal value of ρ by taking $f(\mathbf{x}, \mathbf{w}, \rho) = 0$, i.e., there holds

$$\rho = \sum_{i=1}^{n} \alpha_i T(\mathbf{x}_i, \mathbf{x}_j).$$

In the above expressions, the optimal values $\mathbf{x}_i$, $i = 1, ..., n$, are used.

The next question is how to decide whether a testing observation is abnormal or not when it is interval-valued. In other words, we have a set of functions $f(\mathbf{x}, \mathbf{w}, \rho)$ for every testing observation taking the values $\mathbf{x}$ from $\mathbf{A}$. We again use the minimax strategy. The worst case is when we recognize a new observation as a normal one, but it actually is abnormal. This allows us to formulate the following rule. An interval-valued observation $\mathbf{A}$ is normal if $f(\mathbf{x}, \mathbf{w}, \rho) \geq 0$ for all $\mathbf{x}$ being vertices of the hyper-rectangle $\mathbf{A}$. An interval-valued observation $\mathbf{A}$ is abnormal if $f(\mathbf{x}, \mathbf{w}, \rho) < 0$ for at least one vertex of the hyper-rectangle $\mathbf{A}$.

5   Conclusion

OCC models dealing with interval-valued training data have been proposed in the paper. Classification parameters for every model are obtained by solving a finite set of simple quadratic programming problems whose solution does not meet any difficulties. The simplicity of the OCC models follows from the fact that the optimal solution is achieved at one of the extreme points of a set of probability distributions produced by the underlying interval model. The minimax or pessimistic strategy is used for choosing a single "optimal" probability distribution from the set of distributions produced by interval-valued observations. This strategy has a clear explanation and justification in the framework of decision theory. Therefore, the same clear interpretation have results obtained by means of the proposed model.

Acknowledgement. The reported study was partially supported by RFBR, research project No. 14-01-00165-a.

References

1. Campbell, C.: Kernel methods: a survey of current techniques. Neurocomputing 48(1-4), 63–84 (2002)
2. Campbell, C., Bennett, K.: A linear programming approach to novelty detection. In: Leen, T., Dietterich, T., Tresp, V. (eds.) Advances in Neural Information Processing Systems, vol. 13, pp. 395–401. MIT Press (2001)
3. Cherkassky, V., Mulier, F.: Learning from Data: Concepts, Theory, and Methods. Wiley-IEEE Press, UK (2007)
4. Scholkopf, B., Platt, J., Shawe-Taylor, J., Smola, A., Williamson, R.: Estimating the support of a high-dimensional distribution. Neural Computation 13(7), 1443–1471 (2001)
5. Scholkopf, B., Williamson, R., Smola, A., Shawe-Taylor, J., Platt, J.: Support vector method for novelty detection. In: Advances in Neural Information Processing Systems, pp. 526–532 (2000)
6. Markou, M., Singh, S.: Novelty detection: a review—part 1: statistical approaches. Signal Processing 83(12), 2481–2497 (2003)
7. Bartkowiak, A.: Anomaly, novelty, one-class classification: A comprehensive introduction. International Journal of Computer Information Systems and Industrial Management Applications 3, 61–71 (2011)
8. Khan, S.S., Madden, M.G.: A survey of recent trends in one class classification. In: Coyle, L., Freyne, J. (eds.) AICS 2009. LNCS, vol. 6206, pp. 188–197. Springer, Heidelberg (2010)
9. Hodge, V., Austin, J.: A survey of outlier detection methodologies. Artificial Intelligence Review 22(2), 85–126 (2004)
10. Scholkopf, B., Smola, A.: Learning with Kernels: Support Vector Machines, Regularization, Optimization, and Beyond. The MIT Press, Cambridge (2002)
11. Tax, D., Duin, R.: Support vector data description. Machine Learning 54(1), 45–66 (2004)
12. Tax, D., Duin, R.: Support vector domain description. Pattern Recognition Letters 20(11), 1191–1199 (1999)
13. Bicego, M., Figueiredo, M.: Soft clustering using weighted one-class support vector machines. Pattern Recognition 42(1), 27–32 (2009)
14. Kwok, J., Tsang, I.H., Zurada, J.: A class of single-class minimax probability machines for novelty detection. IEEE Transactions on Neural Networks 18(3), 778–785 (2007)
15. Utkin, L.: A framework for imprecise robust one-class classification models. International Journal of Machine Learning and Cybernetics (2014), doi:10.1007/s13042-012-0140-6
16. Utkin, L., Zhuk, Y.: Robust novelty detection in the framework of a contamination neighbourhood. International Journal of Intelligent Information and Database Systems 7(3), 205–224 (2013)
17. Vapnik, V.: Statistical Learning Theory. Wiley, New York (1998)
18. Walley, P.: Statistical Reasoning with Imprecise Probabilities. Chapman and Hall, London (1991)

19. Utkin, L., Zhuk, Y.: Imprecise prior knowledge incorporating into one-class classification. Knowledge and Information Systems, 1–24 (2014)
20. Pelckmans, K., Brabanter, J.D., Suykens, J., Moor, B.D.: Handling missing values in support vector machine classifiers. Neural Networks 18(5-6), 684–692 (2005)
21. Ishibuchi, H., Tanaka, H., Fukuoka, N.: Discriminant analysis of multi-dimensional interval data and its application to chemical sensing. International Journal of General Systems 16(4), 311–329 (1990)
22. Nivlet, P., Fournier, F., Royer, J.J.: Interval discriminant analysis: An efficient method to integrate errors in supervised pattern recognition. In: Second International Symposium on Imprecise Probabilities and Their Applications, Ithaca, NY, USA, pp. 284–292 (2001)
23. Silva, A., Brito, P.: Linear discriminant analysis for interval data. Computational Statistics 21, 289–308 (2006)
24. Neto, E.L., de Carvalho, F.: Centre and range method to fitting a linear regression model on symbolic interval data. Computational Statistics and Data Analysis 52, 1500–1515 (2008)
25. Angulo, C., Anguita, D., Gonzalez-Abril, L., Ortega, J.: Support vector machines for interval discriminant analysis. Neurocomputing 71(7-9), 1220–1229 (2008)
26. Hao, P.Y.: Interval regression analysis using support vector networks. Fuzzy Sets and Systems 60, 2466–2485 (2009)
27. Wang, J., Lu, H., Plataniotis, K., Lu, J.: Gaussian kernel optimization for pattern classification. Pattern Recognition 42(7), 1237–1247 (2009)
28. Tikhonov, A., Arsenin, V.: Solution of Ill-Posed Problems. W.H. Winston, Washington, DC (1977)
29. Dempster, A.: Upper and lower probabilities induced by a multi-valued mapping. Annales of Mathematical Statistics 38(2), 325–339 (1967)
30. Shafer, G.: A Mathematical Theory of Evidence. Princeton University Press (1976)
31. Halpern, J., Fagin, R.: Two views of belief: Belief as generalized probability and belief as evidence. Artificial Intelligence 54(3), 275–317 (1992)
32. Nguyen, H., Walker, E.: On decision making using belief functions. In: Yager, R., Fedrizzi, M., Kacprzyk, J. (eds.) Advances in the Dempster-Shafer theory of Evidence, pp. 311–330. Wiley, New York (1994)
33. Strat, T.: Decision analysis using belief functions. International Journal of Approximate Reasoning 4(5), 391–418 (1990)
34. Ferson, S., Kreinovich, V., Ginzburg, L., Myers, D., Sentz, K.: Constructing probability boxes and Dempster-Shafer structures. Report SAND2002-4015, Sandia National Laboratories (January 2003)
35. Kriegler, E., Held, H.: Utilizing belief functions for the estimation of future climate change. International Journal of Approximate Reasoning 39, 185–209 (2005)
36. Robert, C.: The Bayesian Choice. Springer, New York (1994)
37. Berger, J.: Statistical Decision Theory and Bayesian Analysis. Springer, New York (1985)
38. Gilboa, I., Schmeidler, D.: Maxmin expected utility with non-unique prior. Journal of Mathematical Economics 18(2), 141–153 (1989)

Manifold Learning in Data Mining Tasks

Alexander Kuleshov[1,2] and Alexander Bernstein[1,2]

[1] Kharkevich Institute for Information Transmission Problems RAS (IITP), Moscow, Russia
kuleshov@iitp.ru
[2] National Research University Higher School of Economics (HSE), Moscow, Russia
abernstein@hse.ru

Abstract. Many Data Mining tasks deal with data which are presented in high dimensional spaces, and the 'curse of dimensionality' phenomena is often an obstacle to the use of many methods for solving these tasks. To avoid these phenomena, various Representation learning algorithms are used as a first key step in solutions of these tasks to transform the original high-dimensional data into their lower-dimensional representations so that as much information about the original data required for the considered Data Mining task is preserved as possible. The above Representation learning problems are formulated as various Dimensionality Reduction problems (Sample Embedding, Data Manifold embedding, Manifold Learning and newly proposed Tangent Bundle Manifold Learning) which are motivated by various Data Mining tasks. A new geometrically motivated algorithm that solves the Tangent Bundle Manifold Learning and gives new solutions for all the considered Dimensionality Reduction problems is presented.

Keywords: Data Mining, Statistical Learning, Representation learning, Dimensionality Reduction, Manifold Learning, Tangent Learning, Tangent Bundle Manifold Learning.

1 Introduction

The goal of Data Mining is to extract previously unknown information from a dataset. Thus, it is supposed that information is reflected in the structure of a dataset which must be discovered from the data. Many Data Mining tasks, such as Pattern Recognition, Classification, Clustering, and other, which are challenging for machine learning algorithms, deal with real-world data that are presented in high-dimensional spaces, and the 'curse of dimensionality' phenomena is often an obstacle to the use of many methods for solving these tasks. To avoid these phenomena in Data Mining tasks, various Representation learning algorithms are used as a first key step in solutions of these tasks. Representation learning (Feature extraction) algorithms transform the original high-dimensional data into their lower-dimensional representations (or features) so that as much information about the original data required for the considered Data Mining task is preserved as possible.

P. Perner (Ed.): MLDM 2014, LNAI 8556, pp. 119–133, 2014.

After that, the initial Data Mining task may be reduced to the corresponding task for the constructed lower-dimensional representation of the original dataset. Of course, the construction of the low-dimensional data representation must depend on the considered task, and the success of machine learning algorithms generally depends on the data representation [1].

Representation (Feature) learning problems can be formulated as various Dimensionality Reduction (DR) problems whose different formalizations depend on the considered Data Mining tasks. Solutions of such DR problems make the corresponding Data Mining tasks easier to apply to the high-dimensional input dataset.

This paper is about DR problems in Data Mining tasks. We describe a few key Data Mining tasks that lead to different formulations of the DR (Sample Embedding, Data Space/Manifold embedding, Manifold Learning as Data Manifold Reconstruction/Estimation). We propose an amplification of the Manifold Learning (ML) problem, called Tangent Bundle ML (TBML), which provides good generalization ability properties of ML algorithms. There is no generally accepted terminology in the DR; thus, some terms introduced below can be different from those used in other works.

We present a new geometrically motivated algorithm that solves the TBML and also gives a new solution for all the considered DR problems.

The rest of the paper is organized as follows. Sections 2 - 5 contain definitions of various DR problems motivated by their subsequent using in specific Data Mining tasks. The proposed TBML solution is described in Section 6.

2   Sample Embedding Problem

One of key Data Mining tasks related to unsupervised learning is Clustering, which consist in discovering groups and structures in data that contain 'similar' (in one sense or another) sample points. Constructing a low-dimensional representation of original high-dimensional data for subsequent solution of Clustering problem may be formulated as a specific DR problem, which will be referred to as the **Sample Embedding** problem and is as follows: Given an input dataset $\mathbf{X}_n = \{X_1, X_2, \ldots, X_n\} \subset \mathbf{X}$ randomly sampled from an unknown nonlinear Data Space (DS) $\mathbf{X}$ embedded in a p-dimensional Euclidean space R^p, find an 'n-point' Embedding mapping

$$h_{(n)}: \mathbf{X}_n \subset R^p \to \mathbf{h}_n = h_{(n)}(\mathbf{X}_n) = \{h_1, h_2, \ldots, h_n\} \subset R^q \qquad (1)$$

of the sample $\mathbf{X}_n$ to a q-dimensional dataset $\mathbf{h}_n$, $q < p$, which 'faithfully represents' the sample $\mathbf{X}_n$ while inheriting certain subject-driven data properties like preserving the local data geometry, proximity relations, geodesic distances, angles, etc.

If the term 'faithfully represents' in the Sample Embedding problem corresponds to the 'similar' notion in the initial Clustering problem, we can solve the reduced Clustering problem for the constructed low-dimensional feature dataset $\mathbf{h}_n$. After that, we can obtain some solution of the initial Clustering problem: the clusters in the initial problem are the images of the clusters discovered in the reduced problem by using a natural inverse mapping from $\mathbf{h}_n$ to the original dataset $\mathbf{X}_n$.

The term 'faithfully represents' is not formalized in general, and in various Sample Embedding methods it is different due to choosing some optimized cost function $L_{(n)}(\mathbf{h}_n|\mathbf{X}_n)$ which defines an 'evaluation measure' for the DR and reflects desired properties of the n-point Embedding mapping $h_{(n)}$ (1). As is pointed out in [2], a general view on the DR can be based on the 'concept of cost functions'.

There exist a number of methods (techniques) for the Sample Embedding, for example, Multidimensional Scaling [3] or Principal Component Analysis (PCA) [4]. Various non-linear techniques are based on Auto-Encoder neural networks [6, 7, 8, 9], Self-organizing Maps [10], Topology representing networks [11], Diffusion Maps [12], Kernel PCA [13], and others.

A newly emerging direction in the fields of the Sample Embedding, which has been a subject of intensive research over the last decades, consists in constructing a family of algorithms based on studying the local structure of a given sampled dataset that retains local properties of the data with the use of various cost functions. Examples of such 'local' algorithms are: Locally Linear Embedding (LLE) [14], Laplacian Eigenmaps (LE) [5], Hessian Eigenmaps (HE) [15], ISOmetric MAPing (ISOMAP) [16], Maximum Variance Unfolding (MVU) [17], Manifold charting [18], Local Tangent Space Alignment (LTSA) [19], and others. Some of these algorithms (LLE, LE, ISOMAP, MVU) can be considered in the same framework, based on the Kernel PCA [13] applied to various data-based kernels [20, 21, 22, 23].

Note that the Sample Embedding algorithms are based on the sample only, and no assumptions about the DS $\mathbf{X}$ are required for their descriptions. However, the study of the properties of the algorithms is based on assumptions about both the DS and a way for extracting the sample from the DS.

3   Data Space (Manifold) Embedding Problem

Another key Data Mining task related to supervised learning concerns the Classification problem in which the original dataset consists of labeled examples: outputs (labels) $\mathbf{\Lambda}_n = \{\lambda_1, \lambda_2, \ldots, \lambda_n\}$ are known for the corresponding inputs $\{X_1, X_2, \ldots, X_n\}$ sampled from the DS $\mathbf{X}$; each label λ belongs to the finite set $\{1, 2, \ldots, m\}$ with $m \geq 2$. The problem is to generalize a function or mapping from inputs to outputs which can then be used to generate an output for previously unseen input $X \in \mathbf{X}$.

In the case of high-dimensional inputs $\mathbf{X}_n$ it is possible to construct low-dimensional features $\{h_1, h_2, \ldots, h_n\}$ (1) by using some Sample Embedding algorithm. After that, we can consider the reduced dataset $\{(h_i, \lambda_i), i = 1, 2, \ldots, n\}$ instead of the sample $\{\mathbf{X}_n, \mathbf{\Lambda}_n\}$. For the possibility of using the solution of the reduced classification problem built for the reduced dataset, it is necessary to construct a lower-dimensional representation for a new unseen (usually called Out-of-Sample (OoS)) input $X \in \mathbf{X} / \mathbf{X}_n$. Thus, it is necessary to consider the following extension of the Sample Embedding problem: Given a dataset $\mathbf{X}_n$ sampled from the DS $\mathbf{X}$, construct a low-dimensional parameterization of the DS which produces an Embedding mapping

$$h: \mathbf{X} \subset R^p \to \mathbf{Y} = h(\mathbf{X}) \subset R^q \tag{2}$$

from the DS **X**, including the OoS points, to the Feature Space (FS) $\mathbf{Y} \subset R^q$, q < p, which preserves specific properties of the DS **X**. This problem can be referred to as the Data Space Embedding problem. The term 'preserves specific properties' is not formalized in general, and can be different due to choosing various cost function reflecting specific preserved data properties.

An 'OoS extension' for the algorithms LLE, LE, ISOMAP and MVU, which are based on the Kernel PCA approach, has been found in [21] with the use of Nyström's eigendecomposition technique [22], [24]. The Cost functions concept [2] allows constructing other Embedding mapping for the OoS points, see [25, 26], etc.

The definition of the Data Space Embedding problem uses values of the Embedding mapping h (2) for the OoS points too. Thus, to justify the problem solution and study properties of the solution, we must define a **Data Model** describing the DS, and a **Sampling Model** offering a way for extracting both the sample $\mathbf{X}_n$ and OoS points from the DS. The most popular models in the DR are **Manifold Data Models**, see [27, 28, 29, 30] and others works, in which the DS **X** is a q-dimensional manifold embedded in a p-dimensional Euclidean space R^p, q < p, and referred to as the **Data Manifold** (DM). In most studies, DM is modeled using a single coordinate chart.

The Sampling Model is typically defined as a probability measure μ on the DM **X** whose support Supp(μ) coincides with the DM **X**. In accordance with this model, the dataset $\mathbf{X}_n$ and OoS points $X \in \mathbf{X} / \mathbf{X}_n$ are selected from the DM **X** independently of each other according to the probability measure μ.

A motivation for the Manifold Data model consists in the following empirical fact: as a rule, high-dimensional real-world data lie on or near some unknown low-dimensional Data Manifold embedded in an ambient high-dimensional 'observation' space. This assumption is usually referred to as **Manifold hypothesis**, or **Manifold assumption**, see [27], [31, 32] and others works.

Non-linear DR problems applied to the data which are described by the Manifold Data Model are usually referred to as the **Manifold Learning** (ML) problem [27, 28, 29, 30]; the above defined Data Space Embedding problem with the Manifold Data Model will be referred to as the **Manifold Embedding** problem which is to construct a parameterization of the DM from a finite dataset sampled from the DM.

In this paper, by the ML we mean a specific DR problem in which an accurate reconstruction of the DM **X** from the sample is required.

4   Manifold Learning Problem as Data Manifold Reconstruction

The Section 4.1 contains the examples of Data Mining tasks, in which reconstruction of the DM **X** from its FS **Y** is required. A necessary of DM reconstruction from the sample arises also when a preserving as much available information contained in the sample as possible is required, see Section 4.2. A strict definition of the ML and its geometrical interpretation are contained in Section 4.3. Generalization ability of arbitrary ML solution is discussed in Section 4.4.

4.1   Data Mining Tasks Requiring Data Manifold Reconstruction

Many Data Mining tasks involve time series prognosis, optimization, etc. In the case of high-dimensional data (multidimensional time series, high-dimensional design space in optimization, etc.), a low-dimensional representation of the original data is used for reducing the initial 'high-dimensional' task.

Example 1. The Electricity price curve forecasting considered in [33] is as follows. Electricity 'daily' prices are described by a multidimensional time series (electricity price curve) $X_t = (X_{t1}, X_{t2}, \ldots, X_{t,24})^T \in R^{24}$, $t = 1, 2, \ldots, T$, consisting of 'hour-prices' in the course of day t. Based on the 'daily-prices' vectors $X_{1:T} = \{X_1, X_2, \ldots, X_T\} \subset R^{24}$ up to day T, it is necessary to construct a forecast X* for X_{T+1}.

In Forecasting algorithm [33], the LLE method [14] embeds the 'daily-prices' vectors $X_{1:T}$ into the feature set $Y_{1:T} = h(X_{1:T}) \subset R^q$, the value q = 4 is selected in [33] as an appropriate dimension of the features. Based on the features $Y_{1:T}$, a forecast $\widehat{Y}_{T+1}$ for Y_{T+1} is constructed by using standard forecasting technique.

The natural inverse mapping from $Y_{1:T}$ to the original dataset $X_{1:T}$ determines a re-construction of the original high-dimensional data $X \in \mathbf{X}$ from its low-dimensional feature y = h(X) for the Feature sample points. Then, it is required to reconstruct the forecast $\widehat{X}_{T+1}$ from the 'low-dimensional' forecast $\widehat{Y}_{T+1}$, which doesn't belongs to the set $Y_{1:T}$ in the general case and can be called Feature-Out-of-Sample (FOoS) point. The forecast $\widehat{X}_{T+1} = g(\widehat{Y}_{T+1})$ for X_{T+1} is constructed in [33] by using the LLE-reconstruction technique [22]. As is pointed out in [33] (citations): 'low-dimensional coordinates to the high-dimensional space is a necessary step for forecasting'; 'the reconstruction of high-dimensional forecasted price curves from low-dimensional prediction is a significant step for forecasting'; 'reconstruction accuracy is critical for the application of manifold learning in the prediction.'

Example 2. Wing shape optimization is one of important problem in aircraft design-ing. Wing shape design variables include a number of high-dimensional vectors X of dimension p which are detailed descriptions of wing airfoils; these vectors consist of coordinates of points lying densely on the airfoils' contours. The dimension p varies usually in the range from 50 to 200; a specific value of p is selected depending on the required accuracy of airfoil description.

Low-dimensional airfoil parameterization [34] is constructed in order to reduce the number of design variables, and DR is one of highly powerful methods for such pa-rameterization [35, 36, 37]. Based on a dataset $\mathbf{X}_n = \{X_1, X_2, \ldots, X_n\}$ consisting of high-dimensional descriptions of airfoils-prototypes, this technique allows describing the wing airfoil X by low-dimensional vector $h(X) \in R^q$, whose dimension q varies in the range from 5 to 10. Thus, the space $\mathbf{X}$ consisting of the airfoil descriptions is a q-dimensional manifold embedded in an ambient p-dimensional space, q << p.

The airfoil optimization problem in the original p-dimensional design space $\mathbf{X}$ can be reduced to the corresponding optimization problem in the Feature space $\mathbf{Y} = h(\mathbf{X})$. Let the vector y* $\in \mathbf{Y}$ be a result of some optimization procedure. Considering this vector as a low-dimensional representation of the 'optimal' airfoil X*, it is necessary to reconstruct X* = g(y*) from the value y* which is the FOoS point in general case.

Thus, in both examples, it is necessary to accurately reconstruct the points $X \in \mathbf{X}$ of the original manifold $\mathbf{X}$ from their low-dimensional features y = h(X).

4.2   Preserving an Information in Manifold Embedding Problem

Manifold Embedding is usually a first step in various Data Mining tasks, in which features y = h(X) are used instead of initial p-dimensional vectors X. If the Embedding mapping h (2) preserves only specific properties of high-dimensional data, then substantial data losses are possible when using a reduced vector y = h(X) instead of the initial vector X. As is pointed out in [38, 39], one objective of the DR is to preserve as much available information contained in the sample as possible.

Under this approach, the term 'faithfully represents' is understood as preserving such information, and this means the possibility of reconstructing high-dimensional points X from low-dimensional representations h(X) with small reconstruction error. Thus, it is necessary to construct a Reconstruction mapping

$$g: \mathbf{Y}_g \subset R^q \to R^p \qquad (3)$$

defined on a domain $\mathbf{Y}_g \supset \mathbf{Y} = h(\mathbf{X})$, which determines a reconstructed value g(h(X)) with Reconstruction error $\delta_{(h,g)}(X) = |X - g(h(X))|$; here $r_{(h,g)}(X) = (g \bullet h)(X)$ is a result of successively applying the embedding and reconstruction mappings to a vector X. Note that the Reconstruction mapping g must be defined not only on the Feature sample $\mathbf{Y}_n = h(\mathbf{X}_n)$ (with an obvious reconstruction), but also on FOoS features y = h(X) ∈ $\mathbf{Y} / \mathbf{Y}_n$ obtained by embedding the OoS points X.

The Reconstruction error can be chosen as an evaluation measure in the Manifold Embedding problem: a small value of $\delta_{(h,g)}(X)$ means that h(X) well preserves information contained in X. The Reconstruction error can be considered as an 'universal quality criterion' [38, 39] in the Manifold Embedding problem describing a measure of preserving the information contained in the high-dimensional data.

4.3   Manifold Learning Problem

In this paper, the ML means the DR under the Manifold assumption, in which a low-dimensional representation of the DM allows accurately reconstructing the DM.

A strict definition of the ML is as follows: Given an input dataset $\mathbf{X}_n$ sampled from a q-dimensional DM $\mathbf{X}$ embedded in an ambient p-dimensional space R^p, q < p, and covered by a single chart, construct an **ML-solution** θ = (h, g) consisting of an Embedding mapping h (2) defined on a domain of definition $\mathbf{X}_h \supset \mathbf{X}$ and a Reconstruction mapping g (3) defined on a domain of definition $\mathbf{Y}_g \supset \mathbf{Y}_h = h(\mathbf{X}_h)$, which ensures the approximate equality

$$r_\theta(X) = r_{(h,g)}(X) = g(h(X)) \approx X \quad \text{for all} \quad X \in \mathbf{X}, \qquad (4)$$

and $\delta_\theta(X) = \delta_{(h,g)}(X)$ is a measure of quality of the ML solution θ at a point X ∈ $\mathbf{X}$.

The solution θ determines a q-dimensional **Reconstructed Manifold** (RM)

$$\mathbf{X}_\theta = \{X = g(y) \in R^p : y \in \mathbf{Y}_\theta \subset R^q\} \tag{5}$$

embedded in R^p and covered (parameterized) by a single coordinate chart g defined on the set $\mathbf{Y}_\theta = h(\mathbf{X})$ called the Feature Space (FS).

The approximate equalities (4) can be considered as the **Manifold proximity** property $\mathbf{X}_\theta = r_\theta(\mathbf{X}) \approx \mathbf{X}$ meaning that the RM $\mathbf{X}_\theta$ accurately reconstructs the DM $\mathbf{X}$ from the sample $\mathbf{X}_n$. Note that the ML solution θ allows reconstructing the unknown DM $\mathbf{X}$ by the parameterized RM $\mathbf{X}_\theta$, whereas the Embedding Manifold solution h (2) reconstructs a parameterization of the DM only.

From the statistical point of view, the defined ML problem may be considered as a Statistical Estimation Problem: Given a finite dataset randomly sampled from a smooth q-dimensional Data Manifold $\mathbf{X}$ covered by a single chart, estimate $\mathbf{X}$ by data-based q-dimensional manifold covered (parameterized) by a single chart also.

There are some (though limited number of) methods for reconstruction of the DM $\mathbf{X}$ from the FS $h(\mathbf{X})$. For a specific linear DM, the reconstruction can easily be made with the PCA. For a nonlinear DM, sample-based Auto-Encoder Neural Networks [6, 7, 8, 9] determine both the embedding and reconstruction mappings. LLE reconstruction, which is done in the same manner as LLE, has been introduced in [22]. LTSA [19] determines an interpolation-like nonparametric reconstruction of the DM.

Note that the defined ML Problem differs from the Manifold approximation problem, in which an unknown manifold embedded in high-dimensional ambient space must be approximated by some geometrical structure with close geometry, without any 'global parameterization' of the structure. For the latter problem, some solutions are known such as approximations by a simplicial complex [40], by finitely many affine subspaces called 'flats' [41], or by other geometrical structure in other works.

4.4   Generalization Ability of Manifold Learning Solution

The Reconstruction error $\delta_\theta(X)$ can be directly computed at sample point $X \in \mathbf{X}_n$; for OoS point $X \in \mathbf{X} \setminus \mathbf{X}_n$ it describes the generalization ability of the considered ML solution θ at a specific point X.

Let $X_0 \in \mathbf{X}$ be some selected point, $U_\varepsilon(X_0) = \{X \in \mathbf{X}: |X - X_0| \le \varepsilon\}$ be the ε-neighborhood of the point X_0 and $\delta_\theta(X_0, \varepsilon) = \max\{\delta_\theta(X): X \in U_\varepsilon(X_0)\}$ be the maximum reconstruction error in this point which characterizes the local generalization ability of the procedure θ in the ε-neighborhood of the point X_0.

Let $L(X)$ and $L_\theta(r_\theta(X))$ be the tangent spaces to the manifolds $\mathbf{X}$ and $\mathbf{X}_\theta$ at the points $X \in \mathbf{X}$ and $r_\theta(X) \in \mathbf{X}_\theta$, respectively. These tangent spaces will be treated as elements of the Grassmann manifold Grass(p, q) composed of q-dimensional linear subspaces in R^p. Let $d_{P,2}(L(X), L_\theta(r_\theta(X))) = \|P_L - P_{L*}\|_2$ be the metric on the Grass(p, q) called the projection metric in 2-norm [45], or simply the projection 2-norm [46, 47]; here P_L and P_{L*} are projectors onto the linear spaces $L = L(X)$ and $L* = L_\theta(r_\theta(X))$. In Statistics, this metric is called the Min Correlation Metric [42], [44].

An asymptotic (as $\varepsilon \to 0$) expansion for the maximum reconstruction error

$$\delta_\theta(X, \varepsilon) = \delta_\theta(X) + \varepsilon \times \delta^*(X) + o(\varepsilon), \tag{6}$$

was obtained in [48]; here the quantity $\delta^*(X)$ was written in an explicit form and the $o(\cdot)$ symbol in the vector case is understood componentwise.

The following upper and lower bounds

$$\gamma_- \times d_{P,2}(L(X), L_\theta(r_\theta(X))) \le \delta^*(X) \le \gamma_+ \times d_{P,2}(L(X), L_\theta(r_\theta(X))) \tag{7}$$

are obtained in [48] for the main term in the asymptotic expansion (6), here γ_- and γ_+ are quantities written in an explicit form and satisfying the inequalities $0 \le \gamma_- \le \gamma_+ \le 1$. It was shown also in [48] that the asymptotic inequalities

$$\eta \times |X-X_0| \times d_{P,2}(L(X), L_\theta(r_\theta(X))) + o(|X-X_0|) \le |(X-X_0) - (r_\theta(X)-r_\theta(X_0))| \le$$

$$|X-X_0| \times d_{P,2}(L(X), L_{\theta(g)}(r_\theta(X))) + o(|X-X_0|); \tag{8}$$

$$|X-X_0| \times \sqrt{1 - d_{P,2}^2(L(X_0), L_\theta(r_\theta(X_0)))} + o(|X-X_0|) \le |r_\theta(X)-r_\theta(X_0)| \le$$

$$|X-X_0| \times \sqrt{1 - \eta^2 \times d_{P,2}^2(L(X_0), L_\theta(r_\theta(X_0)))} \tag{9}$$

hold as $|X-X_0| \to 0$, here $0 \le \eta \le 1$ is some parameter defined in explicit form.

Thus, relation (7) demonstrates a relationship between the local generalization ability and distances between the tangent spaces to the DM $\mathbf{X}$ and RM $\mathbf{X}_\theta$. The inequalities indicate to what extent the mapping r_θ preserves the local structure of the DM $\mathbf{X}$ (relation (8)) and characterizes the local non-isometricity of this mapping (relation (9)).

5   Tangent Bundle Manifold Learning

It follows from relations (6) – (9) that the greater the distances between tangent spaces to the DM $\mathbf{X}$ and RM $\mathbf{X}_\theta$ at the point $X \in \mathbf{X}$ and the reconstructed point $r_\theta(X) \in \mathbf{X}_\theta$, respectively, the lower the local generalization ability of the solution θ becomes, the poorer the local structure is preserved, and the poorer the local isometricity properties are ensured at the considered point $X \in \mathbf{X}$. Thus, it is natural to require that the solution θ ensures not only proximity $\mathbf{X} \approx \mathbf{X}_\theta$ but also Tangent proximity

$$L_\theta(r_\theta(X)) \approx L(X) \quad \text{for all} \quad X \in \mathbf{X} \tag{10}$$

between the corresponding tangent spaces in some selected metric on the Grassmann manifold Grass(p, q). The requirement of Tangent proximity (10) for the ML solution arises also in various applications in which the ML solution is an intermediate step for some Intelligent Data Analysis problem.

The set TB($\mathbf{X}$) = {(X, L(X)): $X \in \mathbf{X}$} composed of points X of the manifold $\mathbf{X}$ equipped by tangent spaces L(X) at these points is known in the Manifold theory

[50, 51] as the Tangent Bundle of the manifold $\mathbf{X}$. Consider an amplification of the ML which consists in accurate reconstruction of the tangent bundle TB($\mathbf{X}$) from the sample and called Tangent Bundle Manifold Learning problem (TBML). A strict definition of the TBML is as follows: Given dataset $\mathbf{X}_n$ sampled from a q-dimensional DM $\mathbf{X}$ embedded in an ambient p-dimensional Euclidean space R^p, $q < p$, and covered by a single chart, construct TBML-solution $\vartheta = $ (h, g, G) consisting of the ML-solution $\theta = $ (h, g) and p×q matrices G(y) smoothly depending on $y \in \mathbf{Y}_\theta$ such that

$$J_g(y) = G(y) \quad \text{for all} \quad y \in \mathbf{Y}_\theta, \tag{11}$$

which provides both the Manifold proximity (4) and the proximity

$$L_G(h(X)) \approx L(X) \quad \text{for all} \quad X \in \mathbf{X}, \tag{12}$$

where the linear space $L_G(y) = \text{Span}(G(y))$ spanned by columns of the matrix G(y) is the tangent space to the RM $\mathbf{X}_\theta$ at the point $g(y) \in \mathbf{X}_\theta$.

The TBML-solution ϑ determines the Reconstructed Tangent Bundle

$$TB_\vartheta(\mathbf{X}_\theta) = \{(r_\theta(X), L_G(h(X))): X \in \mathbf{X}\} = \{(r_\theta(X), L_\theta(r_\theta(X))): X \in \mathbf{X}\} =$$

$$\{(g(y), L_G(y)): y \in \mathbf{Y}_\theta = h(X)\} = \{(X', L_\theta(X')): X' \in \mathbf{X}_\theta\}$$

of the RM $\mathbf{X}_\theta$. The proximities (4) and (12) can be treated as Tangent Bundle proximity $TB_\vartheta(\mathbf{X}_\theta) \approx TB(\mathbf{X})$.

Consider the set $\mathbf{L} = \{L(X), X \in \mathbf{X}\}$ called the Tangent Manifold. The solution ϑ determines the set $\mathbf{L}_\vartheta = \{L_G(y), y \in \mathbf{Y}_\theta \subset R^q\} = \{L_\theta(r_\theta(X)): X \in \mathbf{X}\}$ called the Reconstruction Tangent Manifold, which accurately reconstructs (estimates) the unknown Tangent Manifold $\mathbf{L}$. Note that the Tangent Manifold and the Reconstruction Tangent Manifold are q-dimensional submanifolds in the Grassmann manifold Grass(p, q).

Note that the term 'tangent bundle' is used in the ML for various purposes: as approximation of the manifold shape from the data [40], for geometric interpretation of the Contractive Auto-Encoder algorithm [52] in the manifold Tangent Classifier [32], as the name of the 'Tangent Bundle Approximation' algorithm for the Tangent Space Learning Problem in [53, 54, 55], etc. The above-given TBML definition is new and differs from all known meanings of this term in the ML.

6   Grassmann and Stiefel Eigenmaps Algorithm for TBML

This section describes the solution for the TBML, proposed in [56] and called Grassmann&Stiefel Eigenmaps (GSE) algorithm, which gives new solutions for all the DR problems (Sample Embedding, Data Space/Manifold embedding, Manifold Learning) defined above in Sections 2 - 4.

The GSE-algorithm consists of three successively performed main parts: Part I (approximation of the Tangent manifold), Part II (solving the Manifold Embedding problem) and Part III (Tangent Bundle reconstruction).

In Part I, a sample-based submanifold $\mathbf{L_H} = \{L_H(X) = \mathrm{Span}(H(X)), X \in \mathbf{X_h}\}$ of the Grassmann manifold Grass(p, q), defined by a family $\mathbf{H} = \{H(X), X \in \mathbf{X_h}\}$ consisting of p×q matrices H(X) smoothly depending on $X \in \mathbf{X_h}$, is constructed to approximate the Tangent manifold $\mathbf{L}$. The mappings h (2) and g (3) will be built later in such a way that the Jacobian $J_g(h(X))$ of the mapping g(y) at the point y = h(X) will be close to the matrix H(X) for all the points $X \in \mathbf{X_h}$.

In Part II, given the already constructed $\mathbf{L_H}$, the Embedding mapping h(X) is constructed for $X \in \mathbf{X_h}$. In Part III, given $\mathbf{L_H}$ and the constructed mapping h, the RM $\mathbf{X_\theta}$ (5) approximating the DM $\mathbf{X}$ and the tangent spaces $L_\theta(r_\theta(X))$ to the RM at the point $r_\theta(X)$ are constructed in such a way that the built linear spaces $L_H(X)$ are close to the tangent spaces $L_\theta(r_\theta(X))$, whence comes the Tangent bundle proximity.

Main ideas and justification of the GSE-algorithm are briefly described below.

At first, the tangent space L(X) for an arbitrary point $X \in \mathbf{X_h}$ is estimated by q-dimensional linear space $L_{PCA}(X)$ which is a result of the PCA applied to sample points from an ε-ball in R^p centered at X; the set $\mathbf{X_h}$ consists of the points in which the qth eigenvalue in PCA is positive.

The data-based kernel K(X, X′), X′, $X \in \mathbf{X_h}$, is constructed as a product $K_E(X, X′)$ × $K_G(X, X′)$, where K_E is standard Euclidean 'heat' kernel [5] and $K_G(X, X′)$ = $K_{BC}(L_{PCA}(X), L_{PCA}(X′))$ is the Binet-Cauchy kernel [47, 61] on the Grass(p, q); this aggregate kernel reflects not only geometrical nearness between the points X and X′ but also nearness between the linear spaces $L_{PCA}(X)$ and $L_{PCA}(X′)$, whence comes a nearness between the tangent spaces L(X) and L(X′).

Approximation of the Tangent Manifold. The set $\mathbf{H_n}$ consisting of preliminary values H_i of the matrices H(X) at the sample points $X_i \in \mathbf{X_n}$ that meet the constraints $\mathrm{Span}(H_i) = L_{PCA}(X_i)$, is constructed to minimize the quadratic form

$$\Delta_{H,n}(\mathbf{H_n}) = \frac{1}{2}\sum_{i,j=1}^{n} K(X_i, X_j) \times \left\| H_i - H_j \right\|_F^2 \tag{13}$$

under the normalizing condition $\sum_{i=1}^{n} K(X_i) \times (H_i^T \times H_i) = I_q$ required to avoid a degenerate solution; here $K(X) = \sum_{j=1}^{n} K(X, X_j)$ and $K = \sum_{i=1}^{n} K(X_i)$.

The exact solution of the minimizing problem (13) is obtained as solution of specified generalized eigenvector problem. Given $\mathbf{H_n}$, the matrix H(X) for an arbitrary point $X \in \mathbf{X_h}$ is defined as

$$H(X) = \pi(X) \times \frac{1}{K(X)}\sum_{j=1}^{n} K(X, X_j) \times H_j \tag{14}$$

and minimizes the form $\sum_{j=1}^{n} K(X, X_j) \times \left\| H(X) - H_j \right\|_F^2$ under the condition Span(H(X)) = $L_{PCA}(X)$, here $\pi(X)$ is the projector onto the linear space $L_{PCA}(X)$.

Solving the Manifold Embedding Problem. Given $\mathbf{L_H}$, the TBML solution ϑ will be constructed to provide the equalities $J_g(h(X)) = G(h(X)) \approx H(X)$ and (4). Taylor series expansions $g(y') - g(y) \approx J_g(y) \times (y' - y)$ for near points y, y', under the desired equalities for mappings (2) and (3) specified above, imply the equalities

$$X' - X \approx H(X) \times (h(X') - h(X)) \qquad (15)$$

for near points X, X' $\in$ **X**. Under H_n, these equalities written for near sample points X, X' $\in$ X_n can be considered as regression equations for the embeddings $h(X_i)$. The standard least squares solution, which minimizes the residual $\sum_{i,j=1}^{n} K(X_i, X_j) \times |X_j - X_i - H(X_i) \times (h_j - h_i)|^2$ under natural normalizing condition $h_1 + h_2 + \ldots + h_n = 0$, gives the preliminary values for the embeddings $h(X_i)$. Note that the set h_n gives a new solution for the Sample Embedding Problem.

The value h(X) for arbitrary point X $\in$ X_h (including sample points) is defined as

$$h(X) = h_{KNR}(X) + \left(Q_{PCA}^T(X) \times H(X) \right)^{-1} \times Q_{PCA}^T(X) \times \left(X - \frac{1}{K(X)} \sum_{j=1}^{n} K(X, X_j) \times X_j \right),$$

here $h_{KNR}(X) = \frac{1}{K(X)} \sum_{j=1}^{n} K(X, X_j) \times h_j$ is the Kernel Non-parametric Regression estimator for h(X) based on the values $h_j \in h_n$ of the vector h(X) at the sample points.

Tangent Bundle Reconstruction. The data-based kernel k(y, y') on Y_θ and the linear spaces L*(y) $\in$ Grass(p, q) dependent on y $\in$ Y_θ are constructed in such a way to provide the equalities k(h(X), h(X')) $\approx$ K(X, X') and L*(h(X)) $\approx$ $L_{PCA}(X)$ for the points X $\in$ X_h and X' $\in$ X_n. Denote $\pi^*(y)$ the projector onto the linear space L*(y), and denote by Y_n the set $\{y_i = h(X_i), i = 1, 2, \ldots, n\}$. The function g(y) (3) and matrix G(y) are defined as

$$G(y) = \pi^*(y) \times \frac{1}{k(y)} \sum_{j=1}^{n} k(y, y_j) \times H(X_j)$$

where $k(y) = \sum_{j=1}^{n} k(y, y_j)$, and

$$g(y) = \frac{1}{k(y)} \sum_{j=1}^{n} k(y, y_j) \times X_j + G(y) \times \left(y - \frac{1}{k(y)} \sum_{j=1}^{n} k(y, y_j) \times y_j \right)$$

ensuring the equality (11) and the Tangent Bundle proximity (4), (12).

It was shown in [62], that under an appropriate choice of ball radius $\varepsilon \sim O(n^{-1/(q+2)})$, the relations $|X - r_\theta(X)| = O(n^{-2/(q+2)})$ and $d_{P,2}(L(X), L_G(h(X))) = O(n^{-1/(q+2)})$ hold true with high probability as n $\to \infty$ and. The first rate coincides with the asymptotically minimax lower bound for the Hausdorff distance $H(X_\theta, X)$ between the DM X and RM X_θ, which was set out in [63]. Thus, the RM X_θ estimates the DM X with the optimal rate of convergence.

7   Conclusion

Many Data Mining tasks, such as Pattern Recognition, Classification, Clustering, and others, which are challenging for machine learning algorithms, deal with real-world data presented in high-dimensional spaces, and the 'curse of dimensionality' phenomena is often an obstacle to the use of many methods for solving these tasks. To avoid this phenomena in Data Mining tasks, various Representation learning algorithms are used as a first key step in solutions of these tasks to transform the original high-dimensional data into their lower-dimensional representations such that as much information about the original data required for the considered Data Mining task is preserved as possible. The above Representation learning problems can be formulated as various Dimensionality Reduction problems, whose different formalizations depend on the considered Data Mining tasks.

Different formulations of the Dimensionality Reduction (Sample Embedding, Data Space (Manifold) embedding, Manifold Learning as Data Manifold Reconstruction/Estimation, and newly proposed Tangent Bundle Manifold Learning), which are motivated by various Data Mining tasks, are described. A new geometrically motivated algorithm that solves the Tangent Bundle Manifold Learning and gives new solutions for all the considered DR problems is also presented. If the sample size tends to infinity, the proposed algorithm has the optimal rate of convergence.

Acknowledgments. This work is partially supported by the RFBR, research projects 13-01-12447 and 13-07-12111.

References

1. Bengio, Y., Courville, A., Vincent, P.: Representation Learning: A Review and New Perspectives. arXiv preprint: arXiv:1206.5538v2, 1–64 (2012)
2. Bunte, K., Biehl, M., Hammer, B.: Dimensionality reduction mappings. In: IEEE Symposium Series in Computational Intelligence (SSCI) 2011 - Computational Intelligence and Data Mining (CIDM), pp. 349–356. IEEE, Paris (2011)
3. Cox, T.F., Cox, M.A.A.: Multidimensional Scaling. Chapman and Hall/CRC, London (2001)
4. Jollie, T.: Principal Component Analysis. Springer, New-York (2002)
5. Belkin, M., Niyogi, P.: Laplacian eigenmaps for dimensionality reduction and data representation. Neural Computation 15, 1373–1396 (2003)
6. Hecht-Nielsen, R.: Replicator neural networks for universal optimal source coding. Science 269, 1860–1863 (1995)
7. Hinton, G.E., Salakhutdinov, R.R.: Reducing the dimensionality of data with neural networks. Science 313(5786), 504–507 (2006)
8. Kramer, M.: Nonlinear Principal Component Analysis using autoassociative neural networks. AIChE Journal 37(2), 233–243 (1991)
9. DeMers, D., Cottrell, G.W.: Nonlinear dimensionality reduction. In: Hanson, D., Cowan, J., Giles, L. (eds.) Advances in Neural Information Processing Systems, vol. 5, pp. 580–587. Morgan Kaufmann, San Mateo (1993)
10. Kohonen, T.: Self-organizing Maps, 3rd edn. Springer (2000)
11. Martinetz, T., Schulten, K.: Topology representing networks. Neural Networks 7, 507–523 (1994)
12. Lafon, S., Lee, A.B.: Diffusion Maps and Coarse-Graining: A Unified Framework for Dimensionality Reduction, Graph Partitioning and Data Set Parameterization. IEEE Transaction on Pattern Analysis and Machine Intelligence 28(9), 1393–1403 (2006)

13. Schölkopf, B., Smola, A., Müller, K.: Nonlinear component analysis as a kernel eigenvalue problem. Neural Computation 10(5), 1299–1319 (1998)
14. Saul, L.K., Roweis, S.T.: Nonlinear dimensionality reduction by locally linear embedding. Science 290, 2323–2326 (2000)
15. Donoho, D.L., Grimes, C.: Hessian eigenmaps: New locally linear embedding techniques for high-dimensional data. Proceedings of the National Academy of Arts and Sciences 100, 5591–5596 (2003)
16. Tehenbaum, J.B., de Silva, V., Langford, J.C.: A global geometric framework for nonlinear dimensionality reduction. Science 290, 2319–2323 (2000)
17. Weinberger, K.Q., Saul, L.K.: Maximum Variance Unfolding: Unsupervized Learning of Image Manifolds by Semidefinite Programming. International Journal of Computer Vision 70(1), 77–90 (2006)
18. Brand, M.: Charting a manifold. In: Becker, S., Thrun, S., Obermayer, K. (eds.) Advances in Neural Information Processing Systems, vol. 15, pp. 961–968. MIT Press, Cambridge (2003)
19. Zhang, Z., Zha, H.: Principal Manifolds and Nonlinear Dimension Reduction via Local Tangent Space Alignment. SIAM Journal on Scientific Computing 26(1), 313–338 (2005)
20. Bengio, Y., Delalleau, O., Le Roux, N., Paiement, J.-F., Vincent, P., Ouimet, M.: Learning Eigenfunctions Link Spectral Embedding and Kernel PCA. Neural Computation 16(10), 2197–2219 (2004)
21. Bengio, Y., Delalleau, O., Le Roux, N., Paiement, J.-F., Vincent, P., Ouimet, M.: Out-of-sample extension for LLE, Isomap, MDS, Eigenmaps, and spectral clustering. In: Thrun, S., Saul, L., Schölkopf, B. (eds.) Advances in Neural Information Processing Systems, vol. 16, pp. 177–184. MIT Press, Cambridge (2004)
22. Saul, L.K., Roweis, S.T.: Think globally, fit locally: unsupervised learning of low dimensional manifolds. Journal of Machine Learning Research 4, 119–155 (2003)
23. Saul, L.K., Weinberger, K.Q., Ham, J.H., Sha, F., Lee, D.D.: Spectral methods for dimensionality reduction. In: Chapelle, O., Schölkopf, B., Zien, A. (eds.) Semisupervised Learning, pp. 293–308. MIT Press, Cambridge (2006)
24. Burges, C.J.C.: Dimension Reduction: A Guided Tour. Foundations and Trends in Machine Learning 2(4), 275–365 (2010)
25. Gisbrecht, A., Lueks, W., Mokbel, B., Hammer, B.: Out-of-Sample Kernel Extensions for Nonparametric Dimensionality Reduction. In: Proceedings of European Symposium on Artificial Neural Networks, ESANN 2012. Computational Intelligence and Machine Learning, pp. 531–536. Bruges, Belgium (2012)
26. Strange, H., Zwiggelaar, R.: A Generalised Solution to the Out-of-Sample Extension Problem in Manifold Learning. In: Proceedings of the Twenty-Fifth AAAI Conference on Artificial Intelligence, San Francisco, California, USA, pp. 471–478. AAAI Press, Menlo Park (2011)
27. Cayton, L.: Algorithms for manifold learning. Univ of California at San Diego (UCSD), Technical Report CS2008-0923, pp. 541–555. Citeseer (2005)
28. Huo, X., Ni, X., Smith, A.K.: Survey of Manifold-based Learning Methods. In: Liao, T.W., Triantaphyllou, E. (eds.) Recent Advances in Data Mining of Enterprise Data, pp. 691–745. World Scientific, Singapore (2007)
29. Izenman, A.J.: Introduction to manifold learning. Computational Statistics 4(5), 439–446 (2012)
30. Ma, Y., Fu, Y. (eds.): Manifold Learning Theory and Applications. CRC Press, London (2011)

31. Narayanan, H., Mitter, S.: Sample complexity of testing the manifold hypothesis. In: Lafferty, J., Williams, C.K.I., Shawe-Taylor, J., Zemel, R., Culotta, A. (eds.) Advances in Neural Information Processing Systems, vol. 23, pp. 1786–1794. MIT Press, Cambridge (2010)
32. Rifai, S., Dauphin, Y.N., Vincent, P., Bengio, Y., Muller, X.: The manifold Tangent Classifier. In: Shawe-Taylor, J., Zemel, R.S., Bartlett, P., Pereira, F., Weinberger, K.Q. (eds.) Advances in Neural Information Processing Systems, vol. 24, pp. 2294–2302. MIT Press, Cambridge (2011)
33. Chen, J., Deng, S.-J., Huo, X.: Electricity price curve modeling and forecasting by manifold learning. IEEE Transaction on Power Systems 23(3), 877–888 (2008)
34. Song, W., Keane, A.J.: A Study of Shape Parameterisation Methods for Airfoil Optimisation. In: Proceedings of the 10th AIAA / ISSMO Multidisciplinary Analysis and Optimization Conference, AIAA 2004-4482. American Institute of Aeronautics and Astronautics, Albany (2004)
35. Bernstein, A., Kuleshov, A., Sviridenko, Y., Vyshinsky, V.: Fast Aerodynamic Model for Design Technology. In: Proceedings of West-East High Speed Flow Field Conference, WEHSFF-2007. IMM RAS, Moscow (2007),
    http://wehsff.imamod.ru/pages/s7.htm
36. Bernstein, A., Kuleshov, A.: Cognitive technologies in the problem of dimension reduction of geometrical object descriptions. Information Technologies and Computer Systems 2, 6–19 (2008)
37. Bernstein, A.V., Burnaev, E.V., Chernova, S.S., Zhu, F., Qin, N.: Comparison of Three Geometric Parameterization methods and Their Effect on Aerodynamic Optimization. In: Control with Applications to Industrial and Societal Problems (Eurogen 2011), Capua, Italy, September 14 - 16 (2011)
38. Lee, J.A., Verleysen, M.: Quality assessment of dimensionality reduction based on k-ary neighborhoods. In: Saeys, Y., Liu, H., Inza, I., Wehenkel, L., Van de Peer, Y. (eds.) JMLR Workshop and Conference Proceedings. New Challenges for Feature Selection in Data Mining and Knowledge Discovery, vol. 4, pp. 21–35. Antwerpen, Belgium (2008)
39. Lee, J.A., Verleysen, M.: Quality assessment of dimensionality reduction: Rank-based criteria. Neurocomputing 72(7-9), 1431–1443 (2009)
40. Freedman, D.: Efficient simplicial reconstructions of manifold from their samples. IEEE Transaction on Pattern Analysis and Machine Intelligence 24(10), 1349–1357 (2002)
41. Karygianni, S., Frossard, P.: Tangent-based manifold approximation with locally linear models. In: arXiv preprint: arXiv:1211.1893v1 [cs.LG] (November 6, 2012)
42. Golub, G.H., Van Loan, C.F.: Matrix Computation, 3rd edn. Johns Hopkins University Press, MD (1996)
43. Hotelling, H.: Relations between two sets of variables. Biometrika 28, 321–377 (1936)
44. James, A.T.: Normal multivariate analysis and the orthogonal group. Ann. Math. Statistics 25, 40–75 (1954)
45. Wang, L., Wang, X., Feng, J.: Subspace Distance Analysis with Application to Adaptive Bayesian Algorithm for Face Recognition. Pattern Recognition 39(3), 456–464 (2006)
46. Edelman, A., Arias, T.A., Smith, T.: The Geometry of Algorithms with Orthogonality Constraints. SIAM Journal on Matrix Analysis and Applications 20(2), 303–353 (1999)
47. Hamm, J., Lee, D.D.: Grassmann Discriminant Analysis: a Unifying View on Subspace-Based Learning. In: Proceedings of the 25th International Conference on Machine Learning (ICML 2008), pp. 376–383 (2008)
48. Bernstein, A.V., Kuleshov, A.P.: Manifold Learning: generalizing ability and tangent proximity. International Journal of Software and Informatics 7(3), 359–390 (2013)

49. Kuleshov, A.P., Bernstein, A.V.: Cognitive Technologies in Adaptive Models of Complex Plants. Information Control Problems in Manufacturing 13(1), 1441–1452 (2009)
50. Lee, J.M.: Manifolds and Differential Geometry. Graduate Studies in Mathematics, vol. 107. American Mathematical Society, Providence (2009)
51. Lee, J.M.: Introduction to Smooth Manifolds. Springer, New York (2003)
52. Rifai, S., Vincent, P., Muller, X., Glorot, X., Bengio, Y.: Contractive Auto-Encoders: Explicit Invariance during Feature Extraction. In: Getoor, L., Scheffer, T. (eds.) Proceedings of the 28th International Conference on Machine Learning (ICML 2011), pp. 833–840. Omnipress, Bellevue (2011)
53. Silva, J.G., Marques, J.S., Lemos, J.M.: A Geometric approach to motion tracking in manifolds. In: Paul, M.J., Van Den Hof, B.W., Weiland, S. (eds.) A Proceedings Volume from the 13th IFAC Symposium on System Identification, Rotterdam (2003)
54. Silva, J.G., Marques, J.S., Lemos, J.M.: Non-linear dimension reduction with tangent bundle approximation. In: Proceedings of IEEE International Conference on Acoustics, Speech, and Signal Processing (ICASSP 2005), vol. 4, pp. 85–88. Conference Publications (2005)
55. Silva, J.G., Marques, J.S., Lemos, J.M.: Selecting Landmark Points for Sparse Manifold Learning. In: Weiss, Y., Schölkopf, B., Platt, J. (eds.) Advances in Neural Information Processing Systems, vol. 18. MIT Press, Cambridge (2006)
56. Bernstein, A.V., Kuleshov, A.P.: Tangent Bundle Manifold Learning via Grassmann & Stiefel Eigenmaps. arXiv preprint: arXiv:1212.6031v1 [cs.LG], pp. 1–25 (December 2012)
57. Achlioptas, D.: Random matrices in data analysis. In: Boulicaut, J.-F., Esposito, F., Giannotti, F., Pedreschi, D. (eds.) ECML 2004. LNCS (LNAI), vol. 3201, pp. 1–7. Springer, Heidelberg (2004)
58. Tyagi, H., Vural, E., Frossard, P.: Tangent space estimation for smooth embeddings of riemannian manifold. arXiv preprint: arXiv:1208.1065v2 [stat.CO], pp. 1–35 (May 17, 2013)
59. Singer, A., Wu, H.: Vector Diffusion Maps and the Connection Laplacian. Comm. on Pure and App. Math. (2012)
60. Coifman, R.R., Lafon, S., Lee, A.B., Maggioni, M., Warner, F., Zucker, S.: Geometric diffusions as a tool for harmonic analysis and structure definition of data: Diffusion maps. Proceedings of the National Academy of Sciences, 7426–7431 (2005)
61. Wolf, L., Shashua, A.: Learning over sets using kernel principal angles. J. Mach. Learn. Res. 4, 913–931 (2003)
62. Kuleshov, A., Bernstein, A.: Yanovich, Yu.: Asymptotically optimal method in Manifold estimation. In: Márkus, L., Prokaj, V. (eds.) Abstracts of the XXIX-th European Meeting of Statisticians, Budapest, Hungary, July 20-25, p. 325 (2013), http://ems2013.eu/conf/upload/BEK086_006.pdf
63. Genovese, C.R., Perone-Pacifico, M., Verdinelli, I., Wasserman, L.: Minimax Manifold Estimation. Journal Machine Learning Research 13, 1263–1291 (2012)

Adaptive Multiple-Resolution Stream Clustering

Marwan Hassani, Pascal Spaus, and Thomas Seidl

Data Management and Data Exploration Group,
RWTH Aachen University, Germany
{hassani,spaus,seidl}@cs.rwth-aachen.de

Abstract. Stream data applications have become more and more prominent recently and the requirements for stream clustering algorithms have increased drastically. Due to continuously evolving nature of the stream, it is crucial that the algorithm autonomously detects clusters of arbitrary shape, with different densities, and varying number of clusters. Although available density-based stream clustering are able to detect clusters with arbitrary shapes and varying numbers, they fail to adapt their thresholds to detect clusters with different densities. In this paper we propose a stream clustering algorithm called HASTREAM, which is based on a hierarchical density-based clustering model that automatically detects clusters of different densities. The density thresholds are independently adapted to the existing data without the need of any user intervention. To reduce the high computational cost of the presented approach, techniques from the graph theory domain are utilized to devise an incremental update of the underlying model. To show the effectiveness of HASTREAM and hierarchical density-based approaches in general, several synthetic and real world data sets are evaluated using various quality measures. The results showed that the hierarchical property of the model was able to improve the quality of density-based stream clusterings and enabled HASTREAM to detect streaming clusters of different densities.

1 Introduction

Data streams have become more and more popular over the recent years. The technical advancement allows to create cheap systems that process data with increasing size, dimensionality and speed. The continuous flow of data, i.e. data streams, exists in several fields of application, e.g. health monitoring, economical data, sensor networks. In fact, data streams are generated wherever electronic devices are used to measure, monitor or record something. The interest in online mining of the evolving big data, has signaled the interest in streaming algorithms recently. Increasingly sophisticated stream clustering algorithms were developed for instance, to handle noise, operate on subspaces or adapt better to stream evolution. Thus, stream clustering algorithms have inherited all the challenges that are known when clustering static data, in addition to those challenges which appeared due to the nature of streaming data.

A data stream differs from a static data set in various ways. One of the main differences is that the data of a stream arrives continuously and has to be

P. Perner (Ed.): MLDM 2014, LNAI 8556, pp. 134–148, 2014.

processed at its arrival. Furthermore, due to the huge amount of data, it can not be stored persistently. Another implication of the endless character of the stream, is that multiple runs over the data are not possible, and thus every new object can only be processed once. Every object has to be processed as fast as possible, thus the processing has to be efficient. An additional challenge is the evolution of data streams, thus the algorithm has to recognize these changes in the data and *continuously* adapt to them.

Most stream clustering algorithms follow the online-offline-phases model. In the online phase, a summary of the evolving data is performed and continuously maintained to follow the changing distribution of the data. The offline phase is then performed upon a user request, and uses one of the well known static clustering algorithms to deliver the final clustering. CluStream [1] for example, performs a k-means variant in the offline phase over the data summaries (also called micro-clusters in most algorithms). Motivated by the need to detect arbitrarily-shaped evolving clusters, and being less sensitive to outliers in the stream, density based stream algorithms were developed like DenStream [5] which obtains the final clustering by performing a DBSCAN variant over the micro-clusters. The advantage of density-based clustering algorithms is mainly their ablity to find clusters of arbitrary shape and autonomously detect the number of clusters. However a significant drawback of available density-based approaches, is that they are not able to find clusters of different densities, due to the "static" density threshold parameter that is usually used in such algorithms Figure (1). Having clusters of different densities at the same time is very common in streaming data. Additionally, the same clusters may evolve over the stream such that they will have different densities over the time. Setting the density threshold parameter to a single value all over the stream (like e.g. in DenStream), will lead to missing many clusters (cf. Figure (1), minPts = 5). One approach to counteract this problem is to use hierarchical clustering techniques to enable density-based clustering algorithms to find clusters of different densities, by adapting the density threshold to the characteristics of each considered cluster. In this work, we propose, the first hierarchical density-based stream clustering algorithm based on cluster stability, called HASTREAM. The algorithm is able to detect clusters of different densities and arbitrary shapes. The final *flat* clustering is extracted from the hierarchical clustering. The density

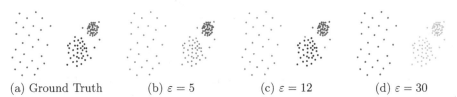

(a) Ground Truth   (b) $\varepsilon = 5$ (c) $\varepsilon = 12$ (d) $\varepsilon = 30$

Fig. 1. Clusters can not be all detected with a single density threshold as in DBSCAN [7]. (b) If ε is too small , only the green cluster is found. (d) However, if ε is chosen too high, the blue and green cluster are detected as a single cluster.

parameters, as well as the number of clusters, are automatically adapted while extracting the clustering, without any need for user intervention.

The remainder of this paper is presented as follows: Section 2 gives an overview on the state-of-the-art. Section 3 presents some needed definitions and the used data structures for extracting the final clustering. In Section 4, we present our HASTREAM algorithm in details. In Section 5, we thoroughly evaluate two variants of HASTREAM against a state-of-the-art algorithm before concluding the paper with an outlook in Section 6.

2   Related Work

DBSCAN [7] has been the most appealing density-based clustering algorithm for a long time. A known problem in DBSCAN however, is detecting clusters with considerably different densities (cf. Figure (1)). **OPTICS:** *Ordering Points To Identify the Clustering Structure* requires two parameters ε and *minPts*, which have the same meaning as in DBSCAN. OPTICS orders the objects such that the closest objects are also neighbours in the ordering. To perform this ordering, the algorithm computes two additional values for each object, called core-distance and reachability-distance. The reachability-distance is crucial for extracting the final clustering. The final cluster ordering can be visualised by the

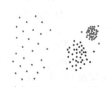

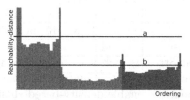

(a) Ground truth   (b) Reachability plot

Fig. 2. The two lines in the reachability plot represent two different thresholds. Using the upper line that represents a higher threshold, leads to two clusters, where the blue and green areas are recognized as a single cluster and the red cluster is correctly recognized. The lower line corresponds to a lower threshold recognizes only the blue and green clusters.

reachability-plot. Figure (2) illustrates a reachability-plot for a 2D data set. The horizontal axis represents the ordering of the objects and the vertical axis their respective reachability-distance. The clusters can be identified by the valleys in the reachability-plot. By setting a threshold, represented by a horizontal line, the clustering can be extracted from the reachability-plot. Each valley the horizontal line crosses represents a cluster. **HDBSCAN:** *Density-Based Clustering Based on Hierarchical Density Estimates.* It is a recent research motivated by the fact that OPTICS only allows to obtain a flat clustering by setting a single density threshold. This approach may not lead to the optimal clustering if the clusters have different densities, as illustrated in Figure (2). Campello et al. [4]

proposed the density-based hierarchical clustering algorithm HDBSCAN. The authors introduced a new cluster stability measure to extract a flat clustering, that contains the most significant clusters from the hierarchy of clusters delivered by HDBSCAN. HDBSCAN builds the hierarchical clustering by computing the $minPts$-core-distance for each object of the data set. Then, a complete graph (called mutual reachability graph) is computed where each object is represented by a vertex and the weight of an edge is equivalent to the mutual reachability distance of the two considered objects. A minimal spanning tree can be extracted from the mutual reachability graph. This minimal spanning tree is used to build a dendrogram that represents the hierarchical clustering, by iteratively removing the largest weighted edge. The newly introduced measure *cluster stability* is used to extract a flat clustering from the dendrogram. The cluster stability measure is an adaptation of the excess of mass concept [11]. The cluster stability of a cluster is directly related to the density of the cluster. Higher densities infer more stability.

Data Stream Clustering algorithms are developed more recently. **CluStream** [1] summarizes the data stream by using the *micro-cluster* structure (Definition (1)). The offline component of CluStream considers the micro-clusters as pseudo-points and performs a k-means variant over the maintained micro-clusters. **DenStream** [5] uses the micro-cluster structure as well. DenStream maintains two lists of micro-clusters, the potential and the outlier micro-clusters. The micro-clusters are restricted by a maximal radius and new streaming data is only merged into an existing micro-cluster if its distance between the new data object and the closest center of a micro-cluster does not exceed the maximal radius threshold. The offline phase is a DBSCAN variant which considers only potential micro-clusters that exceed a certain density threshold. **ClusTree** [8] is an anytime stream clustering algorithm which provides a compact and self-adaptive indexing structure for the micro-clusters to allow handling of fast and slow data streams. Therefore, the authors additionally propose new descent strategies to improve the clustering results. **OpticsStream** [14] is a visualization algorithm which uses a micro-cluster structure to represent the data stream and the offline phase uses the object ordering technique of OPTICS [2] to generate a 3-dimensional surface plot, called reachability surface. The idea is similar to the 2-dimensional reachability-plot of OPTICS, but plotted over time. The plot can be used to obtain insight on how the clusters change over time. **MR-Stream** [15] uses a tree structure to represent the data stream. The data space is partitioned into equally sized cells. At each level of the tree structure, each cell is divided into subcells. E.g. if the number of cell divisions is set to 2, the resulting tree structure is similar to a quadtree. The offline clustering is performed on the cells at a user defined hierarchy level of the tree, by clustering all reachable dense cells to a cluster. Thus the algorithm does not solve the main drawback of density-based stream clustering algorithms, as it introduces a new parameter. Additionally it suffers from the efficiency issues that grid-based stream clustering algorithms usually have. In [6], the **COD** algorithm for clustering of multiple evolving streams. According to speed of the incoming streams, the algorithms selects either a wavelet-based model or a regression-based model for building the summary hierarchy. Different to our algorithm, the COD algorithm does not adapt to the distribution of the underlying data streams. Our algorithm is able to detect clusters with different granualities, and automatically adapts its thresholds to reflect the most stable clustering.

3   Preliminaries and Data Structure

Our algorithm, called HASTREAM, is composed of an online and an offline phase.

3.1   The Online Phase: Micro-Clusters List and Index Structures

The online phase uses the damped window model with an exponentially decaying function, as specified in Equation 1. The parameter λ specifies the importance of previous data. HASTREAM uses the micro-cluster data structure (Definition (1)) to represent the data stream in a compact way.

$$f_b(t) = b^{-\lambda t}, \text{ where } 0 < \lambda < 1 \tag{1}$$

The offline part of the algorithm is built in an abstract way, such that the underlying micro-cluster structure can be easily exchanged. The two used micro-cluster structures are adapted variants of those used by Denstream [5] (list structure) and ClusTree [8] (index structure).

Definition 1 (Micro-Cluster). *Is a tuple* $MC(\overline{\text{CF}^1}, \overline{\text{CF}^2}, w, c, r, t_0)$ *where:* $\overline{\text{CF}^1} = \sum_{j=1}^{n} f_b(t - t_j) p_j$ *is the linear weighted sum of the points,* $\overline{\text{CF}^2} = \sum_{j=1}^{n} f_b(t - t_j) p_j^2$ *is the weighted sum of the squared points,* $w = \sum_{j=0}^{n} f_b(t - t_j)$ *is the weight,* $c = \frac{\overline{\text{CF}^1}}{w}$ *is the center,* $r = \sqrt{\frac{|\overline{\text{CF}^2}|}{w} - \left(\frac{|\overline{\text{CF}^1}|}{w}\right)^2}$ *is the radius and* t_0 *is the creation timestamp.*

If an object p fits into a micro-cluster mc, the statistics are updated as follows: $mc = \left(\overline{\text{CF}^1} + p, \ \overline{\text{CF}^2} + p^2, \ w + 1\right)$. Otherwise, if no object is added to the micro-cluster for a certain time interval δt, then it is updated as: $mc = \left(f_b(\delta t) \cdot \overline{\text{CF}^1}, \ f_b(\delta t) \cdot \overline{\text{CF}^2}, \ f_b(\delta t) \cdot w\right)$.

3.2   The Offline Phase: Density-Based Hierarchical Clustering

This model requires the density threshold parameter $minPts$ merely. To be able to explain the model, the required definitions are presented in the following. A micro-cluster is represented by a vertex when graphs are considered. Thus, when referring to the weight of a vertex, what is in fact referred to is the weight of the micro-cluster represented by the respective vertex.

Definition 2 (Core Distance). *The core distance of a micro-cluster mc_p is defined as the Euclidean distance to its $minPts$-nearest neighbor micro-cluster.*

$$\text{core-dist}_{minPts}(mc_p) = \textit{distance to the } minPts\textit{-th closest micro-cluster}$$

Definition 3 (Mutual Reachability Distance). *The mutual reachability distance between two micro-clusters mc_p and mc_q is defined as the maximum of the Euclidean distance between the centers of the micro-clusters and their respective core-distances.*

$$\text{dist}_{mr}(mc_p, mc_q) = \max\{ \text{dist}_{Euc}(mc_p, mc_q),$$
$$\text{core-dist}_{minPts}(mc_p), \text{core-dist}_{minPts}(mc_q)\}$$

Definition 4 (Mutual Reachability Graph). *The mutual reachability graph $\mathcal{MRG}(V, E)$ is a complete graph. The set of vertices V is represented by the micro-clusters available at timestamp t. The weight of the edge between u and v represents the mutual reachability distance between the two micro-clusters represented by vertices $u, v \in V$.*

$$\mathcal{MRG}(V, E) = \begin{cases} V = \textit{the set of available micro-clusters at time } t \\ E = \{e(u, v) \mid u, v \in V \textit{ with weight } w(e) = \text{dist}_{mr}(u, v)\} \end{cases}$$

Definition 5 (Adjacency List). *Given a connected graph $\mathcal{G}(V, E)$, the adjacency list of a vertex v contains all vertices which are directly reachable from v by one edge.*

$$\text{Adj}_E(v) = \{u \mid u \in V \textit{ and } e(v, u) \in E\}$$

Definition 6 (Connected Component). *A connected component is a set of vertices V, where each vertex has at least one adjacent vertex, for $|V| \geq 2$.*

$$\text{Conn-Component}(V, E) = \left\{v \in V \mid \text{Adj}_E(v) \neq \varnothing \wedge \text{Adj}_E(v) \neq \{v\}\right\}$$

A single vertex is also a connected component.

Definition 7 (Total Weight of a Connected Component). *The total weight of a connected component Conn-Component(V, E) is the sum of the weights of each $v \in V$.*

$$\textit{Total Weight}(\text{Conn-Component}(V, E)) = \sum_{v \in V} \textit{weight of } v$$

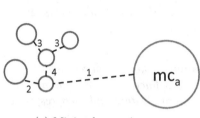

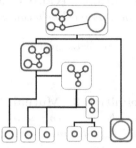

(a) Minimal spanning tree

(b) Dendrogram and extracted flat clustering (red boxes)

Fig. 3. Minimal spanning tree over micro-clusters, the labels of the edges indicate the sequence in which the corresponding edges are removed

Definition 8 (Minimal Spanning Tree). *Given a connected and undirected* $\mathcal{G}(V, E)$, *a minimal spanning tree* $\mathcal{MST}(V, T)$ *is a connected subgraph of* $\mathcal{G}$ *which contains no cycles and minimizes the total weight w.r.t. the edges.*

$$\mathcal{MST}(V', T) = \begin{cases} V' = V \\ T = \left\{ e(u, v) \in E \mid minimize \sum_{e \in E} w(e) \right\} \\ \text{Conn-Component}(V', T) \end{cases}$$

Extraction of the Hierarchical Clusters. HDBSCAN [4] will not be able to extract hierarchical clustering out of the micro-clusters when simply considering them as pseudo points. Figure (3a) illustrates the problem that could appear. If the extraction of the hierarchical clustering is only based on the centers of the micro-clusters, the micro-cluster mc_a would not be recognized as a cluster, even if the micro-cluster covers much of the relevant data. In this case, HDBSCAN would simply consider mc_a as noise, because it is far away and does not belong to the most stable clustering, even though the micro-cluster could form a cluster by itself. Thus, the hierarchical extraction model for the streaming case has to consider the weights of the micro-clusters. The weight of a micro-cluster is compared with the density threshold $minPts$. Following this approach, the micro-cluster mc_a from the example would be recognized as a cluster by itself, since its weight exceeds the density threshold. The model for extracting the hierarchical clusters works as follows:

Given a set of micro-clusters MC, the mutual reachability graph $\mathcal{MRG}$ is computed as defined in Definition (4) and the minimal spanning tree $\mathcal{MST}$ of $\mathcal{MRG}$ is extracted. Since the model is for streaming data, the extraction of the minimal spanning tree is a crucial step for the efficiency of the later algorithm. Thus in this model, the minimal spanning tree is build by using Prim's algorithm [13] with a Fibonacci heap structure. After computing the minimal spanning

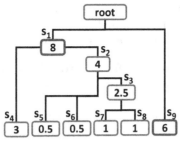

Fig. 4. This figure illustrates a flat clustering extraction w.r.t. the cluster stability measure. The red boxes are selected, i.e. $s_1 = s_9 = 1$. Whereas, the blue boxes are not selected, i.e. $s_j = 0$, $\forall j \notin \{1,9\}$.

tree, the extraction of the clusters can begin. At the start, the root node of the dendrogram contains all micro-clusters, forming one cluster. Starting from a complete minimal spanning tree $\mathcal{MST}$, the largest edge from the $\mathcal{MST}$ is removed. If there exists multiple edges with the same weight, they have to be removed simultaneously. Removing the edges leads to connected subcomponents of the currently considered cluster. A subcomponent can be either a connected component with more than one micro-cluster, or it is a single micro-cluster. In case of a connected subcomponent, the total weight of the component is compared to $minPts$ and if it exceeds the density threshold, the subcomponent can be considered as a cluster at the current hierarchy level. Otherwise, the subcomponent is rejected and is no longer considered as a possible cluster, since the subcomponent can not form a cluster w.r.t. the density threshold. In case of a single vertex, the weight of the micro-cluster, which is represented by the vertex , is compared to $minPts$ and is considered as a possible cluster when it exceeds the density threshold. Otherwise if its weight is lower than the density threshold, the micro-cluster can not form a cluster on its own. This procedure is repeated iteratively until there is no edge left which can be removed. At the end, the hierarchical clustering is represented as a dendrogram. Figure (3b) illustrates the hierarchical clustering in form of a dendrogram of the previous example. This model generates all the clusters which exceed the minimal density threshold $minPts$. This follows from the fact that at each step, only those subcomponents are rejected, which do not exceed the minimal density threshold.

Extraction of the Flat Clustering. In most applications, it is desired to have a flat clustering instead of returning every possible hierarchical cluster. For example in OPTICS [2], the flat clustering is obtained by setting a threshold in the reachability-plot. The resulting clustering is not necessarily optimal. In Section (2), the optimal clustering of the example data set, cf. Figure (2), could not be found by a single threshold. In HDBSCAN [4], a measure was presented to extract the optimal non-hierarchical clustering out of the hierarchical clusters. The referred measure, Cluster Stability, was already explained in more detail in Section (2). An example of a flat clustering extracted from the hierarchical clustering is illustrated in Figure (3b) and Figure (4). The measure for extracting

the flat clustering in this model is an adapted variant of the cluster stability
measure for micro-clusters. The adapted cluster stability takes the weight of the
micro-clusters into account and is defined as follows:

Definition 9 (Cluster Stability for Micro-Clusters). *The cluster stability
CS of a cluster C_i consisting of micro-clusters, is*

$$CS(C_i) = \sum_{mc_j \in C_i} w(mc_j) \cdot \left(\lambda_{max}(mc_j, C_i) - \lambda_{min}(C_i) \right)$$

$$= \sum_{mc_j \in C_i} w(mc_j) \cdot \left(\frac{1}{\varepsilon_{min}(mc_j, C_i)} - \frac{1}{\varepsilon_{max}(C_i)} \right) \quad (2)$$

*where $w(mc_j)$ is the weight of the micro-cluster mc_j. $\lambda_{min}(C_i)$ is the minimum
density threshold at which C_i exists and $\lambda_{max}(mc_j, C_i)$ is the density threshold where mc_j does no longer belong to the cluster C_i. ε_{max} and ε_{min} are the
corresponding ε thresholds.*

ε_{max} and ε_{min} thresholds can be extracted for each cluster from the dendrogram.
Let $HC = \{C_2, \dots, C_k\}$ be the set of all extracted hierarchical clusters from
the dendrogram, except for the cluster C_1 at the root node. In the following,
it is assumed that the number of available micro-clusters is at least $minPts$
and thus the cluster C_1 which contains all micro-clusters is not relevant. A flat
clustering is a non overlapping set of clusters from the hierarchical clustering
such that the most significant clusters are represented in the flat clustering. The
most significant clusters are selected w.r.t. the highest cluster stability measure.
The cluster selection can be formalized as an optimization problem. The goal is
to maximize the sum of stabilities of the extracted clusters. The optimization
problem of extracting the flat clustering is defined as follows:

Definition 10 (Flat Clustering Extraction). *Given a hierarchical clustering
$HC = \{C_2, \dots, C_k\}$, the flat clustering is the set of clusters which maximizes
the sum of cluster stabilities.*

$$\max_{s_2,\dots,s_k} \sum_{i=2}^{k} s_i \cdot CS(C_i)$$

$$\text{with constraints} \begin{cases} s_i \in \{0,1\}, & 2 \leq i \leq k \\ \text{if } (s_i = 1) \Leftrightarrow (s_j = 0), \text{ for all } j \in \mathcal{N}_i \end{cases} \quad (3)$$

where $\mathcal{N}_i$ is the set of all the indices from the nested clusters of cluster C_i.

4   HASTREAM Algorithm

The variant which uses the index structure model, is referred to as $HASTREAM_{IS}$,
whereas the variant that uses the list model, is referred to as $HASTREAM_{LS}$. Algorithm 1 illustrates the main steps of $HASTREAM_i$, with $i \in \{IS, LS\}$.

HASTREAM$_i$ requires the parameter $minPts$, $minClusterWeight$ and the corresponding parameter settings for the online phase. The initialization step is the first phase in the algorithm. During this phase, a set of data stream objects is collected and the initial set of micro-clusters is generated according to the corresponding model. As long as the data stream is ongoing, the new data is processed in Steps 3-4, which represents the online phase. When a clustering is requested, the offline phase is executed, which is represented by the Steps 5-10.

Algorithm 1. HASTREAM$_i$(DataStream ds, $minClusterWeight$)

1: initialization phase
2: **repeat**
3:      get next point $p \in ds$ with current timestamp t_c;
4:      process$_i$(p, online parameter settings);
5:      **if** (t_c mod $updateFrequency == 0$) **then**
6:          compute MRG and corresponding MST
7:          $HC \leftarrow$ extractHierarchicalClusters(MST, $minClusterWeight$); //cf. Alg. 2
8:          $C \leftarrow$ extractFlatClustering(HC);   // cf. Alg. 3
9:          return C;
10:     **end if**
11: **until** data stream terminates

HASTREAM's Online Phase. In the online phase, the new stream object is processed and the micro-clusters are maintained. The function **process**$_i$, $i \in [LS, IS]$ in Algorithm 1 can be replaced by any micro-cluster model.

HASTREAM's Offline Phase. The offline phase of the algorithm is responsible for generating the final clustering over the set of micro-clusters. The offline part can be realized in two different ways. The iterative variant performs a complete computation of the clustering. The incremental variant reuses the previous computations and simply updates the existing minimal spanning tree which is used for the clustering extraction. In the following, the used function from Algorithm 1, Steps 5-10, is explained in more details.

Extraction of the Hierarchical Clusters. Algorithm 2 illustrates the extraction of the hierarchical clusters. Given the minimal spanning tree, the algorithm starts with one connected component, representing one cluster. The largest edge(s), cf. Step 4, are collected and removed from the components which contain any of the edges. This results in connected subcomponents. Either the weight of the subcomponent is high enough w.r.t. $minPts$, and is kept, or otherwise it is rejected, cf. Steps 5-13. At the current hierarchy level of the dendrogram, cf. Step 6, the distance between the components is stored. This allows the computation of the cluster stability in later steps. This procedure is repeated until no more edges are left to process.

Extraction of the Flat Clustering. Algorithm 3 illustrates the extraction of the flat clustering from the dendrogram. In Steps 1-3, the cluster stability for each leaf, as presented in Definition (9), is initialized and each cluster is marked as selected. Then, in Steps 4-13, the cluster stabilities of the inner nodes

---

Algorithm 2. extractHierarchicalClusters(Minimal Spanning Tree MST)

1: $E \leftarrow$ set of edges of MST
2: $all \leftarrow MST$
3: **while** $(E \neq \varnothing)$ **do**
4:     $L \leftarrow$ collect edge(s) with the largest weight
5:     $affected \leftarrow$ components from all containing any edge $e \in L$
6:     set scale value of current hierarchy level to the edge(s) weight
7:     $subcomponents \leftarrow$ removing all edges $e \in L$ from $affected$
8:     **for each** subcomponent c **do**
9:         **if** total weight of $c < minPts$ **then**
10:             remove all edges of component c from E
11:             remove c from $subcomponents$
12:         **end if**
13:     **end for**
14:     $components \leftarrow (components \backslash affected) \cup subcomponents$
15: **end while**

Algorithm 3. extractFlatClustering(Dendrogram den)

1: **for each** cluster $\mathbf{C}_l$ of leaf node l **do**
2:     $CS_p(\mathbf{C}_l) \leftarrow CS(\mathbf{C}_l)$ and $s_l \leftarrow 1$
3: **end for**
4: **while** (current node $i \neq$ root) **do**
5:     **if** $CS(C_i) > \sum_{j \in S_i} CS(\mathbf{C}_j)$ **then**
6:         $CS_p(C_i) \leftarrow CS(\mathbf{C}_i)$
7:         $s_i \leftarrow 1$ and $s_j \leftarrow 0$ for each nested cluster of C_i
8:     **else**
9:         $CS_p(C_i) \leftarrow \sum_{j \in S_i} CS(\mathbf{C}_j)$ and $s_i \leftarrow 0$
10:     **end if**
11:     mark node i and all nodes $j \in S_i$ as visited
12:     update i in a reverse breadth first traversal of only unvisited nodes
13: **end while**
14: **return** all clusters $\mathbf{C}_r$, where $s_r = 1$

---

are computed in a bottom up way. For a faster computation at the current hierarchy level, the computed cluster stabilities are propagated upwards. The current cluster selection, of nested clusters, is rejected when the current stability of the cluster is higher then the selected nested clusters. In that case all nested clusters are unselected and the current cluster is selected. Otherwise the sum of the cluster stabilities of the selected clusters is propagated upwards and the current cluster is rejected. The procedure stops at the root node.

5   Experimental Evaluation

Two variants of HASTREAM are compared to DenStream [5]: HASTREAM$_{LS}$; the variant that uses the index structure to maintain micro-clusters, referred to as Ite-LS and HASTREAM$_{IS}$ the variant that uses the list structure, refereed to as Ite-IS. All algorithms were implemented within MOA [10]. The stream speed is set to 1000 objects per time unit and the first 1000 objects are buffered for the initialization phase of the algorithms. As basis b for the decaying function $f_b(\Delta t)$, cf. Equation (1), b is set to 2 and the decay factor λ is set to 0.25. The window H is set to 1. For DenStream, the parameter β is set to 0.2 and the $\varepsilon_{offline}$ to perform the offline density based clustering is set to $2 \times \varepsilon_{online}$. For HASTREAM$_{LS}$, the online parameter are set equally to those for DenStream, since both use the same structure to maintain the micro-clusters. For HASTREAM$_{IS}$, the maximal height parameter h is set to 4 and the minimal weight threshold ω is set to 0.5. All experiments were performed on a Windows 8 operating system with a 3.5GHz processor and 16GB of memory. The flat clustering quality of all algorithms is evaluated using both the Cluster Mapping Measure (CMM) [9] and the Purity [16]. Furthermore, the runtime in seconds is used to measure the efficiency.

5.1 Data Sets

2D Synthetic Data Set. The 2-dimensional data set was generated by a modified RandomRBFGenerator provided by MOA [10]. Most of the time it consists of 4 drifting clusters with different densities.
KDD CUP'99, Network Intrusion Data Set. The real data set is composed of raw TCP dump data for a local-area network. Each connection has 42 attributes where 34 attributes have continuous and 8 attributes have discrete values. Additionally each connection is labeled as either a normal connection or as an attack. Most of the 494021 TCP connections are labeled as normal connections.
ICML'04, Physiological Data Set. This real data set [12] is a collection of physiological data, collected for the Physiological Data Modeling Contest at ICML 2004. The data set is a collection of different activities such as sleeping or watching TV that were performed by test subjects of different genders. The data set consists of 720792 data objects, where each object has 15 attributes.
Forest Covertype Data Set. This real data set [3] contains 581012 data objects, where each data object has 10 continuous attributes with 7 labels.

5.2 Results

2D Synthetic Data Set. Figure (5) presents ground truths, as well as clusterings at different timestamps for HASTREAM$_{LS}$ and DenStream. The results of DenStream show that the algorithm has difficulties in recognizing clusters with lower densities, due to the single density threshold. However, HASTREAM$_{LS}$ recognizes additionally less dense areas as single clusters, but this can also be a drawback if the stream is currently evolving and old micro-clusters still exist.

    Network Intrusion Data Set. For the network intrusion data set, the parameters for DenStream are set to the same values as used in [5]. The online settings for HASTREAM$_{LS}$ are set equally to the online parameters of DenStream. For HASTREAM$_{IS}$, the maximal height of the indexing tree is set to $h = 3$. For the offline phase of HASTREAM, the density threshold is set to $minPts = 10$.

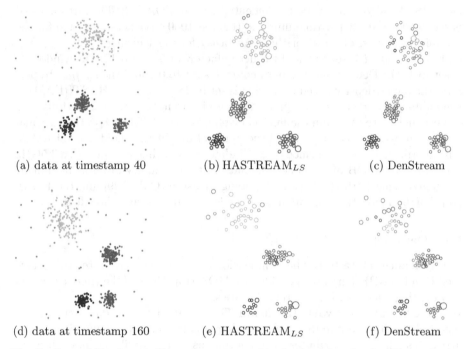

(a) data at timestamp 40   (b) HASTREAM$_{LS}$ (c) DenStream

(d) data at timestamp 160   (e) HASTREAM$_{LS}$ (f) DenStream

Fig. 5. Ground Truth and Clustering output of both algorithm for the 2D synthetic data set at different timestamps

In Figure (6a), the averaged CMM is shown for time intervals of length 100. It can be seen that both HASTREAM variants outperform DenStream. However, HASTREAM$_{IS}$ outperforms HASTREAM$_{LS}$ significantly. Especially in the first time interval $[0, 100]$, HASTREAM$_{IS}$ has a much higher CMM than HASTREAM$_{LS}$. This time interval contained normal connections as well as attacks of each type. One reason for this high difference of CMM could be that the indexing structure is adapting faster to the evolving stream than the list structure. The results for the purity are shown in Figure (6b). The HASTREAM variants have a higher averaged purity on each time interval than DenStream.

Physiological Data Set. For the next real world data set, the online settings for HASTREAM$_{LS}$ are set to $\varepsilon_{online} = 12$, $\mu = 10$ and $\beta = 0.5$. The online parameters for DenStream are set equally to those of HASTREAM$_{LS}$. For HASTREAM$_{IS}$: $h = 3$. For the offline phase of both HASTREAM variants, $minPts = 10$. Figure (7a) illustrates the averaged CMM for time intervals of length 100. HASTREAM$_{LS}$ has slightly higher results than DenStream. HASTREAM$_{IS}$ outperforms both HASTREAM$_{LS}$ and DenStream.

Covertype Data Set. The online parameters of HASTREAM$_{LS}$ are set to $\varepsilon_{online} = 0.2$, $\mu = 10$ and $\beta = 0.5$. The online parameters of DenStream are set equally to HASTREAM$_{LS}$. In HASTREAM$_{IS}$, $h = 3$. The offline parameter of both HASTREAM variants is set to 10. Figure (7b) illustrates the averaged

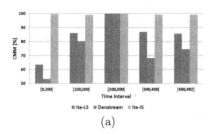 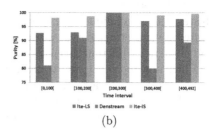

(a)   (b)

Fig. 6. Quality Results for Network Intrusion Dataset: (a) CMM, (b) Purity

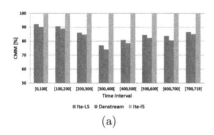 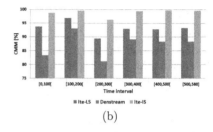

(a)   (b)

Fig. 7. CMM Results for: (a) Physiological Dataset, (b) Covertype Dataset

CMM results for time intervals of length 100. It can be seen that HASTREAM$_{IS}$ outperforms HASTREAM$_{LS}$ and DenStream. The averaged CMM for both HASTREAM variants is over 90% for almost each time interval.

Table 1. Runtime in [ms] and Percentage of each Phase: Physiological Dataset

Algorithm	Timestamps				Runtime [%]	
	200	400	600	719	Online	Offline
Ite-LS	70567	355172	514211	656283	99.04	0.96
DenStream	72209	361969	528425	672252	99.94	0.06
Ite-IS	**17002**	**21793**	**27343**	**30522**	86.71	13.29

Efficiency Evaluation: Covertype Data Set Table (1) shows that both the Ite-LS and Ite-IS are faster than DenStream and that Ite-IS variants has the lowest runtime due to its logarithmic complexity of indexing micro-clusters. Table (1) shows also the percentage of the runtime amount of the online and offline phases respective to the total runtime of each algorithm.

6   Conclusions and Future Work

In this paper, we proposed HASTREAM, a new algorithm for hierarchical density-based clustering on evolving data streams. The presented algorithm is able to detect clusters of different densities, by adapting the density threshold for each cluster, using techniques of hierarchical clustering. The experimental evaluation study on synthetic and real world data sets shows that HASTREAM is able to find clusters of different densities, whereas the competitor DenStream fails to do so. Efficiency and effectiveness experiments shows the superiority of HASTREAM over the state-of-the-art. In the future, the quality of HASTREAM's clusterings could be improved by adding another filtering step to exclude *real* outliers micro-cluster, by discussing e.g. the local outlier factor method.

Acknowledgments. This work has been supported by the UMIC Research Centre, RWTH Aachen University, Germany.

References

1. Aggarwal, C.C., Han, J., Wang, J., Yu, P.S.: A framework for clustering evolving data streams. In: VLDB 2003, pp. 81–92 (2003)
2. Ankerst, M., Breunig, M.M., Kriegel, H.-P., Sander, J.: OPTICS: Ordering Points To Identify the Clustering Structure. SIGMOD Rec. (1999)
3. UCI KDD archive. Data set,
   http://archive.ics.uci.edu/ml/datasets/Covertype
4. Campello, R.J.G.B., Moulavi, D., Sander, J.: Density-based clustering based on hierarchical density estimates. In: Pei, J., Tseng, V.S., Cao, L., Motoda, H., Xu, G. (eds.) PAKDD 2013, Part II. LNCS, vol. 7819, pp. 160–172. Springer, Heidelberg (2013)
5. Cao, F., Ester, M., Qian, W., Zhou, A.: Density-based clustering over an evolving data stream with noise. In: SDM 2006, pp. 328–339 (2006)
6. Dai, B.-R., Huang, J.-W., Yeh, M.-Y., Chen, M.-S.: Adaptive clustering for multiple evolving streams. TKDE 2006 18(9), 1166–1180 (2006)
7. Ester, M., Kriegel, H.-P., Jörg, S., Xu, X.: A density-based algorithm for discovering clusters in large spatial databases with noise. In: KDD 1996, pp. 226–231 (1996)
8. Kranen, P., Assent, I., Baldauf, C., Seidl, T.: Self-adaptive anytime stream clustering. In: ICDM 2009, pp. 249–258 (2009)
9. Kremer, H., Kranen, P., Jansen, T., Seidl, T., Bifet, A., Holmes, G., Pfahringer, B.: An effective evaluation measure for clustering on evolving data streams. In: KDD 2011, pp. 868–876 (2011)
10. MOA. Framework,
    http://moa.cms.waikato.ac.nz/details/stream-clustering/
11. Muller, D.W., Sawitzki, G.: Excess mass estimates and tests for multimodality. Journal of the American Statistical Association, 738–746 (1991)
12. Physiological. Data set,
    http://www.cs.purdue.edu/commugrate/data/2004icml/
13. Prim, R.C.: Shortest connection networks and some generalizations. The Bell Systems Technical Journal, 1389–1401 (1957)
14. Tasoulis, D.K., Ross, G., Adams, N.M.: Visualising the cluster structure of data streams. In: Berthold, M., Shawe-Taylor, J., Lavrač, N. (eds.) IDA 2007. LNCS, vol. 4723, pp. 81–92. Springer, Heidelberg (2007)
15. Wan, L., Ng, W.K., Dang, X.H., Yu, P.S., Zhang, K.: Density-based clustering of data streams at multiple resolutions. ACM Trans. Knowl. Discov. Data, 14:1–14:28 (2009)
16. Zhao, Y., Karypis, G.: Empirical and theoretical comparisons of selected criterion functions for document clustering. Mach. Learn., 311–331 (2004)

Monitoring Distributed Data Streams through Node Clustering

Maria Barouti, Daniel Keren, Jacob Kogan, and Yaakov Malinovsky

University of Maryland Baltimore County, USA and Haifa University, Haifa, Israel
{bmaria2,kogan,yaakovm}@umbc.edu,
dkeren@cs.haifa.ac.il

Abstract. Monitoring data streams in a distributed system is a challenging problem with profound applications. The task of feature selection (e.g., by monitoring the information gain of various features) is an example of an application that requires special techniques to avoid a very high communication overhead when performed using straightforward centralized algorithms.

Motivated by recent contributions based on geometric ideas, we present an alternative approach that combines system theory techniques and clustering. The proposed approach enables monitoring values of an arbitrary threshold function over distributed data streams through a set of constraints applied independently on each stream and/or clusters of streams. The clusters are designed to adapt themselves to the data stream. A correct choice of clusters yields a reduction in communication load. Unlike many clustering algorithms that attempt to collect together similar data items, monitoring requires clusters with *dissimilar* vectors canceling each other as much as possible. In particular, sub–clusters of a good cluster do not have to be good. This novel type of clustering dictated by the problem at hand requires development of new algorithms, and the paper is a step in this direction.

We report experiments on real-world data that detect instances where communication between nodes is required, and show that the clustering approach reduces communication load.

Keywords: data streams, convex analysis, distributed system, clustering.

1 Introduction

In many emerging applications one needs to process a continuous stream of data in real time. Sensor networks [9], network monitoring [6], and real–time analysis of financial data [16], [17] are examples of such applications. Monitoring queries are a particular class of queries in the context of data streams. Previous work in this area deals with monitoring simple aggregates [6], or term frequency occurrence in a set of distributed streams [15]. The current contribution is motivated by results recently reported in [10], [11] where a more general type of monitoring query is described as follows:

P. Perner (Ed.): MLDM 2014, LNAI 8556, pp. 149–162, 2014.
© Springer International Publishing Switzerland 2014

Let $\mathbf{S} = \{\mathbf{s}_1, \ldots, \mathbf{s}_n\}$ be a set of data streams collected at n nodes $\mathbf{N} = \{\mathbf{n}_1, \ldots, \mathbf{n}_n\}$. Let $\mathbf{v}_1(t), \ldots, \mathbf{v}_n(t)$ be d-dimensional, real-valued, time varying vectors derived from the streams. For a function $f : \mathbf{R}^d \to \mathbf{R}$ we would like to monitor the inequality

$$f\left(\frac{\mathbf{v}_1(t) + \ldots + \mathbf{v}_n(t)}{n}\right) > 0 \tag{1}$$

while minimizing communication between the nodes. Often the threshold might be a constant r other than 0. In what follows, for notational convenience, we shall always consider the inequality $f > 0$, and when one is interested in monitoring the inequality $f > r$ we will modify the threshold function and consider $g = f - r$, so that the inequality $g > 0$ yields $f > r$. In e.g. [10,7,8,5] a few real-life applications of this monitoring problem are described; see also Section 2 here.

The difference between monitoring problems involving linear and non-linear functions f is discussed and illustrated by a simple example involving a quadratic function f in [10]. The example demonstrates that, for a non-linear f, it is often very difficult to determine from the values of f at the nodes whether its value evaluated at the average vector is above the threshold or not. The present paper deals with the information gain function (see Section 2 for details), and rather than focus on the values of f we consider the location of the vectors $\mathbf{v}_i(t)$ relative to the boundary of the the the subset of $\mathbf{R}^d$ where f is positive. We denote this set by $\mathbf{Z}_+(f) = \{\mathbf{v} : f(\mathbf{v}) > 0\}$, and state (1) as

$$\mathbf{v}(t) = \frac{\mathbf{v}_1(t) + \ldots + \mathbf{v}_n(t)}{n} \in \mathbf{Z}_+(f). \tag{2}$$

thus, the functional monitoring problem is transformed to the monitoring of a *geometric* condition. As a simple illustration, consider the case of three scalar functions $v_1(t)$, $v_2(t)$ and $v_3(t)$, and the identity function f (i.e. $f(x) = x$). We would like to monitor the inequality

$$v(t) = \frac{v_1(t) + v_2(t) + v_3(t)}{3} > 0$$

while keeping the nodes silent as long as possible. One strategy is to verify the initial inequality $v(t_0) = \dfrac{v_1(t_0) + v_2(t_0) + v_3(t_0)}{3} > 0$ and to keep the nodes silent while

$$|v_i(t) - v_i(t_0)| < \delta = v(t_0), \ t \geq t_0, \ i = 1, 2, 3.$$

The first time t when one of the functions, say $v_1(t)$, crosses the boundary of the local constraint, i.e. $|v_1(t) - v_1(t_0)| \geq \delta$ the nodes communicate, t_1 is set to be t, the mean $v(t_1)$ is computed, the local constraint δ is updated and made available to the nodes. The nodes are kept silent as long as the inequalities

$$|v_i(t) - v_i(t_1)| < \delta, \ t \geq t_1, \ i = 1, 2, 3$$

hold. This type of monitoring was suggested in [13] for a variety of vector norms. The numerical experiments conducted in [13] with the dataset described in Section 5 show that:

1. The number of time instances the mean violates (1) is a small fraction ($< 1\%$) of the number of time instances when the local constraint is violated at the nodes.
2. The lion's share of communications (about 75%) is required because of a single node violation of the local constraint δ.
3. The smallest number of communications is required when one uses the l_1 norm.

We note that if, for example, the local constraint is violated at $\mathbf{n}_1$, i.e. $|v_1(t) - v_1(t_0)| \geq \delta$, and at the same time

$$v_1(t) - v_1(t_0) = -[v_2(t) - v_2(t_0)],$$

while $|v_3(t) - v_3(t_0)| < \delta$ then $|v(t) - v(t_0)| < \delta$, $f(v(t)) > 0$, and update of the mean can be avoided. Separate monitoring of the two node cluster $\{\mathbf{n}_1, \mathbf{n}_2\}$ would require communication involving two nodes only, and could reduce communication load. We aim to extend this idea to the general case – involving many nodes, arbitrary functions, and high-dimensional data.

Clustering in general is a difficult problem, and many clustering problems are known to be NP-complete [4]. Unlike standard clustering that attempts to collect together similar data items [2], we are seeking clusters with *dissimilar data items*, which cancel out each other as much as possible. While sub-clusters of a "classical" good cluster are usually good, this may not be the case when a cluster contains dissimilar objects. These observations indicate that common clustering methods are not applicable to our problem.

A basic attempt to cluster nodes was suggested in [14] with results reported for the dataset presented in Section 5. Clustering together just two nodes reported in [14] reduces communication by about 10%.

In this paper we advance clustering approach to monitoring. The main contribution of this work is twofold:

1. We suggest a specific clustering strategy, and report the communication reduction achieved.
2. We apply the same clustering strategy with l_1, l_2, and l_∞ norms and report the results obtained.

The paper is organized as follows. In Section 2 we present a relevant Text Mining application. Section 3 provides motivation for node clustering. A specific implementation of node clustering is presented in Section 4. Experimental results are reported in Section 5. Section 6 concludes the paper and indicates new research directions.

In the next section we provide a Text Mining related example that leads to a non linear threshold function f.

2   Text Mining Application

Let $\mathbf{T}$ be a textual database (for example a collection of mail or news items). We denote the size of the set $\mathbf{T}$ by $|\mathbf{T}|$. We follow the methodology suggested in [10] and assume that some of the documents are marked as "spam," and a "feature" (word or term for example) is selected. We will be concerned with two subsets of $\mathbf{T}$:

1. $\mathbf{R}$–the set of "relevant" texts (e.g. texts not labeled as "spam"),
2. $\mathbf{F}$–the set of texts that contain a "feature."

We denote complements of the sets by $\overline{\mathbf{R}}$, $\overline{\mathbf{F}}$ respectively (i.e. $\mathbf{R}\cup\overline{\mathbf{R}} = \mathbf{F}\cup\overline{\mathbf{F}} = \mathbf{T}$), and consider the relative size of the four sets $\mathbf{F}\cap\overline{\mathbf{R}}$, $\mathbf{F}\cap\mathbf{R}$, $\overline{\mathbf{F}}\cap\overline{\mathbf{R}}$, and $\overline{\mathbf{F}}\cap\mathbf{R}$ associated with text collection and the "feature" as follows:

$$x_{11}(\mathbf{T}) = \frac{|\mathbf{F}\cap\overline{\mathbf{R}}|}{|\mathbf{T}|}, \; x_{12}(\mathbf{T}) = \frac{|\mathbf{F}\cap\mathbf{R}|}{|\mathbf{T}|},$$

$$x_{21}(\mathbf{T}) = \frac{|\overline{\mathbf{F}}\cap\overline{\mathbf{R}}|}{|\mathbf{T}|}, \; x_{22}(\mathbf{T}) = \frac{|\overline{\mathbf{F}}\cap\mathbf{R}|}{|\mathbf{T}|}. \tag{3}$$

Note that the non negative numbers x_{ij} depend on the "feature",

$$0 \leq x_{ij} \leq 1, \; \text{and} \; x_{11} + x_{12} + x_{21} + x_{22} = 1.$$

The function f is defined on the simplex (i.e. $x_{ij} \geq 0$, $\sum x_{ij} = 1$), and given by

$$\sum_{i,j} x_{ij} \log\left(\frac{x_{ij}}{(x_{i1} + x_{i2})(x_{1j} + x_{2j})}\right), \tag{4}$$

where $\log x = \log_2 x$ throughout the paper. It is well-known that (4) provides the *information gain* for the "feature" (see e.g. [1]).

As an example, we consider n agents installed on n different servers, and a stream of texts arriving at the servers. Let $\mathbf{T}_h = \{\mathbf{t}_{h1}, \dots, \mathbf{t}_{hw}\}$ be the last w texts received at the hth server, with $\mathbf{T} = \bigcup_{h=1}^{n} \mathbf{T}_h$. Note that

$$x_{ij}(\mathbf{T}) = \sum_{h=1}^{n} \frac{|\mathbf{T}_h|}{|\mathbf{T}|} x_{ij}(\mathbf{T}_h),$$

i.e., entries of the global contingency table $\{x_{ij}(\mathbf{T})\}$ are the weighted average of the local contingency tables $\{x_{ij}(\mathbf{T}_h)\}$, $h = 1, \dots, n$.

To check that the given "feature" is sufficiently informative with respect to the target relevance label r, one may want to monitor the inequality

$$f(x_{11}(\mathbf{T}), x_{12}(\mathbf{T}), x_{21}(\mathbf{T}), x_{22}(\mathbf{T})) - r > 0 \tag{5}$$

with f given by (4) while minimizing communication between the servers.

In the next section we provide motivation to node clustering for monitoring data streams.

3 Monitoring Threshold Functions through Clustering: Motivation

In what follows we denote a norm of a vector $\mathbf{v}$ by $\|\mathbf{v}\|$. While the experiments reported in this paper have been conducted with l_1, l_2, and l_∞ norms, the proposed monitoring and node clustering procedures can be applied with any norm. The monitoring strategy proposed in [13] can be briefly described as follows:

Algorithm 31 *Monitoring Threshold Function*

  − *A node is designated as a root* $\mathbf{r}$.
  − *The root sets* $i = 0$.
  − *Until end of stream*
     1. *The root sends a request to each node* $\mathbf{n}$ *for the vectors* $\mathbf{v_n}(t_i)$. *The nodes respond to the root. The root computes the distance* δ *between the mean*
        $$\frac{1}{n} \sum_{n \in N} \mathbf{v}_n(t_i)$$
        and the zero set $\mathbf{Z}_f$ *of the function* f. *The root transmits* δ *to each node.*
     2. *do for each* $\mathbf{n} \in \mathbf{N}$
        If $\|\mathbf{v_n}(t) - \mathbf{v_n}(t_i)\| < \delta$
              the node $\mathbf{n}$ *is silent*
        else
              $\mathbf{n}$ *notifies the root about violation of its local constraint* δ
              the root sets $i = i + 1$
              go to Step 1.
  − *Stop*

An application of the above procedure to data streams generated from the Reuters Corpus RCV1–V2 (see Section 5 for detailed description of the data and experiments) leads to 4006 time instances in which the local constraints are violated, and the root is updated. Results presented in Table 1 show that in 3034 out of 4006 time instances, communications with the root are triggered by constraint violations at exactly one node.

The results immediately suggest to cluster nodes to further reduce communication load. Indeed clustering together, for example, the "longest" vector $\mathbf{v_{n_L}}(t) - \mathbf{v_{n_L}}(t_i)$ with the "shortest" vector $\mathbf{v_{n_S}}(t) - \mathbf{v_{n_S}}(t_i)$ may result in the mean of the two vector cluster being shorter than δ. In such a case communication involving only two rather than all n nodes may prevent mean update for the entire set of nodes in many of the 3034 instances listed in Table 1. Clustering together just two nodes as described above reported in [14] reduces communication by about 10%.

Table 1. Number of local constraint violations simultaneously by k nodes, $r = 0.0025$, l_2 norm, the feature is "bosnia"

# of nodes violators	1	2	3	4	5	6	7	8	9	10
# of violation instances	3034	620	162	70	38	26	34	17	5	0

In this paper we advance the node clustering approach and demonstrate additional communication savings. Each cluster will be equipped with a "coordinator" $\mathbf{c}$ (one of the cluster's nodes). If a cluster node $\mathbf{n}$ violates its local constraint at time t, then the coordinator collects vectors $\mathbf{v_n}(t) - \mathbf{v_n}(t_i)$ from all the nodes in the cluster, computes the mean of the vectors, and checks whether the mean violates the coordinator constraint δ (at this point, node and coordinator constraints are identical). We shall follow [10] and refer to this step as "the balancing process." If the coordinator constraint is violated, the coordinator alerts the root, and the mean of the entire dataset is recomputed by the root (for detailed description of the procedure see Section 4).

A standard clustering problem is often described as "... finding and describing cohesive or homogeneous chunks in data, the clusters" (see e.g. [2]). For the problem at hand we would like to partition the set of nodes $\mathbf{N}$ into k clusters $\Pi = \{\pi_1, \ldots, \pi_k\}$ so that

$$\mathbf{N} = \bigcup_{i=1}^{k} \pi_i, \text{ and } \pi_i \bigcap \pi_j = \emptyset \text{ if } i \neq j.$$

We denote the size of π_i by $|\pi_i|$. If for each cluster π_i one has

$$\frac{1}{|\pi_i|} \left\| \sum_{\mathbf{n} \in \pi_i} [\mathbf{v_n}(t) - \mathbf{v_n}(t_j)] \right\| < \delta, \tag{6}$$

then, due to convexity of any norm, one has

$$\left\| \frac{1}{n} \sum_{\mathbf{n} \in \mathbf{N}} \mathbf{v_n}(t) - \frac{1}{n} \sum_{\mathbf{n} \in \mathbf{N}} \mathbf{v_n}(t_j) \right\| \leq \sum_{i=1}^{k} \frac{|\pi_i|}{n} \left[\frac{1}{|\pi_i|} \left\| \sum_{\mathbf{n} \in \pi_i} [\mathbf{v_n}(t) - \mathbf{v_n}(t_j)] \right\| \right] < \delta.$$

The inequality shows that the "new" mean $\dfrac{1}{n} \sum_{\mathbf{n} \in \mathbf{N}} \mathbf{v_n}(t)$ belongs to $\mathbf{Z}_+(f)$ if the distance from the "old" mean $\dfrac{1}{n} \sum_{\mathbf{n} \in \mathbf{N}} \mathbf{v_n}(t_j)$ to the boundary of this set exceeds δ, and (6) holds for each cluster. We therefore may attempt to define the quality of a k cluster partition Π as

$$Q(\Pi) = \max_i \left\{ \frac{1}{|\pi_i|} \left\| \sum_{\mathbf{n} \in \pi_i} [\mathbf{v_n}(t) - \mathbf{v_n}(t_j)] \right\|, \; i = 1, \ldots, k \right\}. \tag{7}$$

Our aim is to identify k and a k cluster partition Π^o that **minimizes** (7). Our monitoring problem requires to assign nodes $\{\mathbf{n}_{i_1}, \ldots, \mathbf{n}_{i_l}\}$ to the same cluster π so that the total average change within cluster

$$\left\| \frac{1}{|\pi|} \sum_{\mathbf{n} \in \pi} [\mathbf{v_n}(t) - \mathbf{v_n}(t_j)] \right\| \text{ for } t > t_j$$

is minimized, i.e., nodes with **different** variations $\mathbf{v_n}(t) - \mathbf{v_n}(t_j)$ that cancel out each other as much as possible are assigned to the same cluster. Hence, unlike classical clustering procedures, this one needs to combine "dissimilar" nodes together.

The proposed partition quality $Q(\Pi)$ (see (7)) generates three immediate problems:

1. Since the arithmetic mean $\bar{a}$ of a finite set of real numbers $\{a_1, \ldots, a_k\}$ satisfies

$$\min\{a_1, \ldots, a_k\} \leq \bar{a} \leq \max\{a_1, \ldots, a_k\}$$

   the single cluster partition always minimizes $Q(\Pi)$. Considering the entire set of nodes as a single cluster with its own coordinator that communicates with the root introduces an additional unnecessary "bureaucracy" layer that only increases communications. We seek a trade-off which yields clusters with "good" sizes (this is rigorously defined in Section 4).
2. Computation of $Q(\Pi)$ involves future values $\mathbf{v_n}(t)$, which are not available at time t_j when the clustering is performed.
3. Since the communication overhead of the balancing process is proportional to the size of a cluster, the individual clusters' sizes should affect the clustering quality $q(\pi)$ (see (8) below).

In the next section we address these problems.

4   Monitoring Threshold Functions through Clustering: Implementation

We argue that in addition to the average magnitude of the variations $\mathbf{v_n}(t) - \mathbf{v_n}(t_j)$ inside the cluster π, the cluster's size also affects the frequency of updates, and, as a result, the communication load. We therefore define the quality of the cluster π by

$$q(\pi) = \frac{1}{|\pi|} \left\| \sum_{\mathbf{n} \in \pi} [\mathbf{v_n}(t) - \mathbf{v_n}(t_j)] \right\| + \alpha |\pi|, \tag{8}$$

where α is a nonnegative scalar parameter. The quality of the partition $\Pi = \{\pi_1, \ldots, \pi_k\}$ is defined by

$$Q(\Pi) = \max_{i \in \{1, \ldots, k\}} q(\pi_i), \tag{9}$$

When $\alpha = 0$ the partition that minimizes $Q(\Pi)$ is a single cluster partition (that we would like to avoid). When $\max_{\mathbf{n}} \|\mathbf{v_n}(t) - \mathbf{v_n}(t_j)\| \leq \alpha$ the optimal partition is made up of n singleton clusters. In this paper we focus on

$$0 < \alpha < \max_{\mathbf{n} \in \mathbf{N}} \|\mathbf{v_n}(t) - \mathbf{v_n}(t_j)\|. \tag{10}$$

The constant α depends on t and t_j, and below we show how to avoid this dependence.

Computation of $Q(\Pi)$ required for the clustering procedure is described below. In order to compute $Q(\Pi)$ at time t_j one needs to know $\mathbf{v_n}(t)$ at future times $t = t_j + 1, t_j + 2, \ldots$ which are not available (we recall that t_j denotes time instances when the mean of the entire data set is updated). While the future behavior is not known, we shall use past values of $\mathbf{v_n}(t)$ for prediction. For each node $\mathbf{n}$ we build "history" vectors $\mathbf{h_n}(t_j)$ defined as follows:

1. $\mathbf{h_n}(t_0) = 0$
2. if ($\mathbf{h_n}(t_j)$ is already available)
   $$\mathbf{h_n}(t_{j+1}) = \mathbf{h_n}(t_j)$$
   for t increasing from t_j to t_{j+1} do
   $$\mathbf{h_n}(t_{j+1}) = \frac{1}{2}\mathbf{h_n}(t_{j+1}) + [\mathbf{v_n}(t) - \mathbf{v_n}(t_j)]$$

The vectors $\mathbf{h_n}(t_j)$ accumulate the history of changes, with older changes assigned smaller weights. In this paper we arbitrary use $\frac{1}{2}$ as the the weight. It is clear that other values can be used as weights, furthermore the weights do not have to be constants, and may reflect known history of changes.

We shall use the vectors $\{\mathbf{h_n}(t_j)\}$ to generate a node partition at time t_j. We note that normalization of the vector set that should be clustered does not change the induced optimal partitioning of the nodes. When the vector set is normalized by the magnitude of the longest vector in the set, the range for α conveniently shrinks to $[0, 1]$. In what follows we set $h = \max_{\mathbf{n}} \|\mathbf{h_n}(t_j)\|$, assume that $h > 0$, and describe a "greedy" clustering procedure for the normalized vector set

$$\{\mathbf{a}_1, \ldots, \mathbf{a}_n\}, \ \mathbf{a}_i = \frac{1}{h}\mathbf{h_{n_i}}(t_j), \ i = 1, \ldots, n.$$

We start with the n cluster partition Π^n (each cluster is a singleton). If a k cluster partition Π^k, $k > 2$ is already available we

1. identify the partition cluster π_j with maximal norm of its vectors' mean, i.e.,

$$\frac{1}{|\pi_j|}\left\|\sum_{\mathbf{a}\in\pi_j}\mathbf{a}\right\| \geq \frac{1}{|\pi_i|}\left\|\sum_{\mathbf{a}\in\pi_i}\mathbf{a}\right\|, \ i = 1, \ldots, n.$$

2. identify cluster π_i so that the merger of π_i with π_j produces a cluster of smallest possible quality, i.e.,

$$q\left(\pi_j \bigcup \pi_i\right) \leq q\left(\pi_j \bigcup \pi_l\right), \ l \neq j,$$

where cluster's quality is defined by (8).

The partition Π^{k-1} is obtained from Π^k by merging clusters π_j and π_i. The final partition is selected from the $n-1$ partitions $\{\Pi^2, \ldots, \Pi^n\}$ as the one that minimizes Q.

Note that node constraints δ do not have to be equal. Taking into account the distribution of the data streams at each node can further reduce communication.

We illustrate this statement by a simple example involving two nodes. If, for example, there is reason to believe that the inequality

$$2\|\mathbf{v}_1(t) - \mathbf{v}_1(t_i)\| \le \|\mathbf{v}_2(t) - \mathbf{v}_2(t_i)\| \tag{11}$$

always holds, then the number of node violations may be reduced by imposing node dependent constraints

$$\|\mathbf{v}_1(t) - \mathbf{v}_1(t_i)\| < \delta_1 = \frac{2}{3}\delta, \text{ and } \|\mathbf{v}_2(t) - \mathbf{v}_2(t_i)\| < \delta_2 = \frac{4}{3}\delta$$

so that the wider varying stream at the second node enjoys larger "freedom" of change, while the inequality

$$\left\|\frac{\mathbf{v}_1(t) + \mathbf{v}_2(t)}{2} - \frac{\mathbf{v}_1(t_i) + \mathbf{v}_2(t_i)}{2}\right\| < \frac{\delta_1 + \delta_2}{2} = \delta$$

holds true. Assigning "weighted" local constraints requires information provided by (11). With no additional assumptions about the stream data distribution this information is not available. Unlike [12] we refrain from making assumptions regarding the underlying data distributions; instead, we estimate the weights through past values $\|\mathbf{v}_j(t) - \mathbf{v}_j(t_i)\|$.

1. Start with the initial set of weights

$$w_1 = \ldots = w_n = 1 \text{ and } W_1 = \ldots = W_n = 1 \tag{12}$$

(so that $\sum_{j=1}^{n} w_j = \sum_{j=1}^{n} W_j = n$).

2. As new texts arrive at the next time instance t, each node computes updates

$$W_j = \frac{1}{2}W_j + \|\mathbf{v}_j(t) - \mathbf{v}_j(t_i)\|, \text{ with } W_j(t_0) = 1, \ j = 1, \ldots, n.$$

When at time t_{i+1} the root constraint $\delta(\mathbf{r})$ needs to be updated, each node $\mathbf{n}_j$ broadcasts W_j to the root. The root computes $W = \sum_{j=1}^{n} W_j$, and transmits the updated $\delta(\mathbf{n}_j) = w_j\delta(\mathbf{r})$ where $w_j = n \times \frac{W_j}{W}$ (so that $\sum_{j=1}^{n} w_j = n$) back to node j. For a coordinator $\mathbf{c}$ of a node cluster π the constraint $\delta(\mathbf{c}) = \frac{1}{|\pi|}\sum_{\mathbf{n}\in\pi}\delta(\mathbf{n})$.

5   Experimental Results

The data streams analyzed in this section are generated from the Reuters Corpus RCV1–V2. The data is available from http://leon.bottou.org/projects/sgd

and consists of 781, 265 tokenized documents with document ID ranging from 2651 to 810596. We simulate n streams by arranging the feature vectors in ascending order with respect to document ID, and selecting feature vectors for the stream in the round-robin fashion.

In the Reuters Corpus RCV1–V2 each document is labeled as belonging to one or more categories. We label a vector as "relevant" if it belongs to the "CORPORATE/INDUSTRIAL" ("CCAT") category, and "spam" otherwise. Following [10] we focus on three features: "bosnia," "ipo," and "febru." Each experiment was performed with 10 nodes, where each node holds a sliding window containing the last 6,700 documents it received.

First we use 67, 000 documents to generate initial sliding windows. The remaining 714, 265 documents are used to generate datastreams, hence the selected feature information gain is computed 714, 265 times. Based on all the documents contained in the sliding window at each one of the 714, 266 time instances we compute and graph 714, 266 information gain values for the feature "bosnia" (see Figure 1).

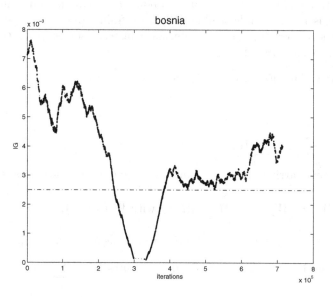

Fig. 1. Information gain values for the feature "bosnia"

For the experiments described below, the threshold value r is predefined, and the goal is to monitor the inequality $f(\mathbf{v}) - r > 0$ while minimizing communication between the nodes.

We assume that new texts arrive simultaneously at each node. The numerical experiment reported in [13] with the feature "febru," and the threshold $r = 0.0025$ are shown in Table 2 where a broadcast is defined as one time transmission of information between different nodes. We run the node clustering monitoring presented in this paper for the same feature and threshold with

Table 2. Number of mean computations, and broadcasts for feature "febru" with threshold $r = 0.0025$, no clustering

norm	mean updates	broadcasts
l_1	2591	67388
l_2	3140	81650
l_∞	3044	79144

Table 3. Number of root and coordinator mean computations, and total broadcasts for feature "febru" with threshold $r = 0.0025$ with clustering

norm	alpha	root mean update	coordinator mean update	total broadcasts
l_1	0.70	1431	0	38665
l_2	0.80	1317	0	35597
l_∞	0.65	1409	0	38093

$\alpha = 0.05, 0.10, \ldots, 0.95$. The best results for l_1, l_2, and l_∞ norms with respect to α are presented in Table 3. The clustering approach in this case is particularly successful – coordinators' constraints are not violated, and the root mean updates are decreased significantly. As a result the number of broadcasts decreases by about 50%.

Next we turn to the features "ipo" and "bosnia." In both cases we run monitoring with clustering, allowing $\alpha = 0.05, 0.10, \ldots, 0.95$, and report results with the lowest number of broadcasts. The results obtained for "ipo" without clustering are presented in Table 4. Application of clustering procedure leads to a significant reduction in the number of broadcasts. Results obtained through clustering procedure are shown in Table 5. The table demonstrates significant inside cluster activity, and a decrease in root mean updates.

Table 4. Number of mean computations, and broadcasts for feature "ipo" with threshold $r = 0.0025$, no clustering

norm	mean updates	broadcasts
l_1	15331	398606
l_2	21109	548834
l_∞	19598	509548

Finally we turn to the feature "bosnia." Application of clustering to monitoring this feature information gain appears to be far less successful. Results obtained without clustering in [13] are presented in Table 6. Application of the clustering procedure leads to a slight decrease in the number of broadcasts in case of the l_2 and l_∞ norms (see Table 7). In case of the l_1 norm, the number of broadcasts increases. Clustering does not offer a universal remedy; in some

Table 5. Number of root and coordinator mean computations, and total broadcasts for feature "ipo" with threshold $r = 0.0025$ with clustering

norm	alpha	root mean update	coordinator mean update	total broadcasts
l_1	0.15	5455	829	217925
l_2	0.10	7414	1782	296276
l_∞	0.10	9768	2346	366300

Table 6. Number of mean computations, and broadcasts, for feature "bosnia" with threshold $r = 0.0025$, no clustering

norm	mean updates	broadcasts
l_1	3053	79378
l_2	4006	104156
l_∞	3801	98826

Table 7. Number of root and coordinator mean computations, and total broadcasts for feature "bosnia" with threshold $r = 0.0025$ and clustering

norm	alpha	root mean update	coordinator mean update	total broadcasts
l_1	0.65	3290	2	89128
l_2	0.55	3502	7	97602
l_∞	0.60	3338	2	91306

cases better performance is achieved with no clustering (by keeping α between 0.05 and 0.95 we force nodes to cluster).

6   Conclusions and Future Research Directions

In this paper we propose to monitor threshold functions over distributed data streams through clustering nodes whose data fluctuations "cancel out" each other. The strategy, if successful, in many cases reduces the communication required to message exchanges within a cluster only, yielding overall communication reduction.

The clustering strategy suggested is based on minimization of a combination of average of a vector associated with a cluster and the cluster size. The nodes are re–clustered each time the entire dataset mean violates its constraint $\delta(\mathbf{r})$. The amount of communication required depends on the "trade off" parameter $0 < \alpha < 1$ selected at the beginning of the monitoring process. While the results obtained show improvement over previously reported ones that do not use clustering [13] it is of interest to introduce an update of α based on the monitoring history each time nodes are re–clustered (see e.g. [3] for feedback theory approach).

Clustering does not provide a universal remedy. It is of interest to identify data streams that benefit from clustering, and those for which clustering does not reduce communication load in any significant fashion. Finally a methodology that measures effectiveness of various monitoring techniques should be introduced, so that different monitoring strategies can be easily compared.

Acknowledgment. The authors thank the reviewers for the valuable remarks that much improved exposition of the results. The research of the second author was supported by Grant No. 2008405 from the United States-Israel Binational Science Foundation (BSF). The work of the fourth author was partially supported by a 2013 UMBC Summer Faculty Fellowship grant.

References

1. Gray, R.M.: Entropy and Information Theory. Springer, New York (1990)
2. Mirkin, B.: Clustering for Data Mining: A Data Recovery Approach. Chapman & Hall/CRC, Boca Raton (2005)
3. Willems, J.C.: The Analysis of Feedback Systems. The MIT Press, Cambridge (1971)
4. Brucker, P.: On the complexity of clustering problems. Lecture Notes in Economics and Mathematical Systems, vol. 157, pp. 45–54 (1978)
5. Burdakis, S., Deligiannakis, A.: Detecting outliers in sensor networks using the geometric approach. In: ICDE, pp. 1108–1119 (2012)
6. Dilman, M., Raz, D.: Efficient reactive monitoring. In: Proceedings of the Twentieth Annual Joint Conference of the IEEE Computer and Communication Societies, pp. 1012–1019 (2001)
7. Gabel, M., Schuster, A., Keren, D.: Communication-efficient outlier detection for scale-out systems. In: BD3@VLDB, pp. 19–24 (2013)
8. Garofalakis, M., Keren, D., Samoladas, V.: Sketch-based geometric monitoring of distributed stream queries. In: PVLDB (2013)
9. Madden, S., Franklin, M.J.: An architecture for queries over streaming sensor data. In: ICDE 2002, p. 555 (2002)
10. Sharfman, I., Schuster, A., Keren, D.: A Geometric Approach to Monitoring Threshold Functions over Distributed Data Streams. ACM Transactions on Database Systems 32, 23:1–23:29 (2007)
11. Sharfman, I., Schuster, A., Keren, D.: A Geometric Approach to Monitoring Threshold Functions over Distributed Data Streams. In: May, M., Saitta, L. (eds.) Ubiquitous Knowledge Discovery. LNCS, vol. 6202, pp. 163–186. Springer, Heidelberg (2010)
12. Keren, D., Sharfman, I., Schuster, A., Livne, A.: Shape Sensitive Geometric Monitoring. IEEE Transactions on Knowledge and Data Engineering 24, 1520–1535 (2012)
13. Kogan, J.: Feature Selection over Distributed Data Streams through Convex Optimization. In: Proceedings of the Twelfth SIAM International Conference on Data Mining (SDM 2012), pp. 475–484. SIAM (2012)

14. Kogan, J., Malinovsky, Y.: Monitoring Threshold Functions over Distributed Data Streams with Clustering. In: Proceedings of the Workshop on Data Mining for Service and Maintenance (held in conjunction with the 2013 SIAM International Conference on Data Mining), pp. 5–13 (2013)
15. Manjhi, A., Shkapenyuk, V., Dhamdhere, K., Olston, C.: Finding (recently) frequent items in distributed data streams. In: ICDE 2005, pp. 767–778 (2005)
16. Yi, B.-K., Sidiropoulos, N., Johnson, T., Jagadish, H.V., Faloutsos, C., Biliris, A.: Online datamining for co–evolving time sequences. In: ICDE 2000 (2000)
17. Zhu, Y., Shasha, D.: Statestream: Statistical monitoring of thousands of data streamsin real time. In: VLDB, pp. 358–369 (2002)

Minimizing Cluster Errors in LP-Based Nonlinear Classification

Orestes G. Manzanilla-Salazar[1], Jesús Espinal-Kohler[2],
and Ubaldo M. García-Palomares[3]

[1] Universidad Simón Bolívar,
Dep. Procesos y Sistemas, Caracas 89000
[2] Universidad Nacional Abierta, Coord. Ingeniería de Sistemas, Caracas 1010
[3] Universidade de Vigo,
Dep. Ingeniería Telemática, Vigo 36310

Abstract. Recent work has focused on techniques to construct a learning machine able to classify, at any given accuracy, all members of two mutually exclusive classes. Good numerical results have been reported; however, there remain some concerns regarding prediction ability when dealing with large data bases. This paper introduces clustering, which decreases the number of variables in the linear programming models that need be solved at each iteration. Preliminary results provide better prediction accuracy, while keeping the good characteristics of the previous classification scheme: a piecewise (non)linear surface that discriminates individuals from two classes with an a priori classification accuracy is built and at each iteration, a new piece of the surface is obtained by solving a linear programming (LP) model. The technique proposed in this work reduces the number of LP variables by linking one error variable to each cluster, instead of linking one error variable to each individual in the population. Preliminary numerical results are reported on real datasets from the Irvine repository of machine learning databases.

1 Introduction

The Binary Classification Problem (BCP) can be defined as the task of building a discriminating function, also denoted as discriminating surface, decision rule or learning rule $h(\cdot) : R^n \to R$ that separates *individuals* belonging to two disjoint classes, say, class $\mathcal{P}$ and class $\mathcal{N}$. BCPs have applications in many areas, like SPAM categorization [8], texture recognition [3], gene expression analysis [4], text categorization [14], as well as in areas related to marketing, accounting, advertising, sales and other domains [6, 7, 13, 15, 19–21], including medical diagnosis [1, 5, 17]. We assume that any individual $x \in R^n$ is identified by its n features $x^1, \ldots, x^n$ and either belongs to $\mathcal{P}$ or $\mathcal{N}$. The focus of our method is on solving non-linearly separable patterns, without any previous knowledge about probability distributions concerning the structure of the data.

This paper proposes a variant of the technique explained in [12], which is in turn based on the Multi Surface Methods (MSMs) developed by Mangasarian *et al* [16–18]. They construct the discriminating function $h(\cdot)$ with a piecewise linear approach where each piece is found by formulating at each iteration a linear programming (LP) model

P. Perner (Ed.): MLDM 2014, LNAI 8556, pp. 163–174, 2014.

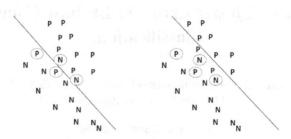

Fig. 1. The hyperplane found by an MSM on the left shows only 4 misclassified elements (encircled). The shifted hyperplane on the right classifies 7 elements that belong to $\mathcal{P}$.

that finds the parameters $(w \in R^n, \theta \in R)$ of an (hyper)plane $w^T x = \theta$. To achieve this result, they must incorporate into the model the constraints

$$
\begin{aligned}
(w^T x_k - \theta) + y^k \geq 0, & \quad \text{for all } x_k \in \mathcal{P}, \\
(w^T x_k - \theta) - y^k \leq 0, & \quad \text{for all } x_k \in \mathcal{N},
\end{aligned}
\tag{1}
$$

and must define an objective function that *somehow* tries to mitigate the errors $y^k, k = 1, ..., m$ that may be present to all individuals. This hyperplane is shifted to ensure that all classified elements belong to the same set and are on the same side of the hyperplane (See Figure 1).

MSMs have several shortcomings: their models need to consider all individuals in R^n, which are each linked to a variable that denotes its classification error. Furthermore the (hyper)plane obtained is not necessarily the optimum. To alleviate these difficulties García and Manzanilla [12] suggested the asymmetric MSM (aMSM), which at the i-th iteration deals with the errors generated by a a subset $\mathcal{S}_i \subseteq R^n$ of individuals of the same class. There are no error variables associated to the rest of the elements. Assuming that we take $\mathcal{S} \subseteq \mathcal{P}$, constraints (1) are reduced to

$$
\begin{aligned}
(w^T x_k - \theta) + y^k \geq 0, & \text{ for all } x_k \in \mathcal{S} \subseteq \mathcal{P}, \\
(w^T x_k - \theta) \leq 0, & \quad \text{ for all } x_k \in \mathcal{N},
\end{aligned}
\tag{2}
$$

which clearly includes fewer constraints and fewer variables than the former system (1). Besides, no hyperplane shifting is needed. In the example depicted in Figure 1 the (hyper)plane on the right is obtained at once from the model.

This paper suggests the asymmetric clustering MSM. The individuals are grouped in clusters and all members of the cluster are forced to share the same variable. This approach can considerably reduce the number of variables in the LP model. Despite that most clustering techniques are computationally demanding, the particular structure of the task allows the use of an LP-based clustering technique, keeping the problem computationally simple. It is inspired on the models proposed in [23] and the ideas exposed in [24]. In order to make this paper self content we end this section with a succinct description of the aMSM.

The rest of the paper is organized as follows: Section 2 describes the optimization model used to find the discriminating surfaces, the clustering method and the algorithm that builds the piecewise surface. In section 3 preliminary numerical results on real databases are shown. Finally, section 4 ends the paper, with additional comments and remarks concerning future work.

1.1   Notation and Description of the aMSM

We stick to the notation given in [12]: R^n denotes the n-th dimensional Euclidean space, lowercase Latin letters are vectors in R^n, x_i^k is the k-th component of x_i, uppercase cursive letters $\mathcal{P}, \mathcal{N}$, etc., denote finite sets in R^n. Given the sets $\mathcal{D}, \mathcal{H}$ the difference set $\mathcal{D} - \mathcal{H}$ consists of those members of $\mathcal{D}$ that do not belong to $\mathcal{H}$. Lower-case Greek letters denote scalars. The rest of the notation is standard. We recall that the subscript i refers to the iteration number, which is dropped when it is clear from the context.

An individual x is described by its n features $x^1, ..., x^n$, and its class indicator will be defined as:

$$c(x) = \begin{cases} 1, x \in \mathcal{P} \\ -1, x \in \mathcal{N}. \end{cases}$$

To simplify the notation we denote $c^k \stackrel{def}{=} c(x_k)$. Also, for convenience we define a class indicator, not only for individuals, but for a given set $\mathcal{S}$,

$$c(\mathcal{S}) = \begin{cases} 1, \text{ if } \mathcal{S} \subseteq \mathcal{P} \\ -1, \text{ if } \mathcal{S} \subseteq \mathcal{N} \\ 0, \text{ otherwise} \end{cases}$$

The BCP can then be defined as the task of finding a discriminating function $h(.)$: $\mathcal{R}^n \to R$, that satisfies

$$h(x) \begin{cases} > 0, \text{ if } x \in \mathcal{P} \\ < 0, \text{ if } x \in \mathcal{N} \end{cases}$$

In aMSM we are given the q basis functions $g^1(\cdot), ..., g^q(\cdot) : R^n \to R$. We define $h_i(x) = w_i^T g(x) - \theta_i$, where $(w_i \in R^q, \theta_i \in R), i = 1, \ldots, t$ are found by sequentially solving model (3). We build $h(\cdot)$ with the non linear pieces $h_i(x) = 0, i = 1, \ldots, t$. These hyper surfaces can be considered as hyperplanes in the q dimensional space spanned by the basis functions. This connotation is of common use with kernels in SVMs and also adopted in this paper.

At the i-th iteration the set $\mathcal{R}_i$ of individuals remaining to be classified is known. The algorithm chooses a subset $\mathcal{S}_i \subseteq \mathcal{R}_i$ of individuals belonging to the same class and $\mathcal{X}_i$ is defined as $\mathcal{X}_i = \{x \in \mathcal{R}_i : c(x) = -c(\mathcal{S}_i)\}$. The model (3) finds a hyperplane $h_i(x)$ and the error variables y_i^k, each associated with a $x^k \in \mathcal{S}_i$. The model attempts to place most individuals from $\mathcal{S}_i$ on one side of the hyperplane, while forcing all $x \in \mathcal{X}_i$

to be on the other side. Finally, The aMSM considers as well classified all individuals in $\mathcal{Q}_i = \{x \in \mathcal{R}_i : c(\mathcal{S}_i)h_i(x) > 1\}$.

$$
\min_{w \in R^q, \, \theta \in R, \, y \in R^{|\mathcal{S}_i|}} \sum_{x_k \in \mathcal{S}_i} y^k
$$
$$
\begin{aligned}
c(\mathcal{S}_i)(w^T g(x_k) - \theta) + y^k &\geq 1, \, x_k \in \mathcal{S}_i \\
-c(\mathcal{S}_i)(w^T g(x_k) - \theta) &\geq 1, \, x_k \in \mathcal{X}_i \\
w \neq 0, \, y &\geq 0 \\
\| w \|_\infty \leq 10, \, |\theta| &\leq 100
\end{aligned}
\tag{3}
$$

An algorithmic scheme of the aMSM is given in Figure 2. The reader is referred to [12] for a complete description and implementation details.

> 1. **Input:** Training set $\mathcal{T}$, basis $g^1(\cdot), \ldots, g^q(\cdot)$
> 2. initial values: $i = 1, \mathcal{R}_1 = \mathcal{T}$
> 3. WHILE $\left((\mathcal{R}_i \cap \mathcal{P} \neq \emptyset) \text{ AND } (\mathcal{R}_i \cap \mathcal{N} \neq \emptyset) \right)$
> 4.   Choose $\mathcal{S}_i \neq \emptyset$:
> $\left(\mathcal{S}_i \subseteq (\mathcal{R}_i \cap \mathcal{P}) \right) \text{ OR } \left(\mathcal{S}_i \subseteq (\mathcal{R}_i \cap \mathcal{N}) \right)$
> 5.   Define $\mathcal{X}_i = \{x \in \mathcal{R}_i : c(x) = -c(\mathcal{S}_i)\}$
> 6.   Let w_i, θ_i, y_i^k be the solution of model (3)
> 7. $\mathcal{Q}_i = \{x \in \mathcal{R}_i : c(\mathcal{S}_i)h_i(x) > 1\}$
> 8. $\mathcal{R}_{i+1} = \mathcal{R}_i - \mathcal{Q}_i$
> 9.   Next iteration: $i = i + 1$
> 10. END WHILE
> 11. **Output:** $c(\mathcal{S}_i), w_i, \theta_i, y_i^k, i = 1, \ldots, t$

Fig. 2. aMSM

Both the aMSM, as well as the algorithm proposed in this paper, may end as soon as a classification accuracy, predetermined by the practitioner, is achieved. We just have to include this termination criterion in line 3 of Figure 2. It is worth mentioning that an explicit close form for the discriminating function $h(\cdot)$ is provided by aMSM and the algorithm proposed in the next section. We assert that

Proposition 1. *Provided $\mathcal{Q}_i \neq \emptyset, i = 1, \ldots, t$, the aMSM described in Figure 2 gives rise to the discriminating function:*

$$
h(x) = \sum_{i=1}^{t} \frac{\mathrm{u}(c(\mathcal{S}_i)h_i(x))}{2^i c(\mathcal{S}_i)} - \frac{1}{2^{t+1} c(\mathcal{S}_t)}, \tag{4}
$$

where $\mathrm{u}(\cdot) : R \to R^+$ is the Heaviside step function, namely: $\mathrm{u}(\alpha) = 1$, for $\alpha > 0$; $\mathrm{u}(\alpha) = 0$, for $\alpha \leq 0$.

Proof. Let us prove that $\left. \begin{array}{l} x \in \mathcal{T}, \\ c(x) = 1 \end{array} \right\} \Rightarrow (h(x) > 0)$.

If $x \notin (\mathcal{Q}_1 \cup \cdots \cup \mathcal{Q}_t)$ then $c(\mathcal{S}_t) = -1$ and $\mathrm{u}(c(\mathcal{S}_i)h_i(x)) = 0, i = 1, \ldots, t$. We conclude that $h(x) = 1/2^{t+1} > 0$.

If $\exists j : x \in \mathcal{Q}_j$ we have that

$$\mathrm{u}(c(\mathcal{S}_i)h_i(x)) = 0, i = 1, \ldots, j-1, \text{ and}$$

$$h(x) > \frac{1}{2^j} - \sum_{i>j} \frac{1}{2^i} > 0.$$

We also need to prove that

$$\left.\begin{array}{l} x \in \mathcal{T}, \\ c(x) = -1 \end{array}\right\} \Rightarrow (h(x) < 0);$$

but the proof is almost identical to the previous case and it is omitted. $\square$

2   Proposed Method

The vector y^k for $x_k \in \mathcal{S}_i$ in model (3) constitutes the error variables and has $|\mathcal{S}_i|$ components. The main goal is to formulate a new LP model for obtaining each piece of the discriminant surface, using fewer error variables. This approach might be render useful for dealing with larger databases.

Let $p \in \{1, \ldots, |\mathcal{S}|\}$ be the number of error variables to be considered. When $p = 1$, the model defines a unique error variable, which is shared by all $x_k \in \mathcal{S}$. On the other hand, when $p = |\mathcal{S}|$, each $x_k \in \mathcal{S}$ is linked with its error $y^k, k = 1, \ldots, |\mathcal{S}|$.

When $1 < p < |\mathcal{S}|$, we identify p clusters and inspired by [24], we force all individuals of the same cluster to share the same error variable. Geometrically, we add up the ∞−norm of the error variables linked to every cluster. When all individuals belonging to the same cluster reside on the same side of the hyperplane generated by the algorithm, the error variable of the cluster drops to zero. We should mention that our model differs from those in [24], as the direct application of those models would lead to a bigger LP model.

Three important questions may arise: (1) how do we formulate an LP that minimizes the ∞-norm of the errors of each individual inside a cluster? (2) how do we choose which individuals are assigned to each cluster? and (3) can the clustering be done in polynomial time, so the overall computational complexity is not severely augmented in an algorithm essentially restricted to solve LP models?

We will elaborate on these questions in the remainder of this section. To begin with, let us graphically show the task described: in Figure 3, assuming $p = 5$ as the number of clusters, the set $\mathcal{S} = \mathcal{P}$ is chosen and each individual in $\mathcal{S}$ is assigned to a cluster, represented inside a geometric figure. The LP optimization model (5) is solved and a surface is generated (see Figure 4). Only members from $\mathcal{P}$ are located on one side of the surface, and will not be considered again by the algorithm.

2.1   LP Model

As SVMs rely on the use of a kernel to separate patterns with non-linear separability, it has become a common interest to reduce the size of the feature space. This issue motivated, among others, the F_∞-norm SVM proposed by Wu and Zhou [24]. They

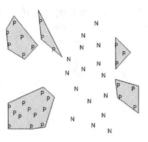

Fig. 3. clusters of individuals in $\mathcal{S} = \mathcal{P}$

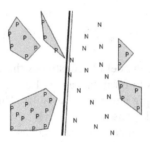

Fig. 4. The separating hyperplane obtained by solving model 5, classifies individuals from $\mathcal{P}$

gather the features by performing a clustering pre-process. In general, a group is formed with features naturally related, and it is assumed that if a feature is not relevant for the classification, the remainder of the features in the same group are as well irrelevant. This can result in the simultaneous elimination of a group of features. Wu and Zhou proposed to minimize the $\infty-$norms of the vector of weights belonging to the same feature-group, instead of the norm of the whole vector of weights, as it is usually done.

Inspired by this approach, we use the same line of thought: if a group of individuals share similar features, they could have a similar error value. This approach leads to a minimization of the ∞-norm of error variables in groups. However, the direct implementation of the LP model in [24] would imply a bigger problem than the one formulated in aMSM. Adapting formulations in [24] to model (3) would add p variables and p constraints. Given that our goal is to achieve a lighter formulation, we instead propose model (5), which has fewer variables than (3).

$$\min_{w \in R^q, \, \theta \in R, \, y \in R^p} \sum_{x_k \in \mathcal{S}_i} y^k$$
$$c(\mathcal{S}_i)(w^T g(x_k) - \theta) + y^1 \geq 1, \, x_k \in \mathcal{G}_i^1$$
$$\vdots \, \vdots$$
$$c(\mathcal{S}_i)(w^T g(x_k) - \theta) + y^p \geq 1, \, x_k \in \mathcal{G}_i^p$$
$$-c(\mathcal{S}_i)(w^T g(x_k) - \theta) \geq 1, \, x_k \in \mathcal{X}_i$$
$$w \neq 0, \, y \geq 0$$
$$\| w \|_\infty \leq 10, \, |\theta| \leq 100,$$
$$\tag{5}$$

where p is chosen according to the LP solver limitations, and the sets of individuals $\mathcal{G}_i^1, \ldots, \mathcal{G}_i^p$ are the clusters, which can be defined by either the clustering procedure to be introduced in Section 2.2, or in a natural way as exemplified in Section 2.3. Note that $\mathcal{G}_i^1, \ldots, \mathcal{G}_i^p$ denote the grouping of individuals, which may be changed at each iteration. We assume that $\mathcal{G}_i^1 \cup \ldots \cup \mathcal{G}_i^p = \mathcal{S}_i$. As y is of size p, the model has only p error variables, which is in general less than $|\mathcal{S}_i|$. The LP model (5) is bounded from below and feasible for appropriate bounds on w and θ. It is therefore solvable.

The task performed by model (5) is graphically shown in figure 5. The figure assumes $p = 4$. Most elements from the first cluster, whose error variable is positive, are correctly classified by the hyperplane $h(x) = w^T g(x) - \theta = 0$, except for one element (marked with a circle), whose distance to the plane $h(x) = w^T x - \theta = 0$ defines the value of y^1, which is shared by the rest of the elements in the cluster. All elements from the second cluster are correctly classified, and therefore its linked error variable drops to zero. The third and fourth clusters have positive error variables $y^3, y^4 > 0$, proportional to the projection of the farthest element (also marked with circles) to the hyperplane $h(x) = w^T x - \theta = 0$. For most elements in clusters with positive errors, the shared error value is bigger than the value needed to make their constraints feasible. In aMSM the error variable y^k is what x_k needs to satisfy its own constraint.

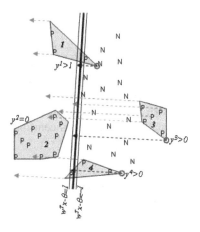

Fig. 5. Minimization of the ∞-norm of errors from each cluster

2.2 Clustering Method

Given p, our task is to form the *clusters* $\mathcal{G}^1, \ldots, \mathcal{G}^p$. Espinal-Kohler [9] ran tests with randomly chosen *clusters* and obtained discouraging results. He concluded that some criteria is necessary to group individuals in a cluster. As the number of clusters in our approach is unrelated to common clustering criteria, we adopt the LP-based clustering technique proposed in [23], which does not represent an excessive computational burden to our method.

This leaves us only with the problem of finding *good* centroids for each cluster. This task is common in the use of radial based classifiers. In general, it is desired to find *good* centroids, in the sense of being *representative* of similar individuals. In [23], the task is done by selecting the p least similar data. This has two shortcomings: it requires to find the maximum *distances* between individuals, which is a computational burden; and the number of clusters for which the approach works well is constrained by the dimensionality of the data.

Fung et al [11] proposed a random choice of centroids with excellent results. In the numerical tests made in [9], random selection in the LP assignment model gave much better results than the least similar test suggested by [23]. Figure 6 describes our approach.

Step 1. Randomly select p centroids among individuals in $\mathcal{S}$.
Step 2. Calculate some measure of distance between each individual and each of the
   p centroids.
Step 3. Solve the LP assignment model in [23], with no upper bound to the number
   of points that can be assigned to each cluster.

Fig. 6. Clustering procedure

The linear assignment model requires $p(|\mathcal{S}| - p)$ variables, and $p + |\mathcal{S}|$ linear constraints [23, Section 4.2]. If the model size exceeds the solver limitations, the feature-space can be split into n-dimensional boxes, making sure each box contains an acceptable number of variables. The LP-based clustering can then be performed on each box. It is highly recommended to assign to each box a number of centroids proportional to the data enclosed by the box. Finally, the number of centroids should not surpass p.

2.3 Natural Clustering

In many examples, datasets are formed by individuals that belong to $p + 1$ mutually exclusive classes $\mathcal{N}_0, \mathcal{N}_1, \ldots, \mathcal{N}_p$. Intuitively, individuals from the same class must have *similar* attributes and may therefore be grouped as a *cluster*. We use this idea in several datasets from the Irvine repository and carried out numerical tests with promising results. We denote $\mathcal{P} = \mathcal{N}_0$ and use the algorithm to find a discriminating function $h(\cdot) : \mathcal{R}^n \to R$ that satisfies

$$h(x) \begin{cases} > 0, x \in \mathcal{P} \\ < 0, x \in (\mathcal{N}_1 \cup \mathcal{N}_2 \cup \cdots \cup \mathcal{N}_p) \end{cases}$$

2.4 Algorithm

Our algorithm inherits many of the aMSM features, and has a very close description. We just have to change lines 1 and 6 of Figure 2 by ac1 and the IF-ELSE instruction ac6. As shown in Figure 7 the algorithm needs the preassigned number of error variables, that is, p, as input and the LP optimization model (3) is solved when $p \geq |\mathcal{S}_i|$,

---

ac1. **Input:** Training set $\mathcal{T}$, basis $g^1(\cdot), \ldots, g^q(\cdot)$,
$\qquad p \in \{1, 2, \ldots, \}$

ac6. IF $p < |\mathcal{S}_i|$
$\qquad$ Define $\mathcal{G}^1, \ldots, \mathcal{G}^p$
$\qquad$ Let (w_i, θ_i, y_i) be the solution to (3)
$\quad$ ELSE
$\qquad$ Let (w_i, θ_i, y_i) be the solution to (5)

Fig. 7. Modifications to the aMSM

which will in general occur in the last iterations. The algorithm returns, as in aMSM, $c(\mathcal{S}_i), w_i, \theta_i, y_i^k, i = 1, \ldots, t$.

Our algorithm requires, at each iteration, the use of a clustering pre-processing. This approach is significantly different from node decomposition or the random selection of p individuals, which are proposed in [12] to deal with large datasets. We group in clusters all individuals of the same class in a multiclass data; otherwise we use the clustering described in Figure 6. In our implementation, Euclidean distances were considered in the assignment LP models.

After the clustering, the algorithm solves model (5) to find a hyperplane, as was shown in the graphical example in Figures 3 and 4. Well classified individuals from $\mathcal{S}_i$ are no longer considered by the algorithm. The process is repeated until no individuals remain to be classified, or the target training accuracy is reached.

We recall that we need to avoid the spurious solution $w = 0$, so we impose the constraint $w \neq 0$, which is implemented by solving two LPs with constraints $\sum_{j=1}^q w^j \geq 1$ and $\sum_{j=1}^q w^j \leq -1$ respectively. We selected the LP solution with greater $|\mathcal{Q}|$. We also implemented an artifice to stop the algorithm when overfitting is detected [12, Section 4.1] but it was never needed.

3 Numerical Results and Analysis

We have coded our program in MATLAB version 7.5 on a laptop with Windows Vista equipped with a 1.8 GHz Intel with 2GB of RAM. We use the COIN-OR clp LP code as suggested by Trötscher [22] for solving (3), (5) and the assignment LP model for the clustering pre-processing described in [23, Section 4.2].

In these initial experiments only the hyperplane form $(g(x) = x)$ is implemented, but we recall that quadratic basis functions are relevant to ensure total classification [12, section 3.1]. As reported in [12] MATLAB linprog code consistently failed and was not used in our reported tests.

For numerical reasons we introduced a tolerance in the definition of $\mathcal{Q}_i$. Specifically,

$$\mathcal{Q}_i = \{x \in \mathcal{R}_i : c(\mathcal{S}_i)h_i(x) > (1 + \epsilon)\},$$

with $\epsilon = 10^{-7}$.

Table 1 shows the results of our algorithm and the aMSM on two datasets from [10]. The *Heart Dataset* is a case where total classification is not easily achieved without

Table 1. Results for *Heart* and Sonar databases

Heart database

Method	p	Accuracy	variables	time (secs)	Planes
aMSM	$|S|$	0.7365	174	2.38	6.9
acMSM	1	**0.7560**	15	6.39	14.8
acMSM	2	**0.7403**	16	4.51	14.6
acMSM	5	0.6916	19	4.00	12.9
acMSM	20	**0.7689**	34	3.34	10.7
acMSM	60	**0.7626**	74	2.81	8
acMSM	80	**0.7388**	94	2.74	7.6

Sonar database

Method	p	Accuracy	variables	time (secs)	Planes
aMSM	$|S|$	0.6710	172	1.07	1.9
acMSM	1	**0.6719**	65	1.02	1
acMSM	2	**0.6780**	63	0.75	1
acMSM	5	**0.6757**	66	0.76	1
acMSM	20	**0.6761**	81	0.80	1
acMSM	60	0.6641	121	0.81	1
acMSM	80	**0.6782**	141	0.83	1

losing the capacity of generalization. The *Sonar Dataset*, is easier to classify without overfitting because of the nature of the problem.

aMSM's performance is compared with the algorithm proposed, with 6 different number of clusters. The accuracy values reported were obtained as the fraction of the testing set that was correctly classified. The table emphasizes the values that gave higher accuracy than aMSM's. Both the time in seconds and the number of planes generated, are reported as averages on 10 runs of the algorithm in each experiment (because of the 10-fold cross-validation). Similar results are shown for the *Sonar* Database. In almost all cases only one plane was needed to classify the training set.

We repeat the experiment on the heart dataset. Healthy individuals were denoted as the $\mathcal{P}$ class and the others were grouped in 4 different classes, that is, $p = 4$, depending on the severity of the heart damage. The mean values found on a 10-fold validation test are: 75.41% prediction ability on the testing set in 5.12 seconds and 16.2 hyperplanes.

The numerical results show that our algorithm has a better prediction ability than aMSM on the datasets above mentioned; however our algorithm needed in general more iterations to achieve total classification and consequently it spent more CPU time.

4   Conclusions and Future Work

We have presented the acMSM, an algorithm devised to decrease the number of variables in the LP models through the use of clustering techniques. It inherits all the attractive features of is predecessor, the aMSM, but outperforms the latter in terms of generalization ability on the numerical experiments performed on two databases taken from the Irvine suppository. It is not clear however, if there is a specific rule to determine the *ideal* number of clusters that gives the best generalization ability.

One question from a practitioner's viewpoint is the following: given that there is no evident relation between generalization ability and number of clusters, shouldn't we simply choose one common variable for all data in S, and get rid of any clustering whatsoever? We never found the best generalization result for $p = 1$; nonetheless if the practitioner has no concerns about generalization ability, it is a valid option. It is still theoretically possible to produce total classification of the training set in a very short time.

Besides the properties shared with the aMSM [12], our algorithm offers the following extra features:

1. The number of clusters may be determined by the size of the LP problems that our computational resources can efficiently handle, so it is appropriate for small and medium range computers.
2. Assignment of individuals to clusters is performed via LP; hence it does not represent a real burden to the algorithm
3. Classification errors are related to the maximum distance of an element of the cluster to the separating hyperplane.
4. In multi class datasets, a cluster defined as all elements of the same class shows competitive results in regards to generalization ability and computational time.

Our implementation is simple, but several variants remain yet to be tested. It would be useful to experiment with the use of some kernel transformation, like the one proposed in [12]. Applications for multi-category classification problems seems a natural line of research. Also, Barros and De Pauda [2] proposed an SVM method in which a pre clustering is performed. When a cluster is formed by elements of the same class, its centroid is the only element considered in the SVM and smaller mathematical programming models are formulated. This scheme may be incorporated into the algorithm proposed, but complete classification might not be possible. Nevertheless, this remains as an open line of research, which could be useful with highly noisy databases.

Acknowledgements. The authors are indebted to Fernando Espinal-Kohler and Juvili Hernández-Monfort for their help and insight in the implementation of our algorithms and models. The third author thanks the support from the Atlantic Research Center for Information and Communication Technologies to his research in progress under project TEC 2010-21405-C02-01.

References

1. Armato-III, S.G., Giger, M.L., MacMahon, H.: Automated detection of lung nodules in ct scans: preliminary results. Medical Physics 28(8), 1552–1561 (2001)
2. Barros de Almeida, M., de Padua Braga, A., Braga, J.P.: SVM-KM: speeding svms learning with a priori cluster selection and k-means. In: Proceedings Sixth Brazilian Symposium on Neural Networks (November 2000)
3. Barzilay, O., Brailovsky, V.: On domain knowledge and feature selecion using a support vector machine. Pattern Recognition Letters 20, 475–484 (1999)

4. Brown, M., Grundy, W., Lin, D., Cristianini, N., Sugnet, C., Furey, T., Ares, M., Hausler, D.: Knowledge-based analysis of microarray gene expression data by using support vector machines. Proceedings of The International Academy of Sciences of The United States of America 97, 262–267 (2000)
5. Buchbinder, S., Leichter, I., Lederman, R., Novak, B., Bamberger, P., Sklair-Levy, M., Yarmish, G., Fields, S.: Computer-aided classifications of bi-rads category 3 breast lesions, radiology. Radiology 280, 820–823 (2004)
6. Chau, M., Chen, H.: A machine learning approach to web page filtering using content and structure analysis. Decision Support Systems 44, 482–494 (2008)
7. Cheung, K., Kwok, J.T., Law, M., Tsui, K.: Mining costumer product ratings for personalized marketing. Decision Support Systems 35, 231–243 (2003)
8. Druker, H., Wu, D., Vapnik, V.: Support vector machines for spam categorization. IEEE Transactions on Neural Networks 10, 1048–1054 (1999)
9. Espinal-Kohler, J.: Método multi-superficies para clasificación binaria con minimización de errores de grupos de datos. Master's thesis. Universidad Simón Bolívar (2012)
10. Frank, A., Asunción, A.: Uci machine learning repository (2010),
    http://archive.ices.uci.edu/ml
11. Fung, G., Mangasarian, O.L., Smola, A.: Minimal kernel classifiers. In: Shawe Taylor, J. (ed.), pp. 312–315 (2002)
12. García-Palomares, U.M., Manzanilla-Salazar, O.G.: Novel linear programming approach for building a piecewise nonlinear binary classifier with a priori accurancy. Decision Support Systems 51, 717–728 (2012)
13. Ince, H., Trafalis, T.B.: A hybrid model for exchange rate prediction. Decision Support Systems 42, 1054–1062 (2006)
14. Joachims, J.: Advances in kernel methods: support vector machines, ch. 11. MIT Press, Cambridge (1998)
15. Li, X.: A scalable decision tree system and its application in pattern recognition and intrusion detection. Decision Support Systems 41, 1–32 (2005)
16. Mangasarian, O.L.: Mathematical programing in neural networks. Journal on Computing 5, 349–360 (1993)
17. Mangasarian, O.L., Setonio, R., Wolberg, W.: Pattern recognition via linear programming: Theory and application in medical diagnosis. In: Coleman, T.F., Li, Y. (eds.) Proceedings of the Workshop on Large-Scale Numerical Optimization, pp. 22–31. SIAM (1990)
18. Nakayama, H., Yun, Y.B., Asada, T., Yoon, M.: Mop/gp mmodel for machine learning. European Journal of Operational Research 166, 756–768 (2005)
19. Peng, Y., Kou, G., Chen, Z.: A multi-criteria convex quadratic programming model for credit data analysis. Decision Support Systems 44, 1016–1030 (2008)
20. Stam, A., Rasgdale, C.T.: On the classification gap in mathematical programming-based approaches to the discriminant problem. Naval Research Logistics 39, 545–559 (1992)
21. Sueyoshi, T.: Extended dea-discriminant analysis. European Journal of Operational Research 131, 324–351 (2001)
22. Trötscher, T.: Linear mixed integer program (September 2009),
    http://www.mathworks.com/matlabcentral/fileexchange/
    25259-linear-mixed-integer-program-solver
23. Wang, J.: A linear assignment clustering algorithm based on the least similar cluster representatives. IEEE Transactions on Systems, Man and Cybernetics 29, 100–104 (1999)
24. Wu, Q., Zhou, D.: The F_∞-norm support vector machine. Statistica Sinica 18, 379–398 (2008)

CBOF: Cohesiveness-Based Outlier Factor
A Novel Definition of Outlier-ness

Vineet Joshi and Raj Bhatnagar

Dept. of Electrical Engineering and Computing Systems
University of Cincinnati, Cincinnati, OH, USA
joshivt@mail.uc.edu
raj.bhatnagar@uc.edu

Abstract. Anomaly detection is an important problem in data mining alongside clustering and classification. The aim of anomaly detection is to detect instances in a dataset that are remarkably different from the rest of the population. Knowledge about such anomalies is of immense practical value, such as for detection of fraud in financial transactions, for finding faulty equipment in industrial plants, and for detection of security breaches in a computer systems etc. In this paper we propose a new technique for identifying anomalies in numerical datasets. It is based on a metric for data points called the *Cohesiveness-Based Outlier Factor (CBOF)*. CBOF determines the outlier-ness score of a data point, which numerically represents how well integrated a data point is within its neighborhood. Through extensive experimentation we demonstrate that our method performs favorably when compared with another popular method for anomaly detection.

Keywords: anomaly detection, outliers, distance-based outlier, detection, local outliers.

1 Introduction

Anomalies are data points that are significantly different from the other points in a given dataset. Hawkins [1] has defined outliers as : "*An outlier is an observation which deviates so much from the other observations as to arouse suspicions that it was generated by a different mechanism.*"

Anomalies are of significant interest because they identify records that are peculiar and indicate unusual and rare phenomena. For example, if an unusual transaction is charged against the account of a credit card holder, it could be because the credit card information of the customer was stolen and is being misused. As another example, if a pressure sensor in an industrial plant monitoring system detects unusual readings, immediate corrective action is required to prevent catastrophic accidents.

By definition, anomalies indicate rare phenomena. While methods such as clustering and classification can be re-purposed to detect anomalies, they are not specifically designed to handle the cases where the records of interest are

P. Perner (Ed.): MLDM 2014, LNAI 8556, pp. 175–189, 2014.

very few. This has fueled research on specialized methods that focus on detecting anomalies.

In this paper we propose a new metric for defining anomalous behavior called the *Cohesiveness-Based Outlier Factor (CBOF)* and compare it with another widely used metric of outlier-ness called the Local Outlier Factor (LOF) [2]. Through extensive experimentation we demonstrate in this paper that in many situations our proposed definition is more effective than LOF in detecting outliers. The outlier-ness of a data point is greatly influenced by the distribution of other data points in its neighborhood. The neighbors of a data point may be scattered around the point in opposite directions or along almost the same direction in the n-dimensional feature space. The LOF and CBOF metrics work equally well for the former case and the CBOF metric works better for the latter case. There are possible data distribution situations where LOF may work better than CBOF. So, our contention is that an anomaly detection scheme that includes both, the CBOF and the LOF metrics to point out the outliers is certainly going to be more robust than the one based on a single metric alone. A data point marked as an outlier by either metric should be considered an anomaly.

CBOF applies primarily to real-valued datasets. The intuition underlying our definition of CBOF is very simple. We assume that anomalous points are relatively isolated from other data points in the population. On the other hand, non-anomalous records are assumed to be centrally located with respect to the points in their neighborhood and distances among such points are assumed to not vary abruptly.

We proceed to make this concept more concrete by examining the k-nearest neighborhood of a data point under consideration. Consider the 2-dimensional dataset depicted in Fig. 1. Essentially, the k-nearest neighborhood means the set of k-nearest neighbors of this data point, though a more precise definition is provided by Breunig [2]. We compute the sum of distances between all pairs of points in this neighborhood. Then we compute the fraction of this sum that is contributed by the distances associated with the data point whose outlier-ness is being investigated. We consider that anomalous points will contribute a higher fraction of this sum than the non-anomalous points.

A high value of this fraction signifies that the data point under investigation does not belong closely enough with its k-nearest neighbors. In other words, its *cohesion* with its neighbors is low. Similarly, a low value of this fraction signifies that the data point belongs within its neighbors and its *cohesiveness* within its neighborhood is high.

The fraction just computed can be considered the outlier-ness score of the data point. Hence, instead of generating a binary label signifying anomalous or normal behavior, our method generates outlier-ness scores for the records in the data set. The analysts can then set thresholds based on their domain expertise and the distribution of the outlier-ness scores of all the records and select all those records that exceed the threshold as anomalies.

LOF is a well-known metric for determining outliers. This method compares the density around any given data point with the density around its neighbors

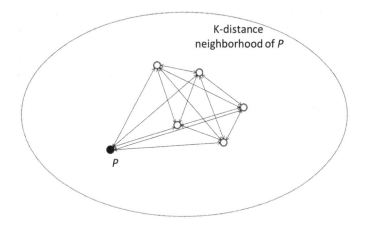

Fig. 1. Example Dataset

and uses them to compute the outlier score for the given point. Higher outlier scores indicate higher likelihood that the data point is an outlier. With this method low-density points lying in the midst of high-density data points incur higher outlier scores and have greater outlier-ness.

While LOF captures the outliers amidst the high density regions, there are scenarios where LOF scores fail to discriminate between cases that display intuitively different outlying behavior. Figures 2 and 3 depict two scenarios where this behavior of LOF becomes apparent.

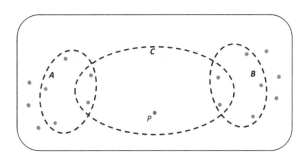

Fig. 2. CBOF vs LOF - Case 1

In Fig. 2 the point P is surrounded by its *4-nearest neighbors* contained within the region C. The regions A and B represent the *4-nearest neighbors* for the neighbors of P. Analogously, in Fig. 3, the point P' has its *4-nearest neighbors* contained within the region C'. Further, regions A' and B' represent the *4-nearest neighbors* of the neighbors of P'.

It should be noted that the density of region A is same as the density of region A', density of region B is same as the density of region B' and the density of region C is same as the density of region C'.

Hence the relative density of region C with respect to the regions A and B is the same as the relative density of region C' with respect to the regions A' and B'. Consequently the LOF score for P in Fig. 2 will be the same as the LOF score for P' in Fig. 3.

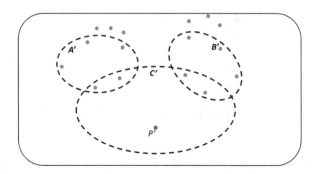

Fig. 3. CBOF vs LOF - Case 2

As is evident from these figures, the neighbors of P are more spread-out than the neighbors of P'. Stated differently, P' is more set-apart from its neighbors than P is from its own neighbors. Hence P' has higher outlier-ness than P. We would expect the outlier scores in these two cases to reflect this reality, which LOF does not capture. We will provide the evidence for this while discussing the experimental results over synthetic datasets representing this scenario in a subsequent section.

CBOF, on the other hand, will produce an outlier score that will reflect the difference between the datasets represented in these two figures. Since CBOF computes the fraction of the sum of distances to the nearest neighbors, it will detect that the point P in Fig. 2 is more *centrally located* with respect to its neighbors when compared with the point P' in Fig. 3. Hence point P' will have higher outlier score than the point P, which is what we expect in this case. Again, we support this with evidence in later sections.

2   Related Work

Proximity among data points forms a foundation of a range of anomaly detection techniques. Three subcategories of anomaly detection techniques that depend on proximity among data points are [3]: Cluster Analysis, Distance-based Analysis, and Nearest Neighbor Analysis.

Clustering based methods implicitly assume that normal data points can be categorized as well-formed clusters and the points that don't cluster well-enough are anomalies [3]. CBOF is quite different from these clustering based techniques as it does not aggregate the data points while determining their outlier-ness. Instead, CBOF considers each data point individually while computing outlier scores.

Distance-based methods use the distance to k-nearest neighbors as a measure of outlier-ness. Ramaswamy et al. presented a definition of outliers that used a fixed value of 'k' while determining the distance to k-nearest neighbors and considered these distances as the outlier scores of the data points[4]. CBOF differs from this method because instead of ranking the points on their distances to k-nearest neighbors, we consider the fraction of distances within the neighborhood that any give point contributes.

Density based methods take into account the variation in the density of data points in different localities of the dataset. Instead of basing the computation on global properties of data points, these methods consider the properties of a data point within its locality while evaluation its outlier-ness. The Local Outlier Factor (LOF) proposed by Breunig et al. is one such definition [2]. This is a widely used definition of outlying data points. In this paper we compare the performance of CBOF with the performance of LOF.

CBOF can be considered another local-outlier detection method like LOF because it considers and compares the distances within the k-nearest neighborhood while determining anomalies. In fact, CBOF can be considered even more local than LOF because unlike LOF, CBOF does not even consider the distances within the neighborhood.

Some outlier-detection methods provide a *yes-no* kind of labels for all the data points in the dataset, identifying them either as an outlier or as a normal data point. The method proposed by Knorr et al. [5] is one such method. However, much of recent work recognizes that anomalous behavior lies on a continuum from completely normal behavior to highly anomalous behavior [3]. These methods provide numerical scores that represent the extent to which any given data point displays anomalous behavior [2].

CBOF also provides outlier scores instead of providing binary labels. This provides the analyst with the flexibility in adjusting the threshold of outlier score to suite the problem domain.

3   CBOF Computation

As described before, CBOF computes the fraction of the sum of distances between all pairs of points in the k-nearest neighborhood that is contributed by a data point. We use this numerical value as a measure of outlier-ness of data point under consideration.

We now describe the steps of the method in greater detail. Breunig [2] has defined the k-distance of a data point P as follows:

Definition 1: $k - distance$

For any positive integer k, the k-distance of object p, denoted by $k - distance(p)$ is the distance $d(p, o)$ between p and another datapoint o such that :

1. For at least k objects, $o' \in D - \{p\}$, it holds that $d(p, o') \leq d(p, o)$
2. For at most $(k - 1)$ objects $o' \in D - \{p\}$, it holds that $d(p, o') < d(p, o)$

Further, Breunig used the above definition of k-distance of a data point to next define its k-distance neighborhood [2]:

Definition 2: $k - distance\ neighborhood$

The $k - distance\ neighborhood$ of a data point p contains every object q whose distance from p is not greater than the $k - distance$ of p. These objects q are called the $k - distance\ neighbors$ of p.

We construct the k-distance neighborhood of a data point based on the above definitions provided by Breunig et al. Consider Fig. 1 in which the dark-shaded point is the data point P whose outlier-ness is being investigated. The k-distance neighborhood of P is marked by the enclosure. Light-shaded points represent the data points within the k-distance neighborhood of P.

Step 1: First we compute the sum of distances between all the data point pairs in the k-distance neighborhood of point P, *including* the point P. We denote this numerical quantity by n_1.

$$n_1 = \left\{ \begin{array}{c} \text{sum of distances between all pairs of points} \\ \text{in } k - distance\ neighborhood \text{ of } P \text{ including} \\ \text{the point } P \end{array} \right\} \qquad (1)$$

Step 2: Next we calculate the sum of distances between all the pairs of points in k-distance neighborhood, but unlike step 1, we *don't include* the data point P among the data points considered. We denote this numerical quantity by n_2

$$n_2 = \left\{ \begin{array}{c} \text{sum of distances between light-shaded pairs} \\ \text{of points} \end{array} \right\} \qquad (2)$$

Step 3: In the final step we compute the Cohesiveness-Based Outlier Factor (CBOF) as follows :

$$CBOF = \left(1 - \frac{n_2}{n_1} \right) \qquad (3)$$

The value of CBOF thus computed has the following properties:

1. $0 \leq CBOF \leq 1$
2. Higher values of CBOF indicate greater outlier-ness displayed by P and can be considered its outlier score.

In our discussion so far we have considered all the attributes of the data set while computing distances. However in a high dimensional dataset often individual subspaces are selected and outlier scores are computed for them. Then, the computed scores are aggregated over the set of selected subspaces, as proposed by Lazarevic et al. [6]. In such methods it helps to have a greater contrast between the outlier scores of anomalous vs non-anomalous points so that the noisy variations in the outlier scores in different subspaces do not drown-out the high outlier-score values that occur in any subspace. The value of CBOF as computed above can be scaled as follows to introduce such a contrast:

$$CBOF' = (CBOF + f)^x \tag{4}$$

Here the value f is a positive parameter chosen such that $(CBOF + f)$ comes out closer to 1 for outliers whereas it remains smaller for normal data points. Further, this sum is raised to an exponent x. Consequently the values for outliers (which are closer to 1) will scale-up much more than the values closer to 0, which represents non-anomalous data points. Quantities f and x are analysis parameters that the analyst needs to provide based on the amount of scaling required. Algorithm 1 describes the steps for computing $CBOF$.

$CBOF'$ is just a scaled version of $CBOF$. Hence the relative ranking of datapoints based on the $CBOF$ scores does not change when using $CBOF'$. $CBOF'$ is just a convenient figure for use in aggregation of outlier scores across subspaces. Hence in our performance evaluation we have only used $CBOF$ and not $CBOF'$.

Algorithm 1. CBOF Score Computation

Input:

    k: the number of nearest neighbors to consider

    D: the dataset containing n records

    f: the fraction to be used in scaling $CBOF$

    x: the exponent to be used in scaling $CBOF$

Output: $C[1..n]$

1 **for** *each point $P \in D$* **do**

2   Compute N_p, the k-nearest neighborhood of P

3   Generate set $N'_P \leftarrow \{N_P \cup P\}$

4   Compute n_1, the sum of distances between all pairs of points in N'_P

5   Compute n_2, the sum of distances between all pairs of points in N_p

6   Compute $CBOF \leftarrow \left(1 - \frac{n_2}{n_1}\right)$

7   Compute $CBOF' \leftarrow (CBOF + f)^x$

8   Set $C[P] \leftarrow CBOF'$

9 **end**

Computational Complexity: The method requires availability of distances between all pairs of points within the neighborhood of the data point whose outlier-ness is being evaluated. The distances beween all pairs of data points can be computed in time $O\left(n^2\right)$ and can be read-off from a lookup table.

For determining the outlier score of each data point, this table will need to be accessed $O\left(k^2\right)$ times. However the value of k is a fixed analysis parameter. It is generally a value much smaller than the number of input records and does not scale as a function of the input size once selected by the analyst. In this case the computational complexity will be determined by the effort required to compute distances between all pairs of data points, and the computation of k-neighborhood of data points. These operations have computational complexity of $O\left(n^2\right)$. Hence this is also the computational complexity of the complete process.

4   Experimental Evaluation

We performed experimental evaluation on synthetic datasets as well as a few real datasets.

4.1   Synthetic Dataset

We first discuss the result of our experiment with the synthetic datasets that represent the scenario described in figures 2 and 3.

Figure 4 displays the CBOF and LOF scores for a 'compact' dataset where the neighbors of the point P are located close together. Figure 5 displays the scores for the dataset derived from compact dataset by spreading-out the neighbors of the point P while retaining the distances among the neighbors. Hence in the compact as well as the spread-out datasets the density among the neighbors is the same. For both CBOF and LOF, the value of k used for both datasets was 4.

CBOF scores of P for the compact and spread-out datasets are 0.45 and 0.32 respectively whereas the LOF scores are 1.83 in the two cases. As expected, the LOF scores do not discriminate between the two cases whereas CBOF scores do bring-out the difference between them and assign higher score to P in the compact dataset where P is more 'left-out' with respect to its neighbors than the spread-out dataset.

Thus we see that as we claimed earlier, CBOF is able to differentiate between the compact and spread-out datasets, whereas LOF assigns the same outlier scores to P in the two cases.

4.2   Real Datasets

We have used the following datasets from the UCI Machine Learning repository [10] to evaluate our method: Email Spam dataset, Sonar dataset, and Thyroid disease dataset.

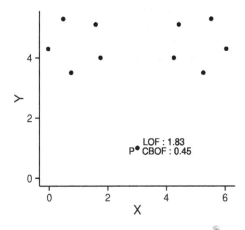

Fig. 4. CBOF and LOF Scores - Compact Dataset

Design of Experiment: It is difficult to obtain datasets with known anomalies for use in evaluating anomaly detection methods. For example the UCI Machine Learning repository contains many datasets for use in classification and clustering etc, but none especially directed towards anomaly detection tasks. A commonly used approach is to use the datasets intended for use with classification tasks for anomaly detection too where records belonging to the smaller class are designated as outliers [11][12].

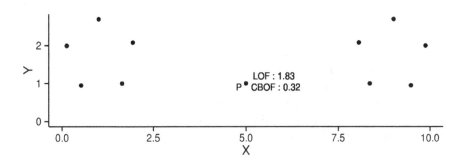

Fig. 5. CBOF and LOF Scores - Spread-Out Dataset

However, the anomaly detection task is fundamentally different from the classification task because the number of anomalies is very small as compared to the non-anomalous points. This renders the datasets intended for classification algorithm testing unsuitable for anomaly detection tasks.

For instance, consider the use of 2-class dataset depicted in Fig. 6 for evaluating an outlier detection algorithm, where records belonging to class B have been a-priori designated as outliers. We should not expect an outlier detection algorithm to report all the records from class B as outliers. The records belonging to class B might be outliers with respect to the records of class A but they appear non-anomalous with respect to other data points of class B. Consequently methods such as CBOF and LOF which examine the neighborhood of a data point while evaluating its outlier-ness will not flag them as outliers and thus fail to meet our expectations with such a dataset.

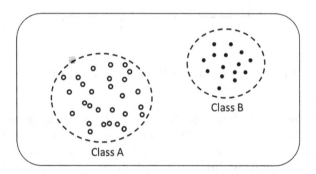

Fig. 6. 2-Class Dataset

Hence a perfectly functional outlier detection algorithm will not report the records in class B as outliers, contrary to our a-priori expectation about the experiment.

This problem arises because we are using an unsuitable dataset for the experiment. An appropriate evaluation dataset should have anomalies that are rare and isolated instead of being numerous and belonging to a well-defined class. With such a dataset we can reasonably expect the outlier detection method to detect the outliers and can then evaluate the methods on their efficacy.

Such a dataset can be obtained if we select a small number of datapoints from class B along with all the points of class A and discard the unselected records from class B. In this *augmented* dataset (Fig. 7), the records from class B will appear rare and isolated, and we can expect the outlier-detection algorithm to detect them. We can expect the outlier detection algorithm to:

- Report all retained records of class B as outliers, and
- Not report any record of class A as an outlier.

If the outlier detection algorithm falls short on any of these goals with the augmented dataset, it will be only because of the inherent limitations of the algorithm instead of the unsuitability of the dataset used for evaluation.

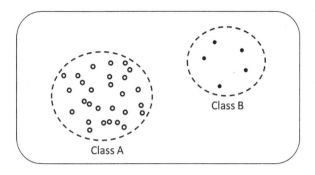

Fig. 7. *Augmented* 2-Class Dataset

With this view, we created the augmented datasets from Email Spam, Sonar and ANN Thyroid Disease datasets taken from the UCI Machine Learning Repository.

Spam dataset : This dataset consists of 4601 records, out of which 2788 records are for non-spam emails and 1813 records are for spam emails. We used a uniformly distributed random number generator to choose 10 records from the spam records and appended them to the set of non-spam records. Further, this was repeated 10 times to generate 10 different augmented datasets.

Sonar dataset : This dataset consists of 208 records out of which 111 records belong to mines and 97 belong to rocks. Again a uniformly distributed random number generator was used to select 10 records from the class of rocks. These were appended to the class of mines to create the augmented dataset. Again, this step was repeated 10 times to generate 10 different augmented datasets.

Thyroid Disease ANN Training dataset : The Thyroid Disease ANN training dataset consists of 3772 records belonging to three distinct classes. We chose to designate the class 3 as the 'normal' class. We then again used the uniformly distributed random number generator to choose 10 records from class 1 and appended them to the records of class 3 to generate the augmented dataset. Again, this was repeated 10 times to generate 10 different augmented datasets.

In all these cases we knew the labels for true outliers a-priori by the virtue of having created the augmented dataset. The outlier scores were computed using both CBOF and LOF. The ROC curves were plotted for both cases and the corresponding *area under curve*(AUC) was computed using the *pROC* package of the R programming environment [13].

We conducted the experiment with all augmented datasets obtained from the three real datasets mentioned earlier. The values of k used were 5, 10, 15 and 20. To provide an overview of the results of all experiments, we report the average

AUC values for all of augmented dataset for these values of k. In the interest of saving space we display the ROC curves for only a few experiments.

Results for Spam Dataset: Figure 8 displays the ROC curve for $k = 5$ for one augmented Spam dataset. Table 1 represents the AUC scores obtained for various values of k and averaged over all the augmented datasets.

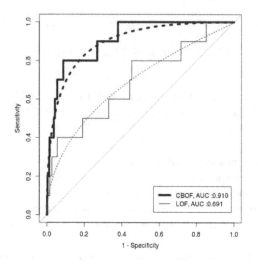

Fig. 8. Spam dataset, $k = 5$

Table 1. Avg. AUC values for Spam dataset

k	$CBOF$	LOF
5	0.81449	0.655301
10	0.809401	0.650872
15	0.808849	0.661611
20	0.808314	0.658328

As is evident from these results, CBOF performs quite well in comparison to LOF for the Spam dataset.

Results for Sonar Dataset: Figure 9 displays the ROC curve for $k = 15$ for one augmented Sonar dataset. Table 2 represents the AUC values obtained for various values of k and averaged over all augmented datasets.

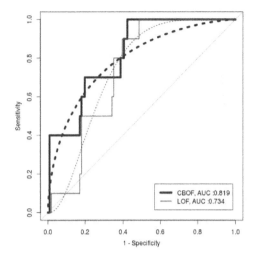

Fig. 9. Sonar dataset, $k = 15$

Thus for Sonar dataset, CBOF outperforms LOF for values 10, 15 and 20 of k, whereas LOF is better for value 5 of k. Hence overall CBOF performs better than LOF.

Table 2. Avg. AUC values for Sonar dataset

k	$CBOF$	LOF
5	0.777838	0.832611
10	0.716035	0.708468
15	0.687478	0.653512
20	0.671983	0.62901

Results for Thyroid Disease ANN Training Dataset: Figure 10 displays the ROC curve for $k = 5$ for one augmented Thyroid Disease ANN training dataset. Table 3 represents the AUC values obtained for various values of k and averaged over all augmented datasets.

For the Thyroid Disease dataset, both CBOF and LOF perform quite well and attain high AUC values. CBOF edges out LOF for values 5 and 10 of k, while the reverse occurs for values 15 and 20 of k. Overall, CBOF and LOF perform equally well for this dataset.

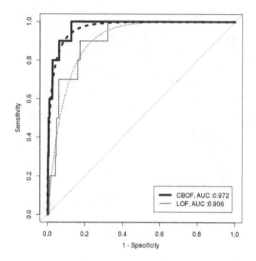

Fig. 10. Thyroid Disease ANN Training dataset, $k = 5$

Table 3. Avg. AUC values for Thyroid Disease ANN Training dataset

k	$CBOF$	LOF
5	0.9709	0.934398
10	0.972945	0.972456
15	0.972811	0.981022
20	0.972175	0.9817

5 Conclusions

In this paper we have proposed a new metric to define outlier-ness of a data point. Some approaches consider distances from cluster centers as measures of outlier-ness [7] and some other methods assume a statistical model and consider deviations from the model as a measure of outlier-ness [8]. Many other methods use distance computations among the data points as the basis for detecting anomalies, such as the Z-score of distances to k-nearest neighbors [9].

CBOF, the novel metric proposed in this paper captures the intent behind Hawkin's definition in a unique way by examining the distances within k-nearest neighborhood of a datapoint. Our extensive experiments demonstrate that CBOF captures facets of outlier-ness that elude LOF. This is exemplified by the results of experiments with the Spam dataset in Fig. 8 and table 1, and the results with Sonar dataset in Fig. 9 and table 2.

However, experiments with the Thyroid Disease ANN training dataset depicted in fig 10 and table 3 demonstrate that there can be instances where outliers are capably identified by LOF too, where it marginally edges out CBOF.

This observation, that perhaps there is no single universal metric of outlier-ness and outcomes from multiple definitions should be combined to produce a final decision about outliers, as has been proposed by Lazarevic et al. [6]. An anomaly detection scheme that uses both, the CBOF and LOF metrics to indicate the outlier-ness of a data point promises to be a more robust scheme. That is, a data point marked as an outlier based on either metric should be considered an anomaly.

Nevertheless, CBOF provides a unique perspective on what constitutes out-lying behavior and in many scenarios is more capable than LOF in detecting anomalies.

References

1. Hawkins, D.: Identification of Outliers. Chapman and Hall (1980)
2. Breunig, M., Kriegel, H.P., Ng, R., Sander, J.: LOF: Identifying Density-based Local Outliers. In: ACM SIGMOD Conference, pp. 93–104. ACM, New York (2000)
3. Aggarwal, C.: Outlier Analysis. Springer, New York (2013)
4. Ramaswamy, S., Rastogi, R., Shim, K.: Efficient Algorithms for Mining Outliers from Large Data Sets. In: ACM SIGMOD Conference, pp. 427–438. ACM, New York (2000)
5. Knorr, E., Ng, R.: Algorithms for Mining Distance-Based Outliers in Large Datasets. In: 24th International Conference on Very Large Data Bases, pp. 392–403. Morgan Kaufmann Publishers, San Francisco (1998)
6. Lazarevic, A., Kumar, V.: Feature Bagging for Outlier Detection. In: 11th ACM SIGKDD International Conference on Knowledge Discovery and Data Mining, pp. 157–166. ACM, New York (2005)
7. Wu, N., Zhang, J.: Factor-analysis Based Anomaly Detection and Clustering. Decision Support Systems 42, 375–389 (2006)
8. Barnett, V., Lewis, T.: Outliers in Statistical Data. John Wiley and Sons, New Jersey (1994)
9. Ebdon, D.: Statistics in Geography: A Practical Approach. John Wiley and Sons, New Jersey (1985)
10. Bache, K., Lichman, M.: UCI Machine Learning Repository. University of California, School of Information and Computer Science, Irvine, CA (2013), http://archive.ics.uci.edu/ml
11. Aggarwal, C., Yu, P.S.: An Effective and Efficient Algorithm for High-Dimensional Outlier Detection. The VLDB Journal 14, 211–221 (2005)
12. Keller, F., Müller, E., Böhm, K.: HiCS: High Contrast Subspaces for Density-Based Outlier Ranking. In: 28th IEEE International Conference on Data Engineering (ICDE), pp. 1037–1048. IEEE Press, New York (2012)
13. Robin, X., Turck, N., Hainard, A., Tiberti, N., Lisacek, F., Sanchez, J.-C., Müller, M.: pROC: An Open-Source Package for R and S+ to Analyze and Compare ROC Curves. BMC Bioinformatics 12, 77 (2011)

A New Measure of Outlier Detection Performance

Kliton Andrea[1], Georgy Shevlyakov[1],
Natalia Vassilieva[2], and Alexander Ulanov[2]

[1] Department of Applied Mathematics, St. Petersburg State Polytechnic University,
Polytechnicheskaya 29 Saint-Petersburg, 195251, Russia
[2] Hewlett-Packard Labs, 1 Artillerijskaya St. Petersburg, 191104, Russia

Abstract. Traditionally, the performance of statistical tests for outlier detection is evaluated by their power and false alarm rate. It requires ensuring the upper bound for false alarm rate while measuring the detection power, which proves to be a difficult task. In this paper we introduce a new measure of outlier detection performance H_m as the harmonic mean of the power and unit minus false alarm rate. The H_m maximizes the detection power by minimizing the false alarm rate and enables an easier way for evaluation and parameters tuning of an outlier detection algorithm.

1 Introduction

In statistics, an outlier is an observation that is numerically distant from the rest of the data [1]. Grubbs [2] defines an outlier as follows: " An outlying observation, or outlier, is one that appears to deviate markedly from other members of the sample in which it occurs ".

A frequent cause of outliers is a mixture of two distributions, which may be two distinct sub-populations, or may indicate "good data" versus "bad data"; this is modeled by a mixture model. Given a data set of n observations $\{x_i\}$, let $\bar{x}$ and s be the sample mean and the standard deviation. A common approach is to declare an observation as an outlier if it lies outside of the interval $(\bar{x} - ks, \bar{x} + ks)$ where the value of k is usually taken between 2 and 3. This approach requires observations to be normally distributed. Another drawback of this method is its poor performance as a result of nonrobust estimates of location and scale.

A better choice is to use the univariate boxplot [3], which is defined by five parameters: the two extremes, the upper UQ and lower LQ quartiles and the sample median. The lower and upper extremes of a boxplot are defined as

$$x_L = max\left\{x_{(1)}, LQ - k_{S_n} S_n\right\}, \quad x_U = min\left\{x_{(n)}, UQ + k_{S_n} S_n\right\}, \quad (1)$$

where $k_{S_n} = 1.5$ and $S_n = IQR$ (interquartile range) for the classical Tukey boxplot. A common practice is to identify the sample elements, which are located beyond the extremes (maximum and minimum), as outliers.

P. Perner (Ed.): MLDM 2014, LNAI 8556, pp. 190–197, 2014.

Parameter optimization is of large importance for data mining. Most parameter settings of learning algorithms are very sensitive, consider, for example, the kernel parameters of Support Vector Machines. Other parameters include preprocessing options, e.g., feature processing and selection. Here, additional parameters come into play which can greatly affect the model quality.

Many outlier detection methods have tunable parameters which can be optimized to improve the performance of these methods. Parameter optimization here is of large importance as for many other data mining tasks. In general, parameter tuning is performed by solving an optimization problem, maximizing performance measures of the method. Traditionally, the performance of statistical tests for outlier detection is evaluated by two measures: detection power and false alarm rate. Performance evaluation of an outlier detection method requires ensuring the upper bound for false alarm rate while measuring the detection power, which proves to be a difficult task. In this paper we propose a new single measure of outlier detection performance, which is easier to use for evaluation and parameter tuning.

The rest of the paper is organized as follows. In section 2 we discuss existing performance measures for outlier detection task. Section 3 presents a definition of the new performance measure H_m. In section 4 we provide an example of application of H_m measure for optimization of outlier detection with boxplots with Monte Carlo experiments. Section 5 concludes the paper.

2   Evaluating Outlier Detection Performance

2.1   Confusion Matrix

Basically, the performance of any classifier [4] is defined by the correctly classified samples in a test set. The confusion matrix collects the statistics about the correctly and incorrectly classified samples as it is shown in Table 1. Most of the metrics used to evaluate classifier's performance are borrowed from Signal Processing and Information Retrieval areas. In this paper, we focus on the outlier detection methods and their performance evaluation. From the statistical point of view the binary classification procedure consists in accepting or rejecting the main (or null) hypothesis H_0. Here, we consider only two cases: the main hypothesis H_0 and its alternative H_1. In our case H_0 is set as: "the sample is an inlier" and its alternative H_1 as: "the sample is an outlier". Table 1 presents a confusion matrix. The rates TP (true positive) and TN (true negative) identify the amount of cases that are correctly predicted, while FN (false negative) and FP (false positive) identify the amount of cases when a wrong prediction takes place.

The confusion matrix of an ideal classification algorithm is the identity matrix with unit values on the main diagonal and zero values otherwise.

True Positive Rate (TPR) and *True Negative Rate (TNR)* are often used to determine the performance of a classifier.

$$TPR = \frac{TP}{TP + FN} \tag{2}$$

Table 1. Confusion matrix

		Reality	
		H_0	H_1
Prediction	H_0	TP	FP
	H_1	FN	TN

$$TNR = \quad Power \quad = \frac{TN}{TN + FP} \tag{3}$$

Or, their complements, *False Negative Rate (FNR)* and *False Positive Rate (FPR)*, which are defined as follows.

$$FNR = \quad Significance\ Level \quad = \frac{FN}{TP + FN} \tag{4}$$

$$FPR = \frac{FP}{TN + FP} \tag{5}$$

We are interested in the statistical significance and the power parameters derived from the confusion matrix.

The performance evaluation of a binary classifier requires the definition of two parameters as we are concerned with optimizing both values on the main diagonal of the confusion matrix. In Information Retrieval the *precision* and *recall* are combined in a single metric (F-measure) [5].

$$\text{Precision} = \frac{TP}{TP + FP} \tag{6}$$

$$\text{Recall} = \frac{TP}{TP + FN} \tag{7}$$

$$F = 2 \times \frac{\text{Precision} \times \text{Recall}}{\text{Precision} + \text{Recall}} \tag{8}$$

2.2   ROC-curve

Assuming the definitions of statistical hypotheses are given by their probability density functions (normal distributions with different parameters of location and scale), the performance of an outlier detection algorithm depends on the optimal choice of the threshold value t, which defines a decision rule of the algorithm. In Fig. 1 the threshold t defined by an outlier detection algorithm defines a following decision rule for a sample: $H_0 : x < t$ and $H_1 : x \geq t$.

In our experimental settings, we generate samples from the mixture of normal distributions (Tukey's gross error model)

$$f(x) = (1 - \varepsilon)\mathcal{N}(x; 0, 1) + \varepsilon\mathcal{N}(x; \mu, s), \tag{9}$$

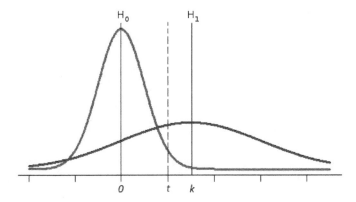

Fig. 1. Threshold value t separating samples described by their probability density functions

where $0 \leq \varepsilon < 1$ is the probability of outliers in the data, μ is the location parameter and $s > 1$ is the scale parameter.

The power and the false alarm rate [6] are used in outlier detection task to evaluate the algorithm performance. The power is related to the outlier detection ability

$$P_D = P(x > t|H_1) = \frac{TN}{TN + FP} \tag{10}$$

and the false alarm rate is related to the case when we fail to accept the main hypothesis

$$P_F = P(x > t|H_0) = \frac{FN}{TP + FN}. \tag{11}$$

By shifting the threshold parameter t, the confusion matrix changes, and as a result the power and false alarm rate values change accordingly. The graphical representation of the power and false alarm values defines the ROC-curve [7] of the method. An ideal ROC-curve is displayed in Fig. 2.

The area under the curve (AUC) is a common metric to compare different classification methods with each other. The closer to 1.0 the value of AUC, the better is the performance of the classification method under study.

3   Definition and Properties of H_m

According to the Neyman-Pearson approach [8], one should prefer the method that has the greatest power given that its false alarm rate is bounded. In many circumstances, it is almost impossible to ensure an upper bound for the false alarm rate. On the other side, it is important to find an optimal balance between these metrics for the method under investigation.

The F-measure (8) is a common performance measure for Information Retrieval tasks, which combines the precision and recall values in a single evaluation

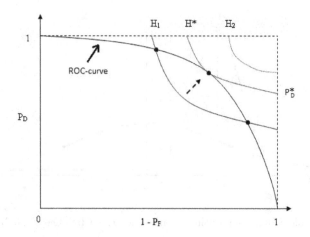

Fig. 2. The relationship between ROC-curve and H_m

estimate. The optimal F-measure value maximizes both the Precision (6) and the Recall (7).

In a similar way, we introduce in this paper a new performance evaluation measure H_m for the outlier detection task that combines the power and false alarm probability. The H_m estimate is defined as follows:

$$H_m = \frac{2\,P_D\,(1 - P_F)}{P_D + (1 - P_F)}, \tag{12}$$

where P_D is defined in (10) and P_F is defined by (11).

There is a relationship between H_m and ROC-curve. For every point on the ROC-curve, it is possible to find the H_m by using (12). An H_m value is represented by a contour line in $(1 - P_F, P_D)$ plane, as shown in Fig. 2. The contour line of an optimal $H_m = H^*$ value tangents the ROC-curve. Given a H_m value, the lower and upper bounds for the power and false alarm rate are established by the following theorem.

Theorem 1. *Given a fixed H_m-value H, the following inequalities hold:*

$$P_D > P_D^{min} = \frac{H}{2 - H}, \quad P_F < P_F^{max} = \frac{2\,(1 - H)}{2 - H},$$

where P_D is the detection power and P_F is the false alarm rate.

Proof. Recall the definition of the H_m in (12). It can be shown that this expression takes the following form:

$$P_D = \frac{(1 - P_F)\,H}{2(1 - P_F) - H}.$$

For a fixed value H, the detection power reaches its minimum value when $P_F \to 0$, which means that $1 - P_F \to 1$. Thus the lower bound for the detection power is

$$P_D^{min} = \frac{H}{2 - H}.$$

The proof for the upper bound of the false alarm rate starts with the following equation:

$$1 - P_F = \frac{P_D H}{2P_D - H}.$$

For a fixed value H, the unit minus false alarm rate reaches its minimum value when $P_D \to 1$. Replacing the $P_D = 1$ and rearranging the terms in the above equation, we get

$$P_F^{max} = \frac{2(1 - H)}{2 - H}.$$

This denotes the upper bound for the false alarm rate and proves the required result.

Using Theorem 1, we exhibit the bound values for the detection power and false alarm rate in Table 2.

Table 2. Lower and upper bounds values for the power and false alarm rate

H	0.5	0.6	0.7	0.8	0.9	0.91	0.92	0.93	0.94	0.95	0.96	0.97	0.98	0.99	0.999
P_D^{min}	0.33	0.43	0.54	0.67	0.82	0.83	0.85	0.87	0.87	0.90	0.92	0.94	0.96	0.98	0.998
P_F^{max}	0.67	0.57	0.46	0.33	0.18	0.17	0.15	0.13	0.11	0.095	0.08	0.058	0.04	0.02	0.002

The optimal threshold value t_0 is defined in terms of H_m by the following way

$$t_0 = \arg\max_t H_m(t). \tag{13}$$

The optimization task for the boxplot case depends on the parameter k_{S_n} defined in (1). Comparing different algorithms by their maximal H_m values is an alternative to the AUC, but the computational complexity burden is much lower.

4 Application of H_m to Optimization of Outlier Detection with Boxplots

Using (1), it is possible to introduce new modifications of boxplots applying highly efficient robust estimates of scale: MAD [9] (median absolute deviation) and FQ_n [10] (fast Q_n [11]). It is necessary to determine the values of the threshold constants k_{S_n} in (1) for each of the modifications. The definition of these values is motivated by the outlier detection task aiming to optimize (13) in terms of k_{S_n}. For this purpose, we conducted Monte Carlo experiments by generating 10 000 samples of observations using (9). The results are displayed in Table 3 for the scale contamination with parameters $\varepsilon = 0.1$, $\mu = 0$ and $s = 3$ and the shift contamination with parameters $\varepsilon = 0.1$, $\mu = 3$ and $s = 3$. The sample size is $n = 100$. Each row of the table contains the optimal value of the parameter k for a given boxplot modification and the values of performance measures F, P_D and P_F for this optimal value of k.

Table 3. The optimal threshold values for the scale and shift contamination in terms of H_m

$\varepsilon = 0.1$	Scale Contamination				Shift Contamination			
	k	H	P_D	P_F	k	H	P_D	P_F
Tukey boxplot	0.36	0.73	0.67	0.20	0.51	0.78	0.74	0.13
MAD-boxplot	0.47	0.73	0.68	0.21	0.77	0.78	0.73	0.11
FQ_n-boxplot	0.5	0.73	0.66	0.19	0.61	0.78	0.75	0.14

Under the same conditions we conducted experiments aiming to maximize the F-measure. The results are displayed in Table 4. The alarm rate values are at a 0 level, which is an advantage, but the detection power is much lower. The explanation is that the F-measure is exploited as a measure of the quality of information retrieval, whereas the H_m is a measure of the detection of outliers quality.

Table 4. The optimal threshold values for the scale and shift contamination in terms of F-measure

$\varepsilon = 0.1$	Scale Contamination				Shift Contamination			
	k	F	P_D	P_F	k	F	P_D	P_F
Tukey boxplot	1.4	0.96	0.35	0.007	1.5	0.97	0.51	0.004
MAD-boxplot	1.9	0.96	0.35	0.007	2	0.97	0.52	0.005
FQ_n-boxplot	1.8	0.96	0.36	0.007	1.9	0.97	0.51	0.003

Traditionally, the threshold constant of Tukey's boxplot is set to the value of 1.5. In our experiments when optimizing F-measure for Tukey's boxplot, we got the same optimal value of the parameter $k = 1.5$, as it is displayed in Table 4. The obtained optimal values for the threshold constants in (1) are much smaller than the conventional ones in terms of H_m. These values raise the false alarm rate in order to reach a greater power of detection, thus requiring a careful usage of the H_m as a performance evaluation estimate. On the other hand, this estimate yields a symmetric confusion matrix.

5   Conclusions

A new performance evaluation measure related to outlier detection methods is proposed. This new measure H_m is based on the harmonic mean of detection power and unit minus false alarm rate of the method under investigation. Although this new metric of performance evaluation is applied to the outlier detection methods, we recommend its use in signal processing applications where the performance of classification tasks is evaluated.

References

1. Barnett, V., Lewis, T.: Outliers in statistical data. Wiley (1994)
2. Grubbs, F.E., Beck, G.: Extension of sample sizes and percentage points for significance tests of outlying observations. Technometrics 14(4), 847–854 (1972)
3. Tukey, J.W.: Exploratory Data Analysis. Addison-Wesley, Reading (1977)
4. Hastie, T., Tibshirani, R., Friedman, J.H.: Elements of Statistical Learning. Springer (2001)
5. van Rijsbergen, C.J.: Information Retrieval. Butterworths, London (1979)
6. Kendall, M.G., Stuart, A.: Kendall's advanced theory of statistics, 2nd edn. Classical inference and the linear model, vol. 2A. Arnold, London (1968)
7. Marzban, C.: The ROC curve and the area under it as a performance measure. Weather and Forecasting 19(6), 1106–1114 (2004)
8. Lehmann, E.L.: The Fisher, Neyman-Pearson theories of testing hypotheses: One theory or two? Journal of the American Statistical Association 88(424), 1242–1249
9. Huber, P.J.: Robust Statistics. Wiley, New York (1981)
10. Smirnov, P., Shevlyakov, G.: Priblizhenie otsenki Q_N parametra masshtaba s pomosh'yu bystrih M-otsenok. Vestnik SibGAU 5(31), 83–85 (2010) (In Russian)
11. Rousseeuw, P.J., Croux, C.: Alternatives to the median absolute deviation. Journal of the American Statistical Association 88(424), 1273–1283 (1993)

A Multi-path Strategy for Hierarchical Ensemble Classification

Esra'a Alshdaifat, Frans Coenen, and Keith Dures

University of Liverpool, Department of Computer Science, United Kingdom
{esraa,coenen,dures}@liv.ac.uk

Abstract. A solution to the multi-class classification problem is proposed founded on the concept of an ensemble of classifiers arranged in a hierarchical binary tree formation. An issue with this solution is that if a miss-classification occurs early on in the process (near the start of the hierarchy) there is no possibility of rectifying this error later on in the process. To address this issue a multi-path strategy is investigated based on the idea of using Classification Association Rule Miners at individual nodes. The conjectured advantage offered is that the confidence values associated with this form of classifier can be used to inform the proposed multi-path strategy. More specifically the confidence values are used to determine, at each node, whether one or two paths should be followed.

Keywords: Hierarchical classification, ensemble classification, multi-class classification, Classification Association Rule Mining.

1 Introduction

Single-label classification (as opposed to multi-label classification) is concerned with learning of classifiers, using a set of training examples, where each example is associated with a single class label c taken from a set of disjoint class labels C. If $|C| = 2$ then we have a simple "binary" classification problem, if $|C| > 2$ it is referred to as a "multi-class" classification problem. In the context of the work described in this paper the focus is on the multi-class single-label classification problem where each example is associated with exactly one element of the class label set C. For simplicity this problem is referred to here as the *multi-class problem*.

An issue with multi-class classification is that when $|C|$ is large the classification accuracy tends to degrade for two reasons: (i) each class is represented by fewer examples in the training dataset than in the case of binary training data, and (ii) a suitable subset of features that can be used to discriminate between large numbers of classes (more than two) is often difficult to identify. One mechanism for addressing this issue is to adopt some ensemble classification technique whereby a set of classification models is used in order to obtain a better composite global model [17]. Much research work has been conducted on ensemble techniques because of the potential benefits of such methods; this research has demonstrated that it is frequently the case that more accurate and

P. Perner (Ed.): MLDM 2014, LNAI 8556, pp. 198–212, 2014.

effective classification results can be produced when using ensemble techniques than when using a single model [17,35,25,13]. With reference to the literature, two main mechanisms for arranging the "base classifiers" within an ensemble model can be identified: (i) concurrent (parallel) [23,8,9], or (ii) sequential (serial) [15,27]. A more recent approach for arranging the base classifiers within an ensemble model involves the creation and utilisation of a hierarchy of classifiers [18,10,32,4,19,24]. One of the most significant advantages of hierarchical classification is its structural flexibility; the ability to modify or adapt the model so as to fit a particular classification problem.

The work presented in this paper is directed at hierarchical ensemble classification using a binary tree representation where nodes near the root of the tree hold classifiers designed to distinguish between groups of classes while the leaf nodes hold classifiers designed to distinguish between individual class labels. Thus classifiers nearer the root of the hierarchy conduct coarse-grain classification with respect to subsets of the available set of classes while classifiers near the leaves of the hierarchy conduct fine-grain classification. More specifically the work is concerned with the resolution of two issues associated with this form of ensemble classification. The first is how best to distribute (group) classes across the nodes in the hierarchy. The second is concerned with the operation of the hierarchy whereby, if a single-path strategy is adopted, a miss-classification early on in the process (near the root of the hierarchy) cannot be rectified later in the process (further down the hierarchy). Three alternative techniques are proposed in this paper to address the first issue: (i) k-means clustering, (ii) divisive hierarchical clustering, and (iii) data splitting. With respect to the second issue the novel idea presented in this paper is to utilise confidence values generated when using Classification Association Rule Mining (CARM) to inform a proposed multi-path strategy that, in certain circumstances, will cause more than one path to be followed through the hierarchy.

The rest of this paper is organised as follows. Section 2 gives a review of related work on multi-class classification. Section 3 describes the process for generating the proposed binary tree hierarchical ensemble classification model, while Section 4 describes its operation. Section 5 presents an evaluation of the proposed hierarchical classification mechanism as applied to a range of different data sets. Section 6 summarises the work and indicates some future research directions.

2   Literature Review

This section reviews some of the previous work directed at the multi-class classification problems. It is widely accepted that multi-class problems can be solved in three ways: (i) using "stand-alone" classification algorithms (Section 2.1), (ii) using a "two-class" classifiers (Section 2.2), and (ii) using ensemble classifiers arranged in some specific form (Section 2.3).

2.1 Using Stand-Alone Classification Algorithms to Solve Multi-class Classification Problems

Some classification algorithms are specifically designed to handle binary classification, for example support vector machines [31]. While other classification algorithms can be used with respect to any number of class labels; examples of the latter include: decision tree classifiers [26], Classification Association Rule Mining (CARM) [12], Neural Networks [34], k-Nearest Neighbors [6], and Bayesian classification [20]. Among these CARM algorithms are of interest with respect to the work described in this paper. The significance is that CARM incorporates the concept of confidence values which in turn, it is argued in this paper, can be use to determine the most appropriate paths through a hierarchical ensemble classification model. CARM integrates Association Rule Mining (ARM) and classification. CARM algorithms works by applying an association rule mining style algorithm, such as Apriori [1] or FPgrowth [16], to produce classification rules from a training set of previously classified data [12] according to user predefined support (frequency)[1] and confidence[2] thresholds, where the focus is to generate association rules that have only a single class label in the consequent. The generated association rules are referred to as Classification Association Rules (CARs) [21], which collectively form the desired classifier. CARM algorithms can be categorised according to how the pruning of low confidence CARs is performed [12]: (i) two stage or (ii) integrated. In the two stage approach all CARs are generated in the first stage and pruned in the second stage. Examples of this approach include Classification based on Multiple Association Rules (CMAR) [21], and Classification Based on Associations (CBA) [22]. Using integrated algorithms the classifier generation is accomplished in a single processing step encompassing both rule generation and pruning. Examples of this latter approach include Classification based on Predictive Association Rules (CPAR) [33] and Total From Partial Classification (TFPC) [12].

2.2 Using Binary Classifiers to Solve Multi-class Classification Problems

With the availability of many effective binary classification algorithms, the multi-class classification problem can be addressed by utilising a "two-class" classifiers; whereby the multi-class problem is decomposed into a binary classification sub-problems addressed using individual binary classifiers. We can, from the literature, identify three commonly referenced methods of using binary classifiers to solve multi-class problems: (i) One-Versus-All (OVA) [28] (ii) All-Verses-All (AVA) [30] (iii) Error-Correcting-Output-Codes (ECOC) [14]. Among these ECOC has often been demonstrated to outperform "stand-alone" multi-class classification algorithms [14,29].

---

[1] The support of a rule describes the number of instances in the training data for which the rule is found to apply [12].

[2] The confidence of the rule is the ratio of its support to the support for its antecedent [12].

2.3   Using Ensemble Classifiers to Solve Multi-class Classification Problems

The use of chains of binary classifiers can be viewed as a form of ensemble classifier. An ensemble classification model is a composite model comprised of a number of classifiers, typically referred to as "base classifiers" or "weak learners", built in order to obtain a better combined global model with more effective classification power than can be acquired from using a single (stand-alone) model. Ensemble methods can be differentiated depending on the relationships between classifiers forming the ensemble, two main types of ensemble can be identified: (i) concurrent ensembles, such as Bagging [23], and (ii) sequential ensembles, such as Boosting [15]. A more recent from of ensemble involves the creation of a hierarchy of classifiers [18,10,32,4,19,24]. A common structure adopted for hierarchical classification is a binary tree constructed in either a bottom-up or top-down manner [18,7]. The top-down model operates in a "divide-and-conquer" manner. The root node contains the entire class label set $\{c_1, c_2, ..., c_n\}$. Starting from the root, the set of class labels at each tree node is recursively divided, and a classifier is trained to discriminate between the two subsets [18]. The process continues until a set of leaf nodes is arrived at. In the bottom-up approach a merging procedure, similar to agglomerative hierarchical clustering, is adopted. At commencement records associate with each class represent the leaf nodes. Nodes separated by the closest "distance" are merged to generate a new node to be associated with a new meta-class [7], and so on until a root node representing the entire class set is arrived at. In this paper an alternative form of generating the desired hierarchies is proposed (see below).

3   Generation of the Hierarchical Model

In this section the generation of the suggested hierarchical model is described. As noted in the introduction to this paper the proposed hierarchical model adopts an ensemble approach founded on the idea of arranging the node classifiers using a binary tree structure. The nature of the classifiers held at each node can be of any form (in previous work the authors have experimented using decision tree and naive bayes classifiers [3,2]), however, with respect to the work described in this paper (and for reasons that will become apparent later in this paper) classifiers generated using Classification Association Rule Mining (CARM) were used. The intuition behind using ensemble classifiers arranged in a hierarchical form is that it could result in better classification accuracy due to: (i) the established observation that ensemble methods tend to improve classification performance [17,35,25,13], and (ii) smaller subsets of class labels are handled at each hierarchy node thus better results might be produced.

In order to group (divide) the input data D during the hierarchy generation process, three different techniques are considered in this paper: (i) k-means clustering, (ii) divisive hierarchical clustering, and (iii) data splitting. Among these k-means is the most commonly used partitioning method where the records are divided into k partitions (in our model $k = 2$ was used because of the binary nature of our hierarchies). The divisive hierarchical clustering approach (top-down) was used to generate a hierarchical decomposition of the given input data. The process starts with all examples in one cluster, in each successive iteration, a cluster is divided into smaller clusters until a "best" cluster configuration is achieved (measured using cluster cohesion and separation measures). The data splitting technique comprises a simple "cut" of the data into two groups so that each contains a disjoint subset of the entire set of class labels. The idea behind the use of clustering algorithms is, at each level and branch of the hierarchy, to group the available class labels into two disjoint groups (clusters) so that the classes within each group share some similar characteristics. Note that, using clustering algorithms for dividing the input data, class labels are considered as a "common" attribute.

The process for generating a hierarchical ensemble binary tree is then as follows. Starting at the root of the hierarchy divide the input data into two groups using one of the proposed clustering or splitting approaches (in acknowledgement of the binary nature of our hierarchies, we refer to these groups as the $leftClassGroup$ and the $RightClassGroup$). Then learn a classifier to distinguish between the two class-groups (in this paper we are using a CARM approach). The process continues recursively until we reach node classifiers that can associate single class labels with records.

4   Operation of the Hierarchical Ensemble Classification Model

Section 3 above explained the process for generating the proposed hierarchical model. After the model has been produced it is ready for usage. The most straightforward mechanism with which to classify a new record is to identify a single path leading through the hierarchy, as dictated by the node classifiers, until a leaf node associated with a single class label is arrived at. In this paper we refer to this as the "Most Confident Path" strategy as it operates by selecting the branch associated with the highest confidence value at each node. The process is as follows. Starting at the root of the hierarchy, the node's CARM classifier classifies the new record as belonging to either the $leftClassGroup$ or the $RightClassGroup$ class. This classification then dictates which branch (left or right) that is followed next. The process proceeds in this manner until a single class label is identified (at a leaf node). This will then be the label to be associated with the record.

However, as already noted, a possible issue with the single path strategy is that if a miss-classification occurs early on in the process their is no opportunity for rectifying this situation later on in the process. To address this problem we propose a multiple path strategy whereby more than one path can be followed within the hierarchy according to a predefined confidence threshold σ, ($0 \leq \sigma \leq 100$). Thus the confidence value $Conf$ associated with class groups ($Conf_{N.leftClassGroup}$ or $Conf_{N.rightClassGroup}$) is used to indicate whether one or two paths (due to the binary structure of our hierarchy) will be followed according to the σ threshold. If the $Conf$ value of the branch associated with the highest confidence is less than σ both branches emanating from the node will be explored further, otherwise the branch with the highest associated $Conf$ value will be selected.

A further issue that results when more than one path is followed through the hierarchy is that more than one final candidate class label may be arrived at, the question then is which class label to select? Two different approaches are suggested to determining the final resulting class label: (i) best confidence and (ii) best normalised accumulated confidence. Using the best confidence approach the "individual" confidence values associated with the identified candidate classes at the leaf nodes are used to select a final class label. When using the best normalised accumulated confidence approach the "normalised accumulated confidence" values associated with the paths that have been followed are used to select a final class label. Thus we have two variations of the multiple class strategy: (i) multiple path with best confidence class label selection ($MultiPathBestConf$) and (ii) multiple path with best normalised accumulated class label selection ($MultiPathNormalisedConf$).

Starting with the $MultiPathBestConf$ approach, Algorithm 1 presents the suggested procedure. The algorithm is similar to the kind of algorithm that might be used to realise the single path strategy (not included in this paper) except with respect to the use of the σ threshold to decide whether to follow a single path or both paths emanating from a node. At the end of the algorithm a list L, which holds all the identified potential class labels with their associated confidence values for the given case, is processed to select the class label c with the highest confidence value. The procedure $MultiPathBestConf(R, N)$ is called recursively as the process progresses. On each recursion the CARM classifier held at the current node is used to produce a confidence value (line 8), with respect to R for the $leftClassGroup$ and the $rightClassGroup$. We then follow one or two paths according the relative nature of the confidence values returned using the CARM classifier at the current node and the σ threshold. Whenever the size of a class group considered at a node is equal to one (lines 11 and 23), indicating that the group comprises a single class label, the class label and associated confidence value are added to L (lines 12 and 24). At the end of the process (line 35) L is processed to identify the class label with the highest associated confidence value.

With respect to the above it should be recalled that a classifier generated using a CARM algorithm comprises a set of CARs whereby the CARs are typically ordered according to confidence value. CARs with the highest confidence are listed first. If two CARs have the same confidence usually the more general rule (that with the smallest antecedent) will appear first, with more specific rules appearing later[3]. Typically the classifier will also include a default rule to be fired when no other rule fits the given example, which will return the most

---

[3] Some authors argue that the more specific rule should be listed first, this remains an open question.

Algorithm 1. *MultiPathBestConf* approach

1: INPUT
2: R a new unseen record
3: N a pointer to the current node in the hierarchy (root node at start)
4: OUTPUT
5: c the predicted class label of the input record R

6: L the set of class labels, together with their associated confidence values, maintained as the procedure progresses, set to {} at start

7: START PROCEDURE $MultiPathBestConf(R, N)$
8: C = Class label set for R with the associated confidence values ($Conf$) generated using classifier held at node N ($C = \{N.leftClassGroup, N.rightClassGroup\}$)
9: **if** ($(Conf(_{N.leftClassGroup}) > Conf(_{N.rightClassGroup})$) **then**
10:     **if** $(Conf(_{N.leftClassGroup}) > \sigma)$ **then**
11:         **if** $(|N.leftClassGroup| == 1)$ **then**
12:             Add class label c_i ($c \in N.leftClassGroup$) to class list L with $Conf_{c_i}$
13:         **else**
14:             $MultiPathBestConf(R, N.leftBranch)$
15:         **end if**
16:     **else**
17:         $MultiPathBestConf(R, N.leftBranch)$
18:         $MultiPathBestConf(R, N.rightBranch)$
19:     **end if**
20:
21: **else**
22:     **if** $(Conf(_{N.rightClassGroup}) > \sigma)$ **then**
23:         **if** $(|N.rightClassGroup| == 1)$ **then**
24:             Add class label $c_i(c \in N.rightClassGroup)$ to class list L with $Conf_{c_i}$
25:         **else**
26:             $MultiPathBestConf(R, N.rightBranch)$
27:         **end if**
28:     **else**
29:         $MultiPathBestConf(R, N.leftBranch)$
30:         $MultiPathBestConf(R, N.rightBranch)$
31:     **end if**
32:
33: **end if**
34: END PROCEDURE $MultiPathBestConf(R, N)$

35: Process L and select class label c with highest confidence

---

frequently occurring class label in the original training set. Given a new record to be classified, the first rule whose antecedent matches the record (or the default rule) is used to classify the record. However, our multi-path strategy requires that we have confidence values for both branches emanating from a node. Thus the CARM classifiers used were modified so that, where possible, the confidence values associated with both branches were returned by finding the first rule in the rule base with respect to both classes (where such rules existed).

Algorithm 2 presents the *MultiPathNormalisedConf* approach. The main difference between the Multiple Path Best Confidence approach and Multiple Path Best Normalised Accumulated Confidence approach is that all confidence values are stored with respect to each path followed (not just the confidence values at the leaf nodes). Consequently a weighting may be derived for each candidate class. We refer to this weighting as a the Normalised Accumulated Confidence (*NormalisedAccumulatedConf*) value. In order to achieve this goal two additional parameters (in addition to the parameters in Algorithm 1) are

Algorithm 2. $MultiPathNormalisedConf$ approach

1: INPUT
2: R a new unseen record
3: N a pointer to the current node in the hierarchy (root node at start)
4: $AccumConf$ the Accumulated summation of the Confidence Value in the followed path (initially 0.0)
5: $ConfCount$ Confidence counter keeping the number of confidence values in the followed path
6: OUTPUT
7: c the predicted class label of the input record R

8: L the set of class labels, together with their associated normalised accumulated confidence values, maintained as the procedure progresses, set to {} at start

9: START PROCEDURE $MultiPathNormalisedConf(R, N, AccumConf, ConfCount)$
10: C = Class label set for R with the associated confidence values $(Conf)$ generated using classifier held at node N ($C = \{N.leftClassGroup, N.rightClassGroup\}$)
11: **if** $(Conf(_{N.leftClassGroup}) > Conf(_{N.rightClassGroup}))$ **then**
12:     $leftAccumConf = Conf(_{N.leftClassGroup}) + AccumConf$
13:     $leftConfCount = ConfCount + 1$
14:     **if** $(Conf(_{N.leftClassGroup}) > \sigma)$ **then**
15:         **if** $(|N.leftClassGroup| == 1)$ **then**
16:             $leftNormalisedAccumConf = leftAccumConf/leftConfCount$
17:             Add class label c_i ($c \in N.leftClassGroup$) to class list L with the
18:             $leftNormalisedAccumConf$.
19:         **else**
20:             $MultiPathBestConf(R, N.leftBranch, leftAccumConf, leftConfCount)$
21:         **end if**
22:     **else**
23:         **if** $(Conf(_{N.rightClassGroup}) \mathrel{!=} Null)$ **then**
24:             $rightAccumConf = Conf(_{N.rightClassGroup}) + AccumConf$
25:             $rightConfCount = ConfCount + 1$
26:         **else**
27:             $rightAccumConf = AccumConf$
28:         **end if**
29:         $MultiPathBestConf(R, N.rightBranch, rightAccumConf, rightConfCount)$
30:         $MultiPathBestConf(R, N.leftBranch, leftAccumConf, leftConfCount)$
31:     **end if**
32: **else**
33:     $rightAccumConf = Conf(_{N.rightClassGroup}) + AccumConf$
34:     $rightConfCount = ConfCount + 1$
35:     **if** $(Conf(_{N.rightClassGroup}) > \sigma)$ **then**
36:         **if** $(|N.rightClassGroup| == 1)$ **then**
37:             $rightNormalisedAccumConf = rightAccumConf/rightConfCount$
38:             Add class label c_i ($c \in N.rightClassGroup$) to class list L with the
39:             $rightNormalisedAccumConf$.
40:         **else**
41:             $MultiPathBestConf(R, N.rightBranch, rightAccumConf, rightConfCount)$
42:         **end if**
43:     **else**
44:         **if** $(Conf(_{N.leftClassGroup}) \mathrel{!=} Null)$ **then**
45:             $leftAccumConf = Conf(_{N.leftClassGroup}) + AccumConf$
46:             $leftConfCount = ConfCount + 1$
47:         **else**
48:             $leftAccumConf = AccumConf$
49:         **end if**
50:         $MultiPathBestConf(R, N.rightBranch, rightAccumConf, rightConfCount)$
51:         $MultiPathBestConf(R, N.leftBranch, leftAccumConf, leftConfCount)$
52:     **end if**
53: **end IF**
54: END PROCEDURE $MultiPathNormalisedConf(R, N, AccumConf, ConfCount)$

55: Process L and select class label c with highest normalised accumulated confidence value

used: *AccumConf* and *ConfCount*, where the *AccumConf* is used to store the summation of the confidence values for the path followed, while *ConfCount* is used to store the number of confidence values for the path followed (so that the final accumulated confidence can be normalised). More specifically, the normalised accumulated confidence value, which will be associated with each candidate class label, is calculated as:

$$NormalisedAccumConf = AccumConf \div Confcount \qquad (1)$$

where $0 \leq NormalisedConf \leq 100$.

As mentioned above our CARM classifiers were modified to return the confidence values associated with both class labels represented by each node. However, in some cases it was not possible to identify both confidence values. In this case only the single identified confidence value was used when calculating the *NormalisedAccumulatedConfidence* for a specific path (See Algorithm 2, Lines 23-27 and 44-48).

5   Experimental Evaluation

In this section we present an overview of the experimental set up used to evaluate the proposed binary tree hierarchical ensemble classification model and the results obtained. Fourteen different data sets (with various numbers of class labels) taken from the UCI data repository [5] were used, pre-processed using LUCS-KDD-DN software [11]. Ten-fold Cross Validation (TCV) was adopted throughout. The evaluation measures used were average accuracy and average AUC (Area Under the receiver operating Curve). Both average accuracy and AUC results are presented here, but for simplicity we will discuss the results in terms of average accuracy.

The objectives of the evaluation were as follows:

1. To compare the operation of the three considered class grouping mechanisms: (i) *k*-means, (ii) divisive hierarchical clustering and (iii) data splitting.
2. To compare the use of the single and multiple path strategies for hierarchical ensemble classification and the two proposed class label selection methods associated with the latter.
3. To compare the operation of the proposed binary tree hierarchical ensemble classification model with stand-alone classification and with alternative established ensemble methods (namely bagging).

The results with respect to the first two objectives are given in Tables 1 and 2. Table 1 gives the TCV classification accuracy values obtained, and Table 2 the AUC results obtained. The results with respect to the third objective are given in Table 3. The results in the context of the above evaluation objectives are discuss in Sub-sections 5.1, 5.2 and 5.3 below.

used: *AccumConf* and *ConfCount*, where the *AccumConf* is used to store the summation of the confidence values for the path followed, while *ConfCount* is used to store the number of confidence values for the path followed (so that the final accumulated confidence can be normalised). More specifically, the normalised accumulated confidence value, which will be associated with each candidate class label, is calculated as:

$$NormalisedAccumConf = AccumConf \div Confcount \tag{1}$$

where $0 \leq NormalisedConf \leq 100$.

As mentioned above our CARM classifiers were modified to return the confidence values associated with both class labels represented by each node. However, in some cases it was not possible to identify both confidence values. In this case only the single identified confidence value was used when calculating the *NormalisedAccumulatedConfidence* for a specific path (See Algorithm 2, Lines 23-27 and 44-48).

5   Experimental Evaluation

In this section we present an overview of the experimental set up used to evaluate the proposed binary tree hierarchical ensemble classification model and the results obtained. Fourteen different data sets (with various numbers of class labels) taken from the UCI data repository [5] were used, pre-processed using LUCS-KDD-DN software [11]. Ten-fold Cross Validation (TCV) was adopted throughout. The evaluation measures used were average accuracy and average AUC (Area Under the receiver operating Curve). Both average accuracy and AUC results are presented here, but for simplicity we will discuss the results in terms of average accuracy.

The objectives of the evaluation were as follows:

1. To compare the operation of the three considered class grouping mechanisms: (i) *k*-means, (ii) divisive hierarchical clustering and (iii) data splitting.
2. To compare the use of the single and multiple path strategies for hierarchical ensemble classification and the two proposed class label selection methods associated with the latter.
3. To compare the operation of the proposed binary tree hierarchical ensemble classification model with stand-alone classification and with alternative established ensemble methods (namely bagging).

The results with respect to the first two objectives are given in Tables 1 and 2. Table 1 gives the TCV classification accuracy values obtained, and Table 2 the AUC results obtained. The results with respect to the third objective are given in Table 3. The results in the context of the above evaluation objectives are discuss in Sub-sections 5.1, 5.2 and 5.3 below.

Algorithm 2. $MultiPathNormalisedConf$ approach

1: INPUT
2: R a new unseen record
3: N a pointer to the current node in the hierarchy (root node at start)
4: $AccumConf$ the Accumulated summation of the Confidence Value in the followed path (initially 0.0)
5: $ConfCount$ Confidence counter keeping the number of confidence values in the followed path
6: OUTPUT
7: c the predicted class label of the input record R

8: L the set of class labels, together with their associated normalised accumulated confidence values, maintained as the procedure progresses, set to {} at start

9: START PROCEDURE $MultiPathNormalisedConf(R, N, AccumConf, ConfCount)$
10: $C =$ Class label set for R with the associated confidence values ($Conf$) generated using classifier held at node N ($C = \{N.leftClassGroup, N.rightClassGroup\}$)
11: **if** ($Conf(_{N.leftClassGroup}) > Conf(_{N.rightClassGroup})$) **then**
12: $\quad$ $leftAccumConf = Conf(_{N.leftClassGroup}) + AccumConf$
13: $\quad$ $leftConfCount = ConfCount + 1$
14: $\quad$ **if** ($Conf(_{N.leftClassGroup}) > \sigma$) **then**
15: $\quad\quad$ **if** ($|N.leftClassGroup| == 1$) **then**
16: $\quad\quad\quad$ $leftNormalisedAccumConf = leftAccumConf/leftConfCount$
17: $\quad\quad\quad$ Add class label c_i ($c \in N.leftClassGroup$) to class list L with the
18: $\quad\quad\quad$ $leftNormalisedAccumConf$.
19: $\quad\quad$ **else**
20: $\quad\quad\quad$ $MultiPathBestConf(R, N.leftBranch, leftAccumConf, leftConfCount)$
21: $\quad\quad$ **end if**
22: $\quad$ **else**
23: $\quad\quad$ **if** ($Conf(_{N.rightClassGroup})$!= Null) **then**
24: $\quad\quad\quad$ $rightAccumConf = Conf(_{N.rightClassGroup}) + AccumConf$
25: $\quad\quad\quad$ $rightConfCount = ConfCount + 1$
26: $\quad\quad$ **else**
27: $\quad\quad\quad$ $rightAccumConf = AccumConf$
28: $\quad\quad$ **end if**
29: $\quad\quad$ $MultiPathBestConf(R, N.rightBranch, rightAccumConf, rightConfCount)$
30: $\quad\quad$ $MultiPathBestConf(R, N.leftBranch, leftAccumConf, leftConfCount)$
31: $\quad$ **end if**
32: **else**
33: $\quad$ $rightAccumConf = Conf(_{N.rightClassGroup}) + AccumConf$
34: $\quad$ $rightConfCount = ConfCount + 1$
35: $\quad$ **if** ($Conf(_{N.rightClassGroup}) > \sigma$) **then**
36: $\quad\quad$ **if** ($|N.rightClassGroup| == 1$) **then**
37: $\quad\quad\quad$ $rightNormalisedAccumConf = rightAccumConf/rightConfCount$
38: $\quad\quad\quad$ Add class label c_i ($c \in N.rightClassGroup$) to class list L with the
39: $\quad\quad\quad$ $rightNormalisedAccumConf$.
40: $\quad\quad$ **else**
41: $\quad\quad\quad$ $MultiPathBestConf(R, N.rightBranch, rightAccumConf, rightConfCount)$
42: $\quad\quad$ **end if**
43: $\quad$ **else**
44: $\quad\quad$ **if** ($Conf(_{N.leftClassGroup})$!= Null) **then**
45: $\quad\quad\quad$ $leftAccumConf = Conf(_{N.leftClassGroup}) + AccumConf$
46: $\quad\quad\quad$ $leftConfCount = ConfCount + 1$
47: $\quad\quad$ **else**
48: $\quad\quad\quad$ $leftAccumConf = AccumConf$
49: $\quad\quad$ **end if**
50: $\quad\quad$ $MultiPathBestConf(R, N.rightBranch, rightAccumConf, rightConfCount)$
51: $\quad\quad$ $MultiPathBestConf(R, N.leftBranch, leftAccumConf, leftConfCount)$
52: $\quad$ **end if**
53: **end if**
54: END PROCEDURE $MultiPathNormalisedConf(R, N, AccumConf, ConfCount)$

55: Process L and select class label c with highest normalised accumulated confidence value

---

Table 1. Average Accuracy for the fourteen evaluation datasets using the three different proposed Hierarchical Ensemble Classification strategies

Data set	Classes	Single Path Stratrgy			Multi. Path Strat. best Conf.			Multi. Path Strat. best Norm. Accum. Conf.		
		K-means	DS	HC	K-means	DS	HC	K-means	DS	HC
WaveForm	3	58.92	57.02	**61.62**	59.02	57.12	61.58	59.02	57.44	61.58
Wine	3	78.99	78.59	**84.28**	78.99	78.59	**84.28**	78.99	78.59	**84.28**
Nursery	5	84.41	**86.81**	53.07	82.53	**86.81**	53.12	80.28	**86.81**	53.06
Heart	5	**53.76**	52.39	53.35	**53.76**	52.39	53.35	**53.76**	52.39	53.35
PageBlocks	5	90.79	89.77	73.73	90.79	89.77	73.88	**91.17**	89.77	73.75
Dermatology	6	74.92	60.28	71.39	75.73	60.28	71.16	**77.34**	60.28	71.16
Glass	7	48.46	**61.56**	52.19	48.46	**61.56**	51.24	48.93	**61.56**	51.71
Zoo	7	**88.00**	85.00	85.00	**88.00**	85.00	85.00	**88.00**	85.00	85.00
Ecoli	8	65.15	**66.57**	52.17	64.55	**66.57**	51.86	64.85	**66.57**	52.17
Led	10	**54.78**	45.53	35.16	52.16	41.38	35.38	52.13	41.53	37.94
PenDigits	10	**61.12**	42.68	50.74	**61.12**	43.89	50.41	60.86	40.12	50.43
Soybean	15	79.01	**88.97**	57.36	79.01	88.43	56.82	79.01	88.43	56.83
ChessKRVK	18	**32.99**	28.13	22.50	32.16	26.58	22.48	31.06	27.28	22.20
LetRecog	26	29.58	**33.12**	21.54	29.17	29.13	21.67	27.87	27.39	21.57
Mean		**64.35**	62.60	55.29	63.96	61.96	55.16	63.81	61.65	55.36

Table 2. Average AUC for the fourteen evaluation datasets using the three different proposed Hierarchical Ensemble Classification strategies

Data set	Classes	Single Path Stratrgy			Multi. Path Strat. best Conf.			Multi. Path Strat. best Norm. Accum. Conf.		
		K-means	DS	HC	K-means	DS	HC	K-means	DS	HC
WaveForm	3	0.59	0.58	**0.61**	0.59	0.58	**0.61**	0.59	0.58	**0.61**
Wine	3	0.81	0.76	**0.86**	0.81	0.76	**0.86**	0.81	0.76	**0.86**
Nursery	5	**0.43**	**0.43**	0.27	0.42	**0.43**	0.27	0.40	**0.43**	0.27
Heart	5	**0.24**	0.20	0.23	**0.24**	0.20	0.23	**0.24**	0.20	0.23
PageBlocks	5	0.24	0.20	0.25	0.24	0.20	0.23	**0.35**	0.20	0.26
Dermatology	6	0.64	0.46	0.66	0.65	0.46	0.66	**0.68**	0.46	0.66
Glass	7	0.29	**0.31**	0.30	0.29	**0.31**	0.30	0.29	**0.31**	0.30
Zoo	7	**0.52**	0.49	0.49	**0.52**	0.49	0.49	**0.52**	0.49	0.49
Ecoli	8	**0.28**	0.20	0.21	**0.28**	0.20	0.20	**0.28**	0.20	0.21
Led	10	**0.55**	0.45	0.35	0.52	0.41	0.35	0.52	0.41	0.38
PenDigits	10	**0.61**	0.42	0.51	**0.61**	0.43	0.50	0.60	0.40	0.50
Soybean	15	0.76	**0.89**	0.52	0.76	**0.89**	0.52	0.76	**0.89**	0.52
ChessKRVK	18	**0.18**	0.14	0.11	0.17	0.13	0.11	0.17	0.13	0.11
LetRecog	26	0.30	**0.33**	0.22	0.29	0.29	0.22	0.28	0.27	0.22
Mean		**0.46**	0.42	0.40	**0.46**	0.41	0.40	**0.46**	0.41	0.40

5.1 Comparison of Class Grouping Mechanisms

As already noted, three techniques were considered for grouping classes at the nodes in the hierarchies: (i) k-means, (ii) data splitting and (iii) divisive hierarchical clustering, indicated by the column headings "K-means", "DS" and "HC" respectively in Tables 1 and 2. Figure 1 shows a comparison between all

the suggested strategies in terms of average accuracy. From the figure it can
be observed that the best method for class distribution between nodes during
the hierarchical model generation was k-means grouping, the second best was
data splitting, while hierarchical clustering produced the worst results. According to the results presented in Table 1, it can be seen that by using the k-means
clustering algorithm to divide classes between nodes within the hierarchy, a best
classification accuracy was obtained for seven of the fourteen datasets considered
in the evaluation (Heart, Page Blocks, Dermatology, Zoo, Led, Pen Digits, Chess
KRvK). While using data splitting produced a best classification accuracy with
respect to five of the datasets (Nursery, Glass, Ecoli, Soybean, Letter Recognition). Using Hierarchical clustering a best classification accuracy was obtained
for only two of the datasets considered (Wine, Wave Form).

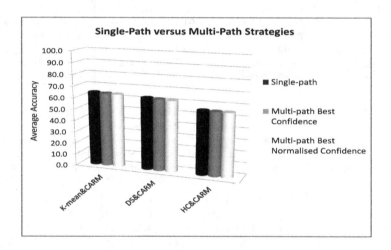

Fig. 1. Comparison between Single-Path and Multi-Path strategies with respect to the
different grouping mechanisms (in terms of average accuracy)

5.2   Comparison of Strategies

With respect to the three proposed hierarchical ensemble classification strategies
(Single path, Multiple path with best confidence class label selection and Multiple path with best normalised accumulative confidence class label selection) it
is interesting to note that their is little difference between the operation of the
strategies. Figure 1 shows three sets of bar charts, one for each strategy, with
respect to the mean accuracy values obtained. With reference to the figure it can
be seen that the Single-path (the most confident path) strategy produced the
best overall result (using k-means grouping). Out of the two multi-path strategies
the best confidence approach tended to produce the best performance.

Regarding the Multiple Path strategies a threshold of $\sigma = 70.0$ was used. A range of alternative σ values were evaluated and it was demonstrated that $\sigma = 70.0$ produced the best results.

From the results presented in Table 1 it can be observed that for six datasets (Wine, Nursery, Heart, Glass, Zoo, and Ecoli) the same classification accuracy was obtained regardless of which strategy was adopted. For another six datasets (Wave Form, Led, Pen Digits, Soybean, Chess KRvK, and Letter Recognition) the single path strategy produced the best classification accuracy, although for one dataset (Pen Digits) the same result was produced as for the multi-path best confidence strategy. For two of the datasets (Page Blocks, and Dermatology) the multi-path normalised accumulated confidence strategy produced the best classification accuracy. The reason behind the weakness of the multi-path strategies is that, in some cases, it was not possible to identify both confidence values for a given node branches. Consequently the unknown confidence values affected the results of multi-path strategies tending it to less value than initially anticipated.

Table 3. Accuracy and AUC Comparison between "stand-alone" CARM, a conventional Bagging ensemble, and the best hierarchical ensemble classification strategy from Sub-setion 5.2

Data set	Classes	CARM		Bagging		Best Hier. Esemb. Class.	
		Acc.	AUC	Acc.	AUC	Acc.	AUC
WaveForm	3	60.04	0.60	60.76	**0.61**	**61.62**	**0.61**
Wine	3	71.88	0.74	61.48	0.61	**84.28**	**0.86**
Nursery	5	73.94	0.36	73.94	0.36	**86.81**	**0.43**
Heart	5	51.70	0.20	45.49	0.24	**53.76**	**0.24**
PageBlocks	5	89.99	0.21	89.95	0.21	**91.17**	**0.35**
Dermatology	6	77.00	0.66	72.12	0.62	**77.34**	**0.68**
Glass	7	**65.05**	**0.43**	53.30	0.31	61.56	0.31
Zoo	7	**94.00**	**0.59**	83.00	0.46	88.00	0.52
Ecoli	8	49.98	0.12	37.90	0.07	**66.57**	**0.28**
Led	10	**67.28**	**0.67**	67.09	**0.67**	54.78	0.55
PenDigits	10	75.99	0.76	**77.09**	**0.77**	61.12	0.61
Soybean	15	84.01	0.86	73.15	0.77	**88.97**	**0.89**
ChessKRVK	18	17.64	0.07	17.43	0.06	**32.99**	**0.18**
LetRecog	26	30.91	0.31	30.72	0.31	**33.12**	**0.33**
Mean		64.96	0.47	60.24	0.43	**67.11**	**0.49**

5.3 Comparison of Hierarchical Ensemble Classification with Conventional Approaches

Table 3 presents the results obtained using a "stand alone" CARM and Bagging together with the best results from Tables 1 and 2. The proposed hierarchical ensemble classification was compared with CARM because we wanted to compare the operation of the proposed ensemble approach with the operation of the more traditional stand-alone form of classification; CARM was used for this purpose because the classifiers held at the individual nodes in our tree ensembles were also generated in this manner (clearly we could also have compared with alternative forms of stand alone classification). We compared with bagging because we also wanted to compare the operation of our hierarchical ensemble classification with an alternative form of ensemble. From Table 3 it can be seen that the proposed hierarchical techniques can significantly improve the classification accuracy with respect to ten of the fourteen datasets considered (Wave Form, Wine, Nursery, Heart, Page Blocks, Dermatology, Ecoli, Soybean, Chess KRvK, and Letter Recognition). In the remaining four cases, the stand-alone CARM classifier produced the best result for three of the datasets (Glass, Zoo, and Led), while the Bagging ensemble classifier produced the best result for only one dataset (Pen Digits).

6   Conclusion and Future Work

In this paper hierarchical classification using a binary tree structure, for solving multi-class problems, has been considered. To generate such a hierarchical model three different techniques to distribute class labels between nodes within the hierarchy were proposed: (i) k-means, (ii) data splitting, and (iii) hierarchical clustering. CARM classifiers were used at each hierarchical node. By utilising the confidence values generated by CARM classifiers two approaches were proposed: (i) Single-Path, and (ii) Multi-Path. The later one coupled with two alternatives to determining the final resulting class label: Best Confidence and Best Normalised Accumulated Confidence. The conjecture here was that a criticism of hierarchical classification ensembles is that a miss-classification made early on in the hierarchy cannot be rectified later on and that this might hinder the effectiveness of the operation of such classifiers.

From the reported experimental results, presented in this paper, it was demonstrated that the proposed hierarchical classification model tends to perform better than stand-alone classifiers and other forms of ensemble classifier such as bagging, with respect to some datasets considered in the evaluation; especially data sets that featured a large number of class labels.

Although best overall result was obtained using the proposed Single-path strategy (the most confident path) the Multi-path strategy coupled with either best confidence or best normalised accumulated confidence improved the classification accuracy with respect to some datasets considered in the evaluation. An issue with Multiple Path strategies is that the confidence values that were used to determine wether one or two paths emanating from a node will be followed, can not be always obtained. Consequently, in many cases, it was not possible to follow more than one path even if this might have been desirable.

With respect to future work the authors intend to investigate alternative forms of hierarchical ensembles using more sophisticated structures than the binary tree structures considered in this paper, such as Directed Acyclic Graph (DAG). The idea is that by using DAGs many paths can be followed because of the many possible combinations of class labels at each level.

References

1. Agrawal, R., Srikant, R.: Fast algorithms for mining association rules. In: VLDB 1994, pp. 487–499 (1994)
2. Alshdaifat, E., Coenen, F., Dures, K.: Hierarchical classification for solving multi-class problems: A new approach using naive bayesian classification. In: Motoda, H., Wu, Z., Cao, L., Zaiane, O., Yao, M., Wang, W. (eds.) ADMA 2013, Part I. LNCS, vol. 8346, pp. 493–504. Springer, Heidelberg (2013)
3. Alshdaifat, E., Coenen, F., Dures, K.: Hierarchical single label classification: An alternative approach. In: Bramer, M., Petridis, M. (eds.) SGAI Conf., pp. 39–52. Springer (2013)
4. Athimethphat, M., Lerteerawong, B.: Binary classification tree for multiclass classification with observation-based clustering. In: 2012 9th International Conference on Electrical Engineering/Electronics, Computer, Telecommunications and Information Technology (ECTI-CON), pp. 1–4 (2012)
5. Bache, K., Lichman, M.: UCI machine learning repository (2013), http://archive.ics.uci.edu/ml
6. Bay, S.D.: Combining nearest neighbor classifiers through multiple feature subsets. In: Proc. 17th Intl. Conf. on Machine Learning, pp. 37–45. Morgan Kaufmann (1998)
7. Beygelzimer, A., Langford, J., Ravikumar, P.: Multiclass Classification with Filter Trees (June 2007), http://hunch.net/~jl/projects/reductions/mc_to_b/invertedTree.pdf
8. Breiman, L.: Bagging predictors. Machine Learning 24(2), 123–140 (1996)
9. Breiman, L.: Random forests. In: Machine Learning, pp. 5–32 (2001)
10. Chen, Y., Crawford, M.M., Ghosh, J.: Integrating support vector machines in a hierarchical output decomposition framework. In: 2004 International Geosci. and Remote Sens. Symposium, pp. 949–953 (2004)
11. Coenen, F.: The LUCS-KDD discretised/normalised arm and carm data library (2003), http://www.csc.liv.ac.uk/~frans/KDD/Software/LUCS_KDD_DN
12. Coenen, F., Leng, P.: The effect of threshold values on association rule based classification accuracy. Journal of Data and Knowledge Engineering 60(2), 345–360 (2007)
13. Dietterich, T.G.: Ensemble methods in machine learning. In: Kittler, J., Roli, F. (eds.) MCS 2000. LNCS, vol. 1857, pp. 1–15. Springer, Heidelberg (2000), http://dl.acm.org/citation.cfm?id=648054.743935
14. Dietterich, T.G., Bakiri, G.: Solving multiclass learning problems via error-correcting output codes. JAIR (1995)
15. Freund, Y., Schapire, R., Abe, N.: A short introduction to boosting. Journal of Japanese Society for Artificial Intelligence 14(5), 771–780 (1999)

16. Han, J., Pei, J., Yin, Y.: Mining frequent patterns without candidate generation. SIGMOD Rec. 29(2), 1–12 (2000), http://doi.acm.org/10.1145/335191.335372
17. Jiawei, H., Micheline, K., Jian, P.: Data Mining: Concepts and Techniques. Morgan Kaufmann (2011)
18. Kumar, S., Ghosh, J., Crawford, M.M.: Hierarchical fusion of multiple classifiers for hyperspectral data analysis. Pattern Anal. Appl. 5(2), 210–220 (2002)
19. Lei, H., Govindaraju, V.: Half-against-half multi-class support vector machines. In: Oza, N.C., Polikar, R., Kittler, J., Roli, F. (eds.) MCS 2005. LNCS, vol. 3541, pp. 156–164. Springer, Heidelberg (2005)
20. Leonard, T., Hsu, J.S.: Bayesian Methods: An Analysis for Statisticians and Inter-disciplinary Researchers. Cambridge University Press (2001)
21. Li, W., Han, J., Pei, J.: Cmar: Accurate and efficient classification based on multiple class-association rules. In: Proceedings of the 2001 IEEE International Conference on Data Mining, ICDM 2001, pp. 369–376. IEEE Computer Society, Washington, DC (2001), http://dl.acm.org/citation.cfm?id=645496.657866
22. Liu, B., Hsu, W., Ma, Y.: Integrating classification and association rule mining. In: Proc. KDD 1998 Conference (AAAI 1998), pp. 80–86 (1998)
23. Machov, K., Bark, F., Bednr, P.: A bagging method using decision trees in the role of base classifiers (2006)
24. Madzarov, G., Gjorgjevikj, D., Chorbev, I.: A multi-class svm classifier utilizing binary decision tree (2008)
25. Oza, N., Tumer, K.: Classifier ensembles: Select real-world applications. Information Fusion 9(1), 4–20 (2008), http://dx.doi.org/10.1016/j.inffus.2007.07.002
26. Quinlan, J.R.: Induction of decision trees. Machine Learning 1(1), 81–106 (1986)
27. Quinlan, J.R.: C4.5: Programs for Machine Learning. Morgan Kaufmann (1993)
28. Rifkin, R.M., Klautau, A.: In defense of one-vs-all classification. Journal of Machine Learning Research 5, 101–141 (2004)
29. Schapire, R.E.: Using output codes to boost multiclass learning problems. In: Machine Learning: Proceedings of the Fourteenth International Conference (ICML1997) (1997)
30. Tax, D.M.J., Duin, R.P.W.: Using two-class classifiers for multiclass classification. In: ICPR (2), pp. 124–127 (2002)
31. Vapnik, V.N.: The Nature of Statistical Learning Theory. Statistics for Engineering and Information Science. Springer (2000)
32. Vural, V., Dy, J.G.: A hierarchical method for multi-class support vector machines. In: Proceedings of the Twenty-First International Conference on Machine Learning, ICML 2004, p. 105. ACM, New York (2004), http://doi.acm.org/10.1145/1015330.1015427
33. Yin, X., Han, J.: Cpar: Classification based on predictive association rules (2003)
34. Zhang, G.P.: Neural networks for classification: A survey. IEEE Transactions on Systems, Man, and Cybernetics-Part C: Applications and Reviews 30(4), 451–462 (2000)
35. Zhou, Z.H.: Ensemble learning. In: Li, S.Z., Jain, A.K. (eds.) Encyclopedia of Biometrics, pp. 270–273. Springer US (2009), http://dblp.uni-trier.de/db/reference/bio/e.html#Zhou09

Label Correction Strategy on Hierarchical Multi-Label Classification

Thanawut Ananpiriyakul, Piyapan Poomsirivilai, and Peerapon Vateekul

Department of Computer Engineering, Chulalongkorn University, Bangkok, Thailand
{Thanawut.A,Piyapan.P}@Student.Chula.ac.th,
Peerapon.V@Chula.ac.th

Abstract. One of the most popular approaches to solve hierarchical multi-label classification problem is to induce Support Vector Machine (SVM) for each class in the hierarchy independently and employ them in a top-down fashion. This approach always suffers from error propagation and yields such a poor performance of classifiers at the lower levels since no label correlation is considered during the construction. In this paper, we present a novel method called "label correction", which takes label correlation into consideration and corrects the results of unusual prediction patterns. In the experiment, our method does not only improve prediction accuracy on data in hierarchical domains, but it also contributes such a significant impact on data in multi-label domains.

Keywords: Hierarchical Multi-Label Classification, Multi-Label Classification, Label Correlation, Support Vector Machine.

1   Introduction

Over the last decade or so, the need for systems capable of automated classification has become more urgent and complex. The most traditional classification assumes that each example can be assigned to only one out of two or more classes. However, a more complex scenario called *"Hierarchical Multi-Label Classification (HMC)"* has recently received numerous attentions from various applications. It is different from the conservative one in two aspects: (*i*) multi-label classification: each example can belong to more than one class simultaneously, and (*ii*) hierarchical classification: classes are organized in the form of hierarchy.

Support Vector Machine (SVM) [1,2] is the state-of-the-art in classification techniques. It is originally used as a binary classifier by constructing a hyperplane to separate between positive and negative classes. In the multi-label domain, the problem is needed to be transformed in order to employ SVM. The most common approach is called Binary Relevance (BR) [3,4] which constructs a one-vs-rest classifier for each class separately without a consideration in a class correlation; thus, it often results in low prediction accuracy. Label Powerset (LP) [5] is an alternative approach which takes a label correlation into account. It induces a set of binary classifiers equal to the number of all possible class combinations; thus, it has prohibitive computational cost and low scalability. Since the number of classes can be very large,

P. Perner (Ed.): MLDM 2014, LNAI 8556, pp. 213–227, 2014.

it is impossible to apply LP, so BR is only an option. It will be referred to as "SVM-BR", as a baseline method.

In the hierarchical domain, a set of classifiers in the BR approach is modified and used in a top-down fashion from more general to more specific classifiers in order to satisfy a hierarchical constraint – if an example belongs to class c_i, it must belong to all c_i's ancestor classes. With this strategy, classifiers at the lower levels always suffer very much from errors generated by those at the upper levels, which is known as "propagated error". The more the levels in the hierarchy, the less the prediction performance.

Since HMC is a composite of multi-label and hierarchical classifications, it inherits the issues from both domains including (*i*) label correlation and (*ii*) error propagation. In this paper, we introduce a novel method called "Label Correction (LC)" to improve prediction accuracy in the HMC domain by focusing on both issues. For the label correlation issue, there is a hypothesis that the same set of class labels should be used to annotate examples throughout the dataset; therefore, a set of class labels in the testing data should be same or close to that in the training data. According to this hypothesis, for a given testing example, if a prediction label set does not match to any label sets occurred in the training data, it should be corrected to the closest training label set. For the error propagation issue, the proposed label correction is enhanced by taking a hierarchical relationship into account. Since subclasses have a strong correlation to their superclasses in the hierarchy, our strategy can correct propagated errors by using such a relationship and regain performance specifically of classifiers at the lower levels of the hierarchy.

To provide a strong evidence of the improvement, the experiments were intensively conducted on several benchmark datasets: 22 multi-label datasets and 10 HMC datasets. The results compared to the baseline SVM show that there is a significant improvement on 18 multi-label datasets and 8 HMC datasets. Moreover, in all HMC datasets, the performance of classifiers at the lower levels is tremendously improved. In the D15_scop_GO dataset, an average performance of the deepest-level classifiers is increased by 250% over the baseline SVM.

The rest of the paper is organized as follows. Section 2 introduces problem transformation and SVM. Section 3 declares relevant performance criteria. Section 4 describes the proposed method in detail. Section 5 shows the results of extensive experiments and discussion. Section 6 is the conclusion of this paper.

2   Related Work

In this section, definition of classification problem is provided. Then, existing classification strategies both in multi-label and hierarchical domains are introduced. Finally, the detail of SVM is mentioned. Note that SVM is one of the most popular classification algorithms which is chosen as a core classification algorithm in this work.

2.1   Problem Statement

Multi-Label Classification is a domain where an example can be associated to multiple classes simultaneously. Let $X \subseteq \mathbb{R}^n$ be an example space of n-dimensional features, and Y be a finite set of classes. Let D be a set of training examples of $(\overrightarrow{x_i}, y_i)$ which contains a feature vector $\overrightarrow{x_i} \in X$ and belongs to multiple classes $y_i \subseteq Y$. The goal of multi-label classification is to find the classifier $g: X \rightarrow 2^Y$ that maps example x to a set of classes y. This kind of tasks can be encountered in various domains, including image [5], text [6], biology [7], audio [8], and video [9]. For example, movie genre classification is multi-label classification because each movie can belong to more than one genre, e.g., both adventure and also fantasy movie simultaneously.

Hierarchical Multi-Label Classification (HMC) refers to a task that follows a multi-label assumption, and classes are organized in a hierarchical structure. In addition, if an example belongs to a class c_i, it must belong to all c_i's superclasses. This property is called "*hierarchical constraint*". Let $H = (C, E)$ be a class hierarchy consisting of a set of class nodes, $C \subseteq Y$, and a set of edges, E. Let $\prec$ represent superclass relationship, and classes $\{c_1, c_2\} \subset C$; $c_1 \prec c_2$ if and only if c_1 is a superclass of c_2. Denote $(\overrightarrow{x_i}, y_i)$ as a tuple of example, which $\overrightarrow{x_i}$ is a feature vector and y_i is a set of classes. Note that y_i must satisfy the hierarchical constraint if $c_1 \prec c_2$ and $c_2 \in y_i$ then $c_1 \in y_i$. The HMC goal is to find the classifier $g: X \rightarrow 2^Y$ that maps example x to a set of classes y such that $g(x)$ follows the hierarchical constraint. Recently this domain has gained a lot of attentions in various applications including functional genomics [10], image annotation [11], and text classification [12].

2.2 Existing Strategies on Multi-label Classification

There are two main approaches [3] including: Algorithm Adaptation (AA) and Problem Transformation (PT). The first approach aims to invent a new algorithm specifically designed for multi-label classification. Many AA algorithms were proposed such as ML-KNN [13] and Ranked-SVM [7]. Conversely, the latter approach tries to apply existing classification techniques by transforming a multi-label classification task into a set of binary classification tasks. In this paper, the proposed technique is a problem transformation, which SVM is a baseline binary classifier. There are two common strategies of how to transform a multi-label problem: Binary Relevance (BR) and Label Powerset (LP).

Binary Relevance constructs a set of binary classifiers which distinguish one single class from all other classes (one-vs-rest). The number of classifiers is equal to the number of classes. For each class y, a binary classifier is constructed from a training data, which examples are labeled as positive if they originally belong to class y, while the rest are labeled as negative. In the testing process, a testing example is predicted by all classifiers, and it belongs to any classes whose score is higher than a predefined threshold. Each classifier is treated separately both in training and testing process, so, BR does not take advantage of label correlation.

Label Powerset is an alternative strategy to transform a multi-label problem into a set of all possible combinations of labels. Then, each of them is considered as a new class label. Although this strategy considers a class relationship, the number of transformed labels grows exponentially from that of the original classes. It, thus, needs a high induction time and cannot be scalable. Since the number of classes in experimental datasets can be large, LP is not included in this paper. Moreover, an issue of imbalanced training sets always occurs when there are a small number of examples described by distinct label sets.

2.3   Top-Down Approach on Hierarchical Multi-Label Classification

In HMC, classes are linked together and form hierarchical structure. To employ BR in HMC, each classifier is constructed separately as in BR but training examples are limited to its immediate superclasses. In prediction process, top-down induction algorithm is used to guarantee hierarchical constraint satisfaction which starts from more general (top-level) to more specific (bottom-level) classifiers.

However, top-down approach always suffers from propagation of error where misclassified examples at the upper level are propagated to lower level classifiers. Propagation of error directly affects overall performance particularly on deep levels.

2.4   Support Vector Machine (SVM)

It is a famous classification algorithm proposed by [1,2]. It was originally developed for dichotomous domains, where each example x_i is assigned to a binary class, y_i (either positive or negative) in a given training data, D. The goal of SVM is to induce a hyperplane function $f(\vec{w}, b) = \vec{w} \cdot x + b$ as follows:

$$\text{Minimize}_{w,b,\xi} \quad \frac{1}{2} w^T w + C \sum_{i=1}^{|D|} \xi_i$$

$$subject\ to \quad y_i(w^T \phi(x_i) + b) \geq 1 - \xi_i \quad , \xi_i \geq 0. \tag{1}$$

Here, weight vector $\vec{w}$ is a vector of the separable hyperplane and b is a bias to determine the offset of the hyperplane relatively to system coordinates. C is a penalty parameter of error due to misclassifications. In a non-linear separable problem, SVM can induce a hyperplane on a higher dimension space using a kernel function below, where $\phi(x_i)$ is a mapping function:

$$K(x_i, x_j) \equiv \phi(x_i)\phi(x_j) \tag{2}$$

In an imbalanced training data, SVM often shows low prediction accuracy since its result tends to be bias towards a majority class. As mentioned in Section 2.2, a training data in LP is more imbalanced than that in BR, so BR is recommended in the HMC domain. It will be referred to as "SVM-BR", as a baseline method for the rest of the paper.

3   Performance Evaluation

3.1   Binary Classification Criteria

In the domain of binary classes (positive and negative), there are 4 based quantities (Table 1) to compute precision and recall as shown in the following equations. Precision (Pr) is a portion of positive prediction examples that are truly positive, and recall (Re) is a portion of truly positive examples that are correctly predicted positive.

Table 1. Prediction criteria

	Truly A	Truly Not A
Predicted A	True Positive (*tp*)	False Positive (*fp*)
Predicted Not A	False Negative (*fn*)	True Negative (*tn*)

Precision, recall and F_β-measure introduced by [14] are shown in (3).

$$Pr = \frac{tp}{tp+fp} \qquad Re = \frac{tp}{tp+fn} \qquad F_\beta = \frac{(\beta^2+1)\times Pr\times Re}{\beta^2\times Pr+Re} \; ; \; \beta > 0 \qquad (3)$$

F_β measure aims to maximize both precision and recall value. When $\beta > 1$, this measure emphasizes precision value. Conversely, when $\beta \in (0,1)$, this measure emphasizes recall value. In this paper, we use $\beta = 1$ which weighs precision and recall equally shown in (4).

$$F_1 = \frac{2\times Pr\times Re}{Pr+Re} \qquad (4)$$

3.2   Multi-label Classification Criteria

There are two ways to combine those classical measures in Section 3.1 [15]: macro-averaging and micro-averaging as in Table 2. Macro-averaging gives an equal weight to every class, whereas micro-averaging weights each class relatively to its example frequency. In an imbalanced circumstance, it is suitable to use macro-averaging over micro-averaging in order to avoid a dominance of majority classes.

Table 2. Macro-averaging and micro-averaging of precision, recall, and F_β, i is a class index

	Macro-averaging	Micro-averaging
Precision	$MaPr = \frac{1}{|L|}\sum_{i=1}^{|L|} Pr_i$	$MiPr = \frac{\sum_{i=1}^{|L|} tp_i}{\sum_{i=1}^{|L|}(tp_i + fp_i)}$
Recall	$MaRe = \frac{1}{|L|}\sum_{i=1}^{|L|} Re_i$	$MiRe = \frac{\sum_{i=1}^{|L|} tp_i}{\sum_{i=1}^{|L|}(tp_i + fn_i)}$
F_β	$MaF_\beta = \frac{1}{|L|}\sum_{i=1}^{|L|} F_{\beta,i}$	$MiF_\beta = \frac{(\beta^2+1)\times MiPr\times MiRe}{\beta^2\times MiPr+MiRe}$

3.3   Hierarchical Multi-Label Classification Criteria

In the HMC domain, a hierarchical measure was introduced by [16]. It is also known as "an example-based measure" because it evaluates a performance for each example along the hierarchical path (Table 3). To evaluate the overall performance (all examples), it is common to employ macro-averaging (*MaHPr, MaHRe, MaHF₁*) and micro-averaging (*MiHPr, MiHRe, MiHF₁*) from the previous section. Note that the measures in Section 3.2 is also referred to as "a label-based measure" since it evaluates a performance for each class label.

Table 3. Example-based precision, recall, and F_1 calculation of j-th example, where $\hat{T}_j$ and $\hat{P}_j$ is set of true classes and set of predicted classes of j-th example

Precision	Recall	F_1
$HPr_j = \dfrac{\left\|\hat{P}_j \cap \hat{T}_j\right\|}{\left\|\hat{P}_j\right\|}$	$HRe_j = \dfrac{\left\|\hat{P}_j \cap \hat{T}_j\right\|}{\left\|\hat{T}_j\right\|}$	$HF_{1,j} = \dfrac{2 \times HPr_j \times HRe_j}{HPr_j + HRe_j}$

Reference [17] proposed a measure called "an example-label-based measure" which uses a harmonic mean to combine those two criteria (per example and per class label) into a single unit. Each entry value of example-label-based is derived from harmonic mean of label-based and example-based. For example, (5) shows macro-averaging example-label-based F_1.

$$Macro\text{-}averaging\ example\text{-}label\text{-}based\ F_1 = \frac{2 \times MaHF_1 \times MaF_1}{MaHF_1 + MaF_1} \qquad (5)$$

4   A Proposed Method

There are three main modules in the proposed system (Fig. 1): (*i*) label correction, (*ii*) performance enhancement, and (*iii*) hierarchical path adjustment. The first module aims to correct a prediction label set to match the closest label set in the training data. Then, the second module ensures that the corrected label sets do not lessen the overall system performance. There is a selection heuristic to correct only some classes; not all classes and excluding classes whose performance is decreased after the correction. Finally, the third module is only applied in the hierarchical domain in order to adjust the result of second module to satisfy the hierarchical constraint.

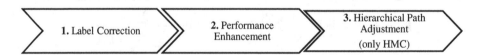

Fig. 1. A process diagram of the proposed system

4.1   Label Correction

First, a set of distinct label sets is extracted from the training data. Each of them is a vector of +1 (positive) and -1 (negative). Then, a distance (cost) between each prediction to the distinct label sets is computed. Finally, if the prediction result does not match to any label sets, it is corrected to match to the closest one (the smallest cost). There are two choices of how to compute the cost: (*i*) constant cost and (*ii*) score cost. In (6), the cost is the total number of labels with different sign between a prediction vector and a training label set vector. In (7), a vector of SVM scores is used instead of a vector of predictions. The cost is a summation of square root of the absolute score of classes with different sign. The range of SVM scores (variance) is varied on different classes and can affect the result, so the square root is chosen in order to alleviate the bias of classes with high variance.

Fig. 2 illustrates an example of how to compute the cost in the label correction module. Assume there are 5 classes, $L = \{A, B, C, D, E\}$, with 4 distinct training label sets, $S = \{A, AB, BE, ACE\}$. Given a testing example x, if a vector of SVM prediction scores is [-0.2, 0.1, 1.5, -0.8, 0.4], a vector of predictions will be [-1, +1, +1, -1, +1] using a sign function which is BCE. Unfortunately, this prediction pattern does not match any training label sets, so this result should be corrected. Using the constant cost, the new prediction is BE with the cost of 1 computed as (6), while the new result is ACE with the score cost of 0.76 computed as (7). Whether distance cost function is preferred will be investigated in the experiment.

$$cost_{constant}\left(vector\ v_i, label\ set\ l_j\right) = \sum_{sign(v_i(k))\neq sign(l_i(k))}^{class\ k} 1 \tag{6}$$

$$cost_{score}\left(vector\ v_i, label\ set\ l_j\right) = \sum_{sign(v_i(k))\neq sign(l_i(k))}^{class\ k} \sqrt{|v_i(k)|} \tag{7}$$

Class = $\{A, B, C, D, E\}$
Distinct Set = $\{A, AB, BE, ACE\}$
SVM Score Vector = [-0.2, 0.1, 1.5, -0.8, 0.4] = $\{BCE\}$

distinct set \ SVM score vector	-0.2	0.1	1.5	-0.8	0.4	$cost_{constant}$	$cost_{score}$
A [+1,-1,-1,-1,-1]	$\sqrt{0.2}$	$\sqrt{0.1}$	$\sqrt{1.5}$	0	$\sqrt{0.4}$	4	2.62
AB [+1,+1,-1,-1,-1]	$\sqrt{0.2}$	0	$\sqrt{1.5}$	0	$\sqrt{0.4}$	3	2.30
BE [-1,+1,-1,-1,+1]	0	0	$\sqrt{1.5}$	0	0	1	1.22
ACE [+1,-1,+1,-1,+1]	$\sqrt{0.2}$	$\sqrt{0.1}$	0	0	0	2	**0.76**

From (6), the lowest $cost_{constant}$ is 1 and the prediction result should be BE.
From (7), the lowest $cost_{score}$ is 0.76 and the prediction result should be ACE.

Fig. 2. An example of label correction process

4.2 Performance Enhancement

In the first process, the prediction results are forced to exactly match one of the distinct label sets. It is possible that the performance of some classes may be dropped. The performance enhancement module is proposed to guarantee that this scenario has never happened; therefore, there is always an improvement in the overall system performance – in terms of F_1.

All the steps of this module are conducted on the training data. First, F_1-score for each class is evaluated on the baseline system, SVM-BR. Second, our label correction (1st module) is applied and, then, F_1-score for each class is reevaluated. Finally, if the performance of any classes is dropped after using the label correction, this module will keep those classes' result of the original system without any correction.

Fig. 3 shows F_1-score for each class of SVM-BR, the traditional system without our label correction. After applying our strategy, F_1-result for each class is shown in Fig. 4. Fig. 4 (a) and Fig. 4 (b) are the results without and with applying the performance enhancement module. In Fig. 4 (b), this module excludes classes A and B from the label correction since their F_1-scores (0.76 and 0.49) in Fig. 4 (a) are dropped comparing to those of the baseline system (0.8 and 0.55) in Fig. 3.

Since some classes are excluded from label correction, the hierarchical constraint may not be satisfied. In that case, the next phase will adjust the result to guarantee the hierarchical constraint.

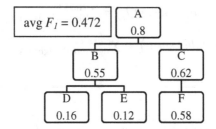

Fig. 3. F_1-score of SVM-BR (without our label correction)

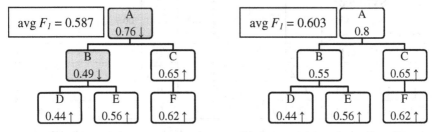

(a) F_1-score without performance enhancement   (b) F_1-score with performance enhancement

Fig. 4. An example of performance enhancement process. ↑ and ↓ refer to as increase or decrease in terms of F_1 comparing to SVM-BR in Fig. 3. The gray node is a class, whose F_1 is dropped.

4.3   Hierarchical Path Adjustment

To adjust the result along the hierarchical path, there are three proposed strategies. First, the decreasing strategy is to follow the superclass' prediction. All subclasses' results are assigned to be negative when the superclassifier predicts as negative. Second, the increasing strategy is to depend on the subclass' result. If the subclassifier predicts as positive, the results of its ancestors must also be positive. Third, the voting strategy is whether adjust the result depending on the majority class of the prediction. On the one hand, if the majority class along the path is negative, the result is adjusted using the decreasing strategy. On the other hand, the prediction is modified using the increasing strategy.

Fig. 5 illustrates an example of the hierarchical adjustment for the class path {A, B, D}. Fig. 5 (a) is the result without the adjustment as {Yes, No, Yes}, where the results of classes B and D disagree and violate the hierarchical constraint. The results of Fig. 5 (b) and Fig. 5 (c) are {Yes, No, **No**} and {Yes, **Yes**, Yes} adjusted using the decreasing and increasing strategies respectively. Since classifier A and D predict as "Yes" and only classifier B predicts as "No", so, the majority of path ABD is "Yes", then, the voting strategy in Fig. 5 (d) gives the same result as increasing strategy, {Yes, **Yes**, Yes}.

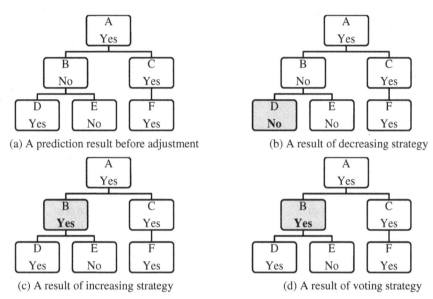

(a) A prediction result before adjustment   (b) A result of decreasing strategy

(c) A result of increasing strategy   (d) A result of voting strategy

Fig. 5. An example of hierarchical path adjustment. The gray node is a class adjusted to follow the hierarchical constraint.

5   Experimental Setup

There are several real-world datasets used in the experiment. The statistical significance is tested on 3-fold cross-validation. A process of data preparation is summarized as follows:

- **Missing Data Handling:** In the case of continuous features, a missing value is replaced by an average of complete data. Otherwise, in the case of categorical features, an imputation is based on the most frequent value.
- **Categorical Data Handling:** Since SVM supports only continuous features, all categorical variables must be transformed to a set of dummy variables. For example, a categorical feature with 3 values, {A, B, C}, is transformed into a set of three binary variables, which $A = (1, 0, 0)$, $B = (0, 1, 0)$, and $C = (0, 0, 1)$.
- **Data Normalization:** All features are normalized into a range of $[0, 1]$.

In the multi-label domain, there are 22 experimental datasets [18] as in Table 4. Let L be a set of classes in the dataset D, $|D|$ be the number of examples, $|L|$ be the number of total classes, and $|Y_i|$ be the number of classes of i-th example. Cardinality [3], density [3], and distinct are calculated using the following equations:

$$Cardinality = \frac{1}{|D|}\sum_{i=1}^{|D|}|Y_i| \quad Density = \frac{1}{|D|}\sum_{i=1}^{|D|}\frac{|Y_i|}{|L|} \quad Distinct = \left|\bigcup_{i=1}^{|D|}\{Y_i\}\right| \quad (8)$$

Table 4. Statistics of multi-label datasets

Dataset	$|D|$	Features	Classes	Cardinality	Density	Distinct
arts	7,484	23,147	26	1.654	0.064	599
bibtex	7,395	1,836	159	2.402	0.015	2,856
business1	11,214	21,925	30	1.599	0.053	233
cal500	502	68	174	26.044	0.150	502
computers1	12,444	34,097	33	1.507	0.046	428
education1	12,030	27,535	33	1.463	0.044	511
emotions	593	72	6	1.868	0.311	27
enron	1,702	1,001	53	3.378	0.064	753
entertainment1	12,730	32,002	21	1.414	0.067	337
genbase	662	1,848	27	1.252	0.046	32
health1	8,867	30,606	32	1.645	0.051	330
mediamill	43,907	120	101	4.376	0.043	6,555
medical	978	1,449	45	1.245	0.028	94
rcv1regions	23,149	47,152	228	1.280	0.006	1,689
rcv1topics	28,596	49,060	22	2.158	0.098	1,341
recreation1	12,828	30,325	22	1.429	0.065	530
reference1	8,027	39,680	33	1.174	0.036	275
scene	2,407	294	6	1.074	0.179	15
science1	6,428	37,188	40	1.450	0.036	457
social1	12,111	52,351	39	1.279	0.033	361
tmc2007	23,149	47,152	101	3.184	0.032	1,787
yeast	2,417	103	14	4.237	0.303	198

In the HMC domain, there are 10 experimental datasets [19,20] as in Table 5.

Table 5. Statistics of HMC datasets

Dataset	$|D|$	Features	Classes	Distinct	Depth
D0_yeast_GO	3,465	5,929	133	737	8
D15_scop_GO	9,843	1,997	572	673	7
D16_struc_GO	11,763	14,804	630	767	7
D17_interpro_GO	11,763	2,815	630	767	7
diatoms	3,119	371	399	706	4
enron	1,648	1,001	57	329	4
imclef07a	11,006	80	97	63	4
imclef07d	11,006	80	47	26	4
reuters	6,000	47,235	103	796	5
wipo	1,710	74,435	189	160	5

6   Results and Discussion

This section contains experimental results of two domains: multi-label classification and hierarchical multi-label classification (HMC). The performance evaluation is based on F_1-measure. To provide the overall performance, a macro-averaging in Section 3.2 is used in the multi-label task, and an example-label-based measure in Section 3.3 is employed in the hierarchical task. For the rest of the paper, the performance of any domains is shorten and called just "F_1".

There are three modules in the proposed system as shown in Fig. 1. The first two modules are used in the multi-label task, and the last module is extended for the hierarchical tasks. Note that the baseline system is SVM-BR. For the first experiment in the multi-label task, there are four alternative systems since there are two cost functions in the label correction module (LC) with two options of using or not using the performance enhancement module (prefixed by "E"). Later the experiment in Section 6.1 will expose the winner, which is "enhanced label correction with a cost function on SVM scores" – its acronym is "ELC". For the second experiment in the HMC task, there are three hierarchical version extended from the winner of the first part. Table 6 shows all of the systems used in the experiments.

Table 6. Systems used in the experiments

Task	Method	Acronym
A Baseline System		SVM-BR
Multi-Label Classification	Constant label correction	Constant LC
	Enhanced constant label correction	Constant ELC
	SVM score label correction	SVM Score LC
	Enhanced SVM score label correction	SVM Score ELC
Hierarchical Multi-Label Classification	SVM score label correction	SVM Score LC
	SVM score ELC with decreasing path adjustment	Decreasing ELC
	SVM score ELC with increasing path adjustment	Increasing ELC
	SVM score ELC with voting path adjustment	Voting ELC

6.1   Result of Multi-label Classification

There are 5 comparison methods as in Table 6. The experimental result on 22 multi-label datasets is summarized and shown in Table 7 and Table 8. The result illustrates that the best strategy is "SVM score label correction with score enhancement (SVM Score ELC)" the method in the last row of Table 7, since it has never lost to any other methods. Comparing to SVM-BR in Table 8, it won on 18 out of 22 datasets with an improvement of [1%, 331%]. It gives the largest F_1-increase on the *scicence1* dataset.

In the label correction (without the enhancement), the third row of Table 7 exhibits that the SVM score function is clearly better that the constant function because it can reflect confidence level of classifier. Moreover, the performance enhancement strategy is shown to be efficient in Table 7. There are 3 datasets, which are significantly better when the enhancement is added to the SVM Score LC.

Table 7. A comparison of SVM-BR and 4 label correction strategies on 22 multi-label datasets. For each cell, *X/Y* means the numbers of winner and loser datasets between the method in row to that in column. They must be significantly different on a paired *t*-test with α=0.05.

	SVM-BR	Constant LC	Constant ELC	SVM Score LC
Constant LC	4/3			
Constant ELC	2/1	3/1		
SVM Score LC	16/0	16/0	16/1	
SVM Score ELC	18/0	16/0	16/0	3/0

Table 8. F_1-result on multi-label datasets. The number in parentheses is a percentage (%) improvement comparing to SVM-BR. * represents a significant difference on a paired *t*-test with α=0.05. The boldface method is a winner on that dataset.

Method Dataset	SVM-BR	Constant LC	Constant ELC	SVM Score LC	SVM Score ELC
arts	0.060	0.063 (+5)	0.060 (0)	**0.143 (+139)***	0.142 (+138)*
bibtex	0.028	0.028 (0)	0.028 (-1)	**0.056 (+101)***	0.053 (+90)*
business1	0.069	0.067 (-3)*	0.069 (0)	0.132 (+91)*	**0.132 (+91)***
cal500	0.073	0.098 (+33)	0.102 (+38)	0.097 (+32)	**0.100 (+37)**
computers1	0.123	0.129 (+5)	0.131 (+7)	**0.180 (+47)***	**0.180 (+47)***
education1	0.074	0.085 (+14)*	0.078 (+5)	**0.165 (+122)***	**0.165 (+122)***
emotions	0.237	0.257 (+8)	0.237 (0)	**0.297 (+25)***	**0.297 (+25)***
enron	0.109	0.113 (+4)	0.111 (+2)	0.112 (+2)	**0.114 (+4)**
entertainment1	0.179	0.188 (+5)*	0.179 (0)	**0.273 (+53)***	0.273 (+53)*
genbase	0.634	0.654 (+3)	0.656 (+3)	**0.675 (+6)**	**0.675 (+6)**
health1	0.093	0.093 (0)	0.094 (+1)*	0.109 (+17)*	**0.110 (+18)***
mediamill	0.028	**0.029 (+5)**	0.029 (+5)	0.028 (+1)	0.028 (+1)*
medical	0.198	0.189 (-5)	0.196 (-1)	**0.274 (+39)***	0.271 (+37)*
rcv1regions	0.203	0.200 (-1)*	0.203 (0)*	**0.361 (+78)***	**0.361 (+78)***
rcv1topics	0.261	0.269 (+3)*	0.270 (+3)*	0.282 (+8)	**0.284 (+9)**
recreation1	0.155	0.159 (+3)	0.155 (0)	**0.329 (+113)***	**0.329 (+113)***
reference1	0.047	0.048 (+4)	0.047 (0)	0.082 (+75)*	**0.082 (+76)***
scene	0.765	0.746 (-3)*	0.765 (0)	**0.804 (+5)***	0.804 (+5)*
science1	0.026	0.029 (+12)	0.026 (0)	**0.112 (+337)***	0.111 (+331)*
social1	0.089	0.091 (+1)	0.089 (0)	**0.108 (+21)***	0.106 (+19)*
tmc2007	0.299	0.299 (0)	0.301 (+1)	0.352 (+18)*	**0.352 (+18)***
yeast	0.299	**0.308 (+3)***	0.308 (+3)	0.303 (+1)	0.305 (+2)*

6.2   Result of Hierarchical Multi-Label Classification

In HMC domain, we compare SVM-BR, SVM score label correction, and three hierarchical path adjustment strategies: decreasing, increasing, and voting in Table 6. There are 10 HMC datasets, and the result is shown in Table 9 and Table 10. Table 9 demonstrates that the best hierarchical adjustment is clearly "the voting strategy". On the *wipo* dataset in Table 10, it has significantly increased from SVM-BR for 75%.

The winner method is further compared to SVM-BR in details. Analyses for each class level are provided on the datasets *D0_yeast_GO* and *D15_scop_GO* in Table 11 and Table 12, respectively. The result shows that our method has such a great impact on classifiers at the lower levels and, thus, improves the overall performance.

Table 9. A comparison of SVM-BR and 4 label correction strategies in HMC. For each cell, X/Y means the numbers of winner and loser datasets between the method in row to that in column. They must be significantly different on a paired t-test with $\alpha=0.05$.

	SVM-BR	SVM Score LC	Decreasing ELC	Increasing ELC
SVM Score LC	7/0			
Decreasing ELC	6/0	0/3		
Increasing ELC	8/0	2/0	6/0	
Voting ELC	8/0	3/0	6/0	2/0

Table 10. F_1-result on HMC datasets. The number in parentheses is a percentage (%) improvement comparing to SVM-BR. * represents a significant difference on a paired t-test with $\alpha=0.05$. The boldface method is a winner on that dataset.

Dataset \ Method	SVM-BR	SVM Score LC	Decreasing ELC	Increasing ELC	Voting ELC
D0_yeast_GO	0.493	0.505 (+2)	0.502 (+2)	0.511 (+4)*	**0.511 (+4)***
D15_scop_GO	0.435	0.504 (+16)*	0.499 (+15)*	0.507 (+17)*	**0.511 (+18)***
D16_struc_GO	0.313	0.362 (+16)*	0.333 (+6)*	**0.378 (+21)***	0.377 (+21)*
D17_interpro_GO	0.159	0.182 (+15)*	0.180 (+13)*	0.182 (+14)*	**0.186 (+17)***
diatoms	0.431	0.641 (+49)*	0.618 (+43)*	**0.644 (+50)***	**0.644 (+50)***
enron	0.336	0.335 (+0)	0.339 (+1)	**0.347 (+3)**	**0.347 (+3)**
imclef07a	0.668	0.743 (+11)*	0.728 (+9)*	**0.745 (+11)***	**0.745 (+11)***
imclef07d	0.722	0.760 (+5)*	0.732 (+1)	**0.764 (+6)***	**0.764 (+6)***
reuters	0.540	0.571 (+6)	0.564 (+5)	**0.574 (+6)**	**0.574 (+6)**
wipo	0.220	**0.386 (+75)***	**0.386 (+75)***	**0.386 (+75)***	**0.386 (+75)***

Table 11. A result of D0_yeast_GO in each level. The number in parentheses is a percentage (%) improvement comparing to SVM-BR. * represents a significant difference on a paired t-test with $\alpha=0.05$. The boldface method is a winner on that level.

Level	#class	SVM-BR	Voting ELC
1	1	1.000	1.000 (0)
2	35	0.366	**0.375 (+3)***
3	37	0.348	**0.381 (+10)***
4	27	0.393	**0.412 (+5)**
5	17	0.410	**0.416 (+2)**
6	9	0.398	**0.452 (+14)**
7	6	0.306	**0.363 (+19)**
Average		0.376	**0.399 (+6)***

Table 12. A result of D15_scop_GO in each level. The number in parentheses is a percentage (%) improvement comparing to SVM-BR. * represents a significant difference on a paired t-test with $\alpha=0.05$. The boldface method is a winner on that level.

Level	#class	SVM-BR	Voting ELC
1	1	1.000	1.000 (0)
2	13	**0.680**	0.630 (-7)*
3	85	0.320	**0.362 (+13)**
4	174	0.294	**0.373 (+27)***
5	292	0.260	**0.355 (+36)***
6	6	0.161	**0.578 (+259)***
Average		0.289	**0.371 (+28)***

7   Conclusion

A problem transformation approach is a strategy to convert a multi-label classification task into a set of multiple one-vs-rest binary classification tasks, such as in SVM-BR. However, it always encounters such a low accuracy since it does not consider a class relationship. This problem is even worse in the hierarchical multi-label classification (HMC) domain due to a propagated error. In this paper, we propose a technique called "Label Correction" to correct an unusual prediction by taking a class correlation into account resulting in F_1-improvement. There are three modules in the proposed system: (*i*) label correction, (*ii*) performance enhancement, and (*iii*) hierarchical path adjustment, where the last module is only employed in the hierarchical domain. Comparing to SVM-BR, our method significantly won on 18 out of 22 multi-label datasets and 8 out of 10 HMC datasets. Moreover, the analyses for each class level in the hierarchy show that there is a huge improvement of classifiers in the lower levels; this can conclude that the number of errors propagated downward is reduced.

References

1. Boser, B.E., Guyon, I., Vapnik, V.: A Training Algorithm for Optimal Margin Classifiers. In: Proceedings of the Fifth Annual Workshop on Computational Learning Theory, pp. 144–152 (1992)
2. Cortes, C., Vapnik, V.: Support-Vector Network. Machine Learning 20, 273–297 (1995)
3. Tsoumakas, G., Katakis, I., Vlahavas, I.: Mining Multi-Label Data. In: Data Mining and Knowledge Discovery Handbook, 2nd edn., pp. 667–686. Springer (2010a)
4. Brinker, K., Fürnkranz, J., Hüllermeier, E.: A Unified Model for Multilabel Classification and Ranking. In: Proceeding of the 17th European Conference on Artificial Intelligence, pp. 489–493 (2006)
5. Boutell, M.R., et al.: Learning Multi-Label Scene Classification. Pattern Recognition 37(9), 1757–1771 (2004)
6. Katakis, I., Tsoumakas, G., Vlahavas, I.: Multilabel Text Classification for Automated Tag Suggestion. In: Proceedings of the ECML/PKDD 2008 Discovery Challenge, Antwerp, Belgium (2008)
7. Elisseeff, A., Weston, J.: A Kernel Method for Multi-Labelled Classification. Advances in Neural Information Processing Systems 14, 681–687 (2001)
8. Briggs, F., et al.: New Methods for Acoustic Classification of Multiple Simultaneous Bird Species in a Noisy Environment. In: IEEE International Workshop on Machine Learning for Signal Processing, pp. 1–8 (2013)
9. Snoek, C.G.M., et al.: The Challenge Problem for Automated Detection of 101 Semantic Concepts in Multimedia. In: Proceedings of the ACM International Conference on Multimedia, pp. 421–430 (2006)
10. Barutcuoglu, Z., et al.: Hierarchical Multi-Label Prediction of Gene Function. Bioinformatics 22, 830–836 (2006)
11. Dimitrovski, I., et al.: Hierarchical Annotation of Medical Images. Pattern Recognition 44, 2436–2449 (2011)
12. Klimt, B., Yang, Y.: The Enron Corpus: A New Dataset for Email Classification Research. In: Boulicaut, J.-F., Esposito, F., Giannotti, F., Pedreschi, D. (eds.) ECML 2004. LNCS (LNAI), vol. 3201, pp. 217–226. Springer, Heidelberg (2004)

13. Zhang, M., Zhou, Z.: ML-KNN: A Lazy Learning Approach to Multi-Label Learning. Pattern Recognition 40(7), 2038–2048 (2007)
14. Van Rijsbergen, C.J.: Information Retrieval, 2nd edn. (1979)
15. Yiming, Y.: An Evaluation of Statistical Approaches to Text Categorization. Information Retrieval 1, 69–90 (1999)
16. Kiritchenko, S., Matwin, S., Nock, R., Famili, A.F.: Learning and Evaluation in the Presence of Class Hierarchies: Application to Text Categorization. In: Lamontagne, L., Marchand, M. (eds.) Canadian AI 2006. LNCS (LNAI), vol. 4013, pp. 395–406. Springer, Heidelberg (2006)
17. Vateekul, P., Kubat, M., Sarinnapakorn, K.: Top-Down Optimized SVMs for Hierarchical Multi-Label Classification: A Case Study in Gene Function Prediction. In: Intelligent Data Analysis (in press)
18. Mulan Multi-Label Dataset, http://mulan.sourceforge.net/datasets.html
19. Schietgat, L., et al.: Predicting Gene Function using Hierarchical Multi-Label Decision Tree Ensembles. BMC Bioinformatics (2010)
20. Dragi, K.: Tree Ensembles for Predicting Structured Outputs. Pattern Recognition 46, 817–833 (2013)

Towards Time Series Classification without Human Preprocessing

Patrick Schäfer

Zuse Institute Berlin, Berlin, Germany
patrick.schaefer@zib.de

Abstract. Similarity search is a core functionality in many data mining algorithms. Over the past decade these algorithms were designed to mostly work with human assistance to extract characteristic, aligned patterns of equal length and scaling. Human assistance is not cost-effective. We propose our *shotgun distance* similarity metric that extracts, scales, and aligns segments from a query to a sample time series. This simplifies the classification of time series as produced by sensors. A time series is classified based on its segments at varying lengths as part of our *shotgun ensemble classifier*. It improves the best published accuracies on case studies in the context of bioacoustics, human motion detection, spectrographs or personalized medicine. Finally, it performs better than state of the art on the official UCR classification benchmark.

1 Introduction

Time series result from recording data over time. The task of analyzing time series data is difficult as the data may be recorded at variable lengths, and are erroneous, extraneous and highly redundant due to repetitive (sub-)structures. Application areas include ECG [4] or EEG signals, human walking motions [6], or insect wing beats [23], for example. The classification of time series has gained increasing interest over the past decade [2,7,8,9,11,14,16,20,26]. It aims at assigning a class label to a time series. For this the features (the model) to distinguish between the class labels are trained based on a labeled train dataset. When an unlabeled query time series is recorded, the trained model is applied to determine the class of the query.

Empirical evaluation suggests that classifiers based on 1-nearest-neighbour Euclidean distance (ED) or dynamic time warping (DTW) are hard to beat [2,7]. These methods calculate the distance between two entire time series to determine their similarity. To make these applicable a lot of time and effort has to be spent by a domain expert to filter the data and extract equal-length, equal-scale, and aligned patterns. Human assistance significantly eases the subsequent data mining task both in terms of the cost of the execution time and the complexity of the algorithm. However, human assistance comes at a high price and is often too time consuming. There is an over-dependence on preprocessed time series data and only few algorithms exists that deal with the data 'as is'. These algorithms are based on matching time series by their structural similarity [8,16].

P. Perner (Ed.): MLDM 2014, LNAI 8556, pp. 228–242, 2014.

As traditional data mining algorithms are not easily applicable to raw datasets, international competitions were staged like identifying whale calls [10], human walking motions [10] and flying insects [23].

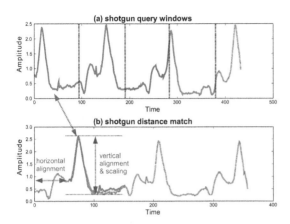

Fig. 1. Shotgun distance consists of segment extraction, horizontal and vertical alignment, and scaling

Our work introduces a novel similarity metric for time series similarity search. It was applied as part of a contest [23]. *Shotgun distance* vertically and horizontally aligns time series segments (subsequences) of a query to a sample time series (Figure 1). Thereby it avoids preprocessing the data for alignment, scaling or length. This is achieved by breaking the query into disjoint subsequences of fixed length first. Next, each query subsequence is slid along a time series sample to find the best matching position in terms of minimizing a distance metric (horizontal alignment). These distances are aggregated. The sample that minimizes this aggregated distance is the 1-nearest-neighbor (1-NN) to a query and most similar. Normalization is applied prior to each distance computation, to provide the same vertical alignment and scaling of each subsequence. The *shotgun ensemble classifier* is based on an ensemble of 1-NN classifiers utilizing the shotgun distance at multiple subsequence lengths. Our contributions are as follows:

- Section 2 presents the motivation and related work on classifying time series.
- We introduce the shotgun distance that provides vertical scaling and horizontal alignment in Section 3.
- We present the shotgun ensemble classifier which represents a time series at multiple subsequences lengths in Section 3.4.
- Two pruning strategies are presented which significantly reduce the computational complexity by one order of magnitude in Section 3.5.
- Our shotgun ensemble classifier is significantly more accurate than state of the art on 5 case studies and the UCR benchmark datasets in Section 4.

2   Motivation and Related Work

The utility of the shotgun distance is related to the observation that a multitude of signals are composed of characteristic patterns (see Section 4.1). Consider human walking motions [6] as a concrete example. The data was captured by recording the z-axis accelerometer values of either the right or the left toe. The difficulties in this dataset result from variable-length gait cycles, gait styles and pace due to different subjects throughout different activities. Figure 2 illustrates the walking motion of a subject, that is composed of 4 gait cycles. Classifying walking motions is difficult, as the samples are not preprocessed to have an approximate alignment, length, scale or number of gait cycles.

Shotgun distance reduces the cost-ineffective human assistance by vertically aligning and horizontally scaling the query to a sample time series. It is an analogy to *Shotgun Sequencing* [24], the process of breaking up a sequence into numerous small segments which are resembled based on overlaps. Figure 2 (bottom) illustrates the result of this process. The distance between the 4 gait cycles in the query and the sample are minimized, even though these differ in scale, have a variable length, and a phase-shift and noise occur. The quality of the shotgun distance is subject to two parameters (Figure 1):

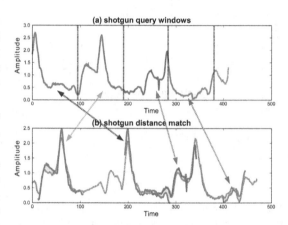

Fig. 2. Matching the gait cycles in the query to the sample is complicated due to different amplitudes, phase-shifts, variable lengths and noise

1. *horizontal alignment* using the *window* length: an integer parameter which is limited by the length of the longest query.
2. *vertical alignment* using the *mean*: a Boolean parameter which defines if the mean should be subtracted prior to the distance calculations. The standard deviation is always normed to 1 to obtain the same scaling. Surprisingly, the mean normalization has not been considered to be a parameter before.

The *window length* parameter controls the length of the segments and depends on the length of the characteristic patterns in the dataset. Furthermore, it regulates how much information on the ordering of the values within the time series is incorporated into the matching-process. For long window lengths the whole query will be treated as a single pattern. This mostly happens with signals which were preprocessed by a human for alignment and length. In contrast, human motions contain repetitive gait cycles. Aligning any gait cycle in the query to any gait cycle in the sample is equivalent. Thus, the ordering information is less relevant, resulting in a window length that should be roughly equal to one gait cycle.

2.1   Related Work

Time series similarity search is a complex task for a computer. It is non trivial to extract a general statistical model from time series as these may show varying statistical properties with time. Classical machine learning algorithms degenerate due to the high dimensionality of the time series and noise [12]. Approaches can be characterized by (a) they try to find a similarity metric that resembles our intuition of similarity in combination with 1-NN classification (*shape-based*) or (b) they transform the data into an alternative data space to make existing data mining algorithms applicable (*structure-based*) [2,14,25]. The UCR time series classification datasets [11] have been established for reference [2,7,11,14,16]. *Shape-based* techniques include 1-NN Euclidean Distance (ED), or 1-NN DTW [17,19] and are used as the reference [7]. However, shape-based techniques fail to classify noisy or long data. *Structure-based* techniques [2,8,14,16,18,26] are based on data mining algorithms such as SVMs, decision trees, or random forests in combination with feature extraction. Feature extraction techniques include DFT [1], PLA [5], SFA [21], SAX [13], or Shapelets. By transforming time series data into an alternative space (i.e. using functional data analysis) the performance of classifiers can be improved [2]. However, the authors failed to show a significant improvement over 1-NN DTW. Shapelets classifiers [16,18,26] extract representative variable-length subsequences (called *shapelets*). A decision tree is build using these shapelets within the nodes of the tree and distance threshold for branching. One algorithms deals with classification on raw data [8]. The shotgun classifier is inspired by shotgun sequencing introduced to find an alignment of two DNA or protein sequences [24]. Shotgun sequencing was used to find the horizontal displacements of steel coils [15]. To find the horizontal displacement the authors use the median on the differences of the calculated starting positions for every pair of subsequences.

3   Shotgun Distance

3.1   Definitions

A time series consists of a sequence of real values:

$$T = (t_1, \ldots, t_n) \tag{1}$$

This time series is split into subsequences (time segments) using a windowing function.

Definition 1. *Windowing: A time series $T = (t_1, \ldots, t_n)$ of length n is split into fixed-length windows $S_w(a) = (t_a, \ldots, t_{a+w-1})$ with length w and offset a in T.*

Two consecutive windows can overlap *within an interval of* $[0, w)$. *Given the overlap, there are* $\frac{(n-w)}{(w-overlap)}$ *windows in T:*

$$windows(T, w, overlap) = \bigcup_{i=0}^{\frac{(n-w)}{(w-overlap)}} S_w(i \cdot (w - overlap) + 1) \qquad (2)$$

To *vertically* align two samples, the query window and the sample window are typically z-normalized by subtracting the mean and dividing by the standard deviation:

$$\hat{\omega}(T, w, overlap) = z\_norms(windows(T, w, overlap)) \qquad (3)$$

However, the mean normalization is treated as a parameter of our model and can be enabled or disabled. For example, heart beats have to be compared using a common baseline but the pitch of a bird sound can be significant for the species. Commonly, the similarity of two time series is measured using a distance metric. The shotgun distance is a distance metric that minimizes the Euclidean distance between each disjoint window in the query Q and the sliding windows in a sample S. For example, each gait cycle is slid along a longer walking motion to find the best matching positions by minimizing the Euclidean distance.

Definition 2. *Shotgun distance: the shotgun distance* $D_{shotgun}(Q, S)$ *between a query Q and a sample S is given by aggregating the minimal Euclidean distance* $D(Q_a, S_b)$ *between each disjoint query window* $Q_a \epsilon \hat{\omega}(Q, w, 0)$ *and each offset b in S, represented by the sliding windows* $S_b \epsilon \hat{\omega}(S, w, w - 1)$:

$$D_{shotgun}(Q, S) = \sum_{a=1}^{len(\hat{\omega}(Q,w,0))} \min \{D(Q_a, S_b) \mid S_b \epsilon \hat{\omega}(S, w, w - 1)\} \qquad (4)$$

This definition resembles the extraction of characteristic patterns (i.e. the gait cycles), and the scaling and aligning of the patterns. The latter provides invariance to the time ordering of the patterns and allows for comparing variable length time series. The shotgun distance is equal to the Euclidean distance for n equal to w.

3.2 Shotgun Distance Algorithm

The shotgun distance in Algorithm 1 makes use of the Euclidean distance, and can be tuned by the two parameters *window length* W_LEN and *mean* MEAN_NORM (the standard deviation of q and s is always normed to 1 regardless of MEAN_NORM). It first splits the query into disjoint windows (line 2) and searches for the position in the sample that minimizes the Euclidean distance (line 4-5). Finally, the distances are accumulated for each query window (line 6).

Algorithm 1. The shotgun distance

```
double ShotgunDistance(query,sample,W_LEN,MEAN_NORM)
(1) totalDist = 0.0
    // for each disjoint query window
(2) for q in disjoint_windows(query,W_LEN,MEAN_NORM)
(3)     qDist = MAX_VALUE
        // search for the position that minimizes the Euclidean distance
(4)     for s in sliding_windows(sample,W_LEN,MEAN_NORM)
(5)         qDist = min(qDist,EuclideanDist(q,s))
(6)     totalDist += qDist
(7) return totalDist
```

Complexity. The computational complexity is quadratic in the length of the time series Q and S: for each query window, all sample windows are iterated and the Euclidean distance for each pair of windows is calculated. There are $\frac{|Q|}{w}$ disjoint query windows and $|S| - w + 1$ sliding windows for window length w:

$$T(\text{Shotgun Distance}) = O\left(\underbrace{\frac{|Q|}{w}}_{\text{disjoint windows}} \cdot w \cdot \underbrace{(|S| - w + 1)}_{\text{sliding windows}}\right) \quad (5)$$

$$\text{for } n = \max(|Q|, |S|) \Rightarrow O\left(n^2 - nw\right) \quad (6)$$

Note that for large window lengths $w \sim n$ this complexity is close to linear in n (like the Euclidean distance). For small window lengths $w \ll n$ the complexity is quadratic in n^2 (like DTW).

3.3   Shotgun Classifier

The shotgun classifier is based on 1-NN classification and the shotgun distance. Given a query, the *predict*-method (Algorithm 2) searches for the 1-NN to a query within the set of samples (line 3-5). Finally, the query is labeled by the class label of the 1-NN *nn*. The *fit*-method (Algorithm 2) uses leave-one-out cross-validation (lines 4–8) to obtain the parameters that maximize the accuracy on the train samples. The accuracies for all window lengths starting from the *maxLen* (the length of the longest time series) down to *minLen* (lines 2–7) are recorded. The *MEAN_NORM*-parameter is a Boolean parameter, which is constant for a whole dataset and not set per sample.

3.4   Shotgun Ensemble Classifier

By intuition every dataset is composed of substructures at multiple window lengths caused by different walking motions, heart beats, duration of vocals, length of shapes. For example, each human may have a different length of a gait cycle. Thus, each sample is represented by a set of window lengths.

Using the *predictEnsemble*-method (Algorithm 2), a label is determined for the best window lengths. Using a constant parameter *factor*ϵ (0, 1] and the best

---

Algorithm 2. The Shotgun Ensemble Classifier

```
String predict(query,samples,W_LEN,MEAN_NORM)
(1) (dist, nn) = (MAX_VALUE, NULL)
(2) for sample in samples
(3)      D = ShotgunDistance(query,sample,W_LEN,MEAN_NORM)
(4)      if D < dist
(5)          (dist, nn) = (D, sample)
(6) return nn.label

String predictEnsemble(query,samples,bestScore,windows,MEAN_NORM)
    // stores for each window length a label
(1) windowLabels = []
    // determine the label for each window length
(2) for (correct, len) in windows
(3)      if (correct > bestScore*factor)
(4)          windowLabels[len] = predict(query,samples,len,MEAN_NORM)
(5) return most frequent label from windowLabels

[(int,int)] fit(samples,labels,MEAN_NORM)
(1) scores = []
    // search for best window lengths in parallel
(2) for len = maxLen down to minLen
(3)      correct = 0
(4)      for query in samples
(5)          nnLabel = predict(query,samples\{query},len,MEAN_NORM)
(6)          if (nnLabel==query.label) correct++
    // store scores for each window length
(7)      scores.push((correct, len))
(8) return scores
```

---

accuracy *bestScore* obtained from the train samples, the best window lengths are given by: *correct* > *bestScore·factor* (line 3). Finally, the most frequent class label is chosen from the set of labels.

While it might seem that we add yet another parameter *factor*, the training of the shotgun ensemble classifier depends solely on the *factor* and *mean* parameters. The shotgun ensemble classifier model is derived from these two parameters using the *fit*-method, which returns the set of window scores. These scores are used as the model and to predict the label of an unlabeled query. In our experiments factors in between 0.95 to 1.0 were best throughout most datasets.

3.5 Pruning the Search Space

The rationale of search space pruning is to early abandon computations, as soon as these can not result in finding a new optimum. Previous work aims at stopping Euclidean distance calculations when the current distance exceeds the best distance found so far [16,18,26].

Early Abandoning: The purpose of the *ShotgunDistance*-method (Algorithm 3) is to accumulate the Euclidean distances for each query window. The Euclidean distance computations are pruned by reusing the best result *qDist* of the previous calculations (line 5). The *ShotgunDistance*-method is executed multiple times for each pair of query and sample. Passing the distance to the current calculation

---

Algorithm 3. Pruning techniques based on early abandoning

```
double EuclideanDist(query,sample,bestDist)
(1) for i = 1 to len(query)
(2)     dist += (sample[i] - query[i])^2
(3)     if (dist > bestDist) return MAX_VALUE            // early abandoning
(4) return dist

double ShotgunDistance(query,sample,W_LEN,MEAN_NORM,bestDist)
(1) totalDist = 0
(2) for q in disjoint_windows(query,W_LEN,MEAN_NORM)
(3)     qDist = MAX_VALUE
(4)     for s in sliding_windows(sample,W_LEN,MEAN_NORM)
            // early abandoning
(5)         qDist=min(qDist,EuclideanDist(q,s,min(qDist,bestDist)))
(6)     totalDist += qDist
(7)     if (totalDist > bestDist) return MAX_VALUE       // window pruning
(8) return totalDist

String predict(q,samples,W_LEN,MEAN_NORM)
[...]
(2) for sample in samples
(3)     D = min(D,ShotgunDistance(q,sample,W_LEN,MEAN_NORM,D))
[...]
```

Algorithm 4. Use an upper bound on the current accuracy

```
[(int,int)] fit(samples,labels,MEAN_NORM)
(1) scores = [], bestCorrect = 0
(2) for len = maxLen down to minLen
(3)     correct = 0
(4)     for q in [1..len(samples)]
(5)         nnLabel = predict(samples[q],samples\{samples[q]},len,MEAN_NORM)
(6)         if (nnLabel==samples[q].label) correct++
(7)         if (correct+(len(samples)-q)) < bestCorrect*factor
(8)             break
(9)     bestCorrect = max(bestCorrect,correct)
[...]
```

---

as *bestDist* allows for pruning calculations as soon as this *bestDist* is exceeded (line 7). Otherwise the sample is a new nearest-neighbor candidate and the distance is used to prune subsequent calls to *ShotgunDistance*. In the best case scenario, we have to compute the distance between one pair of time series and all other distance computations stop after one iteration of the for-loop in line 7.

Upper Bound on Accuracy: While lower bounding on distance computations aims at reducing the complexity in the length n, we present a novel optimization that also aims at reducing the complexity in the number of samples N. For each window length, the best achievable accuracy at any point is given by:

$$\text{correct} \leq (\text{current correct} + \text{remaining samples}) = N \qquad (7)$$

Thus, we do not need to obtain the exact accuracy for a window length in Algorithm 4 (lines 7–8), if the remaining samples will not result in finding a better accuracy (or at least within *factor* to the best accuracy).

Table 1. Test accuracies on the case studies

Dataset	DTW	Best Rival	Shotgun Ensemble Classifier	
Personalized Medicine	62.8%	92.4% [8]	99.3%	*factor* : 1, mean:true
Walking Motions	66.2%	91% [26]	96.9%	*factor* : 0.95, mean:true
Spectrographs	71.3%	72.6% [26]	80.7%	*factor* : 0.95, mean:true
Bio Acoustics	-	93.29% [23]	92.38%	*factor* : 1, mean:true
Astronomy	90.7%	93.68% [18]	95.3%	*factor* : 0.97, mean:true

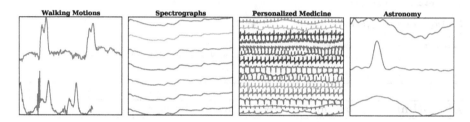

Fig. 3. A sample representing each class of the case studies: abnormal and normal walking motions, wheat spectrographs, ECG signals, and starlight curves

4   Experimental Evaluation

The utility of the shotgun ensemble classifier is underlined by case studies and the UCR time series classification benchmark datasets [11]. Each dataset is split into two subsets: *train* and *test*. By the use of the same train/test splits the results are comparable those previously published [2,3,7,16,18]. **In all experiments we optimized the parameters of the classifiers based on the train dataset. The optimal set of parameters is then used on the test dataset.** Our web page [22] contains a spreadsheet with all raw numbers and source codes. All benchmarks were performed on a shared memory machine running Linux with 8 Quad-Core AMD Opteron 8358 SE and Java JDK x64 1.7.

4.1   Case Studies

Astronomy / Scalability: We test the scalability using the largest dataset in the UCR time series archive [11]. It contains three types of star objects: *Eclipsed Binaries, Cepheids* and *RR Lyrae Variables*. The *Cepheids* and *RR Lyrae Variables* have a similar shape and are hard to separate (Figure 3 top and bottom). To test the scalability of the shotgun *fit*-method, we iteratively doubled the number of samples from 100 to 1000, each of length 1024, and measured the pruning strategies presented in this paper. Figure 4 shows that the time of the *brute force* algorithm grows quadratically to approximately 9 hours for 1000 samples. *Early abandoning* reduces this by a factor of 7 and in combination with the *upper bound* by a factor of 12 to only 41 minutes. Both pruning strategies combined significantly reduce the run-time for training when the number of samples is increased.

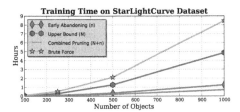

Fig. 4. The time required to execute the shotgun *fit*-method on the *StarLightCurves* dataset using the presented pruning strategies

To the best of our knowledge, the highest reported test accuracy is 93.68% [18] with 52 minutes for training and 1-NN DTW scores 90.7%. The test accuracy of our shotgun ensemble classifier is 95.3% (Table 1).

Personalized Medicine: The BID-MC Congestive Heart Failure Database [4] consists of ECG recordings of 15 subjects, which suffer from severe congestive heart failures (Figure 3). The recordings contain noisy or extraneous data, when the recordings started before the machine was connected to the patient. ECG signals show a high level of redundancy due to repetitive heart beats but even a single patient can have multiple different heart beats. To deal with these distortions a classifier has to be invariant to amplitude, uniform scaling, phase shifts and occlusion. The total size of this dataset is equal to 9 million data points (10 hours sampled at at 250 Hz). We used the train/test split in [8], which selected 150 minutes for training and 450 minutes for testing and search for individual patient heart beats (15 distinct classes). There are 600 samples for training at length 3750 and 600 samples for testing at length 11250.

To the best of our knowledge, the best rivalling approach reported a test accuracy of 92.4% [8] and 1-NN DTW scores 62.8%. The shotgun ensemble classifier obtains a much higher test accuracy of 99.3%. This is a result of the design of the shotgun distance: ECG signals are composed of recurring patterns, which are distorted by all kinds of noise. To obtain this score, training the shotgun ensemble classifier took roughly 2 days as all window lengths have to be evaluated. Prediction on the 600 test samples took roughly 1.5 hours in total.

Human Walking Motions: The CMU [6] contains walking motions of 4 subjects. Each motion was categorized by the labels *normal walk* and *abnormal walk* (Figure 3). The data were captured by recording the z-axis accelerometer values of either the right or the left toe. The difficulties in this dataset result from variable-length gait cycles, gait styles and pace due to different subjects throughout different activities including stops and turns. To deal with these distortions, a classifier needs to be invariant to amplitudes, uniform scaling, phase shifts and occlusions. To make our results comparable to [26], we used the data provided by their first segmentation approach. The dataset contains 40 samples for training and 228 samples for testing, each of variable lengths with an upper

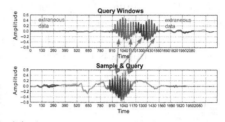

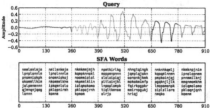

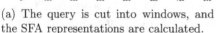

(a) The query is cut into windows, and the SFA representations are calculated.

(b) An insect passes the laser multiple times, causing an *echo*. The shotgun classifier aligns these two signals.

Fig. 5. Insect Classification

limit of 500 points. We search for normal or abnormal walking patterns. Training the shotgun ensemble classifier took less than a minute. This results in a test classification accuracy of 96.9%, due to the repetitive nature of the data. The accuracy is significantly higher than that of the best rivalling approach in [26] with an accuracy of 91% or 1-NN DTW that scores 66.2%.

Spectrographs: *Wheat* [26] is dataset of 775 spectrographs of wheat samples grown in Canada. The data is split into 49 samples of length 1050 for training and 726 samples of length 1050 for testing. The dataset contains different wheat types like *Canada Western Red Spring, Soft White Spring* or *Canada Western Red Winter* (Figure 3). The class labels define the year in which the wheat was grown. This makes the classification problem much more difficult, as the same wheat types in different years belong to different classes. The best rivalling approach [26] reported a test accuracy of 72.6% on this dataset and 1-NN DTW obtains a test accuracy of 72.6%. Our shotgun ensemble classifier obtains a significantly higher test accuracy of 80.69%.

Bioacoustics: Producers set up traps in the field that lure and capture pests, in order to detect and count these. Manual inspection of traps is a procedure that is costly and error prone. Repeated inspection must be carried manually, sometimes in areas that are not easily accessible. A novel recording device has been recently introduced [23]. The core idea is to embed a device in an insect trap to record the fluctuations of light received by a photoreceptor as an insect passes a laser beam and partially occludes light. The samples were recorded at 16 kHz and are 1s long but the insect motion within each sample is typically only a few hundredths of a second long. The bandwidth between 0.2-4 kHz is most characteristic. The dataset D1 was collected from 5 insects, namely: *Aedes aegypti male, Fruit flies mixed sex, Culex quinquefasciatus female, Culex tarsalis female, Culex tarsalis male,* and consists of 5000 recordings for testing and 500 for training.To connect time series analysis with bioacoustics, we use Symbolic Fourier Approximation (SFA) [21]. Its symbolic and thus compact representation of a time series has shown to be capable of exact similarity search and to index terabyte-sized datasets. Our workflow consists of feature extraction and feature matching. SFA is applied to extract features, which are then passed to the shotgun classifier.

The solution utilizes SFA to reduce noise by the use of low pass filtering and quantization. The SFA transformation results in a character string (see Figure 5a). Each symbol represents an interval in the frequency domain. In particular, noise is generated by the angle and the speed of an insect passing the photoreceptor. This affects the recorded intensities. SFA's noise reduction accounts for these differences in the intensity. By introducing the shotgun classifier for feature matching, we obtain invariance to the time of the insect passage. Shotgun distance further deals with outliers like multiple insects passing the laser within a short time frame (see Figure 5b). For the sake of brevity, the interested reader is referred to [20] for details on the approach and SFA. Using a small window length of 130 and a small number of SFA symbols of 22 performed best. Our approach scored within 1 percentage point of the best approach applied in the contest (Figure 6). This proves that time series analysis is applicable to computational bioacoustics.

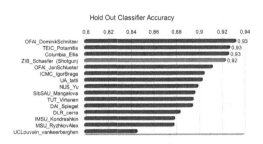

Fig. 6. Classifier accuracies on the test dataset. ZIB_Schaefer represents the shotgun classifier.

4.2   UCR Classification Benchmark Datasets

We used a standardized benchmark [11] to evaluate the classifiers. Each dataset consists of a train and a test subset. The shotgun ensemble classifier is compared to state of the art time series classifiers like shapelets [16], fast shapelets [18], 1-NN classifiers using Euclidean distance or dynamic time warping (DTW) with the optimal warping window, support vector machines (SVM) with a quadratic and cubic kernel, and a tree based ensemble method (random forest). We followed the setup in [16,18]. The authors in [2] used data transformations (i.e. functional data analysis) to improve classifiers performance. However, they failed to show a significant improvement over 1-NN DTW. We omit data transformations prior to classification for the sake of brevity. The scatter plots in Figure 7 show a pairwise comparison of each classifier with the shotgun classifier. Each dot represents the test accuracies of the two classifiers on one concrete dataset. Dots below the diagonal line indicate that the shotgun ensemble classifier is more accurate.

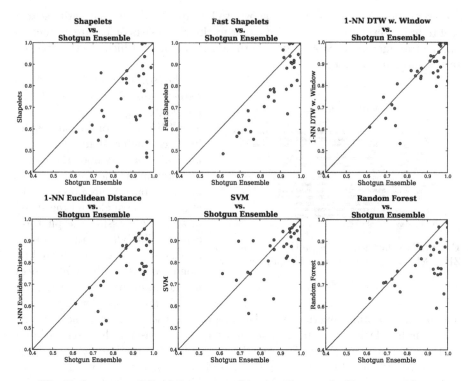

Fig. 7. Accuracy of the shotgun ensemble classifier vs. rivalling approaches

The shotgun ensemble classifier is better than fast shapelets and shapelets on the majority of the datasets. We conclude that this is a result of the sensitivity to the overfitting of the shapelet classifiers and the decision tree in particular, where the difference between the train and test accuracy makes up for up to 50 percentage-points. In contrast the shotgun classifier is more robust towards overfitting (see web page [22]). 1-NN DTW is the established benchmark classifier [7]. Our shotgun ensemble classifier is better than 1-NN DTW or 1-NN Euclidean distance on the majority of datasets by a large margin in terms of accuracy. DTW provides invariance to local scaling (time warping). Shotgun distance does not explicitly provide this invariance. However, the results imply that either (a) most UCR datasets do not require local scaling or (b) the shotgun distance provides some local scaling invariance. This will be part of future work. The shotgun distance is equal to Euclidean distance, if the window length is equal to the query length. Thus, the shotgun ensemble classifier performs better than the 1-NN Euclidean distance. SVMs and the shotgun ensemble classifier complement each other quite well as one classifier is good on a dataset in which the other performs badly. So, at least for datasets which were preprocessed for approximate alignment and fixed length, the choice of the classifier depends on the dataset. When comparing the shotgun ensemble classifier with random forests, the results suggest that the former is more accurate by a large margin. Note that the UCR datasets were preprocessed for approximate alignment and length. Still our shotgun classifier performs significantly better than rivalling state of the art classifiers on a majority of datasets.

5   Conclusion

The time series classification task is complicated by noise, dropouts, subtle distinctions, variable lengths or extraneous data. The shotgun distance is a novel distance measure based on the characteristic patterns in time series. Shotgun distance utilizes time segments which are vertically and horizontally aligned and scaled between the query and a sample, and thereby simplifying the preprocessing. Based on an ensemble of 1-nearest-neighbor classifiers the shotgun ensemble classifier is presented. To deal with the increased complexity, two pruning strategies for the length and the number of time series are presented. This reduces the computational complexity by one order of magnitude. The experimental evaluation shows that the shotgun ensemble classifier performs better than rivalling methods in the context of computational bioacoustics, human motion detection, spectrographs, astronomy, or personalized medicine. This is underlined by the best classification accuracy on the UCR time series classification datasets.

Acknowledgements. The author would like to thank C. Eichert-Schäfer, F. Schintke, F. Wende and S. Dreßler for their comments and the dataset owners.

References

1. Agrawal, R., Faloutsos, C., Swami, A.: Efficient similarity search in sequence databases. In: Lomet, D.B. (ed.) FODO 1993. LNCS, vol. 730, pp. 69–84. Springer, Heidelberg (1993)
2. Bagnall, A., Davis, L.M., Hills, J., Lines, J.: Transformation Based Ensembles for Time Series Classification. In: SDM, vol. 12, pp. 307–318. SIAM (2012)
3. Batista, G., Wang, X., Keogh, E.J.: A Complexity-Invariant Distance Measure for Time Series. In: SDM, vol. 11, pp. 699–710. SIAM/Omnipress (2011)
4. BIDMC, http://www.physionet.org/physiobank/database/chfdb/
5. Chen, Q., Chen, L., Lian, X., Liu, Y., Yu, J.X.: Indexable PLA for Efficient Similarity Search. In: VLDB, pp. 435–446. ACM (2007)
6. CMU Graphics Lab Motion Capture Database, http://mocap.cs.cmu.edu/
7. Ding, H., Trajcevski, G., Scheuermann, P., Wang, X., Keogh, E.: Querying and mining of time series data: experimental comparison of representations and distance measures. Proceedings of the VLDB Endowment 1(2), 1542–1552 (2008)
8. Hu, B., Chen, Y., Keogh, E.: Time Series Classification under More Realistic Assumptions. In: SDM, pp. 578–586. SIAM (2013)
9. Jeong, Y., Jeong, M.K., Omitaomu, O.A.: Weighted dynamic time warping for time series classification. Pattern Recognition 44(9), 2231–2240 (2011)
10. Kaggle: Go from Big Data to Big Analytics, https://www.kaggle.com

11. Keogh, E., Xi, X., Wei, L., Ratanamahatana, C.A.: UCR Time Series Classification/Clustering Homepage, http://www.cs.ucr.edu/~eamonn/time_series_data
12. Keogh, E., Kasetty, S.: On the need for time series data mining benchmarks: a survey and empirical demonstration. In: KDD, pp. 102–111. ACM (2002)
13. Lin, J., Keogh, E.J., Wei, L., Lonardi, S.: Experiencing SAX: a novel symbolic representation of time series. Data Mining and Knowledge Discovery 15(2) (2007)
14. Lin, J., Khade, R., Li, Y.: Rotation-invariant similarity in time series using bag-of-patterns representation. J. Intell. Inf. Syst. 39(2), 287–315 (2012)
15. Lipowsky, C., Dranischnikow, E., Göttler, H., Gottron, T., Kemeter, M., Schömer, E.: Alignment of Noisy and Uniformly Scaled Time Series. In: Bhowmick, S.S., Küng, J., Wagner, R. (eds.) DEXA 2009. LNCS, vol. 5690, pp. 675–688. Springer, Heidelberg (2009)
16. Mueen, A., Keogh, E.J., Young, N.: Logical-shapelets: an expressive primitive for time series classification. In: Proceedings of the 17th ACM SIGKDD International Conference on Knowledge Discovery and Data Mining, KDD 2011, pp. 1154–1162. ACM (2011)
17. Rakthanmanon, T., Campana, B., Mueen, A., Batista, G., Westover, B., Zhu, Q., Zakaria, J., Keogh, E.: Searching and mining trillions of time series subsequences under dynamic time warping, pp. 262–270. ACM (2012)
18. Rakthanmanon, T., Keogh, E.: Fast Shapelets: A Scalable Algorithm for Discovering Time Series Shapelets. In: SDM, pp. 668–676. SIAM (2013)
19. Sakoe, H., Chiba, S.: Dynamic programming algorithm optimization for spoken word recognition. IEEE Trans. Acoust., Speech, Signal Processing (1), 43–49 (1978)
20. Schäfer, P., Dreßler, S.: Shooting Audio Recordings of Insects with SFA. In: AmiBio Workshop, Bonn, Germany (2013) (to appear)
21. Schäfer, P., Högqvist, M.: SFA: a symbolic fourier approximation and index for similarity search in high dimensional datasets. In: Rundensteiner, E.A., Markl, V., Manolescu, I., Amer-Yahia, S., Naumann, F., Ari, I. (eds.) EDBT, pp. 516–527. ACM (2012)
22. Shotgun Distance Webpage, http://www.zib.de/patrick.schaefer/shotgun
23. UCR Insect Contest (2012), http://www.cs.ucr.edu/~eamonn/CE
24. Venter, J.C., et al.: The Sequence of the Human Genome. Science 291(5507), 1304–1351 (2001)
25. Warren Liao, T.: Clustering of time series data—a survey. Pattern Recognition 38(11), 1857–1874 (2005)
26. Ye, L., Keogh, E.J.: Time series shapelets: a novel technique that allows accurate, interpretable and fast classification. DMKD 22(1-2), 149–182 (2011)

Applications of Concurrent Sequential Patterns in Protein Data Mining

Cuiqing Wang[1], Jing Lu[2] and Malcolm Keech[3]

[1] Shenyang University of Chemical Technology, Shenyang China, 110142
[2] Southampton Solent University, Southampton UK, SO14 0YN
[3] University of Bedfordshire, Park Square, Luton UK, LU1 3JU
wangcuiqing@syuct.edu.cn, Jing.Lu@solent.ac.uk,
Malcolm.Keech@beds.ac.uk

Abstract. Protein sequences of the same family typically share common patterns which imply their structural function and biological relationship. Traditional sequential patterns mining has its focus on mining frequently occurring sub-sequences. However, a number of applications motivate the search for more structured patterns, such as protein motif mining. This paper builds on the original idea of structural relation patterns and applies the Concurrent Sequential Patterns (ConSP) mining approach in bioinformatics. Specifically, a new method and algorithms are presented using support vectors as the data structure for the extraction of novel patterns in protein sequences. Experiments with real-world protein datasets highlight the applicability of the ConSP methodology in protein data mining. The results show the potential for knowledge discovery in the field of protein structure identification.

Keywords: protein sequences, data mining, concurrent sequential patterns (ConSP), bioinformatics, ConSP mining, concurrent vector method, PROSITE, knowledge discovery.

1 Introduction

Mining sequential patterns from very large databases is a well-developed technique with important applications which have featured in many business and scientific domains, e.g. analysis of customer transactions to discover frequent buying patterns; processing of web access logs to identify user navigation patterns; and analysis of gene regulatory networks in bioinformatics. The discovery of frequent sequential patterns (such as *motifs*) that occur in many biosequences from a given database (i.e. DNA or protein sequences) could be essential to the interpretation and engineering of the biological data.

One of the most challenging areas of bioinformatics is to understand the genetic regulation mechanisms, as sequence motifs are becoming increasingly important [13]. There are various relationships among motifs and the study of this can lead to the discovery of significant but hidden patterns. Based on sequence databases, Structural Relation Patterns (SRP) mining aims to find and represent more complex information

P. Perner (Ed.): MLDM 2014, LNAI 8556, pp. 243–257, 2014.
© Springer International Publishing Switzerland 2014

which can include concurrent patterns, exclusive patterns and iterative patterns. Concurrent Sequential Patterns (ConSP) mining and modelling methods have already been applied in the analysis of customer behaviour and web usage mining [7].

All of these structures may have corresponding applications in motif identification. Following the previous theoretical research on SRP mining, the focus of this paper will be on applying the methodology in the bioinformatics area, in particular for the analysis of protein sequences. The development of effective ConSP mining techniques in the context of sequence motifs could help to discover new biological information and potentially reduce the cost of experiments.

The remainder of this paper proceeds as follows: some related work is highlighted in section 2 to provide relevant background on knowledge discovery in bioinformatics as well as to frame the concept of concurrent sequential patterns in protein databases. The emphasis is on tailoring the ConSP mining approach to this domain and the method and algorithms are presented in section 3 with a worked example to illustrate their application. An experimental evaluation using several protein datasets is given in section 4, which showcases the effectiveness of ConSP mining at generating new and interesting results. The paper draws to a close by summarising and making brief conclusions, indicating future directions.

2   Related Work

To provide some background and further motivation, the nature of the biological data will be addressed first with the focus on mining patterns from protein sequences. This section will conclude with a concise description of the authors' previous research on post sequential patterns mining as applied to this context.

2.1   Sequential Patterns Mining for Biological Sequences

Biosequences are the primary sequences of DNA, RNA and protein molecules, and they represent the most basic type of biological information [4]. These sequences typically have a very small alphabet, i.e. 4 for DNA sequences and 20 for protein sequences, and many short patterns can occur in most sequences. The goal of pattern discovery in the latter domain is to find new and previously unknown patterns that are common to (or match) all or most of the sequences in the protein dataset.

Frequent patterns mining is useful in biological sequence analysis, where the most frequently occurring patterns are called motifs. It is noted that motif mining is related to the problem of mining sequential patterns and this has been studied extensively. A scalable two-phase approach has been presented to mine frequent patterns in biosequences [14] with segment phase searches for short patterns containing no gaps and pattern phase searches for long patterns containing multiple segments separated by variable length gaps. The biological datasets from the National Center for Biotechnology Information (http://www.ncbi.nlm.nih.gov) have been used to evaluate this approach.

Terai and Takagi proposed a method for detecting element patterns that control gene expressions [10]. The requirement for applying their method is a set of gene expression data and corresponding upstream sequence data. The new patterns discovered from the *S.cerevisiae* genome demonstrate the utility of sequential patterns mining for analysis of eukaryotic gene regulation.

Exarchos et al. presented a method based on protein sequence analysis and classification [1]. Specifically, sequential patterns mining was used for classification of proteins into folds highlighting the important role of data mining in bioinformatics. Gupta and Han reviewed applications of pattern discovery using sequential data mining in [5] which also covers predicting protein sequence functions, analysis of gene expression data, protein fold recognition and protein family detection.

The common feature of the above work is that mining results are sequential. Therefore the question is: are there any other novel patterns that can be discovered based on sequential patterns?

2.2   Concurrence and Concurrent Sequential Patterns

The original method to mine concurrent sequential patterns has been proposed taking sequence databases as input, where each data sequence consists of a list of itemsets [6]. Due to the characteristics of protein databases, i.e. each protein is a linear sequence made up of smaller constituent molecules called amino acids, the following notation is required: let Σ = {A, C, D, E, F, G, H, I, K, L, M, N, P, Q, R, S, T, V, W, Y} be a set of 20 items (e.g. amino acids). A sequence is an ordered list of items and can be written as $X = \langle x_1\ x_2\ \ldots\ x_p \rangle$, where $x_j \in \Sigma$ $(1 \leq j \leq p)$. A sequence $X = \langle x_1\ x_2\ \ldots\ x_p \rangle$ is contained in sequence $X' = \langle x_1'\ x_2'\ \ldots\ x_q' \rangle$ if $p \leq q$ and $x_j = x_k'$ for all j, $1 \leq j \leq p$ and corresponding k, $1 \leq k \leq q$.

A *sequence database* (SDB) is defined as a set {S_1, S_2, ..., S_n}, where each S_r $(1 \leq r \leq n)$ is a sequence. The *support* in SDB of any given sequence S is considered to be the fraction/percentage of sequences in the database that contains S and is defined as sup(S) = |{S_r: $S \subseteq S_r$}|/n, where |...| denotes the number of sequences. Sequence S is called a sequential pattern in SDB with respect to a minimum support threshold *minsup* (0<*minsup*≤1) if sup(S)≥*minsup*. Sequential patterns mining thus discovers a set of patterns from a given SDB under a user-specified *minsup*.

Let α and β be two sequential patterns mined from SDB with minimum support threshold *minsup* and assume that α, β are not contained in each other. With regard to a given sequence $S \in$ SDB, sequential patterns α and β have a concurrent relationship if and only if both of them have occurred in S, i.e. $(\alpha \angle S) \wedge (\beta \angle S)$ is true. This is denoted by $[\alpha + \beta]_S$, where the '+' notation represents the concurrency.

The *concurrence* of sequential patterns α and β is defined as the fraction of sequences that contains both of the patterns. This is denoted by concurrence(α, β) = |{S_r:$(\alpha \angle S_r) \wedge (\beta \angle S_r)$}|/n, where $S_r \in$ SDB, $1 \leq r \leq n$ and n is the total number of data sequences. The user-specified *minsup* provides the threshold for frequency measurement when mining frequent itemsets and sequential patterns. Another fractional value, the minimum *concurrence* threshold, *mincon* (0<*mincon*≤1) is used to check the concurrent relationships of sequential patterns.

Let *mincon* be the user-specified minimum concurrence. The concurrence of sequential patterns sp_1, sp_2, ..., sp_s is defined as concurrence(sp_1, sp_2, ..., sp_s) = |{S_r:[sp_1+sp_2+...+sp_s]}|/n, where $S_r \in SDB$ and $1 \leq r \leq n$. If concurrence(sp_1, sp_2, ..., sp_s)$\geq mincon$ is satisfied, then sp_1, sp_2, ... and sp_s are called *concurrent sequential patterns*. This is represented by $ConSP_s$ = [sp_1+sp_2+...+sp_s], where there is no particular order for the sequential patterns.

Furthermore, the concurrent sequential patterns represented by $ConSP_s$ = [α_1+α_2+...+α_s] are contained in $ConSP_{s+t}$ = [β_1+ β_2+...+β_{s+t}] if $\alpha_i \angle \beta_j$ for all i, $1 \leq i \leq s$ and corresponding j, $1 \leq j \leq (s+t)$. This is denoted by $ConSP_s \angle ConSP_{s+t}$. Concurrent sequential patterns are called *maximal ConSP* if they are not contained in any other concurrent patterns.

Example 1. Given a small protein sequence database PSDB={<HENAC>, <EAEGHASC>, <GAGNAHS>, <GMHSN>, <EGHCNS>}. Table 1 shows the set of all sequential patterns mined with a *minsup* of 60% and this will be used as a running worked example through the paper. Setting *mincon* = 60% as well then three maximal concurrent sequential patterns can be found: [A+H], [GN+GHS] and [EC+HC]. The last two ConSPs can alternatively be represented as G[N+HS] and [E+H]C.

Table 1. Sequential patterns from a sample protein database

Seq ID	Sequence	Sequential Patterns supported by each Sequence (SuppSP), *minsup*=60%
S_1	<HENAC>	A, C, E, H, N, EC, HC, HN
S_2	<EAEGHASC>	A, C, E, G, H, S, EC, GH, GS, HC, HS, GHS
S_3	<GAGNAHS>	A, G, H, N, S, GH, GN, GS, HS, GHS
S_4	<GMHSN>	G, H, N, S, GH, GN, GS, HN, HS, GHS
S_5	<EGHCNS>	C, E, G, H, N, S, EC, GH, GN, GS, HC, HN, HS, GHS

3   Mining Concurrent Patterns Based on Support Vectors

The original method to mine concurrent sequential patterns was proposed by taking general sequence databases as input and first pursuing traditional sequential patterns mining. Protein data mining tailored to the context of bioinformatics will be described in this section through a support vector/matrix-based ConSP mining approach with optimisation. It aims to find potentially valuable ConSPs which are common to all (or mostly all) known protein sequences of a family.

3.1   Support Data Structure

Given the sequence database SDB = {S_1, S_2, ..., S_n} and mined sequential patterns SP={sp_1, sp_2, ..., sp_m}, a compact data structure or **support vector**, SeqVect(sp_i) is defined for every sequential pattern sp_i ($1 \leq i \leq m$) as:

$$SeqVect(sp_i) = [v_1 v_2 ... v_n],$$

where $v_r=1$ $(1 \leq r \leq n)$ if sequential pattern sp_i $(1 \leq i \leq m)$ is contained in data sequence S_r $(1 \leq r \leq n)$, i.e. $sp_i \angle S_r$; otherwise $v_r=0$.

For example, for the PSDB in Table 1 and sequential patterns EC and GH from the set SP, SeqVect(EC)=[11001] and SeqVect(GH)=[01111].

Putting all of these vectors together forms rows of support vectors, SeqVect(SP) in two dimensions with sequential patterns sp_i $(1 \leq i \leq m)$ determining the binary entries v_{ir} corresponding to data sequences S_r $(1 \leq r \leq n)$ across the columns of a matrix:

	S_1	...	S_r	...	S_n
sp_1	v_{11}	...	v_{1r}	...	v_{1n}
...	...	...	...	...	...
sp_i	v_{i1}	...	v_{ir}	...	v_{in}
...	...	...	...	...	...
sp_m	v_{m1}	...	v_{mr}	...	v_{mn}

Consider the intersection of two support vectors for sp_i and sp_j $(1 \leq i \leq m, 1 \leq j \leq m, i \neq j)$ respectively, denoted by $SeqVect(sp_i) \wedge SeqVect(sp_j) = [v_1 v_2 ... v_n]$, where $v_r=1$ $(1 \leq r \leq n)$ if both $SeqVect(sp_i)=1$ and $SeqVect(sp_j)=1$ in their r^{th} columns; otherwise $v_r=0$.

Counting the total number of entries in the resulting vector which have value "1" gives a measure of the support of both sequential patterns across SDB, denoted by

$$w = Count(SeqVect(sp_i) \wedge SeqVect(sp_j)).$$

If $w/n \geq mincon$, then sp_i and sp_j constitute a ConSP = [sp_i+sp_j].

The idea of a support vector and matrix can thus be extended to a group of sequential patterns which will potentially make up a ConSP. The support vector of the concurrent sequential pattern $ConSP_s = [\alpha_1+\alpha_2+...+\alpha_s]$ is defined as:

$$ConVect(ConSP_s) = [v_1 v_2 ... v_n],$$

where $v_r=1$ $(1 \leq r \leq n)$ if $ConSP_s \angle S_r$; otherwise $v_r=0$.

Similarly, putting all the support vectors for ConSPs together forms the rows of a **support matrix**, ConVect(ConSP) with respective concurrent sequential patterns determining the binary entries corresponding to data sequences across columns.

If the intersection of ConSP = [sp_i+sp_j] with another sequential pattern sp_k $(1 \leq k \leq m, k \neq i, k \neq j)$ is considered, then:

$$ConVect(ConSP) \wedge SeqVect(sp_k) = [v_1 v_2 ... v_n],$$

where $v_r=1$ $(1 \leq r \leq n)$ if both ConVect(ConSP)=1 and $SeqVect(sp_k)=1$ in their r^{th} columns; otherwise $v_r=0$.

Counting the total number of "1" entries in the resulting vector again forms w; if $w/n \geq mincon$, then sp_k provides another concurrent sequential pattern which can be added to the existing result, such that ConSP=ConSP$\cup\{sp_k\}$=[$sp_i+sp_j+sp_k$].

3.2   Concurrent Vector Method

The problem of mining concurrent sequential patterns is divided into three phases. A support vector/matrix-based approach will be described using the sample protein database from Example 1 for illustration: PSDB={<HENAC>, <EAEGHASC>, <GAGNAHS>, <GMHSN>, <EGHCNS>}.

Phase 1. Pre-processing. First traditional sequential patterns mining discovers the frequent sub-sequences from a given protein database under a user-specified *minsup*. For those sequential patterns with a specified minimum sequence length, *minlen*, the set of sub/super-sequences is determined and the corresponding support matrix SeqVect(SP) is constructed.

In Table 1 there are 8 sequential patterns of minimum length 2 forming the rows and 5 protein data sequences forming columns, with results as shown in Table 2.

Table 2. A support matrix example

	S_1	S_2	S_3	S_4	S_5
EC	1	1	0	0	1
GH	0	1	1	1	1
GN	0	0	1	1	1
GS	0	1	1	1	1
HC	1	1	0	0	1
HN	1	0	0	1	1
HS	0	1	1	1	1
GHS	0	1	1	1	1

Phase 2. ConSP Generation. As follows:

A. Taking respective sequential patterns sp_i ($1 \leq i \leq m-1$) from SP in turn to provide the initial *seed-sequence*, the next sequential pattern sp_j ($i < j \leq m$) is sought which can constitute an initial ConSP=$[sp_i + sp_j]$, so long as:

$$Count(SeqVect(sp_i) \wedge SeqVect(sp_j))/n \geq mincon.$$

In this case, sp_i and sp_j can form a new seed-sequence for further ConSP checking.

B. Taking the next seed-sequence and corresponding ConSP as input, another sp_k is sought which contributes to forming an expanded ConSP$\cup\{sp_k\}$, if:

$$Count(ConVect(ConSP) \wedge SeqVect(sp_k))/n \geq mincon.$$

In this eventuality, sp_k will be consolidated into the seed-sequence and so on.

C. Revert back to B to determine whether any further sequential patterns can be added to the intermediate ConSP in hand, until all combinations have been checked.

Example 2. Given SP={EC, GH, GN, GS, HC, HN, HS, GHS} from Table 2.
(1) Consider EC first and find the next sequential pattern which can be used to form the initial ConSP. It is HC, giving:

$$Count(SeqVect(EC) \wedge SeqVect(HC)) = Count([11001] \wedge [11001]) = 3;$$

i.e. concurrence(EC,HC)=3/5=60%≥*mincon*, so ConSP_1=[EC+HC].

Using {EC,HC} as a seed-sequence and going through the rest of the sequential patterns in turn, i.e. HN, HS and GHS, none of these will satisfy the minimum concurrence condition. Therefore the ConSP result when taking EC as the initial seed-sequence remains the same and this is shown in the first row of Table 3.
(2) Similarly, taking GH as the next seed-sequence and considering GN to form an initial ConSP, subject to:

$$Count(SeqVect(GH) \wedge SeqVect(GN))/5 = Count([01111] \wedge [00111])/5 = 3/5 \geq mincon.$$

Using {GH,GN} now as the seed-sequence, the remaining sequential patterns are considered in turn and contribute to forming the new concurrent sequential pattern ConSP_2=[GH+GN+GS+HS+GHS], which is shown in the 2nd row of Table 3.
(3) Using GN, GS, HS and GHS respectively as seed-sequences, the result is the same as ConSP_2 above.
(4) Using HC as the seed-sequence, only EC qualifies, and this pair has already made up ConSP_1 above.
(5) Finally, when HN is the seed-sequence, there are no other valid sequential patterns to pair up and form a ConSP under *mincon*.
Therefore, the concurrent sequential patterns generated from this phase are:

ConSP_1=[EC+HC]
ConSP_2=[GH+GN+GS+HS+GHS].

Table 3. Individual seed-sequences, ConSPs and their support vectors

Seed-sequence	ConSPs	ConVect
EC	[EC+HC]	11001
GH	[GH+GN+GS+HS+GHS]	00111
GN	[GN+GH+GS+HS+GHS]	00111
GS	[GS+GH+GN+HS+GHS]	00111
HC	[HC+EC]	11001
HN	-	-
HS	[HS+GH+GN+GS+GHS]	00111
GHS	[GHS+GH+GN+GS+HS]	00111

Phase 3. Concurrent Sequential Patterns Optimisation. Maximal ConSPs can be obtained by deleting those concurrent sequential patterns which are contained by

other ConSPs, then deleting the sequential patterns in particular ConSPs (in turn) which are contained by other sequential patterns within the same ConSP.

For example, for ConSP_2=[GH+GN+GS+HS+GHS], as <GH>, <GS> and <HS> are all sub-sequences of <GHS>, then ConSP_2 can be optimised and reduced to [GN+GHS]. The concurrent sequential pattern [EC+HC] is already maximal.

3.3   Algorithm Development

Following the Concurrent Vector method above, the corresponding techniques have been developed for each phase. The approach to ConSP generation in Phase 2 is illustrated below, invoking the *ConSPgen* algorithm. Recursion is used to demonstrate the completeness of the concurrent sequential patterns mined.

```
i = 1;    ConSPset = Ø;
While i < m        // Use sp_i as initial seed-sequence
For j = i+1, … m
IF Count(SeqVect(sp_i)∧SeqVect(sp_j))/n ≥ mincon
    THEN ConSPstart = [sp_i + sp_j]; DoubleSeed = {sp_i U sp_j};
        IF j = m THEN ConSPset = ConSPset + ConSPstart
        ELSE Call ConSPgen(ConSPstart, j+1, DoubleSeed);
    ENDIF         // Continue with "For j" loop
i=i+1
// Go back up to "While i"
// ConSPset will hold the final set of ConSP's for this
// SDB and minsup/con combination
```

```
Algorithm ConSPgen(ConSP, next, Seed)
For k = next, … m
IF ∃sp_k: Count(ConVect(ConSP)∧SeqVect(sp_k))/n ≥ mincon
    THEN For each such sp_k
ConSPnew = ConSP + [sp_k]; SuperSeed = {Seed U sp_k};
    IF k = m THEN Return ConSPset = ConSPset + ConSPnew
            ELSE Call ConSPgen(ConSPnew, k+1, SuperSeed);
    // Continue with "For each such sp_k" loop
    ELSE Return ConSPset = ConSPset + ConSP
```

The above approach starts with sp_i as the initial seed-sequence before finding the corresponding sp_j which can form a *double-seed* and the initial ConSPstart=[sp_i+sp_j]. The algorithm ConSPgen is then called to find any other sequential pattern sp_k to form a *super-seed* contributing to the continued growth of the current ConSP in hand, i.e. through ConSPnew=[sp_i+sp_j+sp_k]. Sequential patterns from such ConSPs would act as consolidated super-seeds when ConSPgen is invoked recursively.

The fundamental idea of the algorithm is to consider all the possible combinations of sequential patterns to check their concurrence condition. Its actual implementation uses support vectors and matrices with flags and iteration to maximise efficiency while ensuring correctness of the outcomes. In addition, excluding certain types of sequential patterns from the checking loops is important for further improvement of efficiency. This can be achieved based on concurrence properties [6]: (i) if $\alpha \in$ ConSP, then all of its sub-sequences will be contained within the ConSP; (ii) if $\alpha \notin$ ConSP, then none of its super-sequences will be contained within the ConSP; (iii) if $\alpha \in$ ConSP, then all other sequential patterns which share the same *ancestor* (i.e. super-sequence) will be contained within the ConSP.

4   Bioinformatics Experiments

The empirical analysis of the Concurrent Vector method was performed on real-world datasets to test its effectiveness as well as to further illustrate the process of protein data mining. Data pre-processing precedes traditional sequential patterns mining in each case, whereupon ConSP mining is invoked using those SPs with a specified *minlen*. Results are summarised and a range of interesting graphs produced, which are arguably the most stimulating for the *higher level* ConSPs generated.

4.1   PROSITE Database and Data Pre-processing

The datasets for the experiments are exclusively drawn from the PROSITE database (http://prosite.expasy.org/), one of the oldest and most prominent repositories which stores information about protein families, their descriptions and patterns. PROSITE consists of a large collection of biologically meaningful *signatures* described as patterns or profiles [3] which are common to all or nearly all of the members of the family. We seek to determine the membership of novel patterns among these families.

First of all it is necessary to transform the real-world protein sequences into a format suitable for sequential patterns mining (e.g. using PrefixSpan [11]). Table 4 shows a sample of protein sequences in the standard bioinformatics FASTA format, a text-based format for nucleotide sequences or peptide sequences, in which nucleotides or amino acids are represented using single-letter codes.

Table 4. Sample protein sequences

>sp|C1IC47|3FN3_WALAE Three finger toxin Wa-III OS=Walterinnesia aegyptia PE=1 SV=1 MKTLLLTLVVVVTIVCLDLGHTFVCHNQQSSQPPTTTNCSGGENKCYKKQWSDHRGSIT ERGCGCPTVKKGIKLHCCTTEKCNN
>sp|C1IC48|3FN4_WALAE Three finger toxin Wa-IV OS=Walterinnesia aegyptia PE=1 SV=1 MKTLLLTLVVVVTIVCLDLGHTLLCHNQQSSTSPTTTCCSGGESKCYKKRWPTHRGTITE RGCGCPTVKKGIELHCCTTDQCNL
>sp|C1IC49|3FN5_WALAE Three finger toxin Wa-V OS=Walterinnesia aegyptia PE=1 SV=1 MKTLLLTLVLVTIVCLDLGYTLTCLICPKKYCNQVHTCRNGENLCIKTFYEGNLLGKQF KRGCAATCPEARPREIVECCSRDKCNH

The Concurrent Vector method has been implemented in C++ and all experiments have been conducted on a 2.5GHz Intel Core i5 processor with 4GB main memory running under MS Windows 7.

4.2   ConSP Mining from SNAKE_TOXIN Dataset

Our first experiment concerns the SNAKE_TOXIN dataset from PROSITE, a family of eukaryotic and viral DNA binding proteins consisting of cytotoxins, neurotoxins and venom peptides. This bio-dataset is not uncommon in sequential patterns mining experiments – it is dense and can generate many sequential patterns with a medium support threshold [12]. The latest number of hits on proteins which belong to the set under consideration is 389 (http://prosite.expasy.org/PS00272, last accessed 17 February 2014) with an average sequence length of 75.

Table 5 shows the total number of SPs and ConSPs found on this dataset using the Concurrent Vector method with various combinations of *minlen* and *mincon=minsup*. As with increasing *minsup*, the number of patterns reduces as *minlen* increases, e.g. there is no ConSP when *minlen*=13 for *mincon*=95-100%.

Table 5. ConSP mining results summary for SNAKE_TOXIN dataset

mincon =	minlen =9		minlen =10		minlen =11		minlen =12	
minsup (%)	ΣSPs	ΣConSPs	ΣSPs	ΣConSPs	ΣSPs	ΣConSPs	ΣSPs	ΣConSPs
100	8	1	1	0	0	0	0	0
99	68	8	16	2	2	1	0	0
98	234	77	68	25	10	7	0	0
97	504	358	170	77	35	17	4	0
96	874	2072	328	322	77	29	9	6
95	1290	5502	506	994	148	178	25	10

Fig. 1 shows the graphical distribution of sequential patterns and ConSP mining results across the range of *minsup*, *mincon* and *minlen* values. Ostensibly, there are too many patterns to be that useful in general as the thresholds decrease. This trend prevails for *mincon=minsup*=90% – however, for *minlen*=13, there are just 4 ConSPs here from 18 sequential patterns with the highest level a single ConSP$_3$.

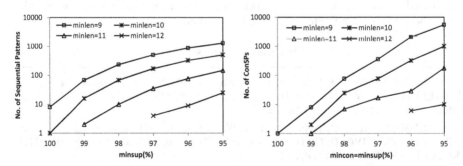

Fig. 1. Pattern distributions from mining SNAKE_TOXIN dataset

Fig. 2 shows the comparison of SP mining and ConSP mining in terms of the minimum length of sequential patterns across different *mincon=minsup* values. It is worth noting that, for example when *minlen=*11, there are single ConSP$_2$, ConSP$_4$, ConSP$_5$, ConSP$_8$ and ConSP$_{16}$ patterns representing respectively the highest levels discovered using the corresponding thresholds (99% down to 95%).

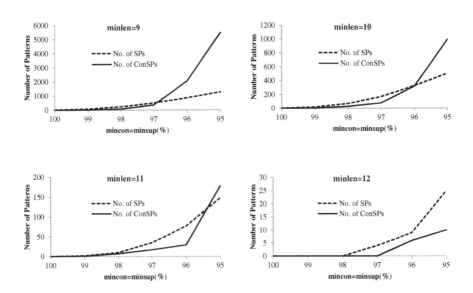

Fig. 2. Comparison between SPs and ConSPs from mining SNAKE_TOXIN dataset

For each *minlen* setting, there is potentially a turning point where the total number of ConSPs exceeds the number of SPs. In the protein data mining domain, the purpose is not to find a large volume of patterns – it is more to discover those ConSPs which are common to all (i.e. *mincon=*100%) or nearly all (e.g. 90%≤*mincon*≤*minsup* <100%) of the members of the family – this broad strategy is applied in the rest of the experiments through the next sub-section.

4.3   ConSP Modelling from Sample Protein Datasets

Several other examples from the PROSITE database are used to illustrate the process and results from ConSP mining. Since the protein alphabet is small, many short patterns that express trivial local similarity may arise. Therefore, longer patterns are expected to imply greater confidence in the similarity of sequences [5]. The parameter strategy thus aims for maximising *mincon=minsup* and *minlen* values here while generating intuitively interesting results.

Table 6 presents a description of protein datasets in the experiments in ascending order of average protein length, which runs from 39 to 580. Protein size is usually measured in terms of the number of amino acids and can range from fewer than 20 to more than 5000 in length, although a typical average is about 350 [2]; the number of sequences reported for the datasets varies between 164 and 1855. The essential nature of the patterns discovered is highlighted using the largest feasible *minlen* values under *mincon=minsup*.

Table 6. Description of protein datasets and patterns mined

Accession ID	Entry Name	# of Seqs	Average Protein Length	*mincon= minsup*	ΣSPs	ΣConSPs
PS00828	RIBOSOMAL_L36	803	39	99%	6 *(minlen=7)*	2 ConSP$_2$
PS00323	RIBOSOMAL_S19	1010	97	99%	5 *(minlen=13)*	4 ConSP$_2$
PS00045	HISTONE_LIKE	603	100	99%	22 *(minlen=11)*	5 ConSP$_2$ 1 ConSP$_3$ 2 ConSP$_4$ 1 ConSP$_6$
PS00352	COLD_SHOCK	164	114	98%	21 *(minlen=10)*	1 ConSP$_5$ 1 ConSP$_8$
PS50889	S4 RNA-BINDING DOMAIN	1855	267	100%	7 *(minlen=9)*	1 ConSP$_7$
PS00012	PHOSPHO-PANTETHEINE	610	545	99%	8 *(minlen=11)*	2 ConSP$_2$ 1 ConSP$_3$
PS50075	ACP_DOMAIN	822	580	99%	161 *(minlen=8)*	··· ··· 5 ConSP$_{11}$

ConSP-Graph was used for the representation of concurrent sequential patterns in [8] and this approach has been adapted in [9] for protein data modelling. A sample of the datasets from Table 6 is thus deployed below to describe the results in a more illuminating way.

HISTONE_LIKE
This bacterial histone-like and DNA-binding protein signature is a set of small and usually basic proteins of about 90 residues that bind DNA. Total number of true positive hits in UniProtKB/Swiss-Prot is 603 among 603 different sequences. Initial ConSP experiments on this dataset show, e.g. when *mincon=minsup=99%* and *minlen=11*, that there are 22 sequential patterns and 9 ConSPs. The highest level is a ConSP$_6$ and this is modelled by the ConSP-Graph in Fig. 3. It shows that protein sub-sequences in the family share a common amino acid "M" at the start and finish up with "K" (i.e. there are no outgoing edges from this final node).

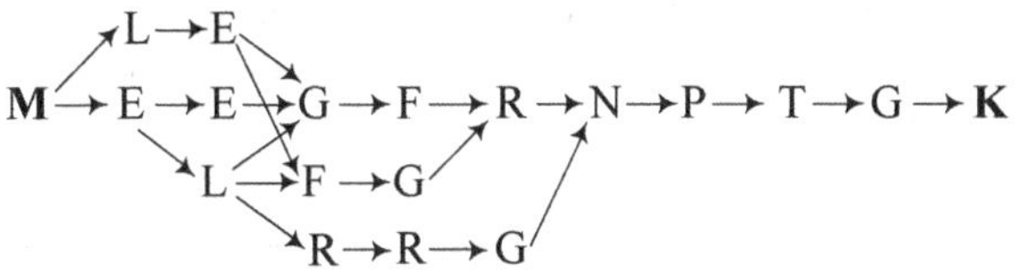

Fig. 3. An example of ConSP$_6$ modelling from HISTONE_LIKE (*mincon=minsup*=99%, *minlen*=11)

COLD_SHOCK

This is a conserved domain of amino acids which has been found in prokaryotic and eukaryotic single-strand nucleic-acid binding proteins. It is known as the 'Cold_Shock domain' and the number of sequences which belong to the set under consideration is 164. When *mincon=minsup*=98% and *minlen*=10, there are 21 sequential patterns with 1 ConSP$_5$ and 1 ConSP$_8$. The latter can be modelled as shown in Fig. 4 and it should be noted that the length of one SP is 11 while all others are 10. The ConSP-Graph shows that protein sub-sequences in this family all finish with either "V" or "G".

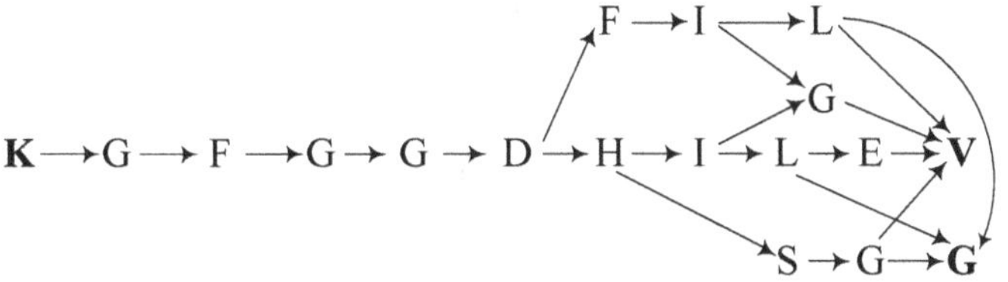

Fig. 4. An example of ConSP$_8$ modelling from COLD_SHOCK (*mincon=minsup*=98%, *minlen*=10)

PHOSPHOPANTETHEINE

This is the prosthetic group of Acyl Carrier Proteins (ACP) in some multi-enzyme complexes where it serves as a 'swinging arm' for the attachment of activated fatty acid and amino acid groups. Total number of hits in UniProtKB/Swiss-Prot is 1052 of which 610 are positive. When *mincon=minsup*=99% there are eight SPs of length 11 and three concurrent patterns, with the highest level a ConSP$_3$. All three ConSPs are

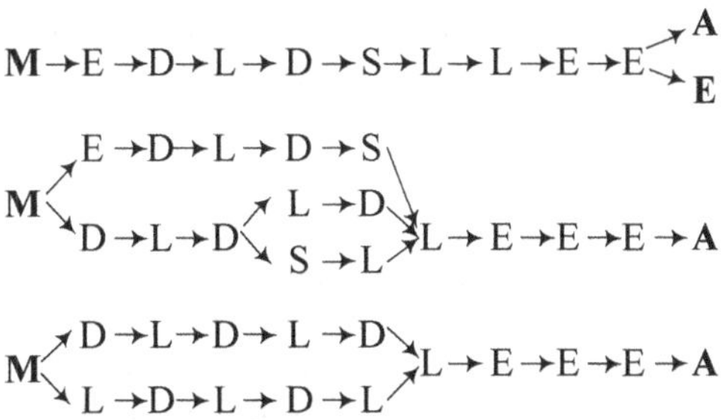

Fig. 5. An example of modelling one ConSP$_3$ and two ConSP$_2$ from PHOSPHOPANTETHEINE (*mincon=minsup*=99%, *minlen*=11)

modelled as shown in Fig. 5, which indicates that protein sub-sequences in this family all start with the common amino acid "M" and finish with "A" or "E".

ACP_DOMAIN

This is an ACP phosphopantetheine domain profile – the number of true positive hits under consideration is 822. Some novel and interesting concurrent sequential patterns can be found from this dataset under *mincon=minsup=99%* and *minlen=8*. While there is a total of 324 ConSPs mined from the 161 SPs, it is instructive to focus on those with greater structure. One of the five highest level $ConSP_{11}$ patterns selected for Fig. 6 shows that protein sub-sequences in this family start with amino acid "M" and finish up with "V", "I" or "E". This is representative of the other four not illustrated here, which all have the same start node and final nodes.

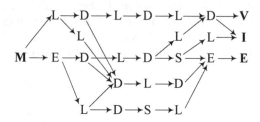

Fig. 6. An example of $ConSP_{11}$ modelling from ACP_DOMAIN (*mincon=minsup=99%*, *minlen=8*)

As can be seen from Figures 3-6, complex structural relations can be discovered and modelled from protein sequence data. In particular the $ConSP_6$, $ConSP_8$ and $ConSP_{11}$ shown convey the increasingly rich connectivity among these progressively higher level patterns. This brings another perspective in pursuit of novel signatures which are common to nearly all of the sequences in a protein family.

5   Conclusion

Motivated by the need to identify more complex protein structures, the ConSP mining approach presented here includes three phases: pre-processing, support vector/matrix-based ConSP generation and ConSP optimisation. Concurrent sequential patterns mining in protein databases aims to discover structures that can re-occur concurrently in protein sequences. This may provide greater insight into the functional role of the proteins which support such structures.

The problem of mining structural motifs has been considered across a range of protein families drawn from PROSITE. In each case the objective is to find ConSPs which can potentially match significant combinations of amino acids, corresponding to functionally or structurally important regions in the proteins. While there are no *consensus* patterns proposed at this stage, it has been demonstrated that novel biological relationships can be predicated upon the results of using the Concurrent Vector method with real-world protein sequences.

This paper has focused on utilising equivalent minimum support and concurrence thresholds with minimum sequence lengths appropriate to each dataset. There is of course the prospect of varying *mincon* a little less than *minsup* and the modelling of this has been taken further in [9]. More especially, an investigation into the applicability of ConSP mining to DNA sequences would be a natural next step for future work. However, the literature suggests that the traditional approach to sequential patterns mining may prove inefficient in this context.

References

1. Exarchos, T.P., Papaloukas, C., Lampros, C., Fotiadis, D.I.: Mining sequential patterns for protein fold recognition. Journal of Biomedical Informatics 41(1), 165–179 (2008)
2. Hofmann, K., Bucher, P., Falquet, L., Bairoch, A.: The PROSITE database, its status in 1999. Nucleic Acids Research 27(1), 215–219 (1999)
3. Hulo, N., Bairoch, A., Bulliard, V., Cerutti, L., De Castro, E., Langendijk-Genevaux, P.S., Pagni, M., Sigrist, C.J.: The PROSITE database. Nucleic Acids Research 34(1), 227–230 (2006)
4. Jonassen, I., Collins, J.F., Higgins, D.G.: Finding flexible patterns in unaligned protein sequences. Protein Science 4(8), 1587–1595 (1995)
5. Kumar, P., Krishna, P.R., Raju, S.B.: Pattern Discovery Using Sequence Data Mining: Applications and Studies. IGI Global, Hershey (2012)
6. Lu, J., Chen, W.R., Adjei, O., Keech, M.: Sequential patterns post-processing for structural relation patterns mining. International Journal of Data Warehousing and Mining 4(3), 71–89 (2008)
7. Lu, J., Keech, M., Wang, C.Q.: Applications of concurrent access patterns in web usage mining. In: Bellatreche, L., Mohania, M.K. (eds.) DaWaK 2013. LNCS, vol. 8057, pp. 339–348. Springer, Heidelberg (2013)
8. Lu, J., Keech, M., Chen, W.R., Wang, C.Q.: Concurrent sequential patterns mining and frequent partial orders modelling. International Journal of Business Intelligence and Data Mining 8(2), 132–154 (2013)
9. Lu, J., Keech, M., Wang, C.Q.: Protein data modelling for concurrent sequential patterns. In: 5th International Workshop on Biological Knowledge Discovery and Data Mining (BIOKDD 2014), Munich (under review 2014)
10. Terai, G., Takagi, T.: Predicting rules on organization of cis-regulatory elements, taking the order of elements into account. Bioinformatics 20(7), 1119–1128 (2004)
11. PrefixSpan source code, http://en.pudn.com/downloads39/sourcecode/math/detail134610_en.html (last access: September 30, 2013)
12. Wang, J., Han, J.: BIDE: Efficient mining of frequent closed sequences. In: 20th International Conference on Data Engineering, pp. 79–90. IEEE (2004)
13. Wang, J., Zaki, M., Toivonen, H., Shasha, D.: Data Mining in Bioinformatics. Springer, London (2010)
14. Wang, K., Xu, Y., Yu, J.X.: Scalable sequential pattern mining for biological sequences. In: 13th International Conference on Information and Knowledge Management, pp. 178–187. ACM (2004)

Semi-supervised Time Series Modeling for Real-Time Flux Domain Detection on Passive DNS Traffic

Bin Yu, Les Smith, and Mark Threefoot

Infoblox Inc., Santa Clara, California, USA
{biny,lsmith,mthreefoot}@infoblox.com

Abstract. Flux domain is one of the most active threat vectors and its behavior keeps changing to evade existing detection measures. In order to differentiate the malicious flux domains from legitimate ones such as content delivery network (CDN) and network time protocol (NTP) services that have similar behavior, a novel time series model is created with a set of features that are not only focused on domain name system (DNS) time-to-live (TTL) but on loyalty and entropy of DNS resource records. An offline system is built with big data technology for training the model in a semi-supervised mode. In addition, an online platform is designed and developed to support large throughput real-time DNS streaming data processing with advanced analytics technologies. The feature extraction, classification, accuracy and performance are discussed based on large amount of real world DNS data in this paper.

Keywords: Network security, Semi-supervised machine learning, Big data analytics, Time series model, DNS flux.

1 Introduction

Flux domain, aka fast flux, is being used as an evasion technique that cyber-criminals and Internet miscreants use to evade identification and to frustrate law enforcement and anticrime efforts aimed at locating and shutting down web sites used for illegal purposes [1].

In a flux network, nodes (typically systems compromised by malware) are used as proxy servers pointed by a flux domain through a DNS server. This allows for very rapid changes to DNS-related data, which helps cyber-criminals and miscreants delay or evade detection and mitigation of their activities. The main purpose of domain flux is to hide true delivery sites used by malware and/or scam operators behind a vast number of relatively short lived internet protocol (IP) addresses that are swapped in and out of a DNS record for a domain. This is often referred as single-flux. However, the same mechanism can be applied to a DNS name server which is referred as double-flux. We will address both problems in this paper.

Domain flux exploits the stability and resilience of the DNS to make it difficult to eliminate systems being used for criminal activities. It can frustrate both administrative remedies and technical remedies. While domain flux isn't a threat to any

P. Perner (Ed.): MLDM 2014, LNAI 8556, pp. 258–271, 2014.

component of the DNS infrastructure, it is a threat to Internet users that are facilitated by DNS. Over the years since the first domain flux was identified in early 2000's, attackers have changed the behavior and made domain flux no longer fast flux. The simple TTL filtering methods will fail because some domain flux attacks are not really fast changing their DNS records. Instead, the DNS records TTL can be as long as up to hours or days. Secondly, many domains used for load balancing such as CDN or NTP have very similar behavior as fast flux.

In this paper, we present a novel time series model based on a set of prominent features that can address behavior changes in domain flux. A semi-supervised training framework is proposed to overcome the difficulties common in traditional supervised machine learning for network securities. Since the amount of network data is extremely large and flows in a fast speed, a horizontally scalable online system is developed to deal with large throughput in a real deployment.

2 Related Work

A number of approaches for detecting flux domains have been studied. Some approaches identify potential flux domain in the URLs found in the body of spam emails that are typically captured by spam traps and filters [2], [3], [4], [5]. The work in [6] proposes to analyze NetFlow information collected at border routers to identify redirection botnets. Hsu et al. proposed a real-time system for detecting flux domains based on anomalous delays in hypertext transfer protocol (HTTP)/hypertext transfer protocol secure (HTTPS) requests from a given client [7]. Perdisci et al. presented a fast flux detection system that is focused on passive analysis of DNS traffic [8]. As many related researches, their work is mainly based on the common characteristics of flux domains [9]: (a) short TTL; (b) high frequency of change of the set of resolved IP addresses returned at each query; (c) the overall set of resolved IP addresses obtained by querying the same domain name over the time is often very large; (d) the resolved IP addresses are scattered across many different networks.

Unfortunately, first of all, botnets are changing the characteristics so that more and more flux domains are no longer returning "short" lived IP addresses. In our observation, some resolved IP address can have a TTL of up to one day. Furthermore, the overall set of resolved IP addresses obtained by querying the same domain name over the time can be as small as 1 or 2. Secondly, some legitimate services, such as legitimate CDN, NTP server pools, internet relay chat (IRC) server pools, etc., are served through sets of domain names that share some similarities with fast-flux domains.

A number of DNS reputation systems mainly targeted at detecting generic malicious domains [10], [11], or malware-specific domains [12] have been proposed. They use large-scale DNS monitoring to detect malicious domains.

3 Data Source

It's always important to get real data for building a real detection system. In this project, we utilize DNS data provided by Internet Systems Consortium (ISC)/Security

Information Exchange (SIE) channel 202 [13], now spun off as part of Farsight Security [14]. This data is collected through its passive DNS technology from more than 80 contributors distributed worldwide. On average, it receives more than 1.8 billion DNS queries per day. We have collected more than six months worth of data for this project. One caveat of using ISC/SIE data is that it doesn't contain cached DNS queries and therefore the end user IP addresses are not provided due to the purpose of privacy protection. Among various DNS query types, type A that is used in this project dominates by 70%. Figure 1 shows the percentage of each DNS query type among the data collection.

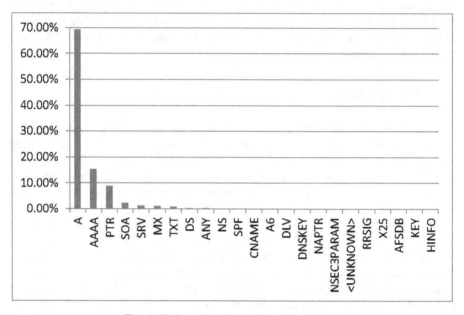

Fig. 1. DNS query distribution by query type

4   Detection Algorithm

The most prominent characteristics that the DNS message of a flux domain may carry include a short TTL value, changing resolved IP addresses and a large set of resolved IP addresses. However, we do observe many flux domain query responses that have very large TTL value and only one resolved IP address. Therefore, it's not reliable to decide if a fully qualified domain name (FQDN) is using fast flux technique or not simply based on each single DNS massage. The proposed detection method is based on a series of DNS messages associated with the underlying FQDN.

Let $m^d(t, P)$ be one DNS message for the FQDN d, where t is the TTL value and

$$P = \{p\}$$

is the set of resolved IP addresses from the DNS response. We need to collect a set of consecutive DNS messages

$$M^d = \{m^d\}$$

against the FQDN d to determine whether d is a flux domain or not. Therefore, we can define

$$U = \bigcup_{m \in M} P$$

as the set of unique IP addresses from set of consecutive messages M.

4.1   Statistical Features

In addition to the features extracted from DNS data, some researchers also rely on third party sources for feature extraction [3] such as WHOIS, ASN and domain registrar. The added latency makes it impossible to detect flux domains in real-time. In this paper, we use features solely extracted from passive DNS data without use of third party databases.

A flux domain tends to set a lower TTL value so that it won't stay in the DNS cache for a long time. The most obvious feature is the average TTL value

$$T = \bar{t} \tag{1}$$

for the set M^d. In addition, a malware wants to frequently change its destination to different IP addresses. In this paper, we only focus on IPv4 addresses and think the same technique can be applied to IPv6 addresses. The number of unique resolved IP addresses is another feature that is

$$N = |U|. \tag{2}$$

One of the challenges in detecting malicious flux domains is to distinguish many legitimate domains owned by CDN, load balancer venders and NTP providers that are also providing a large number of changing destination IP addresses and short TTL value. We call them legitimate flux users.

Compared to the malicious flux domain creators that acquire destination IP addresses from compromised systems randomly distributed everywhere and each has a very short lifetime, the resolved IP addresses provided by the legitimate flux users are distributed in a limited number of subnets and have good shares among the first one, two and three octets of the IPv4 addresses. We can then define an entropy feature E on the first two octets of the addresses of the set U. Entropy is a powerful measure for the uncertainty of a random variable. For a set of elements S, let D_S be the distribution of the unique elements, the entropy of set S is then defined as

$$E_S = 1 - \sum[D_S \log(D_S)], \tag{3}$$

which has a value between 0 and 1. This principle can also be applied to each DNS message. We then have the average entropy calculated on the first three octets of the addresses of the set of resolved IP addresses per each DNS message, which is denoted as

$$F = \bar{e}_c. \tag{4}$$

It is a good indicator of target address neighborhood. The last but the most important feature is the one to measure the loyalty of the resolved IP addresses that is defined as

$$L = \frac{\sum_{m \in M} |P| - |U|}{|M||U| - |U|}. \tag{5}$$

The L value will be higher if the target IP addresses are frequently reused, and vice versa. For legitimate flux users, since the resolved IP addresses come from a reliably managed pool, a destination IP is often reused within a region. On the other side, attackers want to evade the detection systems and accommodate the availability fluctuation of compromised servers by changing resolved IP addresses frequently. The loyalty feature proves to be one of the prominent attributes in lowering false positive rate of the detection.

4.2   Domain Flux Classification

After a thorough feature selection process on a large set of training dataset, we finally focus on the feature vector (T, N, E, F, L), based on equations (1), (2), (3), (4), and (5), that will be used to build a time series model in classification of flux domains. The trivial approach of supervised machine learning is to collect a set of samples with some size and get truth marked by security experts, and then partition the data for training and testing with various machine learning technologies. In the real world of network security, the ratio between the number of malicious messages and number of benign messages in DNS traffic is very tiny. It's very costly to follow this approach and get experts looking through an extremely large set of samples. On the other side, the traditional unsupervised methods will be shy to provide good results on this highly biased network dataset. Therefore, we invent a semi-supervised machine learning method in this paper.

The approach we take in defining the classification system that can detect flux domain queries with a high confidence and, on the other side, won't generate obvious false positive is heuristics based semi-supervised regression. The regression starts from a relatively smaller capture rate and the coverage gets extended to a degree where we start seeing obvious false positive results. In each step, security experts will review the difference of the results which constructs the metrics of coverage and recall rates.

5   Real-Time Detection System

With the increase of network complexity and the number of devices, it's reasonable to expect a throughput up to 1 million DNS messages per second at peak time within a large network. On the other side, the number of unique FQDN to be requested can be around 200 million and this number keeps growing. The real-time detection system needs to be able to deal with both volume and velocity. In order to reduce the workload on flux domain detection, we do conservative benign classification as the first step to drop DNS messages that are not malicious with high confidence. Figure 2 demonstrates the classification diagram.

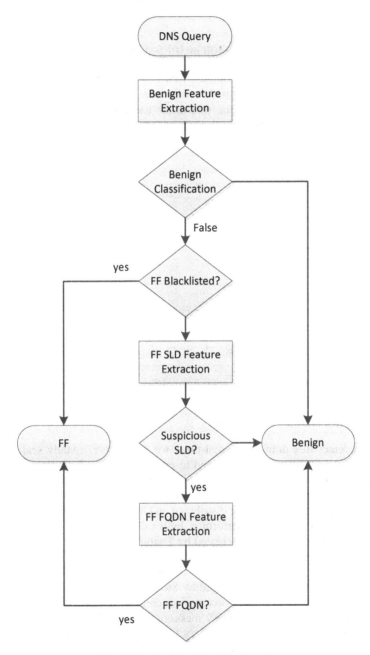

Fig. 2. Fast flux classification diagram

The major component of the benign classifier is a white list of 80 items that are carefully selected from top 100 second level domains or SLDs. Figure 3 shows the coverage of DNS queries by top 100 SLDs. Effectively, 60% of the queries can be safely dropped from fast flux classification.

In addition, all queries

$$\{m^d(t, P): t > 86400 \ or \ |P| > 1 \ and \ E_P < 0.5\},$$

where E_P is the entropy calculated on the first two octets of the resolved IP address set P, can be classified as benign with confidence. Note that when a DNS message m^d is classified as not flux domain, that doesn't mean the underlying FQDN d is benign. It will be classified further when the time series is updated by the new messages.

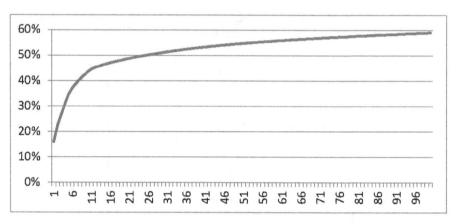

Fig. 3. DNS query coverage by top 100 second level domains

5.1   Analytics Online Architecture

In order to detect flux domains in real-time, we need a horizontally scalable infrastructure that can facilitate real-time end-to-end processing. The time series model also requires a large and fast observation cache that is expected to be in web scale and in DNS speed. Figure 4 shows the architecture of the online platform that is composed of open source components that are distributed and horizontally scalable.

The core part of the online platform is built on the top of a real-time processor cluster that uses Storm technology created by Twitter in open source community [15]. Storm provides a distributed framework that allows applications to run in parallel. User can just build topology networks in application layer based on its API. Each topology that is distributed and managed by Storm framework is for one or more applications. Kafka that is an open source project created by LinkedIn [16] is used as a persistent queue for input of the DNS message stream. HBase is a persistent key-value store in open source [17] that has very fast insertion speed and built-in TTL and versioning features required by our observation cache. It is also used for storing detection results for mitigation and reporting.

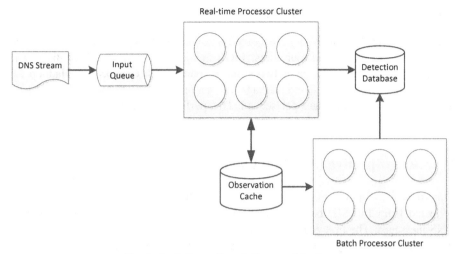

Fig. 4. Analytics online platform architecture

5.2   Asynchronous Detection Mode

Given the large throughput of DNS traffic, to be cost effective, we developed an asynchronous detection mode for the online system. HBase is a key-value store built on Google's BigTable [18]. It is designed to have near in-memory insertion speed. Similar to most database architectures, its random read speed is much slower than write. The system resource requirement is linearly proportional to the data velocity. To some extent, the synchronous detection mode will hit the disk IO bound given limited hardware resource. To overcome this problem, we opt to do online detection in an asynchronous mode to avoid frequent random reads. Therefore, the detection can be done in batch mode by scanning with a MapReduce job that is scheduled for every 10 minutes. The process is illustrated in Fig. 5.

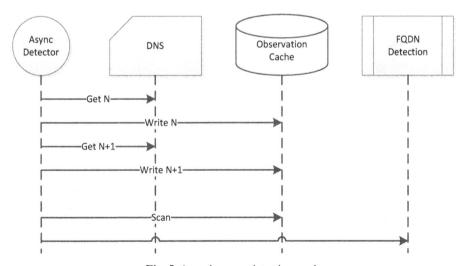

Fig. 5. Asynchronous detection mode

5.3   Online Detection Performance

Our performance benchmarking is executed on a cluster of 9 systems each is equipped with a CPU of 2.9GHz and four cores, 16GB memory, and two 1TB hard disks. The cluster is linked with a 1Gbps network and is loaded with CentOS 6.4, Hadoop 2.0 [19], HBase 0.94, Storm 0.8 and Kafka 0.7. It can reach a throughput of processing up to 2.8 million DNS messages per second. This performance can be further tuned and optimized.

6   Retrospective Detection Eveluation

This study is to analyze more than six month worth of DNS data with a size of 50TB collected from ISC/SIE between November 2012 and June 2013 with 360 billion DNS messages. The first goal of the study is to establish a baseline that can illustrate the detection accuracy with defined metrics which can be used as a reference for future improvement. On the other side, the study on large amount of real data is to uncover some insights about the fast flux malware and its behavior. The analysis is conducted on an analytics offline system shown in Fig. 6. The storage and processor cluster is built on a nine-node Hadoop system. It's mainly used for offline data analysis and semi-supervised training. A MapReduce job is created to simulate online detection including feature extraction, time series modeling, and flux domain detection.

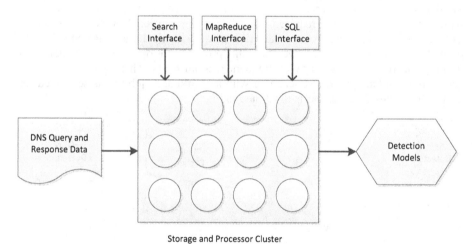

Fig. 6. Offline system

The result shows that, on 200 days worth of DNS data collected from November 2012 to June 2013, there are 10 million of DNS messages for 906 unique domains classified as flux domain. Furthermore, these flux domains come from 527 second level domains and the number of unique resolved IPs reaches a half million. We manually reviewed each of the flux domains and performed cross check. Three of them are less popular sites according to Alexa ranking [20], olendi.com, lodgelocker.com, and loenbun.com. The other five seem to be peer-to-peer seed trackers, dashjr.org, sipa.be, bluematt.me, litecoinpool.org, and xurious.com. What's interesting to us is that the average TTL value of these detected queries can be as high as 14 hours. That means a flux domain can slowly change its target IP address to hide itself from most detection algorithms. On average, 23,892 DNS messages are detected to be of flux domain per day or the rate is 24 every one million messages.

6.1   Botnet Distribution

Assume all flux domains with same SLD are created by the same attacker, we then can list top 10 autonomous system (AS) networks by the number of attackers in Table 1.

Table 1. Top 10 AS networks Compromised by Flux Domains

Number of Flux Domain SLD	Number of Queries	AS Number (ASN)	ASN Name
489	2,109,222	701	UUNET - MCI Communications Services, Inc. d/b/a Verizon Business
475	1,151,573	20,115	CHARTER-NET-HKY-NC - Charter Communications
453	542,027	20,845	DIGICABLE DIGI Ltd.
452	627,966	9,121	TTNET Turk Telekomunikasyon Anonim Sirketi
444	710,814	6,830	LGI-UPC Liberty Global Operations B.V.
438	922,094	812	ROGERS-CABLE - Rogers Cable Communications Inc.
437	509,095	33,491	COMCAST-33491 - Comcast Cable Communications, Inc.
429	272,742	20,001	ROADRUNNER-WEST - Time Warner Cable Internet LLC
428	418,868	10,796	SCRR-10796 - Time Warner Cable Internet LLC
414	303,796	7,015	COMCAST-7015 - Comcast Cable Communications Holdings, Inc

6.2   Detection Latency

As we explained in previous sections, detecting a flux domain based on a single DNS message is often not feasible or accurate. The presented time series model is based on a running history of the DNS messages. Therefore, there is a tradeoff between detection latency and accuracy. On the other side, the latency depends on the data feed rate. In this paper, we define detection latency in two ways, by time and by number of queries between first seen and capture, respectively.

In terms of time, most of the flux domains can be detected in less than a week as illustrated in Figure 7.

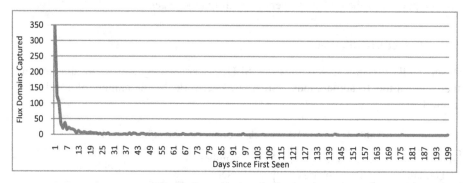

Fig. 7. Detection latency by time

On the other side, most of the flux domains can be captured before seeing the 100th DNS query for the underlying domain name as seen in Figure 8.

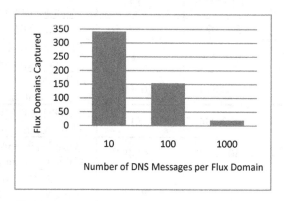

Fig. 8. Detection latency by number of DNS messages

Both metrics tell us that a small portion of flux domains will take very long time or large number of DNS messages to be captured. That may be due to their low activity or because those attacks act more like a normal DNS message in a short period of the time. The evidence for making an accurate decision on them grows slowly. Given a flux domain, the actual latency is determined by the interval set in the online system, e.g., 10 minutes and the DNS traffic on the subject domain to collect enough resolved IP addresses for calculating loyalty and entropy features. Therefore, for the most active flux threats, they can be captured in less than 10 minutes by our online system. Perdisci et al. reported a minimum detection latency of 30 hours and claimed it's days or weeks earlier than other sources publically available [8]. With the use of the methods presented in this paper, detection latency can be proportionally shortened when the data feed rate increases.

6.3   Flux Domain Life Span

With a simple assumption that a flux domain will be active, once it appears, until it gets removed, we count the time between first seen and last seen as the estimate of the life span of a flux domain. Although some flux domains may last longer than six months as shown in Figure 9, most of the flux domains disappear after one month. The short life span of the flux domains make blacklisting techniques less effective and demonstrates a much larger need for online analytical detection platforms that can detect and block new flux domains in a real-time fashion.

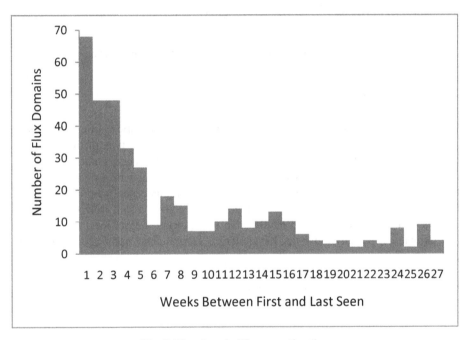

Fig. 9. Flux domain life span estimation

6.4   Effectiveness of Detection Algorithm

This metric is to show how many fast flux queries can be blocked if our online system is deployed on day one, that's November 25, 2012. As the Figure 10 shows, the system can catch and block 27K fast flux queries out of 1.8 billion DNS queries on every day.

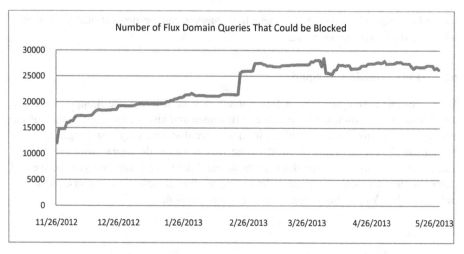

Fig. 10. Mitigation effectiveness based on the proposed detection

7   Conclusion

Domain flux using DNS services is one of the most active threats in computer and network security. It's important to the network users if such a type of threat can be detected and mitigated on DNS transaction that is often the entry point for network connections. Since the early discovery of domain flux, its behavior has changed over the past decade to evade existing detection methods. In this paper, we present a novel time series model with a set of features that are proven by passive DNS traffic of six months in a real-time system to be effective in capturing old and new types of fast flux domains, especially those that have a long TTL and small number of resolved IP addresses.

Due to the difficulty in getting ground truth, it's not trivial to study the false negative rate of the system, which will be addressed in future work by references from some third party works.

References

1. SSAC Advisory on Fast Flux Hosting and DNS, ICANN Security and Stability Advisory Committee (2008)
2. Holz, T., Gorecki, C., Rieck, K., Freiling, F.: Measuring and Detecting Fast-flux Service Networks. In: Proceedings of the Network & Distributed System Security Symposium (2008)
3. Passerini, E., Paleari, R., Martignoni, L., Bruschi, D.: fluXOR: Detecting and monitoring fast-flux service networks. In: Zamboni, D. (ed.) DIMVA 2008. LNCS, vol. 5137, pp. 186–206. Springer, Heidelberg (2008)
4. Nazario, J., Holz, T.: As the Net Churns: Fast-flux Botnet Observations. In: Proceedings of the 3rd International Conference on Malicious and Unwanted Software (2008)

5. Konte, M., Feamster, N., Jung, J.: Dynamics of Online Scam Hosting Infrastructure. In: Moon, S.B., Teixeira, R., Uhlig, S. (eds.) PAM 2009. LNCS, vol. 5448, pp. 219–228. Springer, Heidelberg (2009)
6. Hu, X., Knysz, M., Shin, K.G.: Rb-seeker: Auto-detection of Redirection Botnets. In: Annual Network & Distributed System Security Symposium (2009)
7. Hsu, C.-H., Huang, C.-Y., Chen, K.-T.: Fast-flux Bot Detection in Real Time. In: Jha, S., Sommer, R., Kreibich, C. (eds.) RAID 2010. LNCS, vol. 6307, pp. 464–483. Springer, Heidelberg (2010)
8. Perdisci, R., Corona, I., Giacinto, G.: Early Detection of Malicious Flux Networks via Large-Scale Passive DNS Traffic Analysis. IEEE Transactions on Dependable and Secure Computing 9(5), 714–726 (2012)
9. The Honeynet Project: Know Your Enemy: Fast-flux Service Networks (2007), http://old.honeynet.org/papers/ff/fast-flux.html
10. Antonakakis, M., Perdisci, R., Dagon, D., Lee, W., Feamster, N.: Building a Dynamic Reputation System for DNS. In: Proceedings of the 19th USENIX Conference on Security (2010)
11. Bilge, L., Kirda, E., Kruegel, C., Balduzzi, M.: EXPOSURE: Finding Malicious Domains Using Passive DNS Analysis. In: Proceedings of the ISOC Network and Distributed System Security Symposium (2011)
12. Antonakakis, M., Perdisi, R., Lee, W., Vasiloglou, N., Dagon, D.: Detecting Malware Domains at the Upper DNS Hierarchy. In: Proceedings of the 20th USENIX Conference on Security (2011)
13. ISC Security Information Exchange, http://www.isc.org/
14. Farsight Security, Inc.: https://www.farsightsecurity.com/
15. Storm, http://storm-project.net/
16. Kafka, https://kafka.apache.org/
17. HBase, http://hbase.apache.org/
18. Chang, F., Dean, J., Ghemawat, S., Hsieh, W.C., Wallach, D.A., Burrows, M., Chandra, T., Fikes, A., Gruber, R.: Bigtable: A Distributed Storage System for Structured Data. Google, Inc. (2006)
19. Hadoop, http://hadoop.apache.org/
20. Alexa Internet, Inc.: http://www.alexa.com/

Dictionary Learning-Based Volumetric Image Classification for the Diagnosis of Age-Related Macular Degeneration

Abdulrahman Albarrak[1], Frans Coenen[1], and Yalin Zheng[2]

[1] Department of Computer Science, University of Liverpool, United Kingdom
[2] Department of Eye and Vision Science, University of Liverpool, United Kingdom
{A.Albarrak,coenen,yzheng}@liverpool.ac.uk

Abstract. A discriminative dictionary-based approach to supporting the classification of 3D Optical Coherence Tomography (OCT) retinal images, so as to determine the presence of Age-related Macular Degeneration (AMD), is described. AMD is one of the leading causes of blindness in people aged over 50 years. The proposed approach is founded on the concept of a uniform 3D image decomposition into a set of sub-volumes where each sub-volume is described in terms of a "spatial gradient" histogram, which in turn is used to define a set of feature vectors (one per sub-volume). Feature selection is conducted using the maximum sum of the squared values of each feature vector for each sub-volume. After that, a "coding-pooling" framework is applied so that each image is represented as a single feature vector. The "coding-pooling" framework generates a representative subset of feature vectors called a dictionary, and then use this dictionary as a guide for the generation of a single feature vectors for each volume. Experiments conducted using the proposed approach, in comparison with range of alternatives, indicated that the approach outperformed other existing methods with an accuracy of 95.2%, sensitivity of 95.7% and specificity of 94.6%.

Keywords: Data mining, Image decomposition, Spatial gradient histograms, diactioanry learning, Medical image processing, Optical Coherence Tomography.

1 Introduction

Age-related Macular Degeneration (AMD) is an eye condition that results in vision loss that typically effects people over fifty years of age. AMD can be identified by inspection of retinal imagery. Traditionally this is conducted using 2-dimensional (2D) colour fundus images and a number of techniques for automating this process have been proposed. Of note in this context is the work of Hijazi et al. [1], Zheng et al. [2], Liu et al. [3] and Gossage et al. [4] where data mining techniques (more specifically classification techniques) have been proposed to support 2D retinal image analysis. However, the "traditional" 2D fundus photography for AMD detection has been superseded by three-dimensional

P. Perner (Ed.): MLDM 2014, LNAI 8556, pp. 272–284, 2014.

(3D) Optical Coherence Tomography (OCT) imaging techniques. The use of 3D OCT retinal imagery can provide detailed cross-sectional information for better diagnostic purposes. However, ophthalmologists find that they are now overwhelmed by the large amount of 3D image data in their clinical practice. There is thus a requirement for automated (or semi-automated) decision-support systems for the analysis of 3D OCT retinal image data (with respect to both AMD and other eye conditions).

In order to address the above challenge a new volumetric analysis technique is proposed in this paper for the automated diagnosis (classification) of AMD using 3D OCT retinal image data. The main challenges of this work is how best to extract and represent OCT image features so that a minimal amount of discriminative information will be lost, while at the same time the ensuring compatibility with the generation of effective classifiers. A method for representing 3D OCT images for classification purposes is therefore presented. The proposed method features a 3D decomposition into a collection of sub-volumes such that each sub-volume is defined in terms of a "gradient histogram" vector. We then sum the squared values of each histogram vector and sort them in descending order. We then select the top k vectors. These selected vectors are then translated into a feature vector representation using a coding-pooling framework. The coding part of this mechanism comprises the application of "sparse coding" to form a dictionary (a dictionary is a subset of feature vectors that is representative of a complete set of feature vectors). In the pooling part "spatial max" pooling [5] is applied to the selected feature vectors so as to combined them into a single feature vector to represent a given 3D volume.

The main contributions of the work described are: (i) a mechanism for representing 3D OCT retinal data, using image decomposition and a coding-pooling framework, to support volumetric classification; and (ii) a framework to support automated retinal disease detection using 3D OCT image data.

The remainder of this paper is organised as follows. A brief review of previous research is presented in Section 2. An overview of the application domain is described in Section 3. The design of the proposed approach is described in Section 4. Section 5 then assesses the performance of the proposed approach. Finally, this paper is concluded in Section 6 with a summary of the main findings.

2   Related Work

There has been some previous work directed at feature extraction for 3D image classification. The most common approaches rely on the use of statistical feature extraction and representation such as: (i) Local Phase Quantization (LPQ) [6], (ii) Local Binary Patterns (LBP) [7], and (iii) Scale-invariant feature transform (SIFT) [8]. These methods are used to extract low level image features such as: the frequency of intensity values, changes in the image intensity values and the spatial relationships between neighbouring intensities. LPQ uses the local Fourier transform where by a histogram of the quantised Fourier transform at low frequency is computed [6]. LBPs compute the relationship between each pixel and its immediate neighbours. However, the generation of 3D rotation invariant LBPs are computationally expensive. Zhao and Pietikainen [7] proposed the use of Three Orthogonal Plane LBPs (LBP-TOP). The LBP-TOP representation considers the calculation of LBPs only with respect to neighbouring voxels located in the XY, XZ and YZ planes. SIFT computes the orientation histograms of the image gradient directions. Most of these methods use a global feature vector which is produced by concatenating a range of feature values. This strategy has the disadvantage that some features may be redundant.

In more advanced approaches, feature selection strategies are employed to select only the most discriminative features. For instance, the concept of Dictionary Learning (DL) has drawn the attention of many researchers in computer vision and image classification [9,5,10]. To use DL a subset of a given set of feature vectors is selected to form a dictionary (or codebook). The challenge of DL is the identification of a highly discriminative subset of the available vectors. There are two ways to address this challenge. The first is to use a sparse representation. For instance, in the Maximisation of Mutual Information (MMI) method [11] a Gaussian Process (GP) is combined with k-means Singular Value Decomposition (K-SVD) to optimise the dictionary learning. The second is to use a coding-pooling framework. This framework is argued to be one of the most robust ways to represent images for classification in domains such as face recognition [12]. In the coding part of this framework vectors with similar features are clustered (using, for example, the k-means algorithm), an approach called "vector quantisation" [13]. Alternatively, sparse coding may be used to select a subset of vectors, in this case the selected vectors are said to form a "dictionary" or "codebook". Sparse coding tries to find a vector that represents a group of vectors by measuring the "response" of the vector to the group. The basic idea is to apply sparse coding on a random sample of extracted image descriptors such as SIFT [13], in order to identify a highly discriminative set of features. Then multi-scale spatial max pooling is used to form a single feature vector from a set of sparse feature vectors. These feature vectors can be input into a traditional classifier such as a Support Vector Machine (SVM). An example method that uses the coding-pooling framework is Linear Spatial Pyramid Matching Using Sparse Coding (ScSPM) described in [5] for 2D image classification. Alternatively, Locality-constrained Linear Coding (LLC) may be used instead of sparse coding to achieve the same goal [14]. In this paper, a coding-pooling framework is adopted with respect to the work described in this paper (see below).

3   Application Domain

The work described in this paper is directed at the detection of retinal disease and in particular the identification of Age-related Macular Degeneration (AMD) in 3D OCT images. Analogous to ultrasound, OCT is a relatively new imaging technique that can capture cross-sectional information of the retina. OCT employs low-coherence light and ultrashort laser pulses to detect the spatial position of tissue and resolve depth information. The employment of light waves enables acquisition of images (volumes) with very high resolution that can reveal precise details of internal structures. A series of 2D "slices" are acquired to form a 3D cross-sectional volume.

AMD is a condition, typically contracted in old age, which leads to irreversible vision loss at its advanced stages. This loss of vision for AMD patients is due to the damage to the macula, the centre of the retina that facilitates high level visual activities such as reading and recognition of faces [15]. There are some distinct features of AMD that can be readily identified in OCT image data such as: (i) disturbance of the Retinal Pigment Epithelium (RPE) layer underneath the neuro-retina due to the presence of "drusen" (fatty deposits), pigment epithelium detachment or geographic atrophy; (ii) disruption of layered neuro-retinal tissue; (iii) the presence of intra- and sub-retinal fluid and (iv) retinal thickening.

Two example OCT volumes, one normal and one featuring AMD, are presented in Figure 1. Figure 1(a) shows a 3D image of a normal OCT volume and Figure 1(b) shows a 3D OCT volume that features AMD. From these figures it can be seen that there are notable distinctions between the normal and the AMD volumes. The normal volume features smooth and connected layers. However, the AMD volume features disrupted layers and other abnormal patterns, such as thickening of the RPE layer, the presence of intra-retinal fluid, pigment epithelium detachment and some unusual texture patterns.

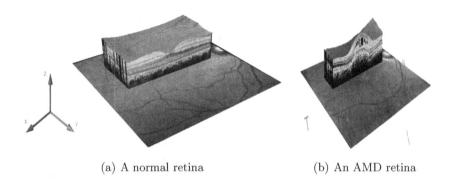

(a) A normal retina   (b) An AMD retina

Fig. 1. The 3D OCT of a "normal" and an AMD retina

Most of the current macular disease diagnosis tools have been directed at 2D images. Instances can be found in [1], [2], [3] and [4]. In [1] and [2], a graph based image representation method was proposed. The image data is decomposed into a set of quadtrees. These quadtrees are analysed using frequent sub-graph mining. Discriminative frequent sub-graphs are selected and used for generating a feature vector for each image. These feature vectors are then used to generate a classifier. In [3] Liu et al. designed a method for automatic detection of retinal diseases including AMD. A Multi-Scale Spatial Pyramid (MSSP) representation, with different levels, was used. The histograms of the LBPs were applied on each sub-block of the MSSP at each level. Dimensionality reduction using Principal Component Analysis (PCA) was also utilised. All the generated LBP's were concatenated together so as to build a global feature vector descriptor. The Radial Basis Function (RBF) kernel based SVM classifier was then used for categorising the feature vectors. In [4] a texture based method was employed using a combination of two methods, namely the Spatial Gray-Level Dependence Matrix (SGLDM) and the Discrete Fourier Transform (DFT) for extracting the OCT image features. Then a statistical method was used to extract the features of the SGLDM such as energy, entropy, correlation, local homogeneity, and inertia. A Mahalanobis distance based method was applied to measure the similarities between image features and then a Bayesian classifier was used to differentiate between features. Using SGLDMs different matrices are typically computed according to selected directions, which leads to long feature vectors with some similar values, which may in turn adversely affect classifier performance.

Recently, a 3D method for the classification of OCT was proposed [16] where a 3D decomposition was employed to support image classification which is similar to the proposed approach in this paper. The image is first decomposed into a set of sub-volumes and then a combination of LBP and Histogram of Oriented Gradients (LBP-HOG) used. First the LBP features were computed for every sub-volume. Following this, the gradient of the LBP features were calculated and the complete set of features normalised. These features were then concatenated forming a single feature vector for every image. Principal Component Analysis (PCA) was used to reduce the dimensionality of the feature vector. A classifier was then applied to categorise the feature vectors. In this research, instead of using PCA as in [16], a coding-pooling framework is adopted. In addition, an extension of the graph based methods for 3D classification described in [17,18] was used.

4   Proposed Approach

The proposed method relies on the coding-pooling framework described in Section 2. The proposed approach comprises the following stages: (i) preprocessing, (ii) image decomposition and feature extraction, (iii) feature vector coding, (iv) pooling, and (v) classification. Each of these stages is discussed in further detail in the following five sub-sections. Note that the input to the proposed process is a set of 3D OCT images $I = \{i_1, ..., i_n\}$, $i_n \in I$, associated with a set of class labels $C = \{c_1, ..., c_n\}$, $c_n \in \{-1, 1\}$. The process of feature vector generation is described in Algorithm 1. In the first stage (line 2) the current volume is enhanced forming a Volume of Interest (VOI) and then (stage 2, line 3) this VOI is decomposed to form a set of sub-volumes. Then feature extraction (low level feature extraction) is applied to form a set of Feature Vectors (FV) for each sub-volume in I (lines 4 to 6). The feature vectors are then ranked, using a "max-sum" calculation, and the top k selected (line 8). A dictionary is then built using sparse coding (stage 3, line 9). Following this, pooling is used to form the final feature representation (stage 4, lines 10 to 12) which is then fed into a classifier generator (stage 5).

Algorithm 1. Pseudocode for the proposed feature vector generation approach

Input: Volumes I
Output: FeatureVector fv_{z_i}
1: **for each** Volume i_n **do**
2:     VOI ← preprocessing(Volume i_n)
3:     Sub-Volumes ← decomposition(VOI)
4:     **for each** Sub-Volume j **do**
5:         FV ← feature-extraction(Sub-Volume j)
6:     **end for**
7: **end for**
8: X_n ← MaxSum(FV)
9: (V, U) ← Coding(X_n) Using algorithm 2
10: **for each** Volume i_n **do**
11:     fv_{z_i} ← pooling $(X_n(i_n)$, V, $U(i_n))$
12: **end for**

4.1   Preprocessing

The quality of OCT 3D images is usually affected by factors such as unwanted structures and alignments. The preprocessing stage is thus intended to improve the quality of the image and reference the images to a uniform coordinate system. Three steps are applied during the preprocessing stage. First of all, in order to remove unwanted structures from the input retinal images, two steps are used. Firstly a Split Bregman Isotropic Total Variation (ITV) algorithm, developed by [19], is applied to every slice of the 3D volume. Secondly, morphological operators are used: (morphological opening is applied in order to remove small objects not connected to the main retina and morphological closing to fill gaps).

The next step is to flatten the retina object. Flattening is applied using a second order polynomial least-square curve fitting procedure [3] according to the nature of the mean surface of the retina (defined according to the top and bottom retina surfaces). In order to do this we select the slice where the top and bottom surfaces of the volume (retina) are furthest apart and consider these two layers in terms of two vectors made up of voxel values. These two vectors are used to define the "middle" vector which is then used as a reference for flattening the entire retinal volume.

Finally, after flattening, a Volume of Interest (VOI) is defined. The basic idea is to select the top and bottom surface for every slice. These surfaces are then used to define the minimum value for the top and the maximum value for the bottom of the VOI. Figure 2 shows two sets of images describing a 3D OCT retina volume before preprocessing (Sub-figure 2(a)) and the VOI after preprocessing (Sub-figure 2(b)).

4.2   Image Decomposition and Feature Extraction

During the image decomposition and feature extraction stage each volume is decomposed into a set of sub-volumes with a "patch size" of $16 \times 16 \times 16$ voxels.

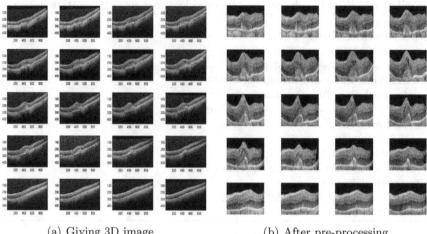

(a) Giving 3D image   (b) After pre-processing

Fig. 2. Example of preprocessed image

For each sub-volume the gradients ∇x, ∇y, ∇z are computed in three dimensions and then normalised to give the sum of the gradients using Equation 1. The "angles" are extracted using Equation 2 [8] to give values of between 0 and 2π, the number of angles was set to 8. For each voxel, an orientation histogram is computed, $hist(angle) = hist(angle) \times magnitude$.

$$magnitude = \sqrt{\nabla x^2 + \nabla y^2 + \nabla z^2} \qquad (1)$$

$$angle = atan^{-1}\left(\frac{\nabla z}{\sqrt{\nabla x^2 + \nabla y^2}}\right) \qquad (2)$$

In order to identify AMD ophthalmologist usually inspect an OCT scan centred on the area around the fovea (the centre of the retina) as this area tends to show more diagnostic information than the remainder of the retina. In order to consider the importance of spatial location, in our approach each sub-volume is given a weight according to its spatial 3D distance from the centre of the retina to the centre of it using Equation 3 where dtc is the sub-volume distance to the centre of the volume and ϵ is a small constant value. A feature vector fv is then formed for each sub-volume: $fv = weights \times hist$. In each case the generated fv is then normalised.

$$weight = \frac{\sqrt{\frac{width}{2} + \frac{height}{2} + \frac{depth}{2}}}{abs(dtc) + \epsilon} \qquad (3)$$

4.3 Feature Vector Coding

During stage 3 of the feature vector generation process, sparse coding is applied to the complete set of feature vectors ($FV = \{fv_1, \ldots, fv_m\}$, where m is the total number of feature vectors for all images) describing the entire image set, produced during the previous stage. The objective is to create a dictionary comprised of a representative subset of FV; the feature vectors in the dictionary should be selected so as to maximise the discriminative power of the eventual image classifier. One of the issues with the sparse coding based methods is that they have a high time complexity associated with them, thus in our approach a subset X_n of the FV for a given image n is used. Sparse coding is then applied to X_n. In [5], the authors proposed to generate the set X_n in a random manner. However, it is conjectured in this paper that random sampling may include feature vectors with less discriminative power which could adversely affect the performance of the eventual classifier. In order to address this problem, in the proposed approach, the sum of the squared value of each feature vector in FV is calculated (Equation 4). The set FV is then ranked, in descending order, and the top k selected to form the set X_n.

$$\max_n \sum_{i=1}^{m} FV_i \qquad (4)$$

Recall that the basic idea behind the concept of sparse coding is to represent a set of feature vectors in terms of a representative sub-set of these vectors. Thus, after the subset of feature vectors X_n has been identified, sparse coding is applied to X_n to form a dictionary V. More specifically the feature vectors in X_n are combined into k representative feature vectors which then defines our dictionary V ($V = \{rfv_1, \ldots, rfv_k\} \subset X_n$). For each $rfv_i \in X_n$ we associate a "feature vector indicator" u_i ($u_i \in U$, where U is the complete set of feature vector indicators) so that each fc_i is linked to the dictionary V (an approach informed by the work of [5]). The problem of optimising the selection of V and U are solved iteratively using a feature-sign search [20]. At the end of this process each feature vector in X_n is associated with its indicator in U and the final dictionary V. Algorithm 2 illustrates how the dictionary V and the complete set of feature vector indicators U are generated. In the algorithm, the dictionary size is set to 1024. The algorithm commences by selecting an initial dictionary randomly $initialV$ from the subset of vectors X_n (line 6). Then the mean values of $initialV$ are subtracted (line 7). In each iteration, in our experiment the number of iterations was set to 5, another random subset of feature vectors $initialB$ is selected (line 12). The gradient is then computed (line 16), which is for measuring the change between the two sets $initialV$ and $initialB$. The gradient is the multiplication of the sparse value of the $initialV$ minus the new set $initialB$. The maximum values of the gradient are selected. The associated indicator U is computed by getting the vectors that minimising the error and maximising the gradient while ensuring the maximum gradient is less than a given threshold (lines 20 to 25). Eventually, dictionary V is generated by considering the best set of feature vectors from the initial dictionaries $initialV$ and $initialB$ (line 31). Each vector in V is linked to U where the minimum error is reached (lines 26 to 33)).

Algorithm 2. Pseudocode for the dictionary generation approach [5,20]

Input: X_n
Output: V, U
 1: $DictSize = 1024$
 2: $\beta = $ 1e-5
 3: $\gamma = 0.15$
 4: $\sigma = $ eye $(DictSize)$.
 5: $itersNumber = 5$
 6: $initialV = $ selected randomaly a subset of size $DictSize$ from X_n
 7: $initialV = initialV$ - mean $(initialV)$
 8: $initialV = initialV \times diag(1/\sqrt{\sum (initialV.*initialV)})$
 9: **for each** i $= 1$ to $itersNumber$ **do**
10:    $A = initialV^T \times initialV + 2 \times \beta \times \sigma$.
11:    $x = $ inital a vector with size of A
12:    $initialB = $ select randomly from X_n
13:    **for each** j $= 1$ to number vectors in $initialV$ **do**
14:       $bb = initialB_j$
15:       $b = -initialVT \times bb$
16:       $grad = A \times sparse(x) + b$;
17:       $ma = \max(abs(grad).*x)$;
18:       $x = (\gamma - grad)/A$
19:       $EPS = $ 1e $- 9$
20:       **while** $ma \leq \gamma \times EPS$ **do**
21:          $U_j = \min(0.5 \times x^T \times A \times x + b^T \times x + \gamma \times |x|)$
22:          $grad = A \times sparse(x) + b$;
23:          $ma = \max(abs(grad).*x)$;
24:          $x = (\gamma - grad)/A$
25:       **end while**
26:       $bu = initialB \times U^T$
27:       $uu = U \times U^T$
28:       $dual = diag(bu - uu)$
29:       $bs = \sum initialB^2$
30:       $uinv = inv(uu + diag(dual))$
31:       $x = mintrace((bu \times uinv \times bu^T) + bs - \sum dual)$
32:       $V = (uu \times diag(x)/bu^T)^T$
33:    **end for**
34: **end for**

4.4 Pooling

The next stage of feature vector generation (stage 4) is data pooling where a single feature vector is generated for each 3D OCT retina volume. Data pooling is a statistical method used to concatenate different feature vectors into a single feature vector. To this end a pooling function is used to combine a collection of different feature vectors into a signal vector. Different factors should be taken into account such as the spatial location and similarities between vectors. Thus, given our subset of feature vectors X_n, and the set of feature vector indicators U linking the elements of X_n with our dictionary V (generated in stage 3 as described in the previous section) the objective is to build a single discriminative feature vector for each volume.

With respect to the work described in this paper spatial max pooling [5] was adopted to encapsulate a set of feature vectors X_i representing a single volume ($X_i \subset X_n$) into a single feature vector guided by the dictionary V. Max pooling recursively computes the histograms of the maximum values for a given set of vectors X_i and their association with elements in V. Equation 5 shows the pooling function used to derived the maximum values with respect to the set U which is then used to generate a new histogram z_r for each region r with M sub-volume. Neighbouring regions are recursively united and then Z_r is applied on each one of them until reaching a final feature vector. On completion of the process a single feature vector fv_{z_i} is obtained that describes each data volume i (retinal volume with respect to the motivation for this paper).

$$z_r = \max\{|u_{1r}|, ... |u_{Mr}|\} \tag{5}$$

4.5   Classification

From the foregoing, a single feature vector (fv_{z_i}) is used to describe each data volume i (retinal volume with respect to the motivation for this paper). For training purposes each feature vector fv_{z_i} was combined with a class label c_i (in our case +1 indicates a retina with AMD, -1 for normal retina, informed by medical retina experts) to create a training set. The resulting representation is compatible with a number of classifier generators. However, linear SVM [21] was used in this paper due to its reasonably good performance in most application. The results of the evaluation are presented in the next section.

5   Results and Discussion

To evaluate the effectiveness of the proposed approach experiments were conducted using 140 3D OCT volumes, 68 "normal" and the remainder AMD. The size of each volume was about (1024×496 $pixels$) $\times$ 19 $slices$ representing a $6 \times 6 \times 2$ mm retinal volume. Ten-fold cross validation was used to evaluate the proposed method. Four metrics were recorded with which to measure the performance of the proposed algorithm: accuracy (Acc) (Equation 6) , sensitivity (Sen) (Equation 7), specificity (Spec) (Equation 8) and the Area Under the receiver operator characteristic Curve (AUC). AUC can reflect the trade-off between sensitivity and specificity.

The experiments were directed at analysing the proposed approach in comparison to approaches using alternative representations taken from the literature, namely: LBP-TOP [7], LPQ [6], SIFT [8], MMI [11], ScSPM [5] and LBP-HOG [16]. We also compared with MSSP which is a 2D based approach described in [4]. With respect to MSSP, the middle slices from the training and test data were used (because these have a high potential for including part of the fovea, the part of the retina where most indicators of AMD are likely to occur). In addition, the LBP-HOG based methods for 3D image classification [16] was included in the experiments. For each different approach, the SVM parameter was tuned for the best performance.

$$Acc = \frac{TP + TN}{TP + TN + FP + FN} \qquad (6)$$

$$Sen = \frac{TP}{TP + FN} \qquad (7)$$

$$Spec = \frac{TN}{TN + FP}. \qquad (8)$$

The results are presented in Table 1. From the table it can be seen that the proposed approach produces better results than the other reported methods. If we compare the accuracy of the proposed method with the other two methods, the proposed method produced a better performance (with a recorded accuracy of 95.2%) while for the LBP-TOP, LPQ, SIFT, MMI, ScSPM, MSSP and LBP-HOG methods accuracy values of 88.6%, 92.9%, 90.4%, 81.3%, 93.8%, 89.8% and 91.4% were recorded respectively. In the context of sensitivity, as shown in Table 1, the proposed approach has a similar result with respect to ScSPM (a recorded sensitivity of 95.7% compared with a recorded sensitivity of 95.3% for ScSPM). MMI performed less well than the proposed approach. In terms of the 3D methods LBP-TOP, LPQ and SIFT, the first two produced a recorded sensitivity of 93.3% compared to 86.0% using SIFT. This sensitivity result for the 2D based method, MSSP, was 87.3%; which was better than the MMI result. In terms of specificity, the proposed approach has a good performance with a recorded specificity of 94.6% with respect to other method: LBP-TOP with 85.0%, LPQ with 91.9%, 3D SIFT with 94.1%, MMI with 79.56%, ScSPM with 92.4% and MSSP with 91.3% and 90.5% for LBP-HOG. The proposed approach also produces better results than the other methods measured by the AUC. The proposed method produced a AUC value of 0.95 while the recorded AUC values with respect to the other methods considered were as follows: LBP-TOP with 0.88, LPQ with 0.92, 3D SIFT with 0.90, MMI with 0.81, ScSPM with 0.93 and MSSP giving 00.89. The AUC value obtained when using LBP-HOG was 0.94.

The results presented in Table 1 also support the idea that by using some guided function such as the sum of the squares with the coding-pooling framework to do image classification, is better than using random selection. In addition it can be noted that discriminative feature extraction methods help to improve the performance of the classifier. Overall LPQ and SIFT, produce a good performance with respect to our dataset. However, the proposed method indicated a better performance.

Table 1. Comparison of the proposed approach with the other methods

Method	Acc	Sen	Spec	AUC
LBP-TOP	88.6%	**93.3%**	85.0%	0.88
LPQ	**92.9%**	**93.3%**	91.9%	0.92
SIFT	90.4%	86.0%	**94.1%**	0.90
MMI	81.3%	83.3%	79.5%	0.81
ScSPM	**93.8%**	**95.3%**	**92.4%**	**0.93**
MSSP	89.8%	87.3%	91.3%	0.89
LBP-HOG	91.4%	92.4%	90.5%	0.94
Proposed method	**95.2%**	**95.7%**	**94.6%**	**0.95**

6   Conclusion

In this paper a new approach to classifying 3D OCT volumes has been proposed with an application in the diagnosis of AMD (AMD vs. non-AMD). More specifically the approach is founded on the use of a coding-pooling framework to identify discriminative features within OCT sub-volumes. The results obtained using the proposed approach demonstrated a better performance in comparison with: LBP-TOP, LPQ, SIFT, sparse based representation methods such as MMI and ScSPM, and the MSSP 2D based method. This research has also shown that using a value, such as the sum of the squares, for the selection of a subset of the available set of feature vectors will improve the performance of the classifier. A further study will be directed at evaluating our approach with respect to different retinal diseases such as diabetic retinopathy. The authors intend to also conduct further research to explore the effect of using different feature selection methods for the coding-pooling framework, and evaluate the proposed approach in larger datasets.

References

1. Hijazi, M. H.A., Coenen, F., Zheng, Y.: Data mining techniques for the screening of age-related macular degeneration. Knowledge-Based Systems 29, 83–92 (2012)
2. Zheng, Y., Hijazi, M.H.A., Coenen, F.: Automated "disease/no disease" grading of age-related macular degeneration by an image mining approach. Investigative Ophthalmology & Visual Science 53(13), 8310–8318 (2012)
3. Liu, Y.Y., Chen, M., Ishikawa, H., Wollstein, G., Schuman, J., Rehg, J.M.: Automated macular pathology diagnosis in retinal OCT images using multi-scale spatial pyramid and local binary patterns in texture and shape encoding. Medical Image Analysis 15(5), 748–759 (2011)
4. Gossage, K., Tkaczyk, T., Rodriguez, J., Barton, J.: Texture analysis of optical coherence tomography images: feasibility for tissue classification. Journal of Biomedical Optics 8(3), 570–575 (2003)
5. Yang, J., Yu, K., Gong, Y., Huang, T.: Linear spatial pyramid matching using sparse coding for image classification. In: IEEE Conference on Computer Vision and Pattern Recognition, pp. 1794–1801 (2009)

6. Päivärinta, J., Rahtu, E., Heikkilä, J.: Volume local phase quantization for blur-insensitive dynamic texture classification. In: Heyden, A., Kahl, F. (eds.) SCIA 2011. LNCS, vol. 6688, pp. 360–369. Springer, Heidelberg (2011)
7. Zhao, G., Pietikainen, M.: Dynamic texture recognition using local binary patterns with an application to facial expressions. IEEE Transactions on Pattern Analysis and Machine Intelligence 29(6), 915 (2007)
8. Scovanner, P., Ali, S., Shah, M.: A 3-dimensional sift descriptor and its application to action recognition. In: Proceedings of the 15th International Conference on Multimedia, pp. 357–360 (2007)
9. Mairal, J., Bach, F., Ponce, J., Sapiro, G.: Online dictionary learning for sparse coding. In: Proceedings of the 26th Annual International Conference on Machine Learning, pp. 689–696 (2009)
10. Yang, M., Zhang, L., Feng, X., Zhang, D.: Fisher discrimination dictionary learning for sparse representation. In: IEEE International Conference on Computer Vision, pp. 543–550 (2011)
11. Qiu, Q., Jiang, Z., Chellappa, R.: Sparse dictionary-based representation and recognition of action attributes. In: IEEE International Conference on Computer Vision, pp. 707–714 (2011)
12. Wang, Z., Feng, J., Yan, S., Xi, H.: Linear distance coding for image classification. IEEE Transactions on Image Processing 22(2), 537–548 (2013)
13. Lazebnik, S., Schmid, C., Ponce, J.: Beyond bags of features: Spatial pyramid matching for recognizing natural scene categories. In: IEEE Computer Society Conference on Computer Vision and Pattern Recognition, vol. 2, pp. 2169–2178 (2006)
14. Wang, J., Yang, J., Yu, K., Lv, F., Huang, T., Gong, Y.: Locality-constrained linear coding for image classification. In: IEEE Conference on Computer Vision and Pattern Recognition, pp. 3360–3367 (2010)
15. Jager, R.D., Mieler, W.F., Miller, J.W.: Age-related macular degeneration. New England Journal of Medicine 358(24), 2606–2617 (2008)
16. Albarrak, A., Coenen, F., Zheng, Y.: Age-related macular degeneration identification in volumetric optical coherence tomography using decomposition and local feature extraction. In: Proceedings of the 17th Medical Image, Understanding and Analysis Conference, pp. 59–64 (2013)
17. Albarrak, A., Coenen, F., Zheng, Y.: Classification of volumetric retinal images using overlapping decomposition and tree analysis. In: IEEE 26th International Symposium on Computer-Based Medical Systems, pp. 11–16 (2013)
18. Albarrak, A., Coenen, F., Zheng, Y., Yu, W.: Volumetric image mining based on decomposition and graph analysis: An application to retinal optical coherence tomography. In: IEEE 13th International Symposium on Computational Intelligence and Informatics, pp. 263–268 (2012)
19. Goldstein, T., Bresson, X., Osher, S.: Geometric applications of the split bregman method: Segmentation and surface reconstruction. Journal of Scientific Computing 45(1-3), 272–293 (2009)
20. Lee, H., Battle, A., Raina, R., Ng, A.Y.: Efficient sparse coding algorithms. In: Schölkopf, B., Platt, J., Hoffman, T. (eds.) Advances in Neural Information Processing Systems, pp. 801–808. MIT Press, Cambridge (2007)
21. Chang, C.C., Lin, C.J.: Libsvm: A library for support vector machine (2001), Software available at http://www.csie.ntu.edu.tw/~{}cjlin/libsvm

A Robot Waiter Learning from Experiences

Bernd Neumann[1], Lothar Hotz[2], Pascal Rost[2], and Jos Lehmann[1]

[1] Department of Informatics,
University of Hamburg, Germany
{neumann,jlehmann}@informatik.uni-hamburg.de
[2] Hamburger Informatik Technology Center, Department of Informatics,
University of Hamburg, Germany
{hotz,7rost}@informatik.uni-hamburg.de

Abstract. In this contribution, we consider learning tasks of a robot simulating a waiter in a restaurant. The robot records experiences and creates or adapts concepts represented in the web ontology language OWL 2, extended by quantitative spatial and temporal information. As a typical task, the robot is instructed to perform a specific activity in a few concrete scenarios and then expected to autonomously apply the conceptualized experiences to a new scenario. Constructing concepts from examples in a formal knowledge representation framework is well understood in principle, but several aspects important for realistic applications in robotics have remained unattended and are addressed in this paper. First, we consider conceptual representations of activity concepts combined with relevant factual knowledge about the environment. Second, the instructions can be coarse, confined to essential steps of a task, hence the robot has to autonomously determine the relevant context. Third, we propose a "Good Common Subsumer" as opposed to the formal "Least Common Subsumer" for the conceptualization of examples in order to obtain cognitively plausible results. Experiments are based on work in Project RACE[1] where a PR2 robot is employed for recording experiences, learning and applying the learnt concepts.

Keywords: Machine Learning, ontology, robot activities.

1 Introduction

Learning capabilities are essential prerequisits for robots to become useful in complex real-world domains. In this paper, we investigate learning of high-level activity concepts represented in the formal knowledge-representation framework OWL 2. Learning will be based on experiences recorded by the robot and, at times, instructions by a human instructor. Examples are taken from the restaurant domain with a robot performing as a waiter. Activities such as serving a guest are typically composed of several levels of subactivities, down to elementary robot capabilities such as moving or grasping.

---

[1] This work is supported by the RACE project, grant agreement no. 287752, funded by the EC Seventh Framework Program theme FP7-ICT-2011-7.

P. Perner (Ed.): MLDM 2014, LNAI 8556, pp. 285–299, 2014.

To convey a first understanding of the learning tasks investigated in this paper, consider the three scenarios sketched in Figure 1. In Scenarios A and B, the robot - here called "trixi" - receives detailed instructions how to serve a coffee to a guest and learns that these activities constitute a "ServeACoffee":

Instructions for Scenario A: "Move to counter1, grasp mug1-A, move to south of table1, place mug1-A at placement area west - this is a ServeACoffee."

Instructions for Scenario B: "Move to counter1, grasp mug1-B, move to north of table1, place mug1-B at placement area east - this is also a ServeACoffee."

In Scenario C, it is assumed that the robot has learnt a concept from the two examples and will serve the coffee to the placement area south right of guest1-C.

Instructions for Scenario C: "Do a ServeACoffee to guest1-C at table2."

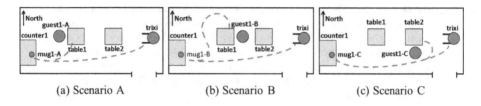

(a) Scenario A   (b) Scenario B   (c) Scenario C

Fig. 1. Scenarios

In all scenarios, we assume that the robot knows the location of the guest and of the placement areas on the table. However, it does not know which placement area to approach for guest1-C. Part of the learning task is therefore to generalize the recorded experiences of Scenarios A and B and create a concept which can be applied to similar but not identical situations. In general, we consider learning scenarios where a service robot, equipped with a repertoire of basic robot operations, is incrementally instructed by examples and expected to autonomously apply its enhanced competence to new situations, possibly requiring further generalizations.

Learning structured conceptual representations from examples is not a new topic. Winston's seminal work on learning block structures such as bridges is a well-known starting point [18]. One can distinguish three major lines of research which have contributed to today's understanding of the field: (i) the development of well-understood and standardized knowledge representation formalisms (see in the following), (ii) research into Cognitive Robotics, connecting symbolic high-level representations with quantitative low-level operations (see [13,7,3] and [15]), and (iii) advances in learning and reasoning models in Cognitive Sciences (see [9,17,10] and [15]).

Our choice of a knowledge representation framework is motivated by previous work on high-level representations of events and activities [3] and the recent development of the standardized knowledge representation language OWL 2 supporting such representations. Furthermore, extensions of OWL 2 in terms of quantitative representations for spatial and temporal information [7] have made it possible to represent robot activities coherently from high-level symbolic activity concepts down to low-level quantitative commands and sensor input. As a drawback, standard OWL representations cannot yet represent the sameness of entities typically occurring in the compositional structures for activity models. Fortunately, recent work by the group of Hitzler [8] has established "Nominal Schemas" as a well-founded way to express sameness by identical variable names, similar to variables in Datalog. We adopt Nominal Schemas for our work.

Learning from examples within a DL framework has been investigated before [6,12], and it is well understood how concept expressions can be generalized or refined if they must be adapted to new examples. In principle, Version Space Learning as introduced in [14] can be applied to define the space of concepts which correctly classify positive and negative examples. In our approach, concept learning has been designed as a cognitively plausible strategy for ontology evolution without bookkeeping of version space boundaries. This is achieved by a learning curriculum essentially based on careful generalizations and avoiding disjunctive expressions. Correspondences between concepts and examples are established following the structure-mapping theory of Cognitive Science [9].

In Section 2, following this introduction, we describe the knowledge representation conventions adopted for our learning work. In Section 3, we present our learning approach, called Ontology-based Learning from Examples (OLE), in detail. This allows the robot to form a new activity concept from examples, to adapt an existing concept to cover a new task, or to refine a concept based on a negative example. In Section 4, we evaluate OLE using several other scenarios in the restaurant domain. It will be shown that the learning procedure can lead to desirable activity models with very few examples, given an appropriate ontology. More examples may be required, however, if learning examples contain many irrelevant details, initially leading to overly restricted concepts. The paper ends with conclusions, summarizing the work and pointing out some open problems for future research. Future work will also cover research on the formal properties of the presented algorithm, such as consistency, completeness, or complexity.

2   Knowledge Representation Conventions

Concepts in OLE are represented in an ontology using a restricted version of the Web Ontology Language OWL 2 and Protégé as an editor. The concept definitions are also the basis for a hierarchical planner as well as other components of the robotic system realized in RACE [16]. For textual concept representations, we use the Manchester Syntax[2] in this paper. Following the conventions of OWL, a concept will be often called 'class' when we deal with OWL representations. As an example of an activity class, the definitions of PlaceObject1-A and classes used as property fillers are listed below.

---

[2] http://www.w3.org/TR/owl2-manchester-syntax/

Listing 1.1. Class definitions of PlaceObject1-A and its property fillers

```
Class:  PlaceObject1-A
SubclassOf: PlaceObject
     that      hasHolding exactly 1 {?Holding1-A}
     and hasOn exactly 1 {?On1-A}
     and hasBefore exactly 1 Before1-A

Class:  Holding1-A
SubClassOf:      Holding
EquivalentTo:   {?Holding1-A}
     that     hasRobot value trixi
     and hasPassiveObject {?Mug1-A}

Class:  On1-A
SubclassOf: On
EquivalentTo:   {?On1-A}
     that     hasPhysicalEntity {?Mug1-A}
     and hasArea value paWest1

Class:  Before1-A
SubclassOf: Before
     that     hasFirst exactly 1 {?Holding1-A}
     and hasSecond exactly 1 {?On1-A}
     and hasBeforeRange exactly 1 TimeRange
```

Class names begin with upper-case, individuals with lower-case letters, property names with the prefix 'has'. The postfix '-A' is part of the class names and used here to mark classes conceptualized from the robot's recording of Scenario A (see Figure 1). In addition to the properties spelled out above, the concepts inherit properties from the ontological ancestor Occurrence:

```
     and     hasStartTime only TimeRange
     and     hasFinishTime only TimeRange
     and     hasDuration only TimeRange
```

The datatype TimeRange is used to express an uncertainty range of a time point and is a shorthand for two separate properties with numerical fillers, e.g.

```
     and     hasStartTimeLowerBound only Int
     and     hasStartTimeUpperBound only Int
```

We currently use only a subset of OWL 2 with the syntax shown in Listing 2. In DL terminology, the syntax corresponds to an Attribute Language with full existential quantification and number restriction (ALEN).

Note the restrictive use of class expressions for property fillers: If a filler requires a more expressive definition, a class name must be introduced and defined in a separate class definition, as shown for Holding1-A and On1-A. The reason for this restrictive grammar is our interest in keeping a simple syntactical correspondence between class definitions and the constituents of episodes, see below. It is apparent that named concepts as property fillers can be replaced by their definitions, creating nested concept definitions and a reduced number of names. Also note the absence of negation. This is motivated by our primary interest in modelling conceptualizations of episodes recorded under an open-world assumption (OWA).

Listing 1.2. Syntax of restricted grammar for concept definitions

```
Class: <className>
SubclassOf: <className>
     [ 'that' [inverse] <propertyName>   <restriction>
     { 'and' [inverse] <propertyName>   <restriction> } ]
<restriction> ::= 'only' <classExpression> |
        'some' <classExpression> |
        'exactly' <integer> [<classExpression>] |
        'min' <integer> [<classExpression>] |
        'max' <integer> [<classExpression>] |
        'value' <individual>
<classExpression> ::= <className> |
        <nominalSchema>
<nominalSchema> ::= '{' <individualVariable> '}'
```

Class definitions can represent hierarchical compositional structures by letting a parent class (an aggregate) refer to its components via partonomical properties. These properties have the common parent property 'hasPart'. We often refer to the root of a compositional hierarchy as root concept or root class.

We now explicate the use of Nominal Schemas as property fillers. A Nominal Schema specifies an individual of a particular class with a variable name which may reoccur as a property filler in several places and must be instantiated with the same known individual [11] which may be, however, any value of its class, different from the usual individuals. As pointed out in the introduction, expressing sameness of individuals is important for compositional hierarchies. For our knowledge representation purposes, the class of the individuals of a Nominal Schema ?X is represented by a class definition with the additional axiom EquivalentTo: ?X. The scope of a Nominal Schema is taken to be the ontology of the domain. In the examples above, ?Holding1-A, ?Mug1-A, and On1-A are Nominal Schemas.

The assertional knowledge of a robot comprises episodes which are stored as experiences in the robot memory. An episode consists of dynamic factual knowledge which describes spatially and temporally coherent occurrences in the restaurant as viewed by the robot, and permanent (or background) knowledge about the robot's environment which is assumed to be valid for all times. Assertional knowledge is stored in triplets relating individuals via properties to other individuals, as customary for DL languages.

3   Concept Formation and Adaption

As illustrated by the scenarios shown in Figure 1, the OLE approach to concept learning includes a supervised learning task where an instructor provides a name for a new concept (ServeACoffee in Scenarios A and B) and its essential components, and an unsupervised learning task where the robot must adapt a concept to a new situation (Scenario C). We believe that learning situations of this kind may play a key role when employing service robots in new domains. Technically, we realize both learning tasks by two procedures: conceptualizing an example and adapting a concept to a conceptualized positive example or, more generally, finding a common subsumer for two corresponding concepts. In effect, this approach allows to formulate learning solely based on conceptual expressions, known as the "single-representation trick" [5]. We also sketch a procedure for refining a concept in order to exclude a negative example.

3.1   Conceptualizing Examples

```
Conceptualize Episode
Input:    - Ontology
          - Episode, background knowledge (recorded by the robot)
          - Robot activities from episode constituting a new concept
            (specified by instructor)
Process:  - Determine relevant assertions from episode
          - Conceptualize assertions
          - Create new class definitions
Output:   - New class definitions updating the ontology
```

Conceptualizing a robot activity carried out in a scenario amounts to establishing a generic description for the ontology of the robot with the purpose that it can be used as a template for future robot activities and other cognitive tasks. The conceptualization procedure is structured as shown above.

In the following, we describe an approach for extracting the assertions relevant for a concept from an episode E. Let A be the high-level robot activities specified by the instructor as constituents of an instance c of the new concept C. We define the *support* of c as the assertions involving A, and all assertions in E connected to A. This can be visualized by a *property graph* with individuals as nodes, c as root node, and properties as directed edges. Figure 2 shows the property graph for placeObject1-A, using definitions of Listing 1, among others. As constituents A, this robot activity comprises a precondition holding1-A with the properties hasRobot and hasPO (hasPassiveObject), and a postcondition on1-A with the properties hasPE (hasPhysicalEntity) and hasArea. Temporal properties of occurrences, specifying start time and finish time, are not included for clarity. The graph also shows some of the observations which the robot has made: guest1-A at the sitting area saWest1, right of the manipulation area maSouth1 and left of the manipulation area maNorth1. Permanent knowledge includes the robot trixi, table1, its sitting areas saWest1 and saEast1, the corresponding placing areas, manipulation areas and premanipulation areas (not all shown). Class information for individuals are not represented in the graph, but are of course available.

We define the *direct support* of c to comprise all nodes which can be reached from the root node via directed partonomical properties (the components of the aggregate hierarchy), and in addition all property fillers of these nodes which have been part of the activity plan and hence included in the episode. In Figure 2, the nodes of the direct support of placeObject1-A are shown in bold, all other nodes are indirect support. In general, an episode may contain many details experienced by the robot and connected to the direct support, e.g. by spatial or temporal relations.

To filter out irrelevant information, we make use of a heuristic which measures *relevance* in terms of the smallest semantic distance of a node to any of the nodes of the direct support. Distance is determined by counting the number of property edges irrespective of their direction. In our experiments we have included nodes within a distance of 2, whereby connections via the two legs of a reified relation (e.g. at, on) where counted as 1. Figure 2 shows that the important instances table1, guest1-A and saWest1 (the sitting area west of table1) are included as a result of the relevance analysis, whereas saEast1 (the sitting area east of table1) is deemed irrelevant.

The next step in the conceptualization procedure is to transform relevant assertional knowledge into conceptual descriptions. We take a strictly conservative approach regarding generalizations and let a conceptualization represent exactly the same activities, however at an arbitrary time (modulo some tolerance regarding durations). Hence given the same environment and the updated ontology, the robot will be able to carry out the learnt activities. Permanent assertional knowledge remains assertional and is associated with the new concept via the relevant property graph. The exact rules for transforming assertional knowledge into conceptual representations are described in [15].

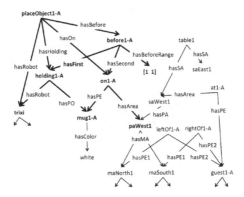

Fig. 2. Property graph for an instance of the concept PlaceObject with related permanent knowledge and observations

3.2 Adapting a Concept to a Positive Example

Modifying an existing concept such that it covers a new example is a frequent step in a learning curriculum. This can be done by conceptualizing the example (as described in the preceding subsection) and then computing the Least Common Subsumer (LCS) of the old concept and the conceptualized example. An effective way to do this for Description Logics (DL) similar to our sublanguage of OWL has been presented in [1]. Computing the LCS of two concepts essentially amounts to determining the product tree of the description trees corresponding to the concepts, and intersecting the subclass memberships. In view of the fact that a formal LCS may not always exist, and following similar ideas as in [2], we will be content to compute a cognitively plausible "good" common consumer, abbreviated GCS.

The adaptation procedure has the following structure:

```
Input:    - Ontology
          - Old concept
          - Positive example (complete or incomplete)
Process:  - Conceptualize example
          - Align example to old concept
          - Compute "Good Common Subsumer" (GCS)
Output:   - New concept subsuming conceptualized example and old concept
          - Updated ontology
```

The scenarios in Figure 1 exemplify two different learning situations. One is the creation of a concept for ServeACoffee after experiencing Scenarios A and B, this can be done based on episodes describing *complete* examples. When the robot is asked, however, to perform a ServeACoffee for a guest in Scenario C, the existing concept has to be adapted on-line to be applicable to partially unfolded scene where only the situation before performing the task is available for concept adaptation. We will refer to this task as adaptation to an *incomplete* example. Both tasks, adaptation of a concept to a complete and adaptation to an incomplete example, can be performed by essentially the same GCS computations, while requiring different alignment procedures.

Alignment. In order to perform component-based generalization of two concepts, they must be structurally aligned. We adopt ideas of analogical reasoning [9] where structure mapping has been investigated in detail in a cognitive context. As one of the key principles, analogical structures require a tight agreement of corresponding relations but little or no agreement between corresponding entities. Accordingly, our alignment process follows the property structure and establishes correspondence mainly based on coinciding property names. Differently from analogy construction, however, classes do play a role, and corresponding classes should not be taxonomically distant.

For adaptation to complete examples, we may assume that both conceptual descriptions have identical single roots which necessarily form a corresponding pair. Further correspondences can then be determined by following the property graphs. For adaptation to incomplete examples, alignment may be more difficult. Roughly, it amounts to searching for large coinciding property structures, similar to searching for graph isomorphisms, except for the specific tolerance requirements. Computational aspects of searching for isomorphisms in conceptual graphs are treated thoroughly in [4].

If a conceptualized example cannot be perfectly aligned with an existing concept because of non-corresponding reified spatial or temporal relations, these are treated as irrelevant and omitted. If a discrepancy is due to non-corresponding robot activities, however, alignment fails and the conceptualized example is made a new concept.

GCS of Corresponding Classes. The GCS computation of two concepts takes the two property graphs of the concepts as input. Hence each concept description typically comprises several class definitions and related permanent knowledge.

Table 1. Key generalization rules for the Good Common Subsumer (GCS)

Property Graph 1	Property Graph 2	Element of resulting property graph (GCS)
<className1>	<className2>	<newClassName>
		subclassOf LNCS (<className1>, <className2>)
<className1>	<individual2>	<newClassName>
		subclassOf LNCS (<className1>, MSC(<individual2>))
<individual1>	<individual2>	if <individual1> = <individual2>: <individual1>
		else: <newClassName>
		subclassOf MSC(<individual1>, <individual2>)
		universal restriction 'only'
'max' <int1>	'max' <int2>	'max' max(<int1>, <int2>)
'min' <int1>	'min' <int2>	'min' min(<int1>, <int2>)
interval	[<int1> <int2>]	interval
[<int3> <int4>]	interval	[min(<int1>, <int3>), max(<int2>, <int4>)]
<propertyName1>	<propertyName2>	LNCS (<propertyName1>, <propertyName2>)

The generalization steps are comparable to LCS computation in DLs, however due to our interest in compact descriptions and the use of restricted concept expressions, there are some differences, manifest in more numerous but shorter class definitions. In Table 1, we list key generalization rules of our GCS. The entities refer to the grammatical structure of properties as described in Section 2. LNCS stands for least named common subsumer, i.e. the closest taxonomical parent. MSC is the most specific concept for an individual.

Restrictions by Nominal Schemas are in principle treated the same way as restrictions by class names. A common new variable name is used for all occurrences of a Nominal Schema. Furthermore, the new class definition corresponding to a Nominal Schema has the conjunct 'EquivalentTo:' <ominalSchema>. Individuals generalized to a class give rise to a new Nominal Schema, if they are property fillers for more than one class definition.

After the generalization phase, a cleaning-up process is carried out to avoid unnecessary new class names. This pertains to class definitions where the generalization has led to a class already existing in the ontology. Illustrating examples for GCS computations are provided in Section 4.

3.3 Adapting a Concept to a Negative Example

Learning from positive examples as described in the preceding section is conservative in the sense that unnecessary taxonomical generalizations are avoided. Nevertheless, conceptualizations may prove too general, and instructions may inform the robot about situations which are negative examples for an existing concept. There may be several reasons:

1. The existing taxonomy is too coarse, preventing a necessary differentiation. For example, there may be no class for standard table items such as pepper, salt and decoration.
2. There are no useful properties which could help to distinguish the positive and negative examples.
3. There are distinguishing features, but the heuristically determined support of a learnt concept has not included this information.

In this section, we sketch two re-learning procedures which resort to recorded episodes in order to adapt a concept to a negative example. This requires, of course, that a learnt concept is linked with its positive examples.

Adding an Affordance Property. It is well-established in robotics to characterize physical or conceptual entities by ways to make use of them, called affordances. For example, if one can sit on an object, it is characterized by the affordance 'sittable'. In the learning situation addressed above in (ii), an affordance property can be added to establish the necessary distinguishing property. To provide a solution where the robot need not invent new names, we introduce the meta-concept 'ActivityName' with all acitivity names of the ontology as possible instances, and postulate that all scene objects have the property 'hasAffordance only ActivityName'. This way, the affordance 'sittable' can be expressed in a concept as the property 'hasAffordance value sit' where 'sit' is an activity name. Hence, if a concept needs refinement, it must receive the appropriate affordance property, and all recorded positive examples must be extended accordingly.

Extending the Support. As a second way of distinguishing positive from negative examples one can search for additional features (properties or scene components) which have not been included in the original conceptualization, but may be recorded in the episodes which have provided the positive examples. For example, if a guest had not been included in the ServeACoffee conceptualization of Scenarios A and B, the impoverished concept would have allowed to place a coffee at any placement area, provoking a negative example. By revisiting Episodes A and B and extending the support to include a guest, the necessary differentiation can be achieved.

4   Experimental Results

In this section, we describe experimental learning results achieved with different learning tasks in various scenarios. Because of the large data volumes we cannot provide a complete coverage but restrict our documentation to the most interesting generalizations resulting from the GCS.

4.1   ServeACoffee Scenarios

The GCS of the conceptualizations of Scenarios A and B produces several generalizations concerning robot destination, placement area, manipulation area, premanipulation area, mug, and spatial relations of the guest. Table 2 shows interesting examples, new concepts are marked with the postfix '-AB'.

Comments to Table 2:

Rows 1 and 2: After applying ServeACoffee to two different premanipulation areas and placing areas of table1, respectively, the concept is generalized to apply to all such areas of table1. The corresponding area concepts are represented by

Table 2. Generalizations by combining conceptualizations of Scenarios A and B

Conceptualization A	Conceptualization B	ServeACoffee-AB
1 ... hasToArea value pmaSouth1	... hasToArea value pmaNorth1	... hasToArea only ?PMA1-AB
2 ... hasArea value paWest1	... hasArea value paEast1	... hasArea only ?PA1-AB
3 maWest1 hasPMA pmaWest1	maEast1 hasPMA pmaEast1	Class: MA1-AB
		EquivalentTo: ?MA1-AB
		SubclassOf: MA
		that hasPMA only ?PMA1-AB
4 Class_Instance: [EastWestTable,	Class_Instance: [EastWestTable,	Class: SA1-AB
table1]	table1]	EquivalentTo: ?SA1-AB
Properties: [hasSA, SA, saWest1]	Properties: [hasSA, SA, saEast1]	SubclassOf: SA
		that inverse hasSA value table1
5 Class: At1-A	Class: At1-B	Class: At1-AB
SubclassOf: At	SubclassOf: At	SubclassOf: At
that hasArea value saWest1	that hasArea value saEast1	that hasArea only ?SA1-AB
and hasPE exactly 1 Guest	and hasPE exactly 1 Guest	and hasPE exactly 1 Guest
6 Class: RightOf1-A	Class: RightOf1-B	Class: RightOf1-AB
SubclassOf: RightOf	SubclassOf: RightOf	SubclassOf: RightOf
that hasFirst value maSouth1	that hasFirst value maNorth1	that hasFirst only ?MA1-AB
and hasSecond exactly 1 Guest	and hasSecond exactly 1 Guest	and hasSecond exactly 1 Guest

Nominal Schemas because the same instances occur as fillers of several properties within the support of ServeACoffee.

Row 4: The new concept for representing all sitting areas of table1 must be related to table1 with the inverse of the existing property hasSA.

Rows 5 and 6: The spatial relations involving a guest now refer to the same sitting areas and manipulation areas of table1 which are also the destination of ServeACoffee.

In Scenario C, the learnt concepts must be applied to a new situation where the guest sits at another table (table2) in a sitting area distinct from any previously encountered sitting areas. The property graph for this situation is shown in Figure 3.

After conceptualization, it is correctly aligned with the property graph of ServeACoffee-AB and the adapted concept SeveACoffee-ABC is determined, with key generalizations shown in Table 3.

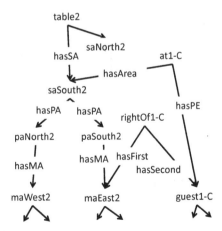

Fig. 3. Property graph of Scenario C before performing ServeACoffee

Table 3. Generalizations by combining the conceptualization of Scenarios A and B with the conceptualization of Scenario C

ServeACoffee-AB	Conceptualization C	ServeACoffee-ABC
1 Class: SA1-AB	Class_Instance: [Table, table2]	Class: SA
EquivalentTo: ?SA1-AB	Properties:	EquivalentTo: ?SA1-ABC
SubclassOf: SA	- [hasSa, SA, saSouth2]	SubclassOf: SA
that inverse hasSA value table1		that inverse hasSA only Table
2 Class: At1-AB	Class: At1-C	Class: At1-ABC
SubclassOf: At	SubclassOf: At	SubclassOf: At
that hasArea only ?SA1-AB	that hasArea value saSouth2	that hasArea only ?SA1-ABC
and hasPE exactly 1 Guest	and hasPE exactly 1 Guest	and hasPE exactly 1 Guest
3 Class: RightOf1-AB	Class: RightOf1-C	Class: RightOf1-ABC
SubclassOf: RightOf	SubclassOf: RightOf	SubclassOf: RightOf
that hasFirst only ?MA1-AB	that hasFirst value maEast2	that hasFirst only ?MA1-ABC
and hasSecond exactly 1 Guest	and hasSecond exactly 1 Guest	and hasSecond exactly 1 Guest

Comments to Table 3:

Row 1: The sitting area saSouth2 of table2 in Conceptualization C causes a generalization of table1 to an arbitrary table.

Rows 2 and 3: The spatial relations involving a guest now refer to sitting areas and manipulation areas of any table which is the destination of ServeACoffee.

ServeACoffee-ABC can be paraphrased as follows: Move to counter1, grasp a mug, move to a premanipulation area belonging to a manipulation area right of the guest, place the mug on the placing area belonging to the sitting area where the guest is located.

4.2   Deal-with-Obstacles Scenarios

In these scenarios, illustrated in Figure 4, it is assumed that the robot has learnt a ServeACoffee as described in the previous subsection. The robot has again the task to serve a coffee to a guest west of table1, but encounters an obstacle in the southern manipulation area from which a coffee is normally served.

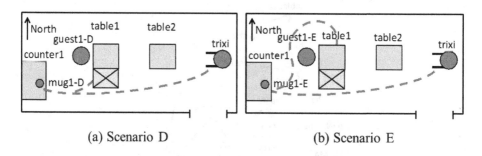

(a) Scenario D   (b) Scenario E

Fig. 4. Scenarios D and E

In Scenario D, this obstacle is a person, and the robot is instructed to wait until the person has moved away. The robot follows the instruction and learns a new concept ServeACoffeeBlocked-D. In Scenario E, a sidetable blocks the manipulation area, and the robot tries to apply the learnt concept, generalizing it to subsume any physical entity as obstacle. As the robot waits for the sidetable to move away, the instructor tells the robot not to wait in this case but to move to the northern premanipulation area and place the coffee on the western placement area.

Both learning situations feature a negative example for an existing concept and require a structurally modified new concept. In Scenario D, the robot first follows a plan based on the existing concept ServeACoffee which fails because of the obstacle. The robot continues following the instructions, records the episode and after conceptualization creates the new concept ServeACoffeeBlocked-D. Parts of the property graph are shown in Figure 5. The temporal relations, omitted for graphical clarity, require that a PutObject1-D is carried out after the blocking event At1-D has terminated.

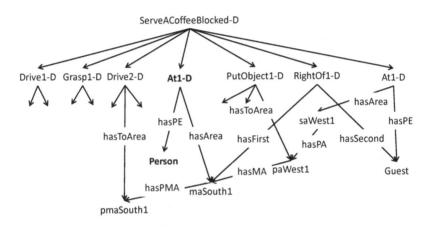

Fig. 5. Conceptualization of Episode D. PutObject1-D begins when At1-D ends

In Scenario E, the robot first applies ServeACoffee unaware of the obstacle and, after failure, adapts ServeACoffeeBlocked-D by generalizing Person to PhysicalEntity (PE) in order to cover the new situation with the sidetable as obstacle (procedure presented in Section 3.2). Following the instructions, the robot does not wait but proceeds to pmaNorth1 at the north of the table and performs the putObject. The conceptualization of this episode results in ServeACoffeeBlocked-E, parts of the property graph are shown in Figure 6. The three concepts, ServeA-Coffee, ServeACoffeeBlocked-D, and ServeACoffeeBlocked-E are preserved, while the concept with the tentative generalization of Person to PhysicalEntity is abandoned.

In a further experimental scenario, Clear-table-smartly, the robot learns to clear all items from a table except for a vase. The initial concept ClearTable lets the robot clear all passive objects (POs). When the robot clears the vase,

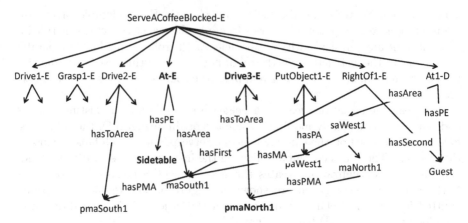

Fig. 6. Conceptualization of Episode E. An additional Drive3-E begins after observing At1-E. PutObject1-E begins after Drive3-E.

the instructor marks this as a negative example. To adapt to the negative example, the robot refines PO in the concept ClearTable by the affordance property 'hasAffordance value clearTable' and extends the positive examples accordingly. For details see [15].

5   Conclusions

We have presented several methods for learning or refining conceptual descriptions based on examples, formalized within an OWL-based knowledge representation framework. While the principles of conceptual learning are well understood for many years, our approach deals with several new aspects. Firstly, our representation formalism is used by an integrated robot system operating in real-world scenarios, collecting experiences in a robot memory, and sharing a common ontology for recording experiences, learning, planning, scene interpretation, and other reasoning tasks. Hence the robot can make immediate use of learning results. Representations include quantitative spatial and temporal information for real world grounding.

Secondly, we have shown that concepts and episodes represented in a restricted OWL 2 dialect can be conveniently transformed into property graphs as a basis for structural matching. Thus, ideas of analogical reasoning can be realized, allowing complex conceptual descriptions to be applied to new situations.

Thirdly, our approach considers realistic learning scenarios where instructions may be vague and the robot may be uncertain about relevant scene components. We have therefore proposed relevance analysis based on the semantic distance between contextual scene components and robot activities.

Finally, due to the experiences stored in the robot memory, the framework can be smoothly extended from learning based on single examples to learning based on a large body of experiences with statistically relevant features.

References

1. Baader, F., Küsters, R., Molitor, R.: Computing Least Common Subsumers in Description Logics with Existential Restrictions. In: Proceedings of the 16th International Joint Conference on Artificial Intelligence (IJCAI-99), vol. 1, pp. 96–101. Morgan Kaufmann (1999)
2. Baader, F., Sertkaya, B., Turhan, A.-Y.: Computing the least common subsumer w.r.t. a background terminology. Journal of Applied Logic 5(3) (2007)
3. Bohlken, W., Koopmann, P., Hotz, L., Neumann, B.: Towards ontology-based real-time behaviour interpretation. In: Guesgen, H., Marsland, S. (eds.) Human Behavior Recognition Technologies: Intelligent Applications for Monitoring and Security, pp. 33–64. IGI Global (2013)
4. Chein, M., Mugnier, M.L.: Graph-based Knowledge Representation. Springer (2009)
5. Cohen, W., Feigenbaum, E.: The Handbook of Artificial Intelligence, vol. 3. William Kaufmann (1982)
6. Cohen, W., Hirsh, H.: Learning the classic description logic: Theoretical and experimental results. In: Proc. Principles of Knowledge Representation and Reasoning (KR-1994) (1994)
7. Günther, M., Hertzberg, J., Mansouri, M., Pecora, F., Saffiotti, A.: Hybrid reasoning in perception: A case study. In: Proc. SYROCO, September 5-7, IFAC, Dubrovnik (2012)
8. Hitzler, P., Krötzsch, M., Rudolph, R., Sure, Y.: Semantic Web (2008)
9. Holyoak, K., Gentner, D., Kokinov, B.: Introduction: The place of analogy in cognition. In: Gentner, D., Holyoak, K., Kokinov, B. (eds.) The Analogical Mind, pp. 1–20. The MIT Press (2001)
10. Keane, M., Costello, F.: Setting limits on analogy: Why conceptual combination is not structural alignment. In: Holyoak, K., Gentner, D., Kokinov, B.N. (eds.) The Analogical Mind, pp. 287–312. The MIT Press (2001)
11. Krötzsch, M., Maier, F., Krisnadhi, A., Hitzler, P.: A better uncle for owl. In: Proc. of the World Wide Web Conference (WWW 2011), pp. 645–654 (2011)
12. Lehmann, J.: Dl-learner: Learning concepts in description logics. Journal of Machine Learning Research (JMLR) 10 (2009)
13. Lutz, C.: Description logics with concrete domains - a survey. Advances in Modal Logic 4 (2003)
14. Mitchell, T.: Generalization as search. Artificial Intelligence 18(2), 203–226 (1982)
15. Neumann, B., Hotz, L., Günter, A.: Learning robot activities from experiences: An ontology-based approach. Tech. Rep. TR FBI-HH-B-300/13, University of Hamburg, Department of Informatics Cognitive Systems Laboratory (2013)
16. Rockel, S., Neuman, B., Zhang, J., Dubba, K.S.R., Cohn, A.G., Konečný, Š., Mansouri, M., Pecora, F., Saffiotti, A., Günther, M., Stock, S., Hertzberg, J., Tomé, A.M., Pinho, A.J., Lopes, L.S., von Riegen, S., Hotz, L.: An ontology-based multi-level robot architecture for learning from experiences. In: AAAI Spring Symposium on Designing Intelligent Robots: Reintegrating AI II. Stanford, USA (2013)
17. Wilson, W., Halford, G., Gray, B., Phillips, S.: The star-2 model for mapping hierarchically structured analogs. In: Holyoak, K., Gentner, D., Kokinov, B. (eds.) The Analogical Mind, pp. 125–160. The MIT Press (2001)
18. Winston, P.: Learning structural descriptions from examples. In: The Psychology of Computer Vision, pp. 157–209. McGraw-Hill, New York (1975)

Parametric Representation of Objects in Color Space Using One-Class Classifiers

Aleksandr Larin[1,*], Oleg Seredin[2], Andrey Kopylov[2],
Sy-Yen Kuo[3], Shih-Chia Huang[4], and Bo-Hao Chen[4]

[1] Moscow Institute of Physics and Technology, Moscow, Russia
ekzebox@gmail.com
[2] Tula State University, Tula, Russia
oseredin@yandex.ru, And.Kopylov@gmail.com
[3] National Taiwan University, Taipei, Taiwan
sykuo@cc.ee.ntu.edu.tw
[4] National Taipei University of Technology, Taipei, Taiwan
schuang@ntut.edu.tw, hd840207@gmail.com

Abstract. Two new approaches to parametrization of specific (flame representative) part of a color space, labeled by an expert, are presented. The first concept is to apply D. Tax's one-class classifier as a steerable descriptor of such a complex volumetric structure. The second concept is based on approximation of the training data by a set of elliptic cylinders arranged along the principal components. Parameters of such elliptic cylinders describe the training set. The efficiency of the approaches has been proven by experimental study which let allowed us to compare the standard Gaussian Mixture Model based approach with the two proposed in the paper.

Keywords: One-class classification, Pixel classifiers, Fire detection, Flame detection, Support Vector Data Description, PCA.

1 Introduction

One of the working stages of computer vision system, solving the flame detection problem, is a description of pixels set that parametrically presents objects of interest in the image. In the course of color analysis any picture pixel can be imagined as a three-dimensional vector, each coordinate of which coincides with the value of corresponding RGB component of the pixel. The construction of mathematical model, based on number of pixels in color space specified by expert, allows the parametric description of the areas of interest in a feature space with given accuracy. This construction is called parameterization problem or Color Data Modeling [1]. Parameterization of flame pixels is an example of such problem and will be covered in the paper.

* This research was supported by grants 14-07-00527 of the Russian Foundation for Basic Research and 12-07-92000 of the Russian Foundation for Basic Research and the Taipei-Moscow Coordination Commission on Economic and Cultural Cooperation.

P. Perner (Ed.): MLDM 2014, LNAI 8556, pp. 300–314, 2014.

The main hypothesis of flame pixels detection in image is based on an expert indication of fragments (rectangles, polygons, explicit labeling of pixels) that cover the flame region with some range of accuracy. It is necessary to provide mathematical model that enables us to classify pixels into one of two following classes: flame pixels - "not-flame" pixels. An example of extempore indication of a region of pixels which acts as training set is presented in Fig. 1. A training set region marked in Fig.1 forms a collection of three-dimensional vectors that can be mapped into RGB space as shown in Fig. 2 (a).

Gaussian Mixture Model is widely used for color data modeling [2]. Various versions of this method were applied to flame detection. For example, the method of experimental data parametrization by spheres was explained in detail in [3]. The spacial arrangement of spheres is based on GMM adjustment in such a way that the center of each sphere is a point represented by mean value of corresponding distribution and the radius of sphere is a doubled standard deviation of the same distribution (see Fig. 2 (b)). For such parametric model, any pixel is considered to be a part of flame in case it falls into at least one approximating sphere. The literature describes various methods of selecting the parameters for a mixture of normal distributions that best describes the original data set, but the most commonly used method based on the EM-algorithm (Expectation-maximization) [5]. Adjustable parameters of the algorithm are the number of normal distributions in the mixture and dispersion threshold for decision.

Fig. 1. The image of flame. The rectangle marks training set region.

The significant restriction of such method is that the number of spheres should be set a priori. The cited paper reports the sufficient quality of flame pixels parameterization can be achieved by usage of ten approximating spheres.

Task formulation, the appearance of the training set in RGB space and our previous experience in pattern recognition inspire us to consider the initial task of color data modeling as a kind of parameterization problem which can be solved by using the methods of machine learning. The specific character of the

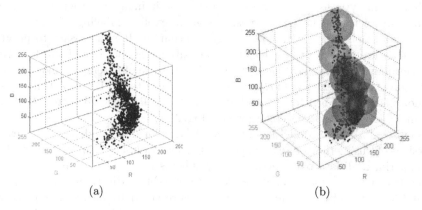

(a)   (b)

Fig. 2. (a) The training set mapping into RGB space, (b) Data approximation by spheres (the center of sphere — distribution mean value, radius — doubled standard deviation

task lays in the fact the training set contains objects of one class only. That is why we suggest to use a specific description of the task for objects identification learning, namely, construction of the one-class classifier.

The first suggested concept is to apply D. Tax's Support Vector Data Description [4] as a steerable descriptor of such a complex volumetric structure. Nevertheless, the computational time needed to form a description based on one-class classifier is hardly predictable and depends on the number of support objects. As an alternative way, we introduce another approximation of expert data in RGB space, namely by a set of elliptic cylinders in the direction of the principal component. First, initial color space is subjected to orthogonal transform by means of Principal Component Analysis (PCA) to reduce the correlation of color components. Next, the transformed space is divided into the set of two-dimensional layers that are orthogonal to the direction of the main component. Then, the ellipse of scattering is constructed for each layer. Thus, the set of parameters of such ellipses, arranged along the main components, describes the training set.

This paper has undertook the study and comparison of three one-class classifiers implemented to the pixels parameterization task in 3D space. In particular, the classifiers, constructed according to Tax's model, Guassian mixture model [5] and model of PCA elliptic cylinders approximation were observed. Algorithms have been compared against three criteria, namely: classification quality, parameterization time of the given data set and time of new objects classification.

2   Support Vector Data Description

Method of one-class classification, which will be applied to solve the given problem, is introduced by Dr. Tax in [4] and is called Support Vector Data Descrip-

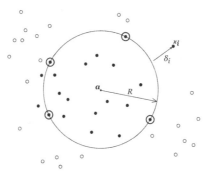

Fig. 3. The spherical model of data description

tion. The peculiarity of this method is in its strong analogy with the V. Vapnik's method of support vectors.

Training set description model ($\mathbf{a} \in \mathbb{R}^n$, n — feature space dimension, $R \in \mathbb{R}$, N — number of objects in training set) in method $\mathbf{x}_i \in \mathbb{R}^n$, $i = 1, \ldots, N$ — hypersphere representing the nearest outer border around data.

Restrictive hypersphere is unambiguously determined by its centre and radius, which are selected in a following way: radius should be minimal but most training total objects should not go beyond its borders (see Fig. 3). At the same time, the objects that fall out of the borders of hypersphere should be fined, and the sum of all fines of all the objects should be minimal as well.

Thus, the structural error of the model should be minimized:

$$\begin{cases} R^2 + C \sum_{i=1}^{N} \delta_i \to \min_{R, \mathbf{a}, \delta}, \\ \|\mathbf{x}_i - \mathbf{a}\|^2 \le R^2 + \delta_i, \ \delta_i \ge 0, \ i = 1, \ldots, N. \end{cases} \tag{1}$$

The given problem is a quadratic programming problem and can be solved by method of Lagrange multipliers. The dual problem is equivalent to the original can be presented as follows:

$$\begin{cases} \sum_{i=1}^{N} \lambda_i (\mathbf{x}_i \cdot \mathbf{x}_i) - \sum_{i=1}^{N} \sum_{j=1}^{N} \lambda_i \lambda_j (\mathbf{x}_i \cdot \mathbf{x}_j) \to \max_{\lambda}, \\ \sum_{i=1}^{N} \lambda_i = 1, \ 0 \le \lambda_i \le C, \ i = 1, \ldots, N, \end{cases} \tag{2}$$

where λ — Lagrange multipliers.

D. Tax suggests using only support objects that are on the borderline of hypersphere to describe training total of decision procedure.

A new object is considered to belong to the area of interest in case the distance between the object and the centre of hypersphere is smaller that its radius.

It follows the function of one-class decision procedure for object recognition will take the form of indicator function:

$$d\left(\mathbf{z}; \lambda, R\right) = I\left(\|\mathbf{z} - \mathbf{a}\|^2 \leq R^2\right), \tag{3}$$

where

$$\|\mathbf{z} - \mathbf{a}\|^2 = (\mathbf{z} \cdot \mathbf{z}) - 2\sum_{i=1}^{N_{SV}} \lambda_i \left(\mathbf{z} \cdot \mathbf{x}_i\right) + \sum_{i=1}^{N_{SV}} \sum_{j=1}^{N_{SV}} \lambda_i \lambda_j \left(\mathbf{x}_i \cdot \mathbf{x}_j\right), \tag{4}$$

$$R^2 = (\mathbf{x}_k \cdot \mathbf{x}_k) - 2\sum_{i=1}^{N_{SV}} \lambda_i \left(\mathbf{x}_k \cdot \mathbf{x}_i\right) + \sum_{i=1}^{N_{SV}} \sum_{j=1}^{N_{SV}} \lambda_i \lambda_j \left(\mathbf{x}_i \cdot \mathbf{x}_j\right), \tag{5}$$

where N_{SV} — number of support objects, and $\mathbf{x}_k$ — any support object.

D. Tax used "kernel trick" method [6,7] to enter rectifying space of larger dimension features to make the data description possible in more "flexible" manner than the sphere. Polynomial and Gaussian kernels are frequently used for this purpose.

$$K\left(\mathbf{x}_i, \mathbf{x}_j\right) = \exp\left(\frac{-\|\mathbf{x}_i - \mathbf{x}_j\|^2}{s^2}\right). \tag{6}$$

Thus, in (2), (4) and (5) it is necessary to change an operation of scalar product of two vectors by calculating the kernel value of two arguments to receive better data description model according to D. Tax's method.

It is evident the Gaussian kernel is more efficient for solving the parametrization problem of flame pixels color representation. Therefore, this function will be further used.

Fig. 1 shows a rectangular area from which pixels range was obtained to train Dr. Tax's classifier. Fig. 10 presents the results of borderline constructed around data, which were received assuming trained classifier. Fig. 4 illustrates application of classifier for flame determination in the image.

It should be noted, the area becomes more "dense" and restricts smaller pixels range around training set in RGB space when Gaussian potential function parameter is reduced. Thus, it becomes necessary to select optimal potential function parameter.

2.1   Selection of Optimal Parameter s for Gaussian Kernel

In his research paper D. Tax suggested the algorithm that allows the selection of an optimal parameter s value of Gaussian kernel. The essence of the algorithm is to search parameter s that will correspond to the given recognition error within the limits of minimum and maximum parameter value. These values can be determined by following formulas:

$$s_{min} = \min_{i,j} \|\mathbf{x}_i - \mathbf{x}_j\|, \; i \neq j, \; s_{max} = \max_{i,j} \|\mathbf{x}_i - \mathbf{x}_j\|.$$

Fig. 4. The results of applying one-class classifier trained on the set of points marked in Fig. 1

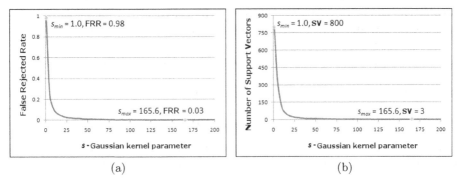

(a)   (b)

Fig. 5. The relations between kernel parameter and (a) classification error in slide control, (b) number of support objects in classifier's decision procedure

Search of parameter should be done according to iterative scheme with the given step until recognition error corresponds to the desired one. It was proposed to estimate recognition error by using the procedure of slide control leave-one-out. Relation between the classification error and Gaussian kernel parameter is shown in Fig. 5 (a).

When attention is drawn to diagram of relation between Gaussian potential function parameter and the number of support objects that take part in recognition decision procedure (see Fig. 5 (b)), it becomes clear the number of support objects decreases with increase of potential function parameter. Thus, the number of support objects in decision procedure is an indirect indicator that helps give rough evaluation of future classification error.

2.2   Optimization of Classifier's Operation Speed

Application of realized program classifier to analyse video stream that works in real time has shown the necessity of speeding-up its operation time at the stage of recognition. The reason is the recognition procedure should be done for each pixel of processing image several times a second.

Focusing attention on the Gaussian kernel structure (6), it becomes evident it inherits two characteristic features from an exponential function: it possesses only positive value at any point of the function domain, and in the case of arguments equality possesses the value of 1. So, the form of the decision rule (3) can be simplified into

$$d\left(\mathbf{z}\right) = I\left(\sum_{i=1}^{N_{SV}} \lambda_i K\left(\mathbf{z}, \mathbf{x}_i\right) - \sum_{i=1}^{N_{SV}} \lambda_i K\left(\mathbf{x}_k, \mathbf{x}_i\right) \geq 0\right),\tag{7}$$

where $\mathbf{x}_k$ — any support vector. As the Gaussian potential function is positive it is enough to partly calculate the first sum in the expression (7). The second sum can be figured out in advance before classification takes place. It becomes clear the smaller number of operations is needed to determine the given expression. It means classification process takes less time.

It is well-known, the estimation of exponent is a difficult operation and we suggested that decrease of this function call can reduce the time of pixels recognition. In our case a feature vector looks as follows:

$$\mathbf{x}_i = (r_i, g_i, b_i).\tag{8}$$

By substituting (8) into initial expression of Gaussian kernel (6) and by using one of exponent's features we have the following expression of Gaussian kernel:

$$K(\mathbf{x}_i, \mathbf{x}_j) = \exp\left(-\frac{|r_i - r_j|^2}{s^2}\right)\exp\left(-\frac{|g_i - g_j|^2}{s^2}\right)\exp\left(-\frac{|b_i - b_j|^2}{s^2}\right).\tag{9}$$

by introducing the following function $f(v)$:

$$f(v) = \exp\left(-\frac{v^2}{s^2}\right)\tag{10}$$

and by changing (9) expression the result is

$$K(\mathbf{x}_i, \mathbf{x}_j) = f(|r_i - r_j|)f(|g_i - g_j|)f(|b_i - b_j|).$$

Note, the value of each RGB color component is restricted by multitude $(0, 1, \ldots, 255)$. That is why the difference module of any two color components will belong to this multitude as well, which means the definition area of the function (10) will be discrete and restricted by $(0, 1, \ldots, 255)$. That is why all values of this function can be figured out before main classification procedure takes place. This allows us to abandon the exponent function evaluation, which takes a lot of time.

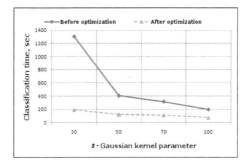

Fig. 6. The results of decision rule optimization

Therefore, the number of operations needed for classification was essentially reduced and the exponent function evaluation was totally abandoned. Comparative results are shown in Fig. 6.

3 PCA-Based Parametric Representation of an Object in Color Space

We introduce here an approximation of expert data in RGB space by a set of elliptic cylinders strung on the principal component. First, initial color space is subjected to orthogonal transform by means of Principal Component Analysis (PCA) to reduce the correlation of color components. Next, the transformed space is divided into the set of two-dimensional layers that are orthogonal to the direction of the main component. Then, the ellipse of scattering is constructed for each layer. Thus, the set of parameters of such ellipses, arranged along the main components, describes the training set.

3.1 Orthogonal Transformation of Color Space

Original RGB color space, which is often used for representation of expert data, is not optimal in terms of computer processing [8], since the color components for real-world objects are strongly correlated, and the space itself is not uniform. Switching to another color space, for example HSV or L*a*b*, cannot refine the situation essentially as it does not allow to take into account the spatial location of the data. We use here a transformation of the training set color space so that the sample variances of the data along each axis have been sequentially maximized while preserving the orthogonality conditions [9], [10]. This transformation can be found by means of PCA [11]. In accordance with that method, three maximum eigenvectors U_1, U_2, U_1 of the empirical sample covariance matrix of the training set form the basis of a new coordinate system. In this paper we use the Jacobi eigenvalue algorithm for calculation of eigenvectors of a covariance matrix. A visual interpretation of such coordinate system is shown in Fig. 7 (a).

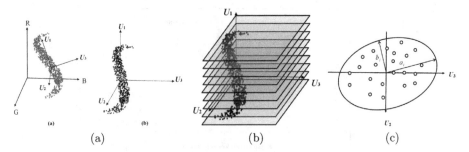

Fig. 7. Algorithm steps of the PCA-based method: (a) transformation of the training set in space of principal components, (b) partitioning the space of principal components by a set of planes, (c) scattering ellipse construction by PCA

The transition to the new coordinate system can be done by superposition of shift and rotation of the coordinate axes.

3.2 Approximation of the Training Set by the Ellipses of Scattering

Cutting the training set in the new coordinate system orthogonally to the first principal component by planes, we get a set of layers (Fig. 7 (b)). Each layer is defined by neighboring planes and contains n_i points, where i is a number of a layer.

Using PCA once again, but now in 2D space, we approximate the set of points at each layer by the scattering ellipse. Two maximum eigenvectors of the covariance matrix of points belonging to the layer i give us the direction of principal components. Now we can define the scattering ellipse in the 2D coordinate system of principal components (Fig. 7 (c)).

Choose, for example, the triple value of the standard deviation $= 3$ as the length of the axis a_i and b_i of the scattering ellipse at layer i such that

$$a_i = \mu\sigma_{u_{i1}} = \mu\sqrt{\frac{\sum_{j=1}^{n_i}\left((u_1)_{ij} - (\bar{u}_1)_i\right)^2}{n_i}},$$

$$b_i = \mu\sigma_{u_{i2}} = \mu\sqrt{\frac{\sum_{j=1}^{n_i}\left((u_2)_{ij} - (\bar{u}_2)_i\right)^2}{n_i}}.$$

It is obvious that standard deviations of components are the square root of the diagonal elements of the covariance matrix. Therefore, the decision rule for recognition of objects, belonging to the area defined by an expert, can be obtained on the basis of the following ellipse equation:

$$\left(\frac{(u_1)_i}{a_i}\right)^2 + \left(\frac{(u_2)_i}{b_i}\right)^2 \leq 1.$$

It is clear that the more objects can be found at the particular layer the more accurate is the scattering ellipse. If the layer has only one point, representing an object, the scattering ellipse cannot be built, since the standard deviations $\sigma_{u_{i1}}$ and $\sigma_{u_{i2}}$ (and the length of the ellipse axes as well) will be zero. But this is fairly common situation in the case of a sparse data. However, some regularization can be applied if we suppose that the area of interest, given by an expert in the color space, is smooth enough, so the scattering ellipses in neighboring layers will have similar parameters. This implies that we have two contradictory requirements: on the one hand, the scattering ellipse must fit the data in corresponding layer as good as possible; on the other hand, the parameters of ellipses in the neighboring layers must have similar values. We apply a simple dynamic programming algorithm [12] to solve a corresponding optimization task and adjust parameters of the neighboring ellipses.

4   Methodology of Experimental Study

A special model of the general population of objects was used for experimental studies. This model is formed by the points located inside the spheres with sinusoidally varying radius and the center located along the spiral curve in the 3D space (Fig.8 (a)). Data generator parameters were adjusted for the general population to include about 100 thousand of objects (all RGB cube contains $256^3 \approx 16.8$ millions objects).

In addition, the part of general population, selected randomly, was performed as a training set in our experiments (Fig.8) and the rest of the cube RGB — as objects to control the recognition quality.

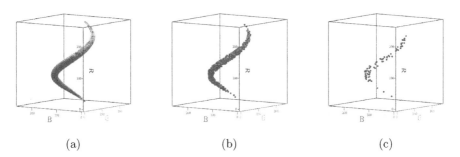

(a)   (b)   (c)

Fig. 8. Imaging of data set that forms experimental training set in RGB space: (a) entire general population (100000 objects), (b) 1 % of the general population (1000 objects), (c) 0.1 % of the general population (100 objects)

In order to avoid inaccuracies when measuring the performance techniques, only C++ implementations of all three algorithms assembled in one test project were used. GMM model realization source code was taken from the site [13].

Experiments with the SVDD method were made using the library, specifically optimized to work with the RGB object space. Method of data approximation with elliptical cylinders (PCA) was independently implemented specifically for this study.

The algorithms were compared against three criteria: classification quality, model training and new object classification time.

To assess the quality of the algorithms with different parameters error estimations of the first and second kind presented in the form of ROC curves were applied [14]. Moreover, both all objects from the general population and remaining set of objects in space of RGB were used as control objects.

The essence of the parameterization classifier speed estimation was in repeated training on the same dataset in order to reduce measurement error. An average parameterization time was then evaluated. In course of training the algorithm parameters were varied as the operating time of algorithms may depend on them.

New RGB object classification time was evaluated as an average time of all RGB cube objects classification: "class" and "non-class" for target classifier.

5   Experimental Study

Fig.9, 10, 11 present results of the methods described above for approximation of source data set shown in Fig.8 a.

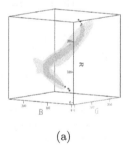   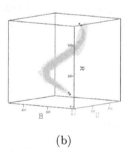

(a)   (b)

Fig. 9. Results of the separating boundary construction that uses the GMM model for different distributions quantity: (a) 3, (b) 5

The numerical estimations of the algorithms rate are shown below in the form of charts demonstrating the dependency between the size of training set and the rate of corresponding run-time operations. The method parameters to be estimated were selected in the way that minimizes the classification error (Fig.14).

The estimation of algorithms quality for training set containing 1 % of general population (Fig.8, a) is presented in the form of ROC curve (Fig.14). ROC curve for training set containing 0.1 % (Fig.8, b) of general population is shown in Fig.15.

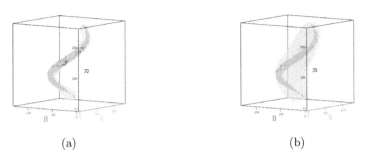

(a)   (b)

Fig. 10. Results of the separating boundary construction that uses the SVDD classifier for different Gaussian kernel parameters s: (a) 50, (b) 100

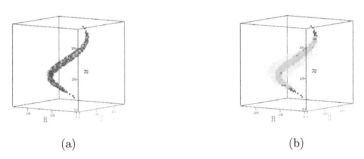

(a)   (b)

Fig. 11. Results of the separating boundary construction that uses the PCA-based classifier for different values of dispersion threshold: (a) 0.5, (b) 1.5

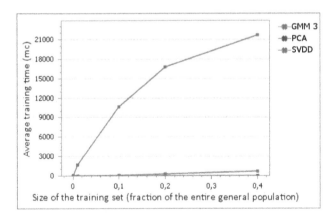

Fig. 12. The diagram showing the dependency between model training time (mc) and the size of the training set (part of the general population)

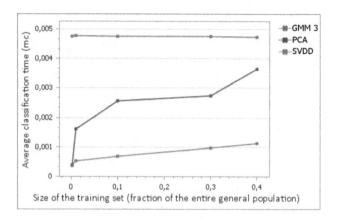

Fig. 13. The diagram showing the dependency between classification time of the new object (mc) and the size of the training set (part of the general population)

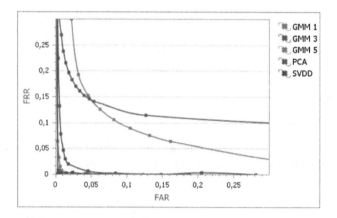

Fig. 14. ROC-curve characterizing the quality of algorithms classification for the training set containing 1 % of the general population

It can be noticed, that the classification error of GMM method with small amount of clusters slightly exceeds the similar error of SVDD method. Classification quality for GMM model can be improved by increasing the amount of the normal distributions that were used. However, when the training set size is small like in the second case (about 100 objects) the model will be the subject to overtraining. Similar effect can be observed for the PCA method, the classification error is dramatically tenfold increased with the decreasing training sample. Therefore, we can conclude that the SVDD method is the most robust in the case of insufficient training material.

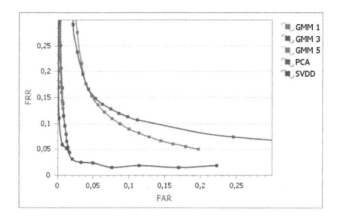

Fig. 15. ROC-curve characterizing the quality of algorithms classification for the training set containing 0.1 % of the general population

6   Conclusion

Our modeling experiments show, that for considered tasks the both SVDD and PCA methods achieve better performance than classical GMMs. PCA method is the fastest of the all methods considered here so it could be used then the training and recognition time is important . But PCA as well as GMM method is very sensitive to the size and especially sparseness of the training set in contrast to the SVDD method, which uses Gaussian kernel. Consequently, this method is the most balanced in terms of speed and quality for solving the task of pixel classification. In addition, the presence of only one tuning parameter in SVDD method can be noted as a significant advantage. This, undoubtedly, greatly simplifies the preset of the best quality SVDD classifier. Further research will be aimed to application of the described classifiers to solve practical problems of image segmentation.

References

1. Rother, C., Kolmogorov, V., Blake, A.: Grabcut: Interactive foreground extraction using iterated graph cuts. ACM Transactions on Graphics (TOG), 309–314 (2004)
2. Ruzon, M.A., Tomasi, C.: Alpha estimation in natural images. In: Proc. IEEE Conf. Comput. Vis. Pattern Recognition, CVPR 2000 (Cat. No.PR00662), vol. 1, pp. 18–25. IEEE Comput. Soc., Los Alamitos (2000)
3. Töreyin, B.: Fire detection algorithms using multimodal signal and image analysis. PhD thesis, Institute of Engineering and Science of Bılkent university (2009)
4. Tax, D.M.J.: One-class classification; Concept-learning in the absence of counter-examples, Ph.D thesis. Delft University of Technology, ASCI Dissertation Series, 146 p. (2001)
5. Bilmes, J.: A Gentle Tutorial of the EM Algorithm and its Application to Parameter Estimation for Gaussian Mixture and Hidden Markov Models. Technical

Report TR-97-021. International Computer Science Institute and Computer Science Division. University of California at Berkeley (April 1998)

6. Vapnik, V.N.: Statistical Learning Theory. Wiley-Interscience, NY (1998)
7. Aizerman, M., Braverman, E., Rozonoer, L.: Method of potential functions in machine learning theory. Nauka, Moscow (1970) (in Russian)
8. Paschos, G.: Perceptually uniform color spaces for color texture analysis: an empirical evaluation. IEEE Transactions on Image Processing 10(6), 932–937 (2001)
9. Tsaig, Y.: Automatic segmentation of moving objects in video sequences: a region labeling approach. Circuits and Systems for Video 12(7), 597–612 (2002)
10. Abdel-Mottaleb, M., Jain, A.: Face detection in color images. IEEE Transactions on Pattern Analysis and Machine Intelligence 24(5), 696–706 (2002)
11. Jolliffe, I.T.: Principal component analysis. Applied Optics 44(30), 64–86 (2005)
12. Mottl, V.V., Blinov, A.B., Kopylov, A.V., Kostin, A.A., Muchnik, I.B.: Variational methods in signal and image analysis. In: Proceedings of the 14th International Conference on Pattern Recognition, Brisbane, Australia, August 16-20, vol. I, pp. 525–527 (1998)
13. Gaussian Mixture Model and Regression (2008),
    http://sourceforge.net/projects/gmm-gmr/
14. Fawcett, T.: ROC Graphs: Notes and Practical Considerations for Researchers. In: Proc. of the 19th International Joint Conference on Artificial Intelligence, pp. 702–707 (2005)

An Automatic Matching Procedure of Ultrasonic Railway Defectograms

Anton Malenichev[1], Valentina Sulimova[1], Olga Krasotkina[1],
Vadim Mottl[2], and Anatoly Markov[3]

[1] Tula State University, Tula, Russia
malenichev@mail.ru, {vsulimova,o.v.krasotkina}@yandex.ru
[2] Moscow Institute of Physics and Technology, Moscow, Russia
vmottl@yandex.ru
[3] Radioavionica Corp, Saint Petersburg, Russia
amarkovspb@gmail.com

Abstract. The paper deals with the problem of the automatic analysis of ultrasonic defectograms, which are presented in the B-scan (brightness diagram) with multiple channels of ultrasonic control. One of the actual problems of ultrasonic railway inspection is presence of uncontrolled regions, taking place due to a bad acoustic contact. This paper proposes simple enough approach for automatic involving the information from the previous ultrasonic rail inspection into the analysis of the results of the current inspection. The proposed approach is based on partial segmentation of defectograms, by selecting most typical rail objects, which are bolt-on rail joints. For comparing fragments of defectograms the special dissimilarity measure is proposed, which is based on the standard Dynamic Time Warping procedure, but takes into account characteristic features of the considered applied task. The experimental results shows good performance of the proposed approach.

Keywords: ultrasonic rail inspection, pairwise signal alignment, partial segmentation.

1 Introduction

The inspection of railway infrastructure components is a very important task in railway maintenance. In particular, the damages on rails are exceptionally dangerous for the operation of rail traffic [1]. Non-destructive rail inspection is the main and often the only possible way to prevent emergency situations [2, 3].

During the last years mobile ultrasonic defectoscopes are actively introduced to inspect rail tracks for damages. These devices register reflected ultrasonic signals at defectograms [1,3].

The defectograms are represented, as a rule, in a form of signals in the coordinate plane "the measured time of ultrasonic signal propagation through the rail – the coordinate of the path along the controlled rail track". Such representation is traditionally called B-scan [3]. The presence of damages or some constructional reflectors (such as

P. Perner (Ed.): MLDM 2014, LNAI 8556, pp. 315–327, 2014.

bolt-on rail joint) leads to appearance of some lines on the defectogram. These lines can has the different form, length and orientation subject to the type of an object on the way of the ultrasonic signal.

It should be noticed, the amount of the registered data is very large. The track length of the Russian Railways (RZD), for example, is about 86 000 km. Every year more than 4,5 million kms of rail track are inspected. The periodicity of the inspection is 2-6 times per month. The result of inspection of each km is a 500 000-length discrete signal. Though there are special softwares for collecting, viewing and interpreting such data [4, 5] and about 50 000 potentially dangerous damages are yearly revealed, about 100-150 rail breaks are occurred [4], each of which can entail serious consequences.

One of the possible reasons of omitting damages is presence of uninspected rail track's sections. They can take the place due to bad contact between defectoscope and an inspected rail in consequence of weather conditions (especially in "cold countries"), high speed of moving ultrasonic device or(and) rail conditions [4, 7]. At that only about 15% of uninspected sections are also uninspected once at the previous inspection acts [6].

So, it is natural to wish involving the information from the previous inspection to compensate missing data at the current inspection. On practice this problem is solved , as a rule, by the reinspection [8], by analyzing the previous data "by hands" [6] or is not considered et all [2, 9]. It is evident, the automatic involving the previous information will allow to essentially reduce expenses of the railway companies due to decreasing a quantity of reinspection acts and amount of hand-made processing.

But it should be noticed, the registered by defectoscope railway coordinates are inaccurate and so there is a problem of finding correspondences between two defectograms. At that the direct matching of the whole signals is impossible because of its too big length.

Instead of that we propose to find a specific elements around the interesting area, belonging to both of defectograms (actually to make its partial segmentation) and then to match the only selected parts of signals.

It is known a number of different methods for segmentation of speech [10], biomedical [11] and other signals [12, 13, 14]. However they have, as a rule, a high computational complexity, solving the more general and difficult problem in contrast to the indicated one.

In this paper we propose to make segmentation on the basis of founding bolt-on rail joints as a specific and repeated elements of the railway.

The proposed strategy requires comparing signals of different length representing fragments of defectograms in process of finding a fragment, which is corresponded to a joint.

There are a number of ways for comparing discrete signals [15]. But the most popular and traditionally used way for comparing signals of different length is based on local warping axis of the compared signals and called the Dynamic Time Warping (DTW) [16]. There are the different modifications of this approach, which allows to restrict the set of considered alignments [17], to increase significance of some important points or to make it more fast [19].

At this paper we use the DTW-based approach, adopted for comparing ultrasonic defectograms by incorporating specially proposed similarity measure of elements of defectograms. It allows to take into account the specificity of the considered applied problem.

2   Dissimilarity Measure of a Defectogram's Elements

It is evident, the comparing of defectograms should be inevitably based on comparing of its elements. At this paper we propose a specific dissimilarity measure for elements of defectograms, which allows to take into account the specificity of the considered applied problem.

In a process of ultrasonic inspection several ultrasonic channels are usually used. They are ultrasonic emitters/ receivers, sending probing impulses of some fixed amplitude and under some angle to a rail roll surface (different channels differs one from another by angle of sending/receiving signal). The signal is registered only in case if the receiver obtain signal with big enough amplitude during some time interval since the sending moment.

Under condition of a good enough acoustic contact and the presence of some reflective surface under $90°$ angle to ultrasonic signal propagation, as a rule, single reflection is occurred. However, when some damage or constructive reflector is located near the rail surface, the ultrasonic signal has no time to fade out and multiple re-reflections can be observed. As a result a number of impulses can be registered with different delays respecting to the initial moment of the emission. In the case of absence of the reflection because of the bad acoustic contact or reflection by some damage turned on the angle differed from $90°$, the impulse will not be registered at all [7].

Each element of the defectogram by the separate channel is represented as impulse signal at the space "delay"-"amplitude" [6]. In accordance with above description it can contain some number $n = 0, 1, 2, ...$ of impulses, characterized by its delay τ_i and amplitude a_i, $i = 1, ..., n$. For convenience of further reasoning the element without impulses we will consider as signal of the length $n = 1$ and with zero delay and amplitude: $\tau_1 = 0$, $a_1 = 0$. Subject to this, each element of the defectogram is represented by two-component n-length signal of pairs $(\tau_i, a_i) \in R^2$, $i = 1, ..., n$.

As it is impossible to direct compare such signals we propose the special mathematical model for its representation in a form which is convenient for comparison. In accordance with this model each element of the defectogram by some channel is described by the sum of normal distributions with the dispersion σ^2 and the mean, equaled to the delay of the respective impulse τ_i:

$$f(\tau \mid x) = \sum_{i=1}^{n} a_i \frac{1}{\sigma\sqrt{2\pi}} \exp\left(-\frac{1}{2\sigma^2}(\tau - \tau_i)^2\right). \tag{1}$$

It should be noticed, that $f(x) = 0$ if zero impulses were registered at the respective point, what is not contradict the common sense.

The graphical interpretation of the proposed model is presented at the Fig. 1.

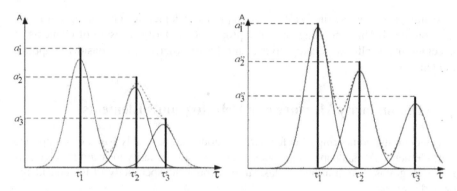

Fig. 1. The representation of the proposed model of description of the defectogram's element by one channel in the form of mix of normal distributions

Let $x' = (\tau'_i, a'_i) \in R^2$, $i = 1,...,n'$ and $x'' = (\tau''_j, a''_j) \in R^2$, $j = 1,...,n''$ - two elements of the defectogram. We propose to measure its dissimilarity as the square of the mismatch area under two compared signals or as the integral of the difference of the respective functions (1):

$$r(x',x'') = \sqrt{\int\limits_{-\infty}^{\infty} [f(\tau \mid x') - f(\tau \mid x'')]^2 \, d\tau} . \tag{2}$$

The convenience of the proposed elements representation (1) consists in that, the integral in (2) is computed analytically. As a result the dissimilarity measure (2) can be written in the next equivalent form:

$$r(x',x'') = \sqrt{\rho(x',x') + \rho(x'',x'') - 2\rho(x',x'')},$$
$$\rho(x',x'') = \sum_{i=1}^{n'}\sum_{j=1}^{n''} a'_i a''_j \exp\left[-\left(\frac{\tau'_i - \tau''_j}{2\sigma}\right)^2\right]. \tag{3}$$

At that the dissimilarity of some element $x' = (\tau'_i, a'_i) \in R^2$, $i = 1,...,n'$ with impulses-free zero element $\phi = (0,0)$, can be calculated by one more simple formula:

$$r(x',\phi) = \sqrt{\sum_{i=1}^{n'}\sum_{k=1}^{n'} a'_i a'_k \exp\left[-\left(\frac{\tau'_i - \tau'_k}{2\sigma}\right)^2\right]} . \tag{4}$$

The Fig. 2 shows the graphical interpretation of the proposed dissimilarity measure.

It is evident, the dissimilarity measure of equal by the form signals will be equal to zero. The most values of the dissimilarity measure will be observed when zero signal will be compared to the signal with big quantity of high-amplitude impulses.

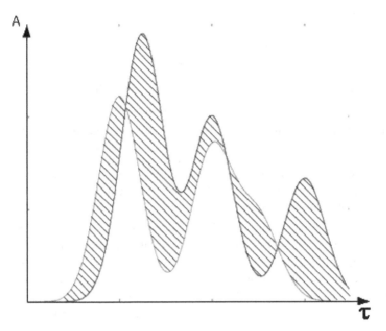

Fig. 2. The graphical representation of the proposed dissimilarity measure of the defectogram's elements

3   Combining Information from Several Channels

The practice of investigations shows, the information, obtained by the only one ultrasonic channel is not enough for making segmentation of a defectogram and identifying possible damages with a required accuracy. In this connection there is a natural desire for joint using information from several channels. Combining information from different sources can be implemented at the classifiers level [21], when separate classifications are combined, and at the detectors level [22], when combining is implemented before making a classifier. Therefore at this work the information from different channels is combined at the stage of comparing elements of the defectograms.

At that each element of a defectogram is considered as m-length vector $\mathbf{x} = [x_1, ..., x_m]^T$, each component of which is obtained from the respective j-th channel and is represented by $f(\tau \mid x_j)$ in accordance with (1).

The resulting dissimilarity measure of two elements $\mathbf{x}' = [x_1', ..., x_m']^T$ and $\mathbf{x}'' = [x_1'', ..., x_m'']^T$ is computed as a linear combination of partial dissimilarity measures, computed in accordance with (2) with nonnegative weights $\alpha_i > 0$:

$$\tilde{r}(\mathbf{x}', \mathbf{x}'') = \sum_{j=1}^{m} \alpha_j r(x_j', x_j''), \quad \sum_{j=1}^{m} \alpha_j = 1. \tag{5}$$

4   Comparing Defectograms

Each defectogram is presented at the form of m -component discrete signal $\mathbf{X} = (\mathbf{x}_1, ..., \mathbf{x}_{N_\mathbf{X}})$ of some length $N_\mathbf{X}$. In accordance with the previous sections each t -th element $\mathbf{x}_t = (x_{t,1}, ..., x_{t,m})$, $t = 1, ..., N_\mathbf{X}$ for each j -th channel $j = 1, .., m$ is represented by continuous signal $f(\tau | x_{t,j})$ of the form (1).

It should be noticed, defectograms obtained even for one and the same part of the railway will in general case will have different length because of inconstant defectoscope's velocity. Moreover, some part of the signal can be lost, particularly, because of bad contact. So the problem of comparing two defectograms is formulated as the problem of finding the optimal pairwise correspondences between elements of these signals.

Let we have to discrete signals $\mathbf{X} = (\mathbf{x}_1, ..., \mathbf{x}_{N_\mathbf{X}})$ and $\mathbf{Y} = (\mathbf{y}_1, ..., \mathbf{y}_{N_\mathbf{Y}})$. One of them (let it be $\mathbf{X}$) to take as the basis signal and the other $\mathbf{Y}$ - as the reference one.

Let also the metric $\tilde{r}(\mathbf{x}, \mathbf{y})$ be defined at the set of its elements in accordance with (5). It is required to match each element of the basis signal to some of the elements of the reference signal and to memorize the result in the form of the reference table $T = (\theta_t, t = 1, ..., N_\mathbf{X})$, where $\theta_t \in \{1, ... N_\mathbf{Y}\}$ - is the absolute reference (the elements' number of the reference signal corresponding to the respective element t at the basic one).

Let search the optimal pairwise correspondences, giving the minimum for the optimality criterion: $\hat{T} = \arg\min_T J(\mathbf{X}, \mathbf{Y}, T)$:

$$J(\mathbf{X}, \mathbf{Y}, T) = r(\mathbf{x}_1, \mathbf{y}_1) + \sum_{t=2}^{N_\mathbf{X}} \gamma(\theta_{t-1}, \theta_t),$$

$$\gamma(\theta_{t-1}, \theta_t) = \begin{cases} \beta + r(\mathbf{x}_t, \mathbf{y}_{\theta_t}), & \theta_t = \theta_{t-1}, \\ \beta|\theta_t - \theta_{t-1} - 1| + \sum_{j=\theta_{t-1}+1}^{\theta_t} r(\mathbf{x}_t, \mathbf{y}_j), & \theta_t > \theta_{t-1}, \\ \infty, & \theta_t < \theta_{t-1}, \end{cases} \tag{6}$$

where $\beta > 0$ - is the penalty for nonparallel references, corresponded to the local warping (compressing and stretching) of signal axes.

Let us use the notation $J(T)$ at the next text as short denotation of the criterion $J(\mathbf{X}, \mathbf{Y}, T)$ (4) for alignment of two signals $\mathbf{X} = (\mathbf{x}_1, ..., \mathbf{x}_{N_\mathbf{X}})$ and $\mathbf{Y} = (\mathbf{y}_1, ..., \mathbf{y}_{N_\mathbf{Y}})$.

The criterion (6) provides taking into account all the elements of the basis and the reference signals. It supplements additional references, connecting the t -th element of the basis signal with the elements $\theta_{t-1} + 1, ..., \theta_t$ of the reference signal in the case of $\theta_t > \theta_{t-1}$. Besides this criterion is not allowed for presence of so-called "cross-references" by infinite penalizing. So, if the element $(t-1)$ of the basic signal refers to the element θ_{t-1} of the reference signal, the next element t of the basic signal can refer only to the element with biggest number $\theta_t \geq \theta_{t-1}$.

Schematically the idea of the pairwise alignment is presented at the Fig. 3. Dotted lines show references, added for taking into account all elements of the reference signal.

The function of type (6) is called the separable function with the chain-shaped variable adjacency graph 20, because of it can be represented as the sum of more simple functions, each of which depends only on two neighbor variables θ_{t-1} and θ_t. Its extremum can be found by dynamic programming procedure, which is implemented in two passes.

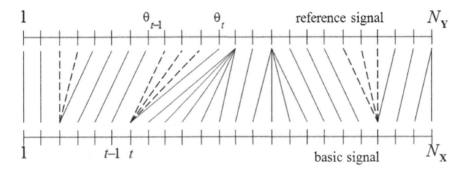

Fig. 3. An example of alignment of two signals

At the forward pass the moving along elements of the basis signal from the first to end is implemented. At that the incomplete criterion values $\tilde{J}_t(\theta_t)$ are computed by the next recurrent scheme:

$$t = 1: \quad \tilde{J}_1(\theta_1) = J(0,\theta_1), \quad \theta_1 \in \{1,...,N_Y\}$$
$$t = 2,...,N:$$
$$\tilde{J}_t(\theta_t) = \min_{\theta_{t-1}\in\{1,...,\theta_t\}} \left[\gamma(\theta_{t-1},\theta_t) + \tilde{J}_{t-1}(\theta_{t-1}) \right], \quad \theta_t \in \{1,...,N_Y\}.$$

Besides, the return recurrent relations are stored:

$$t = 2,...,N:$$
$$\tilde{\theta}_{t-1}(\theta_t) = \arg\min_{\theta_{t-1}\in\{1,...,\theta_t\}} \left[\gamma(\theta_{t-1},\theta_t) + \tilde{J}_{t-1}(\theta_{t-1}) \right], \quad \theta_t \in \{1,...,N_Y\}.$$

At the return way the back moving along elements of the basic signal is implemented. At that for each element the optimal value of the corresponded reference is defined as $\hat{\theta}_{N_X} = N_Y$, $\hat{\theta}_{t-1} = \hat{\theta}_{t-1}(\hat{\theta}_t)$, for $t = N_X,...,2$. The result of performing this procedure is a table of references $\hat{T}$ and a reached minimum of the criterion $J(T) = \tilde{J}_{N_X}(\hat{\theta}_{N_X})$, which can be considered as a dissimilarity measure of two defectograms. The reached minimum $J(T)$ and references disposition not depends on that, which of signals was taken as the basic signal and which – as the reference one.

But for playing the role of a dissimilarity measure the reached minimum should be divided in the total number of disposed references.

The Fig. 4 shows the example of the optimal pairwise alignment of defectogram's fragments, containing bolt-on rail joint. The alignment was made only by "the red channel". The signal at this channel has the form for right-tilted lines.

Fig. 4. An example of the optimal alignment of two signals

It is clear, the presented example confirms the correctness of working the procedure of the optimal warping of the defectograms.

It should be noticed, the proposed in the section III way of combining information from several channels has a general character and allows to take into account an information from the only one or from several channels at once without changing the way of comparing defectograms, but the only by setting appropriate values of the linear combination in (5), what is very convenient in practice.

5   The Main Principle of Finding the "Region of Interest" at the Previous Defectogram

The proposed principle is based on a simple scheme of partial segmentation of defectograms with finding regions, corresponded to bolt-on joints.

The proposed procedure of finding joints consists in a sequential passing along analyzed signal of the length L by a sliding window $\mathbf{w}_i = (\mathbf{x}_i,...,\mathbf{x}_{i+k-1})$, $i = 1,...,L-k+1$ of the fixed width k and comparing fragments of the defectogram $(\mathbf{x}_i,...,\mathbf{x}_{i+k-1})$, cut out by this window, with some signal $\mathbf{Y} = (\mathbf{y}_1,...,\mathbf{y}_k)$ of the length k , respected to the bolt-on rail joint.

Dissimilarity of two fragments of defectograms is computed in accordance with the section IV of this paper.

The result of passing the sliding window along the analyzed defectogram is a signal $\mathbf{d} = (d_1,...,d_{L-k+1})$, each element of which corresponds to the dissimilarity $d_i = d(\mathbf{w}_i,Y)$ of fragment, cut out by the window $\mathbf{w}_i$ and the etalon joint $\mathbf{Y}$. The Fig. 5 shows an example of such a signal. This signal has three main local clearly expressed minimums, corresponded to joints and a number of additional "noising" local minimums.

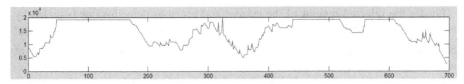

Fig. 5. An example of a signal with 3 joints, obtained by the sliding window

Let's consider the i-th fragment as like for a joint if the value at the respected point is the minimal value in some Δ- neighbourhood:

$$d_i = \min_{j \in [i-\Delta,...,i+\Delta]} d_j\left(\mathbf{w}_j, Y\right).$$

Acceptance (or rejection) of the classification decision about belonging the respective fragment to a class of joints is making on the basis of comparing the computed dissimilarity with some threshold value.

This principle is used twice while finding the "region of interest" (uncontrolled region from the current inspection) at the defectogram for the previous inspection of the same railway (Fig. 6).

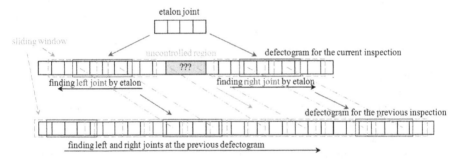

Fig. 6. The idea of finding the "region of interest" at the defectogram for the previous inspection of the same railway

At first the left and the right joints surround the region of interest are found by moving sliding window and comparing the respective fragments with some typical (etalon) joint.

At second, each of the left and the right joints, which have been found at the current defectogram, are found at the defectogram from the previous inspection on the basis of the same principle of sliding window, but by comparing with concrete joints instead of the etalon.

To exclude the situation when some joint is founded incorrectly (i.e. to exclude false acceptance) we use here a big enough threshold for acceptance the hypothesis that the considered fragment corresponds to the required joint. If a joint can't be found with the required confidence level, the first stage is repeated and another left or right joint is found.

Finally the selected regions of the current and previous defectograms bounded by the respective left and right joints are aligned in accordance with the procedure from the section IV. The obtained table of references $\hat{T}$ allows to easily find the region of interest at the defectogram from the previous inspection.

For increasing computational efficiency of the proposed approach, the alignment is not performed for so-called low-significance fragments of defectograms. We define a fragment as low-significance one if its most part (more than 50%) contains the only "bottom" signal, reflected from bottom of the rail and so don't contains constructional reflectors and joints.

Such heuristic in accordance with applying of fast realization of the proposed procedure, based on 23, allows essentially reduce the amount of computations without decreasing a quality.

6   Experimental Results

It is evident, a quality of a decision of the considered problem of compensation the missing data by the information from the previous inspection, essentially depends on an accuracy of finding joints. So at the experimental study we first of all check performance of finding joints.

For this purpose a set of defectograms was used, which contains 530 joints at total. For each joint a center point was marked by experts.

In accordance with the described principle of the sliding window, described in the section V, 10600 different fragments were obtained. The length of each fragment is 1001 elements (2 meters).

At the first stage of experiment, each of these fragments was classified into two classes (a joint / not a joint) by comparing with etalon joint. All 100% of these fragments at using information from all channels were classified correctly.

At the second stage an accuracy of finding joints was computed. For this purpose, for each of 530 fragments containing a joint and obtained at the previous stage, we perform alignment of it for the etalon fragment. The center of each joint was computed as the element of the respective signal, matched to the central element of the etalon fragment as the result of the alignment.

The computations were performed 4 times: for each of 3 different ultrasonic channels separately and for all 3 channels at once with combining information from the different channels in accordance with (5) with equal coefficients values.

For each of these 4-th cases founded center's deviations from the true expert values were computed. The main statistics for these computations are presented at the Table 1.

Table 1. Errors of segmentaition

channel	min	max	mean	std	median
1	0.000	29.600	5.443	8.632	1.200
2	0.000	27.600	4.921	8.397	1.600
3	0.000	35.800	6.814	10.241	1.400
1+2+3	0.200	5.400	1.343	1.346	0.800

The table 2 shows percents of joints (from 530), whose computed centers are differed from the marked by experts on the value of the indicated range.

Table 2. Percents of joints with the stated accuracy of segmentaition

channel	Error of segmentation			
	0 sm	<1 sm	<3 sm	<6 sm
1	3.57%	42.86%	71.43%	82.14%
2	7.14%	42.86%	71.43%	85.71%
3	3.57%	35.71%	67.86%	78.57%
1+2+3	0.00%	57.14%	92.86%	100.00%

Tables 1 and 2 shows, the using information, obtained from all channels allows essentially increase the segmentation quality. As the result the proposed approach allows to perform segmentation of joints with the error which is not exceed 5.4 sm. At that at 92.86% of cases the error is less than 3 sm.

7   Conclusions

The article deals with the problem of the automatic analysis of ultrasonic defectograms, which are presented in the B-scan (brightness diagram) view with multiple channels of ultrasonic control. One of the actual problems of ultrasonic railway inspection is presence of uncontrolled regions, taking place due to a bad acoustic contact. However the respective information is as a rule contained at the defectogram, obtained at the previous inspection.

This paper proposes simple enough approach for automatic involving the information from the previous ultrasonic rail inspection into the analysis of the results of the current inspection.

The proposed approach is based on partial segmentation of defectograms, by finding most typical rail objects, which are bolt-on rail joints. For comparing fragments of defectograms the special dissimilarity measure is proposed, which is based on the standard Dynamic Time Warping procedure, but takes into account characteristic features of the considered applied task by introducing special dissimilarity measure for elements of defectograms.

The experimental results shows the proposed approach allows to detect rail joint placement with good median accuracy of 0.8 sm and at 92.86% of cases the error is less than 3 sm. Such characteristics allows to successfully automatically find the region of interest at the previous defectogram and to involve the respective information into the analysis of the current inspection. So, it allows to essentially reduce expenses of the railway companies due to decreasing a quantity of reinspection acts and amount of hand-made processing.

Acknowledgment. We are very appreciate Russian Foundation For Basic Research for financial support (projects № 12-07-13141 and №12-07-13142).

References

1. Jimenez–Redondo, N., Bosso, N., Zeni, L., Minardo, A., Schubert, F., Heinicke, A.: Simrothhubert Automated and Cost Effective Maintenance for Railway (ACEM-Rail). In: Procedia: Social & Behavioral Sciences, vol. 48, pp. 1058–1067 (2012)

2. Jemec, V., Grum, J.: Automated Non-Destructive Testing and Measurement Systems for Rails, ECNDT 2010 (2010)

3. Markov, A.A., Shpagin, D.A.: Ultrasonic rail defectoscopy. In: Education – Culture, 2nd edn., pp. 284–441. St-Petersburg (2013) (in Russian)

4. Shilov, M.N.: Methodological and algorithmical founation and software for registration and analysis of defectograms at ultrasonic rail inspection. PhD thesis, St-Petersburg, 153 p. (2007)

5. Heckel, T., Thomas, H., Kreutzbruck, M., Rühe, S.: High Speed Non-Destructive Rail Testing with Advanced Ultrasound And Eddy-Current Testing Techniques. In: Indian National Seminar & Exhibition on Non-Destructive Evaluation, NDE 2009, December 10-12 (2009)

6. Markov, A.A., Kozyakov, A.B., Kuznetsova, E.A.: Decryption of defectogram of ultrasonic rail control. Practical Aid, p. 206. St-Petersburg (2006) (in Russian)

7. Fedirenko, D.B.: Problems of automatization of multichannel ultrasonic rail signals decryption. In: Radioelectronnye Kompleksy Mnogocelevogo Naznacheniya: Procedings, pp. 117–120. St-Petersburg (2011) (in Russian)

8. Foreing rail transport. Series 4. Railway and railway equipment. Engineering and building. Express-information. Issue 1 and 2, Moscow (2002) (In Russian)

9. Papaelias, M.P., Roberts, C., Davis, C.: A review on nondestructive evaluation of rails: state-of the art and future development. Proc. IMechE. Part F J. Rail and Rapid Transit. 222, 367–384 (2008)

10. Wang, D., Lu, L., Zhang, H.J.: "Speech segmentation without speech recognition'. In: International Conference on Multimedia and Expo (ICME 2003), vol. 1, pp. 405–408 (2003)

11. Kehagias, A., Nidelkou, E., Petridis, V.: A dynamic programming segmentation procedure for hydrological and environmental time series. Stochastic Environmental Research and Risk Assessment 20, 77–94 (2006)

12. Fearnhead, P.: Exact Bayesian curve fitting and signal segmentation. IEEE Transactions on Signal Processing 53(6), 2160–2166 (2005)

13. Lavielle, M.: Optimal segmentation of randomprocesses. IEEE Transactions on Signal Processing 46(5), 1365–1373 (1998)

14. Sasan, M., Sharif, B.S.: A nonlinear variational method for signal segmentation and reconstruction using level set algorithm. Signal Processing 86(11), 3496–3504 (2006)

15. Wang, X., et al.: Experimental comparison of representation methods and distance measures for time series data. Data Mining and Knowledge Discovery, 1–35 (2010)

16. Martens, R., Claesen, L.: On-line signature verification by dynamic time-warping. In: ICPR, pp. 38–42. IEEE (1996)

17. Myers, C.S., Rabiner, L.R.: A comparative study of several dynamic time-warping algorithms for connected word recognition. The Bell System Technical Journal 60(7), 1389–1409 (1981)

18. Feng, H., Wah, C.C.: Online signature verification using a new extreme points warping technique. Pattern Recognition Letters 24, 2943–2951 (2003)

19. Al-Naymat, G., Chawla, S., Taheri, J.: SparseDTW: A Novel Approach to Speed up Dynamic Time Warping. In: The 2009 Australasian Data Mining, vol. 101, pp. 117–127. ACM Digital Library, Melbourne (2009)

20. Bellman, R., Kalaba, R.: Dynamic programming and the modern control theory, p. 118. Nauka, Moscow (1969)
21. Ruta, D., Gabrys, B.: An overview of classifier fusion methods. Computing and Information Systems 7, 1–10 (2000)
22. Tatarchuk, A., Sulimova, V., Windridge, D., Mottl, V., Lange, M.: Supervised Selective Combining Pattern Recognition Modalities and Its Application to Signature Verification by Fusing On-Line and Off-Line Kernels. In: Benediktsson, J.A., Kittler, J., Roli, F. (eds.) MCS 2009. LNCS, vol. 5519, pp. 324–334. Springer, Heidelberg (2009)
23. Salvador., S., Chan, P.: FastDTW: Toward Accurate Dynamic Time Warping in Linear Time and Space. In: KDD Workshop on Mining Temporal and Sequential Data, pp. 70–80 (2004)

Open Issues on Codebook Generation in Image Classification Tasks

Luca Piras and Giorgio Giacinto

Department of Electrical and Electronic Engineering University of Cagliari
Piazza D'armi, 09123 Cagliari, Italy
{luca.piras,giacinto}@diee.unica.it

Abstract. In the last years the use of the so-called bag-of-features approach, often referred to also as the codebook approach, has extensively gained large popularity among researchers in the image classification field, as it exhibited high levels of performance. A large variety of image classification, scene recognition, and more in general computer vision problems have been addressed according to this paradigm in the recent literature. Despite the fact that some papers questioned the real effectiveness of the paradigm, most of the works in the literature follows the same approach for codebook creation, making it a standard *"de facto"*, without any critical investigation on the suitability of the employed procedure to the problem at hand. The most widespread structure for codebook creation is made up of four steps: dense sampling image patch detection; use of SIFT as patch descriptors; use of the k-means algorithms for clustering patch descriptors in order to select a small number of representative descriptors; use of the SVM classifier, where images are described by a codebook whose vocabulary is made up of the selected representative descriptors. In this paper, we will focus on a critical review of the third step of this process, to see if the clustering step is really useful to produce effective codebooks for image classification tasks. Reported results clearly show that a codebook created according to a purely random extraction of the patch descriptors from the set of descriptors extracted from the images in a dataset, is able to improve classification performances with respect to the performances attained with codebooks created by the clustering process.

Keywords: Bag of Word, Visual codebook, Descriptor sampling.

1 Introduction

The difficulty in capturing the complex semantics of images prompted the researchers in the fields of computer vision, image retrieval, and image classification to propose new and more effective low-level representation of images (e.g., color features, texture, shapes, local descriptors, etc.) over the years. It is easy to understand that no single representation by itself is capable of capturing all the semantics in the best possible way. In some cases, for example, one can be interested in finding several images of the same category (e.g., car), while, in other cases, one may be interested in searching an image archive to find all images of the same object [35].

P. Perner (Ed.): MLDM 2014, LNAI 8556, pp. 328–342, 2014.
© Springer International Publishing Switzerland 2014

In the field of visual concept detection, many global features have been proposed to describe image contents such as color [7,14], and texture [8] histograms. Indeed, global features are not useful to distinguish different parts of the images, e.g., the foreground from the background, or near-duplicate images. A trade-off between global and local features is the use of the so-called bag-of-features approach, that has extensively exhibited high levels of performance [13,43].

This approach, originally developed in the field of text classification [17], is based on the idea that images that contain the same or similar objects share specific areas of the image. These areas, identified as patches or interest points, are described through the use of low-level descriptors such as scale invariant features (e.g., SIFT [27], or SURF [2]), or normalized pixel values. These patches form the so-called vocabulary of *"visual words"*, i.e., the basic building blocks that allow describing the images in the dataset. The number of descriptors extracted from each image can vary from a few tens to several hundreds, and this number clearly depends on the image itself, and on the way in which the patches are sampled within the figure, i.e., densely, by using multi-scale resolution, or by focusing on particular points of interest. As the total number of visual words extracted form images of an archive can be quite large, to create a descriptor of uniform length it is necessary to select a subset of the descriptors that will be used to form the 'vocabulary' or 'codebook' [18] used to represent all images.

Codebooks are usually constructed by clustering local descriptors from a set of training images, and selecting the centroids of the clusters as representative of the clusters. An image is thus represented by a vector where the i-th component can represent either the number of local descriptors that falls in the i-th cluster ("hard" assignment) or how close are the local descriptors to the different cluster centroids ("soft" assignment). As the number of points can be quite large, clustering is usually performed by either sampling densely or sparsely the set of local descriptors extracted from the set of training images.

Images are then assigned to one of the data classes by a supervised classification step that is usually called *post-supervised* as the supervision is applied only after the unsupervised codebook creation. It is also possible that the information of the image class is used to construct the codebook, i.e. by selecting different vocabularies for different classes, and then creating the codebook by stacking all the visual vocabolaries. In this case the classification process is called *pre-supervised*. While the clustering step is the most widely used approach to create codebooks, its effectiveness in producing a discriminant representation is not clearly proven [47].

As it can be clearly seen by reading the papers published in recent years in this field, despite the approach for creating codebooks is composed of several parts (i.e., detection of interest points, extraction of effective descriptors, vocabulary coding, classification), and each part can be implemented in several different ways, the most widespread structure is the following [39]:

- dense sampling image patch detection
- use of SIFT as patch descriptors
- use of k-means as clustering technique
- use of SVM as classifier

In this article we will focus primarily on the third step of this process, by analysing the reasons why this approach is so widely used, and whether this choice is supported by either theoretical motivations or by experimental results. In fact, despite the popularity of the k-means algorithm for generating the codebook in the vast majority of research papers published in the last years in the fields of computer vision and image classification [39], few authors have focused on the reasons *why* one approach works better than another, and how the representativeness of a method does not necessarily correspond to an effective ability to discriminate. In this article, we are going to discuss if the use of k-means is able to represent and encode the visual words in an effective way, or if a codebook whose visual words are selected at random is still able to provide good classification performances.

This paper is organized as follows. Section 2 briefly reviews how the codebook creation process has been addressed in the literature. In particular, in Section 2.3 we deeply analyze the third step of this process, i.e., the "clustering phase", while in Section 2.4 we discuss more in details the usefulness of this approach. In Section 3 we argue that random selection of visual words can be as effective as the use of clustering. Experimental results are reported in Section 4, and show that random selection of visual words is actually veru effective, providing better performances than those attained by the usual technique based on clustering. Conclusions are drawn in Section 5.

2   Codebook Creation

2.1   Image Patch Detection

The easiest approach that allows extracting a large number of patches from an image is based on exhaustively sampling different sub-parts of the image at each location and scale [12]. Even if this method has proven to be very effective [18], it suffers from severe drawbacks that lies in its high computational cost. In order to overcome this problem, a fixed grid of image patches has been proposed in [23] in the context of scene classification, where sampling is performed at each n-th pixel, and at fixed multiple scales. Instead of focussing on a more or less dense exhaustive pixels sampling, other approaches focus on the search of a set of a few but informative interest points. In [46] the authors propose a distinction of corner detectors, blob detectors, and region detectors. Among all, the Harris Laplace corner detector [30], and the Hessian-Laplace blob detector [31] have proved successful in many applications. These approaches to detect interesting patches in an image, despite their success, are still under investigation in order to better understand how the classification results are related to the type of application in which they are used [22,18].

2.2   Patch Descriptors Extraction

Intensity-based descriptors have been widely used so far to extract distinctive invariant features from interest points. The Scale Invariant Feature Transform

(SIFT) descriptor proposed in [26] is one of the most used approaches. SIFT describes the patches of an image by edge orientation histograms. In order to try to cope with the main disadvantage of this method, i.e., the 'curse of dimensionality' because each point is described by a 128 dimensional vector, in [19] the dimensionality of the feature space was reduced from 128 to 36 by principal components analysis (PCA), even if in [32] it has been proven that this descriptor is less discriminative than SIFT. These descriptors are not invariant to changes in light color, because the intensity channel is a combination of the R, G and B channels. In order to add color invariance, and increase its discriminative power, several color descriptors have been proposed [41]. More recently, [2] proposed the "Speeded-up Robust Features (SURF)" that are based on Hessian matrix, by using a very basic approximation called 'Fast-Hessian' detector.

2.3   Codebook Generation by Clustering Patch Descriptors

Clustering is the third step in the codebook generation process and is usually performed by taking into account the interest point/patch descriptors of all the images in the training set. The clustering process aims at detecting the so-called *vocabulary* or *codebook*, i.e., the basic low-level building blocks that allows detecting the concepts of interests. The underlying assumption is that each concept can be detected by representing the images in terms of the number of interest points/patches falling in each cluster that represent the *visual words* of the codebook. It is easy to see that the clustering process depends on the way interest points/patches are detected, and on the employed descriptor.

Usually, from a dataset of a few thousands of images, it is possible to extract from a few millions to several dozens of million of interest points depending on the employed detector. To reduce the computational cost of the clustering process, the set of interest points is usually under-sampled up to a few hundreds of thousand of elements. However, this means that, on average, one is actually considering less than ten points for each image in datasets containing some dozens of thousands images.

In [18] the authors argue that the key-points extraction, in spite of a more compact coding and lower computational cost, has not been designed to select the most informative regions for classification, and dense sampling gives significantly better results. In addition, they claim that standard unsupervised clustering algorithm such as k-means works well for texture analysis on images containing only a few homogeneous regions. On the other hand, they are suboptimal for recognition tasks where dense patches from natural scenes are extracted, because they create suboptimal codebooks as most of the cluster centres fall near high density regions, thus under-representing equally discriminant low-to-medium density regions. In that paper the authors proposed a strategy that combines on-line clustering [29] and mean-shift [10]. The algorithm produces an ordered list of centres, and the under-sampled descriptor vectors [15] are assigned to the first centre in the list that lies within a fixed radius r of them, or otherwise it is left unlabelled if there is no such a centre.

In [33] the authors propose to use an ensemble of randomly created clustering trees (Extremely Randomized Clustering Forests) instead of the k-means algorithm to quantize the large numbers of high-dimensional image descriptors. The reported experimental results show that the approach proves to outperform k-means based coding in terms of the training time, the memory occupancy, and the classification accuracy.

In [47], the authors experimentally compare techniques for selecting histogram codebooks for the purpose of image classification. They study some unsupervised clustering algorithms (k-means [4], Linde-Buzo-Gray (LBG) algorithm [24], Self-Organising Map (SOM) [20], and Tree-Structured SOM (TS-SOM) [21]) in a task of histogram codebook generation when post-supervised classification is performed. For comparison, they include in the experimental results those obtained by using a random selection of descriptors as codebook. They also consider several methods for supervised codebook generation that exploit the knowledge of the image classes to be detected. According to the reported experimental results, the authors conclude that at least in the dataset used in the experiments, the k-means and the LBG algorithms produced the best performing codebooks, but they also noticed that, even if the other two algorithms generate good codebooks, their classification performance is worse than that attained by randomly selected codebooks. It is their opinion that this fact demonstrates the non-existence of a direct link between the quality of clustering and the quality of the resulting codebooks.

In the past years, some authors have explored other paradigms than clustering, such as [49] where the author proposed a histogram intersection kernel technique to form the vocabulary. More recently, a dictionary learning method based on manifolds identification has been presented in [25].

In [28] and [3], the authors go as far as to completely eliminate the step of codebook creation using directly the set of descriptors using a random forest of trees and the Naive Bayes Nearest Neighbour (NBNN) [5] approach, respectively. Also in [16], the authors follow the idea that codebooks generated by clustering the descriptors are not sufficiently flexible to model heterogeneous datasets, so they propose an image representation based on a multivariate Gaussian distribution estimated over the extracted local descriptors. Also, in [51], the authors do not code the images using a codebook, but they exploit a sparse coding with locality constraints approach in order to create a raw image representation [48], thus overcoming other similar approaches as NBNN [5].

2.4   "Is Clustering Useful for Codebook Definition?"

Some papers such as [34], and, afterwards, [39], and [9], review and compare several techniques to sample image patches from the images, and cluster or reduce them in some way in order to create an effective short codebook. Unfortunately, reported results do not allow drawing sound conclusions on the best technique to be employed to create codebooks, for a number of reasons. First, experimental results heavily depend on the values of the parameters used in the various steps of codebook generation that we mentioned in the Introduction. Secondly, the values of the parameters have to change depending on the dataset, and the application at hand. Indeed, by reviewing all the papers that have been published over the years on this topic, it is possible to see that they share the same basic framework as shown in Section 1 without any critical discussion on its validity. It seems like researchers have "surrendered", more or less consciously, to the idea that there is not much room for improvement in this field, and that the above algorithm should be taken for granted.

From our point of view, a number of questions still remain open: Why, despite of the number of papers providing different solutions, researchers typically rely on the same approach based on k-means clustering?

Is it possible that this behavior is motivated by the ease with which nowadays it is possible to find open-source libraries that implement the basic framework as a black box?

Is it possible that the alternative solutions proposed in the literature, while providing modest improvements in performance, exhibit a very complicated implementation that prevents their wide adoption by the scientific community?

It is also worth noting that there is also one paper that proposed a very much simpler approach whose results can be compared to those attained by much more sophisticated methods [9] but, to the best of our knowledge, this proposal has not gained much attention so far. It is our opinion that the use of a codebook allows attaining an effective representation of images, but the question is: how to make this representation discriminative enough to justify the effort to implement it?

3 Random Selection of Interest Points for Codebook Creation

The clustering process is the step that transforms the set of interest point descriptors of a set of training images into "bags of visual words" and codebooks. The process of creating a codebook in this way is quite straightforward:

- m interest points are extracted from all the training images (for details see Section 2.1);
- points are clustered by using an unsupervised clustering algorithms (see Section 2.3). Typically, k-means clustering is used, where $k << m$;
- each image is mapped into a codebook of size k by assigning the interest points of the image to the nearest centroid;
- the new feature vector representation is used for image classification.

To say the truth the use of the approach as is, it is not feasible; in fact to ensure that the clustering process is effective some *tricky* settings are needed. First of all it is necessary to estimate the "optimal" number of clusters (i.e., the codebook's dimension), and then the number of interest points to be clustered.

Let's examine the issue of the number of interest points to be clustered. From a dataset of a few thousands of images, it is possible to extract up to several dozens of million of interest points depending on the technique that is used.

Thus, the process of extracting the subset of points that will form the vocabulary is computationally quite expensive for a clustering task, so the set of interest points is usually under-sampled up to a few hundreds of thousand of elements (Step 2, Figure 1(a)). Clustering is then applied to this reduced set of points. This means that, on average, we are actually considering less than ten points for each image.

A common procedure to set the total number of interest points to be clustered is based on a random extraction of a subset of points of the totality of points belonging to all images in the training set. This number is usually fixed regardless of the number of images in the dataset in order to limit the computational complexity of the clustering algorithm. Consequently, the number of points per image automatically decreases as long as the number of images increases. For example, the *Color Descriptors* toolkit [41] (i.e., one of the most popular toolkits available to generate codebooks) employs the k-means clustering algorithm to produce a "bag of visual words" representation by randomly selecting 250,000 points from the set of points of the images of the input dataset, regardless of the dimension of the dataset. Even if this approach can be considered a reasonable choice with respect to the representativeness of the interest points, it is not possible to say the same with regard to its ability to discriminate.

Thus, it is hard to claim that the output of the clustering process actually summarize the most discriminative point of the image archive. In order to overcome this limitation, in [38] a larger number of interest points are taken into account by resorting to the ensemble paradigm. Different codebooks are created by extracting different subsets by random or pseudo-random under-sampling techniques, then classification results obtained by using each different codebook are combined. In fact, it has been shown that the codebook generated by considering different subsets of interest points provide diverse classification results.

Thus, the first issue we want to point out is that the random extraction of a subset of points for further clustering heavily affect the classification performances. Depending on the number of points extracted, and on the way points are randomly sampled from each image, the resulting codebook may exhibit a diverse degree of discriminative power from a classification point of view.

The choice of the dimension of the codebook is the second parameter that affects the discriminative power in a classification task. Usually, the dimension is chosen so that the computational cost of the clustering phase is constrained within specified limits. We recall that the clustering phase is the most time consuming operation in the pipeline of codebook generation. According to the specifications of the *Color Descriptors* toolkit [41] the *"clustering performed on 250,000 descriptors on 384 (ColorSIFT) dimensions take at least 12 hours per iteration of k-means"*. In addition, the number of clusters strictly depends on the dataset at hand (e.g., number of classes, variability in each class, etc.) so that it is not usually possible to set this parameter in advance. Thus, usually different codebook sizes are tested in order to find the most suitable solution. It is worth recalling that this result is also affected by the number of points we are considering in the clustering process, and thus different codebook sizes my provide similar performances as long as they are computed from a different subset of the set made up of all the interest points extracted from the training images. Finally, let us also recall that the output of the k-means clustering algorithm may depend on the choice of the initial cluster centroids (Steps 3 and 4, Figure 1(a)).

Summing up, the performances of image classification based on the bag-of-visual-words paradigms depends on two random processes

- the initial sampling of interest points that are used in the clustering process
- the initialization of the cluster centroids in the k-means algorithm

Motivated by the above considerations, in this paper we *suggest* to skip the clustering step at all, and generate the codebook by direct random sampling a fixed number of descriptors from the whole set of descriptors, and use them as visual words of the vocabulary to build the codebook (Step 2, Figure 1(b)). Our proposal is based on the idea that a random sub-sampling of the interest points of the images followed by a random initialization of the cluster centroids produces a randomness that can in some way be imitated by an purely random extraction of the codebook's "words" from the dozens of million of interest points of all images in the dataset.

Reported results show how a codebook created in this way is able to generate feature histograms able to attain some improvements in a classification task with respect to the performance obtained with a codebook created by the clustering process. In other words, we will show that by completely neglecting the clustering phase we can also attain better performances, and greatly reduce the time required to build the vocabulary.

1. m interest points are detected from all the training images;
2. a subset of n (where $n << m$) interest points is randomly extracted from the previous set;
3. k points (where $k << n$) are randomly selected as initial cluster centroids;
4. k-means algorithms is used to cluster the n points in k clusters;
5. each image is mapped into a codebook of size k by assigning the interest points of the image to the nearest k^{st} centroid;
6. the new feature vector representation is used for image classification.

1. m interest points are detected from all the training images;
2. a subset of k (where $k << m$) interest points is randomly extracted from the previous set;
3. each image is mapped into a codebook of size k by assigning the interest points of the image to the nearest k^{st} point;
4. the new feature vector representation is used for image classification.

(a) k-means approach   (b) Random approach

Fig. 1. k-means and random approach schemes

4 Experimental Results

Experiments have been carried out using two datasets, namely a subset of the MIRFLICKR[1] collection proposed for the ImageCLEF 2012 Photo Flickr Annotation Task [44] and the MICC-Flickr101 dataset [1]. The first dataset comprises 25 thousand multi-labelled images that have been manually annotated using 94 concepts. The MICC-Flickr101 dataset is based on the 101 object categories of the Caltech101 dataset, while the images were obtained by downloading them from Flickr in January 2012. This dataset is made up of 7348 single labelled images with at least about 40 images per class, the median of the number of elements per class being equal to 70. Images are at high resolution, 1024 x 768 pixels on average, and depict objects in daily-life real scenarios. The ImageCLEF 2012 dataset was originally subdivided into a training and a test set, while we randomly subdivided the MICC-Flickr101 dataset.

A dense sampling strategy for interest point detection has been used: features are extracted at 4 scales (0.5, 1, 1.5, and 2) with a regular grid spaced 10 pixels. For extracting the SIFT descriptors, the ISIS Color Descriptors[2] toolkit has been used [41].

In order to run different experiments we performed four random extractions of 512 points to create four different codebooks according to the usual approach based on k-means clustering. The choice of the dimension of the codebook has been aimed to limit the overall computational cost related to all the phases involved in obtaining the "traditional" bag-of-visual-words representation, and the later parameter estimation for each SVM that has to be trained for each representation. It is well known that all these phases are highly time consuming. We compared the proposed approach to the common procedure, i.e. randomly selecting 250,000 points from the set of points extracted from the entire training set. The clustering has been performed on 250,000 128D (SIFT) vectors using two Intel Xeon E5-2630 2.3Ghz with 64 GB of RAM and took on average 8 hours for each of the five iterations of k-means. Also, in this experiments the dimension of the codebooks, i.e. the number of clusters k in the k-means approach, has been fixed to 512. In several preliminary experiments, we also used different values for the k parameter, and the comparison between the proposed mechanism and the usual one exhibits the same behaviour as the one reported in the paper.

The Support Vector Machine with RBF kernel has been used as the base classifier for its good performance on various image classification tasks [11,6]. In particular, we implemented a multi-label classifier by independently training a SVM classifier for each class [42,45]. SVM parameters have been set by exploring the set of parameters in order to select the detector with the highest performance in the training set. For each classifier, a different decision threshold has been set according to the threshold optimization approach proposed in [37] aimed at maximizing the overall classification performance when, for each pattern, a score value is available for each class.

---

[1] http://press.liacs.nl/mirflickr/

[2] http://koen.me/research/colordescriptors/

Table 1. Results in terms of F1 micro and macro for the ImageCLEF dataset

	k-means	Random
F1u	49.51%	50.62%
F1M	23.07%	24.88%

Table 2. Results in terms of F1 micro and macro for the MICC-Flickr101 dataset

	k-means	Random
F1u	28.12%	30.86%
F1M	24.15%	26.67%

4.1 Results

Performance are evaluated in terms of Macro-averaged and micro-averaged F-measures [42,45]. In multi-label classification tasks, the F-measure [40] over all classes can be defined in terms of empirical averages in two different ways. Macro-averaging (denoted with the capital 'M') consists of averaging over all the classes the corresponding class-related measure. It equally weights each class, and thus tends to be dominated by the performance on rare classes, which is usually lower than that attained for common ones [50]. Micro-averaging (denoted with 'u') consists of computing the measure with respect to the sum of the true positive, false positive and false negative values over all classes.

The results of the experiments on the ImageCLEF and the MICC-Flickr101 datasets are reported in the following Tables. Each value in the table refers to the average performance of four concept detectors, each one trained on different codebooks created either by independently random selected points (Random) or performing k-means clustering on 250,000 descriptor randomly extracted from the set of all image descriptors (k-means).

Reported results clearly show that the different ways of selecting the visual words for the creation of the codebooks influences the performance, depending on the number of the images in the dataset. Table 1, that is related to the Image-CLEF dataset, shows a limited improvement of the proposed random approach if compared to the improvement that the random approach exhibited over the k-means approach on the MICC dataset (see Table 2). In the MICC dataset the average improvement reach 2.5%, while in the ImageCLEF dataset the improvement is around 1.5%. However, it is quite remarkable that by just creating a codebook at random we are able to perform better than with a codebook generated by a structured approach.

The experiments run over the two dataset also show that the same improvement trend is observed by computing the F1M and F1u measures. It is worth to note that the two measures work in a different way. While the F1u measure computes the sum of the true positive values, the false positive values, and the false negative values over all classes, and then evaluates a "global" average, the F1M measure averages the corresponding class-related measure over all the classes, with equal weights for each class. Consequently, the F1M tends to be dominated by the performance of the classes containing a tiny fraction of patterns.

The same trend of improvements show how the (RANDOM) approach works well also for the classes with a small number of patterns.

This behavior can be translated into the following observation. When the dataset contains a large number of images, and, consequently, a very large number of interest points are extracted, the (RANDOM) procedure create a codebook that is able to outperfrom the usual approach based on clustering that suffers from the clustering of a so large number of points. On the other hand, when the number of images in the dataset is not so large, the (RANDOM) procedure still allows extracting those points that permit to create codebooks that are able to produce larger increment of the improvement. The reader might think that as the number of images increases, then the variability of image content becomes higher, and it should be better to increase the dimension of the codebook. It is worth to note that such a direct relationship has not been assessed in the literature. However, while some works [33] claim that within certain limits this relation exists, other authors argue that this relation can hold just for some dataset and not for others [34,9].

In order to better understand the behavior of the random procedure with respect to the usual k-means approach, let us analyze the the correlation (i.e., the diversity) of the codebooks. To this end, Figure 2 shows the degree of correlation among the scores of the 8 concept detectors, four of them trained with the codebooks obtained by randomly extracting the "visual words" of the vocabulary from the set of the descriptors of the images ($\mathbf{R}$x), the other four trained with the codebooks obtained by following the common approach ($\mathbf{T}$x). It is easy to see that, regardless the technique used to produce different codebooks, the output scores tend to exhibit high correlation values (higher than 0.8 for ImageCLEF and higher than 0.7 for MICC-Flickr), and this support our idea that an offhanded extraction of the "words" of the codebooks exhibits the same randomness of the common procedure where a random sub-sampling of the interest points of the images is followed by a random initialization of the cluster centroids.

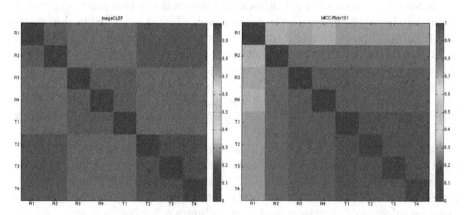

Fig. 2. Correlation matrix between the score obtained using the proposed approach ($\mathbf{R}$x) and the score obtained using the common clustering method ($\mathbf{T}$x) in the Image-CLEF and MICC-Flickr101 dataset

5   Conclusion

By extensively reviewing the papers that have been published over the past years in the fields of image classification, scene recognition, and more in general computer vision, it can be clearly seen that they share the same basic framework, i.e., the use of the bag-of-visual-words paradigm. In this framework, a very important step is the clustering phase that usually is performed using one of the most common unsupervised clustering approach, the k-means algorithm. Despite its wide use to create codebooks, its effectiveness in producing a discriminant representation is not clearly proven, whereas its drawbacks are well known. First of all, the clustering step is the most time consuming operation in the pipeline of codebook generation, and, secondly, it is hardly scalable with the number of points to be clustered. For the latter reason, often this step is not performed over all the points extracted by the training images, but a random sub-sampling step is needed to reduce the total number of interests points, thus reducing the number of points for each image. As a consequence, it is hard to claim that the output of the clustering process actually summarize the most discriminative point of the image archive, as clustering is usually performed over a random selection of a few points per image. In addition, it is also worth noting that the output of the k-means clustering algorithm depends on the choice of the initial cluster centroids. In this light, it is possible to say that the performances of image classification based on the bag-of-visual-words paradigms depends on two random processes.

For these reasons, in this paper we investigated if codebooks generated through a direct random sampling of a fixed number of descriptors from the whole set of descriptors is able to generate feature histograms able to attain similar performances with respect to the usual procedure that involves the use of the k-means clustering algorithm. Surprisingly, we found out that the generation of codebook by randomly selecting the visual words actually produced significant improvements in classification performances. Reported results showed that by completely neglecting the clustering phase it is possible to attain better performances, and greatly reduce the time required to build the vocabulary.

References

1. Ballan, L., Bertini, M., Del Bimbo, A., Serain, A.M., Serra, G., Zaccone, B.F.: Combining generative and discriminative models for classifying social images from 101 object categories. In: Proc. of International Conference on Pattern Recognition (ICPR), Tsukuba, Japan (November 2012) (Poster)
2. Bay, H., Ess, A., Tuytelaars, T., Gool, L.J.V.: Speeded-up robust features (surf). Computer Vision and Image Understanding 110(3), 346–359 (2008)
3. Becker, J.H., Tuytelaars, T., Gool, L.J.V.: Codebook-free exemplar models for object detection. In: WIAMIS, pp. 1–4. IEEE (2012)
4. Bishop, C.M.: Pattern Recognition and Machine Learning (Information Science and Statistics). Springer (October 2006), http://www.worldcat.org/isbn/0387310738
5. Boiman, O., Shechtman, E., Irani, M.: In defense of nearest-neighbor based image classification. In: CVPR. IEEE Computer Society (2008)

6. Chang, C.C., Lin, C.J.: Libsvm: A library for support vector machines. ACM TIST 2(3), 27 (2011)
7. Chang, S.F., Sikora, T., Puri, A.: Overview of the mpeg-7 standard. IEEE Trans. Circuits Syst. Video Techn., 688–695 (2001)
8. Chatzichristofis, S.A., Boutalis, Y.S.: Fcth: Fuzzy color and texture histogram - a low level feature for accurate image retrieval. In: Proceedings of the 2008 Ninth International Workshop on Image Analysis for Multimedia Interactive Services, pp. 191–196. IEEE Computer Society (2008)
9. Chavez, A., Gustafson, D.: Building an effective visual codebook: Is k-means clustering useful? In: Bebis, G., et al. (eds.) ISVC 2012, Part II. LNCS, vol. 7432, pp. 517–525. Springer, Heidelberg (2012)
10. Comaniciu, D., Meer, P.: Mean shift: A robust approach toward feature space analysis. IEEE Trans. Pattern Anal. Mach. Intell. 24(5), 603–619 (2002)
11. Cristianini, N., Shawe-Taylor, J.: An Introduction to Support Vector Machines and Other Kernel-based Learning Methods. Cambridge University Press (2000)
12. Crowley, J.L., Sanderson, A.C.: Multiple resolution representation and probabilistic matching of 2-d gray-scale shape. IEEE Trans. Pattern Anal. Mach. Intell. 9(1), 113–121 (1987)
13. Csurka, G., Dance, C.R., Fan, L., Willamowski, J., Bray, C.: Visual categorization with bags of keypoints. In: Workshop on Statistical Learning in Computer Vision, ECCV, pp. 1–22 (2004)
14. Deselaers, T., Keysers, D., Ney, H.: Features for image retrieval: an experimental comparison. Inf. Retr. 11(2), 77–107 (2008)
15. Estabrooks, A., Jo, T., Japkowicz, N.: A multiple resampling method for learning from imbalanced data sets. Computational Intelligence 20(1), 18–36 (2004)
16. Grana, C., Serra, G., Manfredi, M., Cucchiara, R.: Image classification with multivariate gaussian descriptors. In: Petrosino (ed.) [36], pp. 111–120
17. Joachims, T.: Text categorization with suport vector machines: Learning with many relevant features. In: Nédellec, C., Rouveirol, C. (eds.) ECML 1998. LNCS, vol. 1398, pp. 137–142. Springer, Heidelberg (1998)
18. Jurie, F., Triggs, B.: Creating efficient codebooks for visual recognition. In: ICCV, pp. 604–610. IEEE Computer Society (2005)
19. Ke, Y., Sukthankar, R.: Pca-sift: A more distinctive representation for local image descriptors. In: CVPR (2), pp. 506–513 (2004)
20. Kohonen, T.: The self-organizing map. Neurocomputing 21(1-3), 1–6 (1998)
21. Koikkalainen, P., Oja, E.: Self-organizing hierarchical feature maps. In: 1990 IJCNN International Joint Conference on Neural Networks, vol. 2, pp. 279–284 (1990)
22. Lazebnik, S., Schmid, C., Ponce, J.: Beyond bags of features: Spatial pyramid matching for recognizing natural scene categories. In: CVPR (2), pp. 2169–2178. IEEE Computer Society (2006)
23. Li, F.F., Perona, P.: A bayesian hierarchical model for learning natural scene categories. In: CVPR (2), pp. 524–531. IEEE Computer Society (2005)
24. Linde, Y., Buzo, A., Gray, R.: An algorithm for vector quantizer design. IEEE Transactions on Communications 28(1), 84–95 (1980)
25. Liu, B.D., Wang, Y.X., Zhang, Y.J., Shen, B.: Learning dictionary on manifolds for image classification. Pattern Recognition 46(7), 1879–1890 (2013)
26. Lowe, D.G.: Object recognition from local scale-invariant features. In: ICCV, pp. 1150–1157 (1999)
27. Lowe, D.G.: Distinctive image features from scale-invariant keypoints. International Journal of Computer Vision 60(2), 91–110 (2004)

28. Martínez-Muñoz, G., Delgado, N.L., Mortensen, E.N., Zhang, W., Yamamuro, A., Paasch, R., Payet, N., Lytle, D.A., Shapiro, L.G., Todorovic, S., Moldenke, A., Dietterich, T.G.: Dictionary-free categorization of very similar objects via stacked evidence trees. In: CVPR, pp. 549–556. IEEE (2009)
29. Meyerson, A.: Online facility location. In: FOCS, pp. 426–431. IEEE Computer Society (2001)
30. Mikolajczyk, K., Schmid, C.: Indexing based on scale invariant interest points. In: ICCV, pp. 525–531 (2001)
31. Mikolajczyk, K., Schmid, C.: Scale & affine invariant interest point detectors. International Journal of Computer Vision 60(1), 63–86 (2004)
32. Mikolajczyk, K., Schmid, C.: A performance evaluation of local descriptors. IEEE Trans. Pattern Anal. Mach. Intell. 27(10), 1615–1630 (2005)
33. Moosmann, F., Triggs, B., Jurie, F.: Fast discriminative visual codebooks using randomized clustering forests. In: Schölkopf, B., Platt, J.C., Hoffman, T. (eds.) NIPS, pp. 985–992. MIT Press (2006), http://eprints.pascal-network.org/archive/00002438/01/nips.pdf
34. Nowak, E., Jurie, F., Triggs, B.: Sampling strategies for bag-of-features image classification. In: Leonardis, A., Bischof, H., Pinz, A. (eds.) ECCV 2006, Part IV. LNCS, vol. 3954, pp. 490–503. Springer, Heidelberg (2006)
35. Penatti, O.A.B., Silva, F.B., Valle, E., Gouet-Brunet, V., Torres, R.d.S.: Visual word spatial arrangement for image retrieval and classification. Pattern Recognition 47(2), 705–720 (2014)
36. Petrosino, A. (ed.): ICIAP 2013, Part II. LNCS, vol. 8157, pp. 2013–2017. Springer, Heidelberg (2013)
37. Pillai, I., Fumera, G., Roli, F.: Threshold optimisation for multi-label classifiers. Pattern Recognition 46(7), 2055–2065 (2013), http://www.sciencedirect.com/science/article/pii/S0031320313000320
38. Piras, L., Tronci, R., Giacinto, G.: Diversity in ensembles of codebooks for visual concept detection. In: Petrosino (ed.) [36], pp. 399–408
39. Ramanan, A., Niranjan, M.: A review of codebook models in patch-based visual object recognition. Journal of Signal Processing Systems 68(3), 333–352 (2012)
40. van Rijsbergen, C.J.: Information Retrieval. Butterworth (1979)
41. van de Sande, K.E.A., Gevers, T., Snoek, C.G.M.: Evaluating color descriptors for object and scene recognition. IEEE Trans. Pattern Anal. Mach. Intell. 32(9), 1582–1596 (2010)
42. Sebastiani, F.: Machine learning in automated text categorization. ACM Comput. Surv. 34(1), 1–47 (2002)
43. Sivic, J., Zisserman, A.: A text retrieval approach to object matching in videos. In: ICCV, pp. 1470–1477. IEEE Computer Society (2003)
44. Thomee, B., Popescu, A.: Overview of the imageclef 2012 flickr photo annotation and retrieval task. Tech. rep., CLEF 2012 working notes, Rome, Italy (2012)
45. Tsoumakas, G., Katakis, I., Vlahavas, I.P.: Mining multi-label data. In: Maimon, O., Rokach, L. (eds.) Data Mining and Knowledge Discovery Handbook, pp. 667–685. Springer (2010)
46. Tuytelaars, T., Mikolajczyk, K.: Local invariant feature detectors: A survey. Foundations and Trends in Computer Graphics and Vision 3(3), 177–280 (2007)
47. Viitaniemi, V., Laaksonen, J.: Experiments on selection of codebooks for local image feature histograms. In: Sebillo, M., Vitiello, G., Schaefer, G. (eds.) VISUAL 2008. LNCS, vol. 5188, pp. 126–137. Springer, Heidelberg (2008)

48. Wang, J., Yang, J., Yu, K., Lv, F., Huang, T., Gong, Y.: Locality-constrained linear coding for image classification. In: IEEE Conference on Computer Vision and Pattern Recognition (CVPR), pp. 3360–3367 (2010)
49. Wu, J., Rehg, J.M.: Beyond the euclidean distance: Creating effective visual codebooks using the histogram intersection kernel. In: ICCV, pp. 630–637. IEEE (2009)
50. Yang, Y.: A study on thresholding strategies for text categorization. In: ACM (ed.) Proceedings of the International ACM SIGIR Conference on Research and Development in Information Retrieval, pp. 137–145 (2001)
51. Zhang, C., Wang, S., Liang, C., Liu, J., Huang, Q., Li, H., Tian, Q.: Beyond bag of words: image representation in sub-semantic space. In: Jaimes, A., Sebe, N., Boujemaa, N., Gatica-Perez, D., Shamma, D.A., Worring, M., Zimmermann, R. (eds.) ACM Multimedia, pp. 497–500. ACM (2013)

A Neuro-Genetic System for Cardiac Arrhythmia Classification

Elina Maliarsky[*], Mireille Avigal, and Maya Herman

The Open University of Israel, Raanana, Israel
meliana18@yahoo.com, {miray,maya}@openu.ac.il

Abstract. Electrocardiography (ECG) is a medical test used to measure the heart's electrical conduction system. Many cardiac abnormalities can be detected through ECG analysis. Various computerized techniques have been applied to assist physicians in an accurate diagnosis, among them Artificial neural networks (ANNs), Genetic algorithms (GAs) and their combinations (GANNs). A cardiac arrhythmia computerized diagnostics is a good example of such an application. Many different ANN approaches to arrhythmia classification appear in the literature, but we couldn't find any work related to GANN for this problem. In this paper we are closing this gap by presenting classification system of cardiac arrhythmia using GANN classifier. ANN is the base of the system, while GAs are used to evolve ANN architecture and weights. In addition, another GA is used to perform a feature selection task. The system is trained and tested for online UCI Machine Learning data set for cardiac arrhythmia. The classification performance of the system is evaluated by means of classification accuracy. The proposed classifier gives best classification results of 90.23%, 100%, 94.98%, 99.46%, 97.88% and 86.5% on classifying ischemic changes, old anterior myocardial infarction, old inferior myocardial infarction, sinus tachycardia, sinus bradycardia and right bundle branch block respectively. These results are competitive with the state of the art results in the field; that proves the effectiveness of our application. In addition the tool is enough generic to be used in solving of a wide range of problems. Also we have investigated effectiveness of GA as a training method. Exhaustive experiments demonstrate that classification accuracy of GA-trained classifiers is inversely proportional to the number of classification cases and depends on the content and size of the feature set for the classifier that is built by help of it.

Keywords: artificial neural networks, cardiac arrhythmia classification, genetic algorithms, Gradient Descent back-propagation, Levenberg-Marquardt back-propagation, multi-class classification, primary feature selection, simultaneous classification.

1   Introduction

An arrhythmia is a problem with either the rate or the rhythm of the heartbeat [17],[32]. There are a number of arrhythmia types that can be roughly divided into

---

[*] Corresponding author.

P. Perner (Ed.): MLDM 2014, LNAI 8556, pp. 343–360, 2014.
© Springer International Publishing Switzerland 2014

two categories: *bradyarrhythmias* (slow heart rate) and *tachyarrhythmias* (fast heart rate) [32]. Sustained over a long period of time some arrhythmias may cause damage to the heart or even a sudden death [7]. An in-time correct diagnosis of cardiac arrhythmia allows choosing appropriate anti-arrhythmic measures. Therefore, it is important to constantly improve the diagnosis procedures/abilities.

The computerized electrocardiography (ECG) is currently a main test for discovering the heart rhythm irregularities. During ECG electrical impulses generated by the heart are picked up at the surface of the body and then translated into a waveform that can be interpreted by physicians [32]. Various Machine Learning and data-mining techniques have been applied to assist physicians in an accurate arrhythmia diagnosis. Good results have been achieved when applying artificial neural networks (ANNs).

R.D. Raut and S.V. Dudul [26] presented a classification system for cardiac arrhythmias using multi-layered perceptron (MLP) architecture and standard gradient descent back-propagation (BP) learning algorithm. They examined it on the UCI Machine Learning Data Base for cardiac arrhythmias [9]. From the comparative analysis of obtained results, it turned out that the estimated MLP with BP algorithm operates as an excellent classifier for given task: while testing the average classification accuracy between 90 to 100% for about seven arrhythmias classes has been achieved. B.Anuradha and V.C Veera Reddy [1] proposed a method to accurately classify cardiac arrhythmias through a combination of wavelets and ANN. ANN has been used as a classifier. In this study accuracy of 90.56% has been achieved. S.M. Jadhav, S.L. Nalbalwar and A.A. Ghatol [16] proposed classification system for cardiac arrhythmia from standard 12 lead ECG recordings data, using a Generalized Feed-Forward neural network (GFNN) classifier. The GFNN classifier has been trained on UCI Machine Learning Data Base for cardiac arrhythmias [9] to classify arrhythmia cases into normal and abnormal classes. The abnormal class has been merged from the records of 15 classes, representing 15 types of arrhythmia. The experimental results presented in their paper show that up to 82.35% testing classification accuracy can be achieved. In the study [17] Jadhav et al. used ANNs with MLP architecture and BP learning algorithm to classify different arrhythmia types using the same data set. For this classification one arrhythmia class has been used against normal arrhythmia class. After careful and exhaustive experimentation, the researchers concluded that proposed classifier gave the best classification results in terms of classification accuracy of 100 % for old inferior myocardial infarction and, 94.25%, 98.72%, 97.4%, 91.5%, 92.1% for ischemic changes, old anterior myocardial infarction, sinus tachycardia, sinus bradycardia and right bundle branch block respectively.

ANNs are able to learn and generalize input patterns into output patterns usually through learning algorithm like BP [13]. BP is generally used for training Multi-layer Perceptron (MLP) [13]. MLP is the most popular type of ANNs; it is frequently used in biomedical data processing [3]. Unfortunately, MLP trained by BP has three significant drawbacks: (1) BP learning procedure is very time-consuming; (2) building an MLP requires some prior decisions regarding its topology. Those decisions are made based on application-specific heuristics, which may cause sub-optimal results; (3) BP algorithm may be caught by local minima, which may decrease network performance.

BP with Levenberg-Marquardt algorithm [13], which comes instead of gradient descent, significantly improves the running time, but the problems with local minima and ANN configuration still present.

Recently there have been rather successful attempts to improve the performance of ANNs by genetic algorithms [11],[20],[23],[27]. Genetic algorithms (GA) are stochastic search techniques inspired by natural evolution through survival of the fittest [22].They are widely used in various optimization problems, where deterministic methods are not suitable. A number of approaches called Genetic Algorithm Neural Networks (GANNs) have been applied to incorporate GAs into the ANNs building/training process. In the most popular GANNs the role of GA is limited by the search for optimal ANN architecture [2]. Any learning algorithm can then be used to evolve weights. GA itself can be such an algorithm. Using GA for evolving weights may help to overcome the local minima problem and decrease the run-time [23].

A number of GANN classifier for ECG data processing can be found in literature. Mansouria Sekkal, Mohamed Amine Chikh, and Nesma Settouti [28] investigated in the effectiveness of a GANN classifier and its application to the classification of premature ventricular contraction (PVC) beat. K.S.Kavitha and Manoj Kumar Singh [18] described a new system for detection of coronary artery disease based on feed forward NN architecture and GA learning algorithm. Stanislaw Osowski, Krzysztof Siwek and Robert Siroic in [25] have presented the application of GA for the integration of neural classifiers combined in the ensemble for the accurate recognition of heartbeat types on the basis of ECG registration. Zumray Dokur and Tamer Olmez in [8] comparatively investigated three neural networks for ECG beat classification: MLP, restricted Coulomb energy (RCE) and a novel hybrid NN trained by GA. In all the experiments above GANN classifiers over-performed their classical ANN competitive.

We couldn't find any work related to GANN for arrhythmia classification problem. In this paper we are closing this gap by presenting GANN classifier with MLP architecture, partially evolved by GA for cardiac arrhythmia. During the series of experiments we are investigating the effectiveness of GA as training algorithm while trying to separate each type of arrhythmia from normal state one by one and to classify simultaneously different arrhythmia types. The effectiveness of GA-trained classifiers is measured in classification accuracy and compared to accuracy of the same MLPs trained by gradient-descent BP (GDBP) and by Levenberg-Marquardt BP (LMBP) respectively. To evaluate the performance of MLPs, we are using the online UCI Machine Learning Data set for cardiac arrhythmia that contains 452 records (instances) with 279 attributes [9]. The proposed tool firstly preprocesses the data set and performs feature selection (FS) in order to make the data suitable for processing by ANN. FS has been done as follows: the data set has been firstly cleaned from the features for those more than half of the values have been equal to zero and then GA-based FS method has been applied to perform additional FS step.

This paper is organized as follows: Section 2 describes the classification system. Section 3 presents the experiments with their results. In section 4 the analysis of the results is done. Section 5 presents the conclusion and the future scope.

2   Methods

The classification system, presented in this paper, consists of four basic processes, shown at Fig. 1.

The processes are: data pre-processing, feature selection, developing of the classifier and classification.

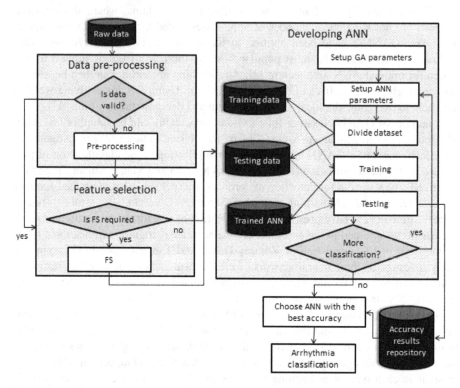

Fig. 1. Flowchart of our arrhythmia classification system. (The abbreviation FS means Feature selection).

2.1   Data Pre-processing

The data used in this work have been taken from the UCI cardiac arrhythmia ECG signal data set [9] that contains 452 records with 280 features in each record. This data set is well known and widely used by researchers [10],[1617]. The first 279 features stored in data set, are age, sex, weight and information, extracted from ECG. The 280-th feature is the decision of experienced cardiologist, belonging to one of 16 diagnostic classes, presented in the Table 1. The classes are coded by two digits, starting from 01, which refers to "normal" ECG. Classes 02 to 15 refer to different classes of arrhythmia and class 16 refers to the unclassified arrhythmias.

ANNs cannot receive just the data of any sort and produce a meaningful result [15]. Initial pre-processing usually should be done. It includes incomplete data handling, imbalanced data handling and normalization. These steps have been sequentially performed.

Table 1. Arrhythmia classes distribution in the online UCI Machine Learning Data set for cardiac arrhythmia [9]

Class code	Class name	Number of instances
01	Normal	245
02	Ischemic changes (Coronary Artery Disease)	44
03	Old Anterior Myocardial Infarction	15
04	Old Inferior Myocardial Infarction	15
05	Sinus tachycardia	13
06	Sinus bradycardia	25
07	Ventricular Premature Contraction (PVC)	3
08	Supraventricular Premature Contraction	2
09	Left bundle branch block	9
10	Right bundle branch block	50
11	1.degree Atrioventricular block	0
12	2. degree AV block	0
13	3. degree AV block	0
14	Left ventricular hypertrophy	4
15	Atrial Fibrillation or Flutter	5
16	Others	22

Incomplete Data Handling
The original data set contains empty values, represented by "?". ANNs are unable to work with incomplete data, so we need to handle missing values. The original data set contains empty values for the following features: vector angles on front plane of T, P, QRST and J[5] waves (features 11-14 respectively), and heart rate (feature 15). The values of feature 14 are missing significantly for almost all the classes. We decided to remove the feature from the data set. The missing values for the rest of the features have been imputed by averages of the remaining feature values per class [29].

Imbalanced Data Handling
According to the Table 1, the original data set contains 3 records for class 07, 2 records for class 08, 9 records for class 09, 4 records for class 14 and 5 records for class 15. Classes 11-13 have no records at all. ANN model requires sufficient number of records for training and testing. Therefore the classes with low amount of records, namely classes 07-09 and 11-15, are removed from the data set. Even after the elimination the data set remains imbalanced: there are hundreds of records for class 01, but only tens for each other class. To overcome this problem we've created synthetic records using SMOTE algorithm [6].

Data Normalization

When using ANN as classifier, the input values should often be normalized to some restricted numeric range for maximum efficiency. All the linear values have been scaled into [0..1] interval, the values, representing classes have been normalized by one-of-n method [14], i.e. replaced by n-length binary vectors, where n – is the total number of classes.

After the data preprocessing has been done we've obtained a normalized and complete data set with 8 balanced classes and 1691 instances.

2.2   Feature Selection

The data in [9] contains a huge amount of attributes (279), whose values ranges vary significantly. It has been observed, that inclusion of all the features in the input vector may decrease the performance of ANNs in different aspects [31], so the original feature set should be reduced via FS procedure. There is a lot of FS methods mentioned in literature [19]. In this paper we've used hybrid method of FS, combining empirical analysis and GA. The basis for our choice was the result obtained by this FS method in our research regarding feature selection.

At the first stage of FS we've analyzed the data and it has been found that there are a lot of features for those more than half of the values are '0'. The main task of the selection process is to choose these features which are correlated in the best way with the class, but uncorrelated with each other [12] and their values differ significantly for different classes [30], so that features can be safely removed, reducing the feature set to 122 features. At the second stage we've applied GA on the reduced feature set. The problem of FS via GA can be considered as a case of feature weighting, where the numerical weights for each of the features have been replaced by binary values. A value of 1 means the inclusion of the corresponding feature into the subset, while a value of 0 means its absence. GA is wrapper FS method [19], and like in any wrapper method, it uses classifier like a black box in order to evaluate the subsets of the features. GA is also used to evolve the number of hidden neurons of ANN.

GA Settings

Encoding. We've used GA not only for FS, but also for evolving the number of hidden neurons of MLP. Therefore the solutions have been represented as integer strings of the n+1 size, where n is the total number of features. The first n values of these strings were binary, with the value of 1 representing the presence of the feature and the value 0 – absence of it. The n+1's value was integer value in the range of [10..20], that represents the number of hidden neurons.

Crossover and Mutation. We have used scattered crossover [4], that creates random binary string and then the genes are selected from the first parent where the vector is 1, and from the second parent, where the vector is 0, and combines the genes to form the first child, and vice versa to form the second child. The crossover probability Pc was 0.8. Gaussian mutation [22] with the probability of the mutation Pm =0.001 has been used. The values above are empirically chosen.

Fitness. For each integer vector a new MLP has been created and trained. The MLP has the same number of input nodes as the number of 1-s in the integer vector. There is only one hidden layer with a number of hidden nodes as the value of the last element of the chromosome. LMBP training function has been used, because this training method is robust and converges quickly. Each MLP has been trained three times, as in related research [11]. Average accuracy a(x) has been calculated, where x is an integer string. In addition to a(x), the cost s(x) was calculated by dividing the number of 1's in the string by n.

Both a(x) and s(x) contributed to the multi-objective fitness function, represented by the following equation:

$$f(x) = -\left((2 - a(x)) - \left(\frac{s(x)}{2 - a(x)} \right) \right)$$

$$(1)$$

This equation is the slight modification of the function, proposed by Taskaran [31] and described in details in his study. We both try to improve the accuracy a(x) and to reduce the cost s(x) of the network, while in the original function author tried to reduce the error e(x) of ANN and the cost s(x).

Other Parameters. Pool size: 40, the number of generations: 100. Initially we trained the MLP for 200 generations, but we empirically established that 100 generations is enough to ensure the convergence of the process.

After hybrid FS we obtained feature set with 21 features instead of 279. GA has evolved the number of hidden neurons equal to 20. The classification accuracy of the resulting classifier is 86.5%, that is competitive with the state of the art results in the field [16,17].

ANN Settings

Three-layered MLP has been chosen as architecture; the number of input and hidden neurons is defined by GA and the number of output neurons is 8, as the number of arrhythmia classes. Levenberg-Marquardt Back-propagation (LMBP) has been chosen as a learning algorithm because it proved to be fast and robust [24]. The data set has been divided to the train set and the test set by using Stratified Random Sampling [21], where the random division is performed for each class separately, in order to sustain the balanced number of records in each class. We've selected 70% of the records for training and 30% for the testing. This ratio is the most popular in the literature because of the following reason: greater training set may cause *over-fitting* (the network will memorize the patterns instead of learning from them and will show poor generalization later) while smaller training set may cause *under-fitting* (there are no enough records for ANN to learn properly) [13].

2.3   Classifier Developing

We started from three-layered MLP architecture. As the result of FS, the number of input neurons is 21 and number of hidden neurons is 20. The number of output neurons is 8 as the number of arrhythmia classes. The number of neurons at each level will be changed during experiments. We've used the following notation for MLPs: *i-h-o*, where *i* is the number of input nodes, *h* is the number of hidden nodes and *o* is the number of output nodes.

In all our experiments MLPs have been trained once via GDBP, once via LMBP and once via GA. Like at the FS stage, the data set has been randomly divided into the train set (70% of the records) and the test set (30% of the records) via Stratified Random Sampling. According to the paper [11], we satisfied by averaging of the accuracy of each MLP after three sequential runs on test set. This average accuracy has been recorded for the later comparative analysis. The accuracies are presented as the numbers with floating point between 0 and 1.

At the end of the experiments the classifier with the highest accuracy has been chosen; it is done by empirical comparative analysis of the best results obtained in each experiment. The best classifier is chosen for the classification stage.

In all our experiments we use a single performance measure such as classification accuracy. We consider using additional performance measures in future scope.

GA Settings
Encoding. MLP is scanned layer by layer and connection weighs are inserted into array of numbers with floating point. The length of array is the same as the number of connections.

Crossover and Mutation. We used scattered crossover with the probability Pc=0.8, and Gaussian mutation with the probability Pm=0.001, as in the GA for feature selection.

Fitness. A multi-objective optimization function that tries to reduce an error as well as to improve the accuracy, has been presented. The function we are going to minimize is *f(x)=e + 10*(1-a)*, where *e* is the error of the network and *a* is the accuracy of the network on training set and *10* is the penalty (the value of 10 has been chosen after a number of trials).

Other Parameters. Reproduction: rank-based with the elitism (the two best binary-strings from the population are moved to the next generation), Pool size: 100, the number of generations: 200.

2.4   Classification

At this stage the classification has been done: given feature vector, the classification system maps it to one of arrhythmia classes. This stage can be considered as diagnostics.

3   Experiments and Results

The objective of this stage was developing of ANN-based tool for classification of arrhythmia with as the highest accuracy as possible and investigation of the effectiveness of GA as a learning algorithm. It has motivated us to perform a sequence of experiments that can be classified into the following types:

Simultaneous Multi-class Classification: As the name implies, the classification is done simultaneously to the defined number *num_of_classes* of classes and the MLPs built for this task have *num_of_classes* output neurons. The obvious disadvantage of such model is the complexity: the more classes we have, the more difficult is the classification and it may affect the accuracy.

Binary Multi-class Classification: In this model the classifiers have only two outputs, normal and abnormal. The abnormal class is combined from the records of one or many arrhythmia classes, depending of what we want to do. For example, if we want to separate sick persons from healthy ones, as in [16], the abnormal class contains the records of all the arrhythmia classes. If we want to perform multi-class classification, we have to build *num_of_classes* binary classifiers, where *num_of_classes* is the number of arrhythmia classes that we want to classify, and to run them separately. Binary multi-class classification is very popular in the literature because it is simple and maintains a good accuracy of classification [17].

Constrained Simultaneous Multi-class Classification: This model is the mix of the models above; we still perform simultaneous classification, but constrain the classes by grouping them together according to some common feature or features.

Binary multi-class classification with the most significant features: in this model we are performing binary multi-class classification, but for each classifier we perform an additional step of FS, with the purpose to select the most significant features for the classifier.

Classification with the Most Significant Feature Subset: In this model we are building the most significant features subset by combining the features obtained after additional FS steps in the binary multi-class classification with the most significant features experiment. Further classification is performed with this features subset. This classification may be binary and simultaneous.

We started from simultaneous multi-class classification, and then we argued that binary multi-class classification significantly improves the accuracy. Then we tried binary multi-class classification with the most significant features. Then we moved to constrained simultaneous multi-class classification, combining similar arrhythmia classes together. We have finished our experiment series by classification with the most significant feature subset; this classification has been done both in simultaneous and binary mode.

3.1   Simultaneous Multi-class Classification

Objective: To develop a classifier for simultaneous classification of different arrhythmias and to compare effectiveness of learning algorithms, such as LMBP, GDBP and GA.

Experiments and Results: We've built the MLP with the configuration 21-20-8 [2.3 explains the notation]. The MLP has been trained once via GDBP, once via LMBP and once via GA. After each training procedure MLP was tested on the records of the test set. The procedure of training and testing repeated three times [2.3] and average accuracy result has been collected. The following accuracies have been achieved: 0.65817 for MLP trained GDBP, 0.86405 for MLP trained by LMBP and 0.40915 for MLP trained by GA. The highest accuracy has been received as a result of LMBP training; the significantly poor accuracy of GA training results could be considered as a failure.

Class 16, representing the *unclassified arrhythmia* records (Table 1), refers to the patients for whom even expert cardiologists couldn't make a distinct diagnosis. This was the motive for our second experiment, related to simultaneous multi-class classification. We disregarded this problematic class, adapted the MLP architecture appropriately (21-20-8 → 21-20-7) and performed training and testing as in previous experiment. The following accuracies have been achieved: 0.75362 for GDBP, 0.90313for LMBP and 0.48207for GA. Relatively to the first experiment the accuracy has been improved for all considered cases, especially for MLPs trained by GDBP and GA.

Conclusion: From the experiments above we can conclude that while performing simultaneous multi-class classification, it is preferred to use LMBP learning algorithm. GA fails to train multi-class MLP; but we can see that the accuracy of GA-trained MLP with 7 outputs is significantly better than the accuracy of GA-trained MLP with 8 outputs. This motivated us to continue with experiments in order to find optimal conditions for GA to be used as an efficient learning algorithm.

3.2   Constrained Multi-class Classification

Objective: To check dependency of GA-trained MLPs' classification ability on the number of classification cases.

Experiments and Results: We've reviewed the data set from the medical point of view and grouped together related arrhythmia classes [7],[32]. Each group was treated as a separate classification case. We've built 9 random multi-classed MLPs with the configuration 21-20-o, where o is the number of classification groups. Each MLP has been trained by LMBP, GDBP and GA three times [2.3]. Average accuracies have been recorded. Figure Fig. 2 summaries the comparative accuracy results of all these MLPs, trained by GDBP, LMBP and GA. The notation of each MLP is as follows. We regard only the output layer, where groups are separated by "-". Grouping of classes is noted by chaining them together with the removing of leading '0'. For example, we want to note MLP, separating three groups, the first is composed of the records of classes 02,03 and 04, the second is composed from the records of classes 05 and 06, and the third presents "normal" classes, coded by 01. The notation will be 234-56-1.

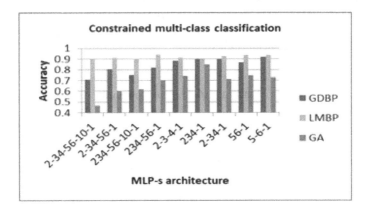

Fig. 2. The accuracy results for constrained multi-class classification. 9 MLPs are presented with their comparative accuracies. The notation for each MLP is as follows: MLP is presented by its output layer, each group is encoded by its classes' codes without leading zero, different groups are separated by "-".

Conclusion: From Fig. 2 we can conclude the following. The accuracies of MLPs trained by LMBP are almost invariant. This may be due to the method of FS: recall that we performed FS by GA, where the part of the fitness function has been the accuracy of MLP trained by LMBP. The accuracies of MLPs trained by GDBP and GA change mostly proportionally and depend on the following factors:

- The number of classification classes/groups
- The medical relationship between separated classes/groups (tachycardia and brady-cardia are related, infarct and ischemia are related, but infarct and tachycardia are not related [32]).

The dependency from these factors is as follows: classification accuracy is inversely proportional to the number of unrelated classification cases.

The results of this experiment lead us to the next assumption: if we try to classify each arrhythmia class versus "normal" class separately, the accuracy results for MLPs, trained by GA will be close to the accuracy results of MLPs, trained by canonical BP algorithms. This assumption motivated us to the additional experiment.

3.3   Binary Multi-class Classification

Objective: To prove the assumption that accuracy results of binary MLPs, trained by GA are close to accuracy results obtained by training the same MLPs with canonical BPs.

Experiments and Results: We've performed multi-class classification, but via binary classification: we've built six separate binary classifiers; each one has been used to separate one arrhythmia class from the normal class. For each arrhythmia class MLP with the configurations 21-20-2 [2.3] have been created and trained by GDBP, LMBP and GA and tested as in previous experiments. The accuracy results for all six MLPs are presented in the Fig. 3. Each MLP is named according to the arrhythmia that separates it from the healthy class.

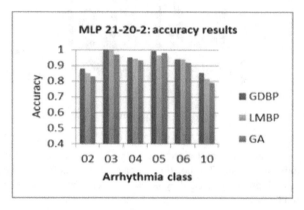

Fig. 3. The accuracy results for binary multi-class classification. Each one of 6 arrhythmia classes (coded by 02-06, 10) is separated from the normal class. For each class binary MLP is built and trained by GDBP, LMBP and GA.

Conclusion: The best results have been achieved by GDBP that significantly improves when training binary MLPs. Binary multi-class classifier, trained by GDBP, over-performs its competitive algorithm that performs simultaneous classification; binary MLPs trained by GA are still not as good as the MLPs with the same configurations, trained by BP methods, but very close to them. This motivated us to focus on binary multi-class classification.

3.4   Binary Multi-class Classification with the Most Significant Features

Objective: To check the dependency of classification accuracy from the size specificity of feature set. The assumption leading to this experiment was as follows: when performing multi-class classification by binary method, we actually need fewer features for each classifier. For example, if we want to identify bradycardia or tachycardia, usually heart rate is quite enough [7],[32].

Experiments and Results: For each binary MLP classifier we performed additional run of FS by GA. We used the same parameters, as while performing primary feature selection, except pool size: it has been empirically chosen to be 20. After finishing the work of all feature selectors, we have got 6 feature sets. As the result, we need only 2 features to train the classifier to separate classes 03 and 10, one feature is enough for classifying individuals with *tachycardia* and *bradycardia* (classes 05 and 06 respectively), separating of classes 04 and 10 requires 4 and 3 features respectively.

With these feature sets we've built 6 MLPs with significantly reduced amount of input (as number of features) and hidden nodes. The number of hidden nodes has been chosen as 6: in the first FS runs the number of hidden nodes always converged to 6, so it has been chosen for all other runs. These MLPs have been trained and tested as in previous experiments. On Fig. 4 (a) we can see their accuracy results. On Fig. 4 (b) the results of GA-trained MLP are compared to the best results obtained by binary multi-class classification (GDBP-trained 21-20-2 MLP).

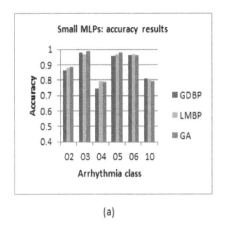
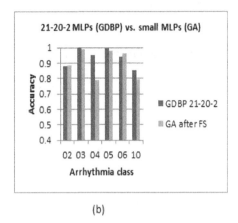

(a)   (b)

Fig. 4. The comparison of accuracy results for small MLPs, trained by GDBP, LMBP and GA (a) and the comparison of best results in the previous and current experiments (b)

Small GA-trained MLPs are at least as good as for MLPs trained by other methods and even over-perform other methods at the half of the cases. For a half of the cases small GA-trained MLPs shows the better result that GDBP-trained 21-20-2 MLPs.

Conclusion: The obvious conclusions from this experiment are the following: GA learning algorithm competes BP methods when performing binary multi-class classification on small and specific feature sets. But there is a slight regression when separating the classes 04 and 10. This regression can be explained by sticking at the local minima while performing GA during additional step of FS.

3.5   Classification with the most Significant Feature Subset

Objective: To perform additional reduction of the initial feature set with 21 features without loss in accuracy. The assumption leading to his experiment was as follows: after the combination of all the features, obtained after secondary FS, we'll be able to classify all arrhythmia classes with as good accuracy as if we'll take all 21 features.

Experiments and Results: We've taken six feature sets, obtained in the previous experiment, combined them together and as the result we've got the set with 9 features. For this experiment we performed simultaneous multi-class classification with 8 and 7 output classes and binary multi-class classification as in previous experiments. Table 2 presents a comparison of the results, obtained in this experiment, to the results, obtained by multi-class classification with 21-inputed MLPs. Fig. 5(a) demonstrates the comparison of the accuracy results for 9-inputed MLPs trained by GDBP, LMBP and GA. Fig. 5(b) shows the comparative accuracy of 21-20-2 MLPs trained by GDBP, small MLPs trained by GA and 9-inputed MLPs trained by LMBP. By this we are comparing the results of the most accurate MLPs, obtained in this and the previous experiments.

Table 2. The accuracy results for 8-classed and 7-classed MLPs with 21 and 9 inputs

Learning algorithm	21-20-8	9-9-8	21-20-7	9-8-7
LMBP	0.86405	**0.72418**	0.90313	**0.80168**
GDBP	0.65817	**0.55556**	0.75362	**0.66133**
GA	0.40915	0.42222	0.48207	0.50267

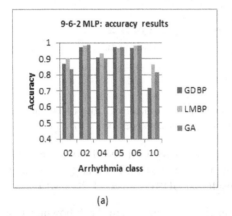

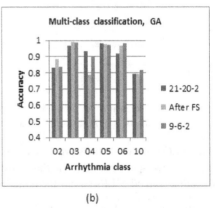

(a)   (b)

Fig. 5. The accuracy results for binary multi-class classification by 9-inputed MLPs (a) and by the best classifiers (b)

Conclusion: From Table 2 we can see that there is a regression in the results of 9-inputed MLPs trained by LMBP and GDBP, but the results of MLPs trained by GA are improved. It proves again that the best results for GA-based classification are obtained for small and specific MLPs.

From Fig. 5(a) we can see that in the half of the cases 9-inputed MLPs trained by GA are at least as accurate as MLPs trained by other methods. In general 9-inputed MLPs, trained by LMPB give the better accuracy, than 9-inputed MLPs trained by GA and GDBP. From Fig. 5(b) we can see that 9-inputed MLPs, trained by LMBP gives the best accuracy result at least at the half of the cases.

During the experiments above we have developed effective approach for cardiac arrhythmia classification problem.

4   Discussion

We've developed a classifier that simultaneously classifies seven arrhythmia classes. This classifier is trained and tested on the data set with 21 features, obtained after primary FS procedure. The accuracy of this classifier is 0.9013. Simultaneous multi-class classifier can be build according to the flowchart on Fig. 1. The obvious disadvantage of simultaneous multi-class classification is the complexity: the more classes we have, the more difficult is the classification and as we've seen it affects the accuracy for generic learning algorithms like GA.

We've also developed a classifier that can separate 6 different arrhythmia classes from "normal" class, while performed binary multi-class classification. This classifier has been trained and tested on the data set with 21 features, obtained after primary FS procedure, data sets, obtained after an additional FS step and data set with 9 features, obtained by combining all the features after additional FS step.

The best accuracy results are summarized in the Table 3. In this table we also compare our best results to the results, obtained by other researchers, performing the similar experiments [16, 17].

Table 3. Summary of the base accuracy result for each classification (each arrhythmia class vs. Healthy), compared to the results obtained by other reasearches in the similar experiments

Classification	MLP	Learning algorithm	Accura-cy	Accuracy in literature
02 vs. 01	9-6-2	LMBP	0.90238	0.9425
03 vs. 01	21-20-2	GDBP	1	1
04 vs. 01	21-20-2	GDBP	0.94987	0.9872
05 vs. 01	21-20-2	GDBP	0.99467	0.974
06 vs. 01	9-6-2	GA	0.97887	0.915
10 vs. 01	9-6-2	LBMP	0.86567	0.921

The results, obtained by other researchers, can be considered as validation results. We can see that our results are rather close to validation results. Also we can see that we've got the same or better results, while detecting arrhythmias of classes 03, 05 and 06. Our best results for separating of the classes 02, 04 and 10 are not as good as in competitive research, especially for class 10. It can be explained by sticking at the local minima while performing primary feature selection. The accuracy of simultaneous multi-class classifier is comparable to validations results too. From the above mentioned description we can conclude that our classification system is effective for solving arrhythmia classification problem. From Table 3 we can conclude that the final classification system should use the benefits of all the learning algorithms and feature sets.

5   Conclusions and Future Scope

In this paper we've proposed a novel neuro-genetic approach for cardiac arrhythmia classification. From the comparative analysis of the obtained results, it has been found that the classifier based on the principle of binary multi-class classification better suits to classify given cardiac arrhythmia ECG data. The proposed classification system gives the best classification results of 90.23%, 100%, 94.98%, 99.46%, 97.88% and 86.5% on classifying ischemic changes, old anterior myocardial infarction, old inferior myocardial infarction, sinus tachycardia, sinus bradycardia and right bundle branch block respectively. These results are comparable to the results, obtained by the similar researches and even partially over-perform them. That proves the effectiveness of our application.

We've also investigated the effectiveness of GA as learning algorithm and found out that the classification abilities of MLPs trained by GA are inversely proportional to the number of unrelated classes/groups and to the specificity of feature set.

We've seen that the accuracy result, obtained by our simultaneous multi-class classifier is about 90% that is competitive with the state of the art results in the field. We should try to improve these results by matching more suitable architecture. Regarding binary multi-class classifier, we should improve the classification ability of each binary MLP, especially of this one with the accuracy less than 90%. In the future we also should consider using additional performance measures.

Our application is not restricted to a specific task; it suits to any classification problem that can be solved by supervised learning. Further works will investigate the possibility to use the application for other problems in medical field and also for different problems in various other fields such as bioinformatics, stock trading, natural language processing, robotics etc..

References

1. Anuradha, B., Veera Reddy, V.C.: ANN for classification of cardiac arrhythmias. ARPN Journal of Engineering and Applied Sciences 3(3) (2008)
2. Arifovic, J., Gencay, R.: Using genetic algorithms to select architecture of a feed-forward artificial neural network. Physica A 289, 574–594 (2001)
3. Begg, R., Kammruzzaman, J., Sarker, R.: Neural networks in healthcare, potential and challenges. IDEA Group (2006)
4. Bocko, J., Nohajová, V., Harčarik, T.: Application of methods of selection and crossover to identification of parameters of Borden-Partom model. Modeling of mechanical and mechatronic system (2011)
5. Bronis, K., Kappos, K., Manolis, A.S.: Early repolarization not benign anymore J-wave syndrome. Hospital Chronicles 7(4), 215–228 (2012)
6. Chawla, N.V., Bowyer, K.W., Hall, L.O., Kegelmeyer, W.P.: SMOTE: Synthetic Minority Over-sampling Technique. Journal of Artificial Intelligence Research 16, 321–357 (2002)
7. Cox, A.: SADS – sudden arrhythmic death syndrome, Cardiac Risk in the Young. Produced by Cardiac Risk in the Young (2003)
8. Dokur, Z., Olmez, T.: ECG beat classification by a novel hybrid neural network. Computer Methods and Programs in Biomedicine 66, 167–181 (2001)

 9. Güvenir, A.H., Acar, B., Muderrisoglu, H.: Arrhythmia Data set. Center for Machine Learning and Intelligent Systems. University of California, Irvine
10. Güvenir, H.A., Acar, B.: Feature selection using a genetic algorithm for the detection of abnormal ECG recordings. In: Proceedings of the World Conference on Systemics, Cybernetics and Informatics (ISAS/SCI 2001), Orlando, Florida, pp. 437–442 (July 2001)
11. Fiszelew, A., Britos, P., Ochoa, A., Merlino, H., Fernández, E., García-Martínez, R.: Finding optimal network architecture using genetic algorithms. Research in Computing Science 27, 15–24 (2007)
12. Hall, M.A., Smith, L.A.: Feature Selection for Machine Learning: Comparing a Correlation-based Filter Approach to the Wrapper. American Association for Artificial Intelligence (1998)
13. Haykin, S.: Neural Networks: a comprehensive foundation. Prentice Hall, Upper Saddle River (1999)
14. Heaton, J.: Introduction to neural networks for C#, 2nd edn. First Printing. Heaton Research (2008)
15. Heaton, J.: Programming neural networks with Encog 2 in C#. Heaton Research (2010)
16. Jadhav, S.M., Nalbalwar, S.L., Ghatol, A.A.: Generalized feed-forward neural network based cardiac arrhythmia classification from ECG signal. In: 2010 6th International Conference on Advanced Information Management and Service (IMS), pp. 351–256 (2010)
17. Jadhav, S.M., Nalbalwar, S.L., Ghatol, A.A.: Arrhythmia disease classification using artificial neural network model. In: 2010 IEEE International Conference Computational Intelligence and Computing Research (ICCIC), pp. 1–4 (2010)
18. Kavitha, K.S., Ramakrishnan, K.V., Singh, M.K.: Modeling and design of evolutionary neural network for heart disease detection. International Journal of Computer Science Issues (IJCSI) 7(5) (2010)
19. Ladha, L., Deepa, T.: Feature Selection Methods and Algorithms. International Journal on Computer Science and Engineering (IJCSE) 3(5) (2011)
20. Miller, G.F., Todd, P.M., Hegde, S.U.: Designing neural networks using genetic algorithms. In: Schaffer, J.D. (ed.) Proceedings of the Third International Conference on Genetic Algorithms, pp. 379–384. Morgan Kaufmann (1989)
21. Milley, A.H., Seabolt, J.D., Williams, J.S.: Data Mining and the Case for Sampling. SAS Institute Inc. (1988)
22. Mitchell, M.: An introduction to genetic algorithms (complex adaptive systems). MIT Press (1996)
23. Montana, D.J.: Neural network weight selection using genetic algorithms. Intelligent Hybrid Systems (1995)
24. Nandy, S., Sarkar, P.P., Das, A.: An Improved Gauss-Newtons Method based Backpropagation algorithm for fast convergence. International Journal of Computer Applications (0975 – 8887) 39(8) (2012)
25. Osowski, S., Siwek, K., Siroic, R.: Neural system for heart beats recognition using genetically integrated ensemble of classifiers. Computers in Biology and Medicine 41, 173–180 (2011)
26. Raut, R.D., Dudul, S.V.: Arrhythmia classification using MLP neural network and statistical analysis. Emerging Trends in Engineering and Technology (ICETET 2008), 553–558 (2008)
27. Schaffer, J.D., Whitley, D., Eshelman, L.J.: Combinations of genetic algorithms and neural networks: a survey of the state of the art. International Workshop on COGANN 92, 1–37 (1992)

28. Sekkal, M., Chikh, M.A., Settouti, N.: Evolving neural networks using a genetic algorithm for heartbeat classification. Journal of Medical Engineering and Technology 35(5), 215–223 (2011)
29. Somasundaram, R.S., Nedunchezhian, R.: Evaluation of three simple imputation methods for enhancing preprocessing of data with missing values. International Journal of Computer Applications 21(10), 14 (2011)
30. Staroszczyk, T., Osowski, S., Markiewicz, T.: Comparative Analysis of Feature Selection Methods for Blood Cell Recognition in Leukemia. In: Perner, P. (ed.) MLDM 2012. LNCS, vol. 7376, pp. 467–481. Springer, Heidelberg (2012)
31. Taskiran, H.D.: A genetic algorithm approach to feature subset selection for pattern classification using neural networks. Intelligent Systems and their Applications (2005)
32. University of Pittsburg medical center, Arrhythmia (2006)

Finding Multi-dimensional Patterns in Healthcare

Andreia Silva and Cláudia Antunes

Department of Computer Science and Engineering
Instituto Superior Técnico – University of Lisbon
Lisbon, Portugal
{andreia.silva,claudia.antunes}@tecnico.ulisboa.pt

Abstract. The amount of healthcare data is increasing at a rapid pace, and with that is also increasing the need for better and automated analyzes that are able to transform these data into useful knowledge. In turn, this knowledge may bring huge benefits to the healthcare management and leverage the healthcare system in all aspects. In the last decade, data mining has been widely applied to this domain and several approaches, technology and methods were developed for improving decision support. Despite the advances in healthcare mining, the characteristics of these data – complexity, volume, high dimensionality, etc. – still demand more efficient and effective techniques. In this work we present a case study on the healthcare domain. We propose to use a multi-dimensional model of the hepatitis dataset and to apply a multi-dimensional data mining algorithm to find all patterns in the database. These patterns are then used to enrich classification and results show a significant improvement on prediction.

Keywords: Pattern Mining, High Dimensionality, Healthcare Domain, Classification.

1 Introduction

The analysis of healthcare data is mandatory, not only because huge amounts of data are continuously being generated, but also because it may help in many areas of healthcare management, such as evaluating treatment effectiveness, understanding causes and effects, anticipating future demanded resources, predicting patient's behaviors and best treatments, defining best practices, etc. [9,8]. Due to the nature of this information, results of these analysis may make the difference, not only by decreasing healthcare costs and, but also, at the same time, by improving the quality of healthcare services and patients' life.

Healthcare data are usually massive, too sparse and complex to be analyzed by hand with traditional methods. In the last decade, data mining has begun to address this area, providing the technology and approaches to transform huge and complex data into useful information for decision making [9]. Data mining (DM) [6] has been successively applied to many different subfields of healthcare management, which results proved to be very useful to all parts involved [9,8].

P. Perner (Ed.): MLDM 2014, LNAI 8556, pp. 361–375, 2014.

One of the characteristics of the data collected in the healthcare domain is their high dimensionality. They include patient personal attributes, resource management data, medical test results, conducted treatments, hospital and financial data, etc. Thus, healthcare organizations must capture, store and analyze these multi-dimensional data efficiently.

Multi-Relational Data Mining, or MRDM [5] is an area that aims for the discovery of frequent relations that involve multiple tables, in their original structure, i.e. without joining all the tables before mining. In recent years, the most common mining techniques have been extended to the multi-relational context. However, there are few capable of dealing with a massive number of records, and of taking into consideration all dimensions at the same time [13].

In this paper, we propose a case study on the healthcare domain, based on the use of the Hepatitis dataset, created by Chiba University Hospital, containing information about 771 patients having hepatitis B or C, and more than 2 million examinations dating from 1982 to 2001. In this study, we consider a multi-dimensional model of the dataset that facilitates the analysis, and we then make use of a MRDM algorithm to find inter-dimensional and aggregated patterns that are able to characterize patient exam behaviors, which in turn may be used as classification features to predict if a patient has hepatitis or not, which type, or even the state of the hepatitis. Experimental results reveal improvements in prediction and on the classification model built, when compared with the baseline models.

Section 2 describes the Hepatitis dataset and section 3 presents the multi-dimensional model for the Hepatitis data, in order to promote their analysis for decision making. Section 4 describes how can we apply data mining to a multi-dimensional model and use the discovered patterns to enrich classification data and improve prediction results. Section 5 shows and discusses our goals, approaches and results. And finally, section 6 concludes the work.

2   The Hepatitis Case Study

The Hepatitis dataset[1] contains information about laboratory examinations and treatments taken on the patients of hepatitis B and C, who were admitted to Chiba University Hospital in Japan. There are 771 patients, and more than 2 million examinations dating from 1982 to 2001, from about 900 different blood and urine types of exams. The dataset also contains data about the biopsies (about 695 biopsy results) and interferon treatments (about 200) performed to patients. Biopsies reveal the true existence of hepatitis and respective fibrosis state. However, they are invasive procedures, and therefore there is an interest in finding other indicators that allow the detection of hepatitis in a more friendly way. Interferon treatments have also been seen and used as an effective way to treat hepatitis C, although it has tough side effects, and its efficacy is not yet proved. Hence there is the need to understand the impact of this treatment.

---

[1] The Hepatitis dataset was made available as part of the ECML/PKDD 2005 Discovery Challenge: http://lisp.vse.cz/challenge/CURRENT/

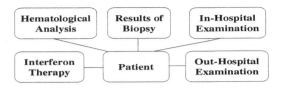

Fig. 1. Hepatitis relational model [11]

The hepatitis dataset is composed of several data tables, modeled in a relational schema centered on the patient. This model is shown in figure 1. Each patient may have performed some biopsies, several hematological analysis, in-hospital and out-hospital exams, and may also be under interferon therapy. Each one of these aspects is stored in one different table and is independent of the others.

Despite being modular, this schema does not facilitate the analysis of these data, for several reasons: (1) the various exams – both in-, out-hospital and hematological analysis – are not directly related, although the same type of exams may be present in more than one table; (2) relating both exams, or exams and biopsies or interferon therapy requires joining the tables for a common analysis. This process of joining the tables is time consuming and non trivial, and the resulting table hinders the analysis, since it will contain lots of redundant data, as well as lots of missing values; and (3) time is not directly modeled, and therefore there is no easy way to understand the interconnection between co-occurring events (e.g. exam results during interferon therapy), neither the disease evolution. Moreover, most data are distributed irregularly, either in time, as well as per patient, making a direct analysis unfeasible.

The work presented in [11] is the first step on the multi-dimensional analysis of the hepatitis data. The authors use a multi-relational algorithm to connect biopsies and urinal exams, and to generate association rules that estimate the stage of liver fibrosis based on lab tests. However, they are only able to mine two dimensions of the relational model at a time, and therefore they cannot relate the biopsies with, for example, both the blood tests (Hematological Analysis) and the other tests (In- and Out-Hospital Examinations).

3   The Multi-Dimensional Model

As stated before, one of the characteristics of the data collected in the healthcare domain is their high dimensionality. In the case of the Hepatitis dataset, we have administrative data such as patient's features (sex and date of birth), pathological classification of the disease (given by biopsy results), duration of interferon therapy, and temporal data about the blood and urine tests performed to patients. Note that we could have more data, such as treatment and tests' cost, hospital data related to out-hospital exams, information about doctors in

charge of patients, etc., which would increase the dimensionality of the dataset
and the complexity of the relational model.

One efficient way to store high-dimensional data is through the use of a multi-
dimensional model – a star schema, in particular. A star schema clearly divides
the different dimensions of a domain into a set of separated data tables, in-
terrelated by a central table, representing the occurring events. In the case of
the Hepatitis data, we can identify several dimensions – *patient*, *biopsy*, possible
exams and *date* information – and events correspond to patient examinations.

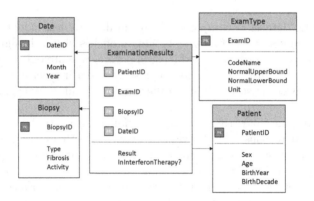

Fig. 2. Hepatitis star schema

In this sense, one of the possible star schemas that can be defined is presented
in figure 2. The star schema is composed of 4 dimensions (*Patient*, *Biopsy*, *Exam
Type* and *Date*) and one central fact table that corresponds to the *Examination
Results*. Each dimension is independent and contains the respective characteristics (*Patient* contains patients' features, and *Exam Type* contains data about
possible exams, like upper and lower bounds and units). By analyzing the central table, we can understand the relation between all dimensions: one patient
P, with active biopsy B, performed exam E on date D. The result of this event
was r (given by attribute *Result* in the central table), and at the moment of this
examination, it was (or not) being administrated interferon therapy (attribute
$InInterferonTherapy?$).

Adding new dimensions to this star schema is straightforward. For example,
we could add dimensions Hospital and Doctor just by adding the respective keys
into the central table, and each event in that table would correspond to one
exam E, performed to patient P, with active biopsy B, on date D, in hospital
H with doctor Doc.

In this case study, we only considered the most significant exams, based on the
report carried out by [15]. These exams are GOT, GPT, ZTT, TTT, T-BIL, D-
BIL, I-BIL, ALB, CHE, T-CHO, TP, WBC, RBC, HGB, HCT, MCV and PLT.
Exam results were categorized into 7 degrees: extremely, very or simply *high* (UH,
VH, H), *normal* (N), *low*, very or extremely low (L, VL, UL). The thresholds

and categories for each of the selected exams are described in [15]. The exam results of patients with more than one result for the same type of exam in one day were averaged.

Fibrosis are considered stable 500 days before and 500 days after a biopsy [15]. Therefore, for each examination in the central table, the corresponding active biopsy is the most recent one performed for the patient, within the 500 days interval (or none).

The resulting star schema contains almost 600 thousand examinations performed, for 722 patients (the other 50 patients have not performed none of the most significant exams). From all patients, only 27.5% were diagnosed with hepatitis B, at some point in time, 40% with hepatitis C, and the rest 32.5% have no biopsy performed, which means that they were not diagnosed with any type of hepatitis, yet. In average, each patient performed about 500 - 700 exams. Also, there are more cases of hepatitis in their early states than in severe states.

4   Multi-Relational Data Mining

In order to deal with multiple tables, data mining has to join somehow the different tables, creating the tuples to be mined. An option that allows the use of the existing single-table algorithms, is to join all the tables in one before mining (a step also known as *propositionalization* or *denormalization*). At a first glance, it may seem easy to join the tables into one, and then do the mining process on the joined result [10]. However, when multiple tables are joined, the resulting table will be much larger and sparser, with many repetitions of the same values, and the mining process more expensive and time consuming. Moreover, these repetitions of values may cause distortions in the calculations of the measures of interest and therefore hinder the discovery of really interesting patterns.

Multi-Relational Data Mining, or MRDM [5] is an area that aims for the discovery of frequent relations that involve multiple tables, in their original structure, i.e. without joining all the tables before mining.

In recent years, the most common mining techniques have been extended to the multi-relational context [4,10,16,12,13]. However, there are few capable of dealing with high number of records, and of taking into consideration all dimensions at the same time [13].

4.1   Multi-dimensional Pattern Mining

Frequent pattern mining is a sub-area of data mining that aims for enumerating all frequent patterns that conceptually represent relations among entities. A well-known example of a transactional pattern is a market-basket: the set of items that are bought together frequently by a customer. In this healthcare domain, an example of a transactional pattern is a set of frequent examinations performed and respective results. These frequent patterns can then be exploited in various ways: for further direct analysis, for creating association rules [14], expressing tendencies, and also for improving classification results and predictions.

Problem Definition. Following the example of the star schema in figure 2, dimensions (e.g. *Patient*) are composed of a primary key (e.g. *PatientID*) and a set of attributes describing the dimension (e.g. *Sex* and *Age*). Each dimension can be seen as a simple set of pairs *(attribute, value)*, corresponding to the characteristics of the elements of that dimension. For example, patient 1 = $\{(Sex, M), (Age, 30), ...\}$. In the context of data mining, an *item* is one of those pairs, and an *itemset* is just a set of pairs.

The central fact table (*ExperimentationResults*) contains one foreign key for each dimension, and a set of measurement/indicator fields. Conceptually, this table contains all business events (the actual occurrences). In this case, our fact table contains all performed examinations.

The *support* of an itemset is defined as the number of its occurrences in the database. In the case of a database modeled as a star schema, the number of occurrences of one item of some dimension (one pair $attribute - value$) depends on the number of occurrences of the respective transactions in the central table. A pattern P is a frequent itemset, i.e. an itemset whose support is greater or equal than a user defined minimum support threshold, and the problem of multi-dimensional frequent pattern mining over star schemas is to find all patterns in a star.

Patterns can be:

Intra-dimensional: if they only contain items of the same dimension (e.g. $\{(Sex, M), (Age, 30)\}$, i.e. most male patients have 30 years old);

Inter-dimensional: if they contain items from more than one dimension (e.g. (Sex, M), (BiopsyType, B), i.e. most male patients have hepatitis B);

Events in the fact table can also be aggregated according to some entity or entities, so that we can discover frequent behaviors or profiles. For example, if we aggregate the facts in the ExaminationResults star, per patient and biopsy, we can discover the sets of exam results that patients usually have, corresponding to each biopsy (i.e. diagnosis). In this sense, patterns can also be:

aggregated: if they result from the aggregation of events of the fact table, i.e. if they contain combinations of items with the same attribute (e.g. $\{(ExamCodeName, GOT), (Result, GOT = H), (ExamCodeName, GPT), (Result, GPT = VH)\}$, i.e. patients have a high result in the GOT exam and very high in the GPT exam frequently).

Related Work. There are many stand-alone algorithms to mine different types of patterns in traditional databases, with *FP-growth* [7] one of the most efficient. This algorithm follows a pattern-growth philosophy and represents the data into a compact tree structure to facilitate counting the support of each set of items and to avoid expensive, repeated database scans. It then uses a *depth-first search* approach to traverse the tree and find the patterns.

Some of the traditional algorithms have been extended to the multi-relational case. In this work we focus on frequent pattern mining over star schemas. The first multi-relational methods have been developed by the *Inductive Logic*

Programming community about ten years ago [4], but they are usually not scalable with respect to the number of relations and attributes in the database and they need all data in the form of prolog tables. Nevertheless, there has been some efforts to create new ILP algorithms that are able to deal with large datasets [1]. An apriori-based algorithm was introduced in [3], which first generates frequent tuples in each single table, and then looks for frequent tuples whose items belong to different tables via a multi-dimensional count array; [10] proposed an algorithm that mines first each table separately, and then two tables at a time; [16] presented *MultiClose*, that first converts all dimension tables to a vertical data format, and then mines each of them locally, with a closed algorithm.

The algorithm *StarFP-Growth* [12] is a pattern-growth method based on FP-Growth. It is able to mine all dimensions without physically join them, by using the central fact table to define and inter-relate the tuples to be mined. The idea is simple and consists in finding first all frequent patterns in each dimension, based on the support given by the central table, and then finding all multi-dimensional patterns by using the central table again to know how and what local patterns co-occur. The algorithm uses an efficient tree structure to store the frequent itemsets, as well as an efficient pattern-growth strategy to generate and propagate the co-occurrences in a divide and conquer manner, and without the need for candidate generation.

In this work we decided to use *StarFP-Growth*, since it works directly on the whole star, and it is an efficient multi-dimensional algorithm.

4.2   Classification Enriched with Multi-Relational Patterns

Classification is a data mining task widely used for predicting future outcomes. As an example, in this healthcare domain, it can be used to predict if a patient is infected with hepatitis or not, as well as to predict the type and stage of hepatitis.

Classification algorithms create a prediction model based on the existing data (*training data*), for which we know a set of features and also the outcomes, and then use this model to predict the unknown outcomes of new data, based on their observed features.

There are several algorithms and models proposed for classification, that have been applied to vast number of different domains. However, results are far from being satisfactory. One of the reasons for the low levels of performance may be the fact that the relations between attributes are not being considered. In fact, in a multi-relational domain there are implicit relations between data that are easily modeled through a relational schema, but that cannot be easily modeled in one singular training data table. Naturally, if we could somehow incorporate these relations into classification, prediction results would be likely to improve.

If we consider the multi-relational (MR) patterns described above, both inter-dimensional and aggregated patterns represent the relations between entities. This means that, if a record (or event) in data satisfies a MR pattern, we can say that this record encloses the relations represented by the pattern.

In this sense MR patterns can be seen as a compact way to model the relationships in data, and can be used as features to enrich data for classification. The simplest way to incorporate these patterns is to pre-process individual records for verifying the satisfaction of each pattern identified, and extend these records with one boolean attribute per pattern, corresponding to the satisfaction (or not) of each pattern by the respective record. In this manner, what is multi-relational by nature becomes tabular, without loosing the dependencies identified before, and traditional classifiers are applicable without the need for any adaptation.

In this paper we propose the use of multi-relational patterns to enrich classification data in the healthcare domain. We tackled several experiments, and we show that running classification over these enriched data improves not only the accuracy of the predictions, but also the classification models built.

5   Results and Discussion

In this section we first describe the goals of this analysis and our approach, and then we discuss some experimental results.

5.1   Goals

For the analysis of the hepatitis star schema, we decided to address two topics of interest, suggested for this dataset: (1) Discover the differences between patients with hepatitis B and C; and (2) Evaluate whether laboratory examinations can be used to predict if a patient is infected with hepatitis B or C (or not), as well as to estimate the stage of liver fibrosis.

This second topic is of particular importance, because biopsies are invasive to patients, and therefore doctors try to avoid them.

By using the star in figure 2 we are able to relate the exams (and other dimensions) with the type of hepatitis, as well as with the fibrosis state. We can look for what examination results are common (frequent) along with hepatitis B or/and C, and see the differences (goal 1). Similarly, we can look for frequent exam results for each hepatitis and fibrosis state (goal 2), and then use those patterns to help classifying other patients with similar results.

5.2   Approach

In order to analyze the hepatitis dataset and achieve our goals, we decided to follow the next methodology: (1) run multi-relational pattern mining over the Hepatitis star schema; (2) filter the best inter-dimensional and aggregated patterns; (3) enrich the classification data (baseline) with these patterns; and (4) run classification over both the baseline and this enriched dataset, and compare the results (the average of the predictions, and the size of the models built).

For the first step, we decided to run the algorithm *StarFP-Growth* over the *Examination Results* star schema. So that we could understand the behavior of patients and discover frequent sets of exam results, the algorithm aggregates into one singular record all the exams of the same patient while each particular biopsy is valid, i.e. each pair *(patient, biopsy)* of the central fact table is considered as only one event. By doing this we are able to discover, not only frequent exam results (like traditional pattern mining algorithms), but also sets of results that co-occur frequently. We can find, for example, that patients with hepatitis B frequently have, at the same time, high results in exams *GOT* and *GPT* but low results in *PLT*.

After finding the patterns, the next step is to filter them and chose the best ones. In theory, using all MR patterns with at least two items to enrich the classification training data should achieve the best results. However, this will eventually lead to overfitting of the models found, which also may lead to poor results when classifying new instances. In this sense, we should choose somehow only those patterns that achieve an higher information gain.

Therefore, we tested our approach with several filters and different number of patterns selected (let it be N). These filters are:

Support: Since the support of a pattern is the number of times it occurs in data, the higher the support, the more patients share the same characteristics, and the more likely it is to cover more patients. Thus, this filter chooses the N patterns with higher support. However, patterns with higher support are the smallest ones, and therefore those that represent smaller relations. Also, in this domain, if they are shared by a high number of patients, it may mean that they are not discriminant of the type of hepatitis or its stage;

Size: On another side, the largest patterns model more relations than smaller ones, and hence may be more interesting. So, this filter choses the N largest patterns. The downside of this is the fact that these patterns tend to have smaller supports and cover a very small part of the data;

Closed: One characteristic of patterns is that, if one is frequent, all of its subsets are also frequent (anti-monotonicity), and this means that some might be redundant. A pattern is closed if none of its immediate supersets have the same support (otherwise, this patterns is not interesting). In this filter, we consider only the closed MR patterns, and choose those with higher support;

Rough independence: According to probability theory, two events A_1 and A_2 are independent if $P(A_1 \cap A_2) = P(A_1)P(A_2)$. And if two events are independent, the occurrences of one do not influence the probability of the other, and therefore patterns that contain these two events are not interesting. For more than two events, they can be mutually independent if $P(A_i \cap A_{i+1} \cap ... \cap A_n) = P(A_i)P(A_{i+1})...P(A_n)$, for all the power set of those events. Taking this into consideration, for this work, we use a rough independence measure:

$$RInd(\{A_1, A_2, ..., A_n\}) = \frac{P(A_1 \cap A_2 \cap ... \cap A_n)}{P(A_1)P(A_2)...P(A_n)}$$

If $RInd$ is 1, it means that the elements of the pattern are rough independent, and therefore less important. The higher the value of $RInd$, the more dependent are the elements, and more important is the pattern. Therefore, using the *rough independence* filter, patterns are ordered in a decreasing order of their value for $|RInd - 1|$, and only the N patterns with highest difference are chosen.

Rough Chi-square (χ^2): Chi-square is an interesting measure that evaluates the correlation between variables [2]. Generally, the more correlated, the more interesting are the relations. The chi-square of two variables is defined as:

$$\sum_{i,j=1}^{n,m} (observed_{ij} - expected_{ij})^2 / expected_{ij}$$

in which $observed_{ij}$ is the observed support of values i and j, and $expected_{ij}$ the expected probability of those values if the variables were independent. In this work we use a rough chi-square measure to evaluate the correlation of elements in a pattern:

$$R\chi^2(\{A_1, ..., A_n\}) = \frac{(support(A_1 \cap ... \cap A_n) - P(A_1)...P(A_n))^2}{P(A_1)...P(A_n)}$$

The higher the value of $R\chi^2$, the more rough correlated are the elements in patterns, and therefore more interesting. In this sense, using a *rough chi-square* filter, the N patterns with highest value of $R\chi^2$ are chosen.

For this case study, since or goal is to predict the type or the stage of hepatitis based on exam results, the baseline used for classification is a table composed of the patient information (*Patient* dimension), and the results for all the 17 most significant exams discriminated in section 3. Then, we defined two similar baselines, **Type** and **Fib**, and applied this methodology to both. The first to predict if a patient is infected with some type of hepatitis, and the second to predict the state of the hepatitis, if present. In this sense, the class of the baseline **Type** is the type of hepatitis (B, C or $None$), and the class of the baseline **Fib** is the stage of the fibrosis (from 0 to 5).

Once we have the best N patterns, we extend the baseline table by adding N boolean attributes. Each value for these attributes is *true* if the patient satisfies the pattern, or *false* otherwise.

Finally, we applied the classification algorithm *C4.5* on these enriched datasets, and compared the results over the same algorithm applied to the baselines. Classification results presented are the average of several 10-cross fold validations.

We used an implementation of *StarFP-Growth* made in Java (JVM version 1.6.0 37) and the *C4.5* implementation available in Weka.

5.3   Multi-Relational Patterns

Table 1 presents a subset of the frequent patterns found. For simplicity, we only present the patterns in which the exam results have abnormal values (i.e. low or high results).

Table 1. Some examples of the multi-relational patterns found in the hepatitis dataset

	Pattern	Support
1	(Result=GOT_H)	975
2	(Result=WBC_L)	536
3	(Result=I-BIL_H)	495
4	(Result=CHE_VL)	491
5	(Type=C)	290
6	(Result=ZTT_H,Result=D-BIL_H,Result=TTT_H)	577
7	(Result=GOT_H,Result=GPT_H,Result=ZTT_H,Result=MCV_H)	586
8	(Result=GOT_H,Result=GPT_H,Result=D-BIL_H,Result=CHE_VL)	386
9	(Result=GOT_H,Result=GPT_H,Result=PLT_L,Result=WBC_L)	362
10	(Result=GOT_VH,Result=GPT_VH,Result=ZTT_H,Result=WBC_L)	355
11	(Sex=M,Result=MCV_H)	579
12	(Sex=M,Result=MCV_H,Result=HCT_H)	469
13	(Type=None,Fibrosis=None,Result=D-BIL_H)	534
14	(Type=None,Result=GOT_H,Result=GPT_H)	508
15	(Type=C,Result=GOT_H,Result=GPT_H,Result=ZTT_H)	236

The first five patterns are intra-dimensional, and contain only one item. The first, for example, means that exam named *GOT* is frequent and appears 975 times in these data with an higher value (*H*). Also, the data contains 290 diagnosis of hepatitis C (pattern 5).

The next five patterns are aggregated patterns, since they represent frequent sets of examination results, and they are discovered because we aggregated the data in the star schema per pair (patient, biopsy). We can see that exams *GOT* and *GPT* frequently appear together, and with very similar (and high) results. We can also observe that, e.g. more than 300 patient diagnosis have high results in both *GOT* and *GPT*, and at the same time, low results in *PLT* and *WBC* exams.

Patterns 11 to 15 are inter-dimensional patterns, that contain items from more than one dimension. From these, the first 2 patterns relate the *Patient* and *Exam* dimensions, and the rest relates the *Biopsy* with the examinations and respective results. In these examples, we note that, from the 800 biopsy diagnosis of male patients, almost 600 correspond to examinations with an high value in exam *MCV*.

The last 3 patterns relate the type of hepatitis and the stage of the fibrosis with examination results. As an example, an high value for the *D-BIL* exam was frequently associated with no hepatitis cases. The last patterns suggest that an higher value for both the *GOT* and *GPT* tests, along with *ZTT*, indicate that the patient has hepatitis C, since more than 80% of the cases of hepatitis C show these values. However, we can see in pattern 14 that these values are also associated with not having hepatitis. This may evince that these values or tests are not discriminant. We also have to note that, in these data, not having information about a biopsy does not say that a person do not have hepatitis. It says only that the person has not been diagnosed yet. However, it may be an indicator for finding relations by which doctors think there is no need for a biopsy.

5.4   Enriched Classification

Figures 3 and 4 show the accuracy of the classification step, over the baseline Type and Fib, respectively, and corresponding datasets enriched with the multi-relational patterns from the pattern mining step.

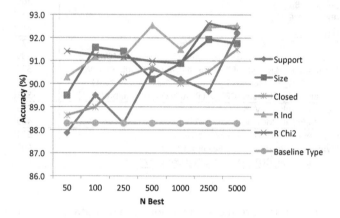

Fig. 3. Accuracy for baseline Type and respective extensions with MR patterns

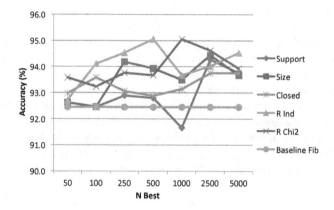

Fig. 4. Accuracy for baseline Fib and respective extensions with MR patterns

When we add the patterns that represent patient exam behaviors, we can see in the figures that the accuracy improves in both cases, as expected. Although small, the improvements indicate that patterns are chosen instead of specific exams, and this may result in models with less over fitting, and therefore more accurate when predicting new instances. Also, results show that, in general, the more N best patterns are chosen, the better the accuracy.

When analyzing the different filters, there are small fluctuations, but both the *rough independence* and *rough chi-square* filters revealed to achieve better results, in both baselines. Choosing the patterns with higher support is the approach that brings less improvements, because they are the smallest ones, and they might not be discriminant of patients with different type or stages of hepatitis. The *closed* filter is similar to the *support*, while the *size* filter achieves intermediate results. These tendencies happened in both baselines.

Figures 5 and 6 analyze the size of the trees created by the classifier (i.e. the size of the models).

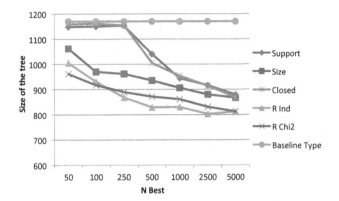

Fig. 5. Size of the trees for baseline Type and respective extensions with MR patterns

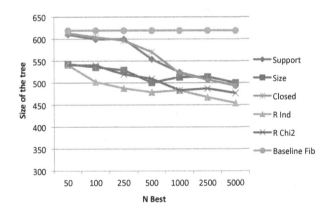

Fig. 6. Size of the trees for baseline Fib and respective extensions with MR patterns

We can see that for both baselines, also as expected, the trees resulting from classifying the enriched datasets are smaller than the base tree (they can have less 300 nodes for $N = 500$ patterns when we are predicting the type of hepatitis).

The tendencies of the different filters are the same. Both *rough independence* and *rough chi-square* filters result in the smallest trees, which means that they choose the patterns that bring more information gain to the models (therefore they are chosen instead of individual examination results). Again, on the contrary, filters *support* and *closed* are the ones that achieve less improvements on the size of the models.

6   Conclusions

In this work we presented a case study on the healthcare domain. Using the Hepatitis dataset, we showed how these data can be modeled and explored in a multi-dimensional model to promote decision support. We also discussed the use of multi-relational data mining algorithms to mine this model, as well as the use of the results to improve classification.

Results show that we can discover structured patterns from the multi-relational model, and find frequent sets of examination results that are common to some type of hepatitis or that lead to some fibrosis stage.

Classification experiments show that, by enriching the training data with the discovered multi-relational patterns, it is possible not only to improve the accuracy of classification, but also to create better and smaller models.

As future work, and in order to surpass the difficulties of this dataset, other paths must be taken. One of the problems comes from the lack of data and their quality. The hepatitis dataset contains more than 30% of patients that never performed any biopsy, and therefore are undiagnosed, and more than 75% of examinations for which there is no information about an active biopsy. To have a better understanding about why these patients have not performed a biopsy requires domain knowledge, and may help partitioning of the data as well as improve the results. In line with the above, this dataset contains a very low number of instances for each type and stage of hepatitis. There is the need for the integration and analysis of more data in this domain.

The use of different approaches may also result in better outcomes, such as infrequent pattern mining [17], for finding rare patterns; or sequential and temporal pattern mining, for the analysis of the evolution of the disease.

We can also try to understand the use of the interferon therapy, and aggregate the data per patient in (and out of) interferon therapy, and find the differences in frequent patterns that happen before, during and after the administration of that therapy (we can also apply the same algorithm, *StarFPGrowth*).

Acknowledgment. This work is partially supported by FCT – Fundação para a Ciência e a Tecnologia, under research project D2PM (PTDC/EIA-EIA/110074/2009) and PhD grant SFRH/BD/64108/2009.

References

1. Appice, A., Ceci, M., Turi, A., Malerba, D.: A parallel, distributed algorithm for relational frequent pattern discovery from very large data sets. Intell. Data Anal. 15(1), 69–88 (2011)
2. Brin, S., Motwani, R., Silverstein, C.: Beyond market baskets: Generalizing association rules to correlations. In: Proc. of ACM SIGMOD Int. Conf. on Manag. of Data, pp. 265–276. ACM, USA (1997)
3. Crestana-Jensen, V., Soparkar, N.: Frequent itemset counting across multiple tables. In: Terano, T., Liu, H., Chen, A.L.P. (eds.) PAKDD 2000. LNCS (LNAI), vol. 1805, pp. 49–61. Springer, Heidelberg (2000)
4. Dehaspe, L., Raedt, L.D.: Mining association rules in multiple relations. In: Džeroski, S., Lavrač, N. (eds.) ILP 1997. LNCS, vol. 1297, pp. 125–132. Springer, Heidelberg (1997)
5. Džeroski, S.: Multi-relational data mining: an introduction. SIGKDD Explor. Newsl. 5(1), 1–16 (2003)
6. Frawley, W.J., Piatetsky-Shapiro, G., Matheus, C.J.: Knowledge discovery in databases: an overview. AI Mag. 13(3), 57–70 (1992)
7. Han, J., Pei, J., Yin, Y.: Mining frequent patterns without candidate generation. In: Proc. of the 2000 ACM SIGMOD, pp. 1–12. ACM, New York (2000)
8. Kaur, H., Wasan, S.: Empirical study on applications of data mining techniques in healthcare. Journal of Computer Science 2(2), 194–200 (2006)
9. Koh, H., Tan, G.: Data mining applications in healthcare. Journal of Healthcare Information Management 19(2), 64–71 (2005)
10. Ng, E.K.K., Fu, A.W.-C., Wang, K.: Mining association rules from stars. In: ICDM 2002: Proc. of the 2002 IEEE Intern. Conf. on Data Mining, pp. 322–329. IEEE, Japan (2002)
11. Pizzi, L., Ribeiro, M., Vieira, M.: Analysis of hepatitis dataset using multirelational association rules. In: ECML/PKDD 2005 Discovery Challenge, Porto, Portugal (2005)
12. Silva, A., Antunes, C.: Pattern mining on stars with FP-growth. In: Torra, V., Narukawa, Y., Daumas, M. (eds.) MDAI 2010. LNCS, vol. 6408, pp. 175–186. Springer, Heidelberg (2010)
13. Silva, A., Antunes, C.: Finding patterns in large star schemas at the right aggregation level. In: Torra, V., Narukawa, Y., López, B., Villaret, M. (eds.) MDAI 2012. LNCS, vol. 7647, pp. 329–340. Springer, Heidelberg (2012)
14. Srikant, R.: Fast algorithms for mining association rules and sequential patterns. PhD thesis, The University of Wisconsin, Madison (1996); Supervisor-Naughton, J.F.
15. Watanabe, T., Susuki, E., Yokoi, H., Takabayashi, K.: Application of prototypelines to chronic hepatitis data. In: ECML/PKDD 2003 Discovery Challenge, Cavtat, Croatia (2003)
16. Xu, L.-J., Xie, K.-L.: A novel algorithm for frequent itemset mining in data warehouses. Journal of Zhejiang University - Science A 7(2), 216–224 (2006)
17. Zhou, L., Yau, S.: Efficient association rule mining among both frequent and infrequent items. Computers and Mathematics with Applications 54(6), 737 (2007)

A Generalized Relationship Mining Method for Social Media Text Data

Tuhin Sharma and Durga Toshniwal

Department of Computer Science and Engineering
Indian Institute of Technology Roorkee
Roorkee - 247667, Uttarankhand, India
{tuhinsharma121,durgatoshniwal}@gmail.com

Abstract. Increasing popularity of Social Media has resulted in the creation of a huge amount of user generated documents. A large number of research works have focused on inferring relationship in certain specific social network domains. Few have considered structured data to establish syntax based relationship. In this work, we develop a two-step syntax based and semantic based relationship mining approach. Here we generalize the concept of relationship mining for all structured as well as unstructured unsupervised text documents from all social network domains. At first, we choose suitable features from individual document and store them in graph structure. Then we establish relationships in the graph generated to obtain Reduced node Social Graph with Relationships (RSGR). Our empirical study on various social media document validates the effectiveness of our approach and suggests its generality in finding relationships irrespective of the type of text documents and the social network domains.

Keywords: Social network analysis, Relationship mining, Concept, Wordnet, Freebase, Social graph, Visualization.

1   Introduction

Web consists of billions of documents, with an increasing rate of growth of 7.3 million pages per day. A document could be a facebook post, a tweet, a blog, a review or even a video. With this rapid growth of information in the web, there is an important need for identifying the relationships between various documents, since all these documents are not useful unless manually read.

In the current scenario, every document is fed individually in social media analysis [1][2]. Here, no relationships exist between documents because every document is considered as a single isolated entity. So we have knowledge of every document individually but not in the presence of other documents. So certain queries cannot be answered like who is the most active Author who talks about data mining, what are the other similar documents the Author is talking about. So, if we can introduce some relationship links or edges we can find out the importance of certain documents as well as we shall be able to give corresponding output according to such queries.

P. Perner (Ed.): MLDM 2014, LNAI 8556, pp. 376–392, 2014.

In this work, we develop a hybrid of syntax and semantic based relationship mining approach for all structured as well as unstructured unsupervised text documents from all social network domains. At first, suitable features are selected and extracted from individual documents and are stored in the form of graph structure. In the second step we establish relationships in the context of the graph generated and merge duplicate nodes. The relationships are represented in the form of edges. The strengths or weights of the relationships represented in numerical form are also stored as part of those relationship edges. Thus finally we obtain a Reduced node Social Graph with Relationships (RSGR).

2   Related Work

Earlier research works in relationship mining mainly focus on applying text mining technique on structured data. Bonaventura Coppola et al. propose a machine learning framework to mine relations automatically, where the target objects are structurally organized in a tree[3]. But the approach is syntax based and semantic meaning related to the extracted dimensions is not used.

Current solutions for social relationship mining are generally statistical learning-based approaches[4]. Christopher P. Diehl et al. propose a supervised learning model, which treats links as features and labeled pairs as training data[5]. Wenbin Tang et al. proposed a Partially-labeled Pairwise Factor Graph Model to infer the type of social ties[6]. But only advisor-advisee, manager-subordinate relations and friendships from mobile network are mined for domain specific data. Tang et al. propose a transfer-based factor graph to incorporate social theories into a semi-supervised learning framework[7]. Lei Tang and Huan Liu propose a clustering-based approach which differentiates latent social dimensions from the social network[8]. But the challenges in social media analysis are to handle the dynamic nature of social network and its multiple entities[9]. Each day, huge number of new members join the social network as well as new connections occur among the existing members. How to efficiently update the relationship model accordingly remains a challenge.

Chi Wang et al. develop a time-constrained probabilistic factor graph model (TPFG) that takes a research publication network as input to find advisor-advisee relationships[10]. But the problem is, how to generalize the approach of relationship mining to enable semi-supervised learning and to cope with multi-typed nodes and links are not done yet.

In this work, we have done the following innovations. We establish relationship between different Authors of different documents on the basis of common interest and user-id. For Domain we establish similarity on the basis of popularity. We have connected different documents on the basis of common topic and reference to other documents. We even connect a document to other Domains apart from its source Domain. The next sections describe the work we have done. In section 3. the proposed work is elaborately discussed. Section 4 represents experimental results along with different possible applications to support the effectiveness of our work.

3   Proposed Work

The flow chart for the proposed work is shown in Fig.1.

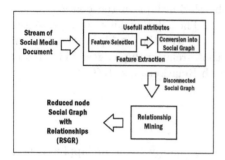

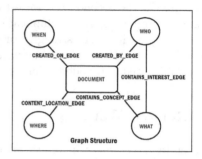

Fig. 1. Proposed process for Relationship Mining

Fig. 2. Each document converted into graph structure

3.1   Feature Extraction

Each of the incoming documents can be categorized into 4 types which are Boards (e.g. large blogs, facebook posts etc.), Reviews (e.g. tripadvisor review), Microblogs (e.g. tweets) and Videos (e.g. youtube, dailymotion videos). Each of them has 4 types of information regarding its Author, Timestamp, Domain and Text associated with it. So for individual documents we create nodes corresponding to these 4 types and we segregate corresponding information and store them accordingly in graph structure as shown in Fig.2. We additionally create WHAT nodes to store different concepts or keywords of text documents and to store interest of Authors.

The edges shown are not assigned any weight. For DOCUMENT, WHO, WHERE and WHEN nodes we just read and store the attribute values for respective documents. For the creation of WHAT nodes we perform natural language processing. The significance of these nodes is as follows:-

DOCUMENT Node: It keeps track of the actual document. It contains the *Url* of the document, the *Type* of the document, i.e. whether it is a Boards, Microblogs, Reviews, Videos, etc (categorized by us). It also contains *Texthtml, Subjecthtml, Review rating*, etc.

WHO Node: It contains information about the Author who has created the document like *Real name, Username, Location, Gender, Profile-url,Author description*, etc.

WHERE Node: It contains information about the place where the document has been created like *Domain, Geo-location*, etc. Additionally, we store the Pagerank score. We have used Quantcast dataset[15] for determining the Pagerank score of a particular domain. Pagerank score of the WHERE node is calculated as the difference between the pagerank of the domain and the lowest possible rank in the Quantcast dataset.

WHEN Node: It contains information about timestamp like *date of creation* of the document, *date of registration* of the Author etc.

WHAT Node: These are basically concepts. Concepts are important keywords which solely determine the meaning of a document or the interest of an Author. The concepts are fetched from unstructured fields which can be either the *Texthtml* and *Subjecthtml* or the *Author description*. We do not use classical TF-IDF measure while mining concepts as it is more appropriate when we need to eliminate frequently occurring stopwords. But we deal with documents irrespective of their category. So it may frequently happen that an important concept (not a stopword) occurs in every or most of the documents. In these cases, if we take TF-IDF measure then that concept will be lost and we shall encounter error. So, we use Apache Opennlp library[14] for mining concepts. From corresponding fields, plain text data is extracted using HTML parser. The resulting plain text is tagged using opennlp POS tagger[14]. Then the plain text is tokenized and stopwords and punctuations are removed from those respective tokens. Here we store the top 10 concepts out of which at most 3 concepts will be proper nouns according to the frequency of occurrence in respective documents. If there exist two different concepts having the same term frequency then first we select the concept which occurred first in the text and then we select the other concept. We are not eliminating any verbs or adverbs because even they can correspond to a concept. For example, *I love to **swim (verb)** for long time.* All the concepts are stemmed using Porter stemmer library included in [14]. Then the stemmed and the actual concepts are stored in the corresponding WHAT nodes. The WHAT nodes mined from *Texthtml* or *Subjecthtml* are connected with the DOCUMENT node and those mined from *Author description* are connected with the corresponding WHO node as shown in Fig.2. So, in this step we create atomost 10 WHAT nodes for every document and each of them contains a single concept along with its stemmed form.

3.2   Relationship Mining

Depending on the type of nodes different approach is followed to establish relationships and to remove duplicates.

Approach for WHO Nodes. For establishing relationships between WHO nodes 2 types of relations are considered as shown in Fig.3. They are as follows:

Explicit Relationship. It represents whether one Author is a friend or follower of another Author in social media like Facebook and Twitter. To mine this relation facebook api and twitter api have been used. It is called explicit as it is explicitly defined by the social network provider. We connect such WHO nodes via EXPLICIT_RELATIONSHIP_EDGE(*ERE*).

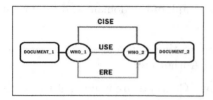

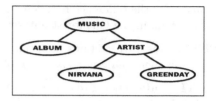

Fig. 3. Relationship between WHOs   **Fig. 4.** MUSIC domain's Ontology tree

Common interest relationship. It represents whether two Authors have any similar interests. To mine this relation the WHAT nodes associated with one WHO node are compared to those associated with the other WHO node using Wordnet[16] and Freebase[17]. The reason for using the Wordnet dictionary is to figure out the similarity between two WHAT nodes on the basis of synonym and hyponym-hypernym relations. The reason for the use of Freebase is to figure out the similarity between WHAT nodes which belong to the same category. If the concepts are Proper nouns then we cannot find synonyms. For example,

---

Algorithm 1. Algorithm for creating CISE

INPUT: Disconnected social graph
OUTPUT: Social graph with CISE
  for all pair of WHO nodes **do**
    for all pair of WHAT nodes containing Proper noun concepts i.e. one WHAT node associated with one WHO node and other WHAT node associated with the other one **do**
      Calculate the Wu-Palmer similarity (WPS) value according to Freebase
    end for
    Say,there are such m pairs of Proper noun concepts
    for all WHAT nodes containing concepts other than Proper noun associated with any one of the WHO nodes **do**
      synonyms from Wordnet are fetched for each actual concept and stored in respective WHAT nodes and associated with those actual concepts
      for each of the WHAT nodes associated with the other WHO node **do**
        Compare actual concept against each of the actual concepts from the other set and against the synonyms associated using string comparison
        Compare stemmed concept against each of the stemmed concepts from the other set using string comparison
      end for
    end for
    Say,there are such n matched concepts which are not Proper nouns
    $match\_value = \sum_{i=1}^{m} WPS_i + n$
    $CISV = match\_value/avg\_length.$
    if $CISV > (THRESHOLD\_CISV\_CONNECT = 0.5)$ **then**
      Connect the two WHO nodes by a CISE and store $CISV$ in it
    end if
  end for

---

let WHAT_1 and WHAT_2 represent the interests *Nirvana* and *Greenday* respectively. Here, there exists a relationship between WHAT_1 and WHAT_2 because both the interests are related to the music category. So to detect this type of relationship we use Freebase. For WHAT_1 as well as for WHAT_2 we shall get an ontology path starting from the root node to the desired node in the ontology tree of Freebase. Now, if these two paths share more common nodes then they will be more similar.

For example, let us consider the ontology tree for the Music domain as per Fig.4. Here, there exists a relationship between *Album* and *Artist* as well as between *Nirvana* and *Greenday*. But *Nirvana* and *Greenday* are more similar as compared to *Album* and *Artist*. It is because *Nirvana* and *Greenday* appear at greater depths in the ontology tree as compared to that of *Album* and *Artist*. So additionally we have to consider the depth in the ontology tree also. For this, we use Wu-Palmer similarity measure[11] while considering the similarity based on Freebase. The algorithm for calculating common interest similarity value ($CISV$) and creating COMMON_INTEREST_SIMILARITY_EDGE (CISE) is depicted in Algorithm.1. Here, avg_length is the total number of WHAT nodes associated with those two WHO nodes divided by 2.

User-id similarity relationship. It represents whether two Authors are similar or not on the basis of their user-ids. To mine this relation 5 fields i.e. *Username, Gender, Author description, Location* and *Original name* are compared. Generally when a person gets registered in two different social sites he usually chooses Usernames which starts with a matching subsequence of characters (eg. *Alex Maxwell* is present in facebook as *alexm* and in twitter as *alexmax*). To capture this type of similarity both the number of characters matched and the length of starting subsequence matched are to be considered. So, for *Username* Jaro-Winkler distance[12] is used. For *Gender, Location* and *Original name* Jaccard coefficient[13] is calculated to find out the similarity. For *Author description* we use $CISV$. The priorities of the similarity values corresponding to these 5 fields in descending order are as follows:

Gender>Location>Author description>Original name>Username.

Algorithm 2. Algorithm for creating USE

INPUT: Disconnected social graph
OUTPUT: Social graph with USE

    for all pair of WHO nodes **do**
        Calculate $S_i \forall \ 1 \leq i \leq m$
        $W_i = \frac{m-i+1}{\sum_{i=1}^{m} i}$
        $USV = \sum_{i=1}^{m} W_i S_i$
        if $USV > (THRESHOLD\_USE\_CONNECT = 0.8)$ **then**
            Connect the two WHO nodes by a USE and store USV in it
        end if
    end for

The reason for such priority ordering is very simple. If two WHO nodes have different gender then irrespective of other fields they cannot be the same person. If gender is same for two WHO nodes but they belong to two different countries then they cannot be the same person and so on. So, the similarity due to *Gender* should get highest weightage and that due to *Username* should get the lowest weightage when we use those similarity values to calculate the user-id similarity value. As we are considering the intra social network relationship as well as inter social network relationship so it is quite often that we find some attributes are applicable to certain documents and some are not. As for example, for a *facebook* document we shall find an attribute named *number of friends*, but in case of *tripadvisor* document we cannot find such attribute. So, while calculating the similarity we shall consider only those attributes for which the values are present in both of the documents.

Let, A_i represents the ith attribute and there are m number of attributes for which values are present in both WHO nodes. The priorities in descending order are as shown in Eq.1.

$$A_1 > A_2 > \ldots\ldots\ldots > A_m [1 \leq m \leq 5] \tag{1}$$

For each of the attributes we get respective the similarity values (S_i) and we assign weights W_i to those S_i to calculate userid similarity value (USV). The process to calculate USV and to establish USERID_SIMILARITY_EDGE (USE) is described in Algorithm.2.

Approach for WHERE Nodes. For establishing relationships between WHERE nodes we create PAGERANK_SCORE_SIMILARITY_EDGE (PSSE) based on pagerank score as shown in Fig.5. It is described in Algorithm.3.

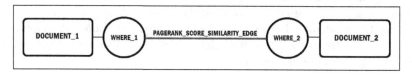

Fig. 5. Relationship between WHERE nodes

Algorithm 3. Algorithm for creating PSSE

INPUT: Disconnected social graph
OUTPUT: Social graph with PSSE
  for all pair of WHERE nodes **do**
    Calculate $diff$ as the difference between pagerank scores of those 2 WHERE nodes

    if $diff > (THRESHOLD\_PSSE\_CONNECT = 50)$ **then**
      Connect the two WHERE nodes by a PSSE
    end if
  end for

Approach for DOCUMENT Nodes. For establishing relationships between DOCUMENT nodes 2 types of relations are considered as shown in Fig.6. They are as follows:

Document Concept similarity relationship. We check if two documents have any similar concepts. To mine this relation Algorithm 1 is followed except we consider DOCUMENT nodes instead of WHO nodes and we calculate document concept similarity value (DCSV) and store them in DOCUMENT_CONCEPT_SIMILARITY_EDGE (DCSE) connecting the two DOCUMENT nodes.

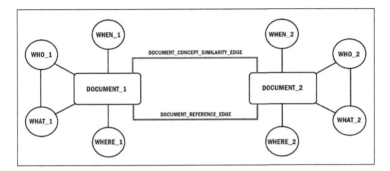

Fig. 6. Relationship between DOCUMENT nodes

Document Reference relationship. It represents if a DOCUMENT is referred by another DOCUMENT. To find this relationship we parse the *Texthtml* using html parser and find the links to other document if present in it. If we find any such document then we connect those documents using DOCUMENT_REFER-ENCE_EDGE (DRE).

Relationship between DOCUMENT node and WHERE Node. We establish 1 type of relation between DOCUMENT and WHERE node other than CONTENT_LOCATION_EDGE as shown in Fig.7.

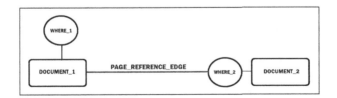

Fig. 7. Relationship between DOCUMENT and WHERE nodes

This relationship represents the domain or the WHERE node from which the documents referred in the current document comes from. We analyze the

Texthtml of the DOCUMENT and find out different domain names and connect those corresponding WHERE nodes with the DOCUMENT node via PAGE-REFERENCE_EDGE (PRE). So this type of edge shows the other domains a particular document is connected to apart from the source domain.

After all these steps, finally we get the Reduced node Social Graph with Relationships (RSGR) as shown in Fig.8.

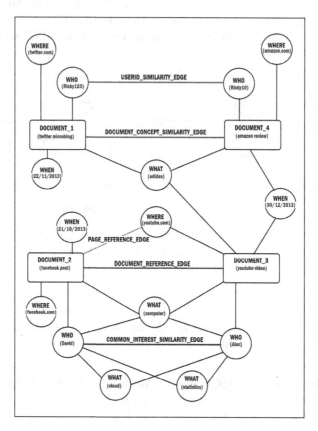

Fig. 8. Reduced node Social Graph with Relationships (RSGR)

4   Experimental Results

In this section we show various experiments that support the effectiveness and the accuracy of our approach.

4.1   Experimental Setup

Data Sets. We use real life data from various social sites, especially from facebook, twitter, tripadvisor, amazon, youtube, dailymotion, hotel.com etc. It consists of 60,370 unsupervised documents. After the feature extraction step the resulting graph database consists of $8,01,350$ nodes. For the purpose of experimental analysis we take a random sample of 600 documents (resulted into 8229 nodes) consisting of 150 documents from each category, i.e. Boards, Microblogs, Reviews and Videos. These 4 types of categories are associated with the documents by us to explain the experimental results. To test the accuracy of the different types of discovered relationships, we adopt two data sets. In the first one (D_1), those selected 600 documents are manually labeled by looking into the actual document in which the best concepts of the *Texthtml* and the *Author description* are selected. In the other one (D_2), the same process of labeling is adopted, but only top 10 concepts are chosen for the documents. In both the cases, the labeling i.e. selecting the concepts is done by 10 unbiased persons. These persons are not expert in any domain and as the data used also do not belong to any specific category, we assume the labeling as the ground truth.

Method. At first we try to find out the behavior of accuracy and the redundancy of our algorithm for finding the concepts with respect to D_1 and D_2. The accuracy and redundancy is calculated as follows:

Let us consider a document for which, A represents the set of concepts that we have got applying our algorithm and B represents the set of concepts that we have got from the labeled dataset. $N(A)$ and $N(B)$ represent the number of concepts in the sets A and B respectively. Then the accuracy and redundancy are calculated as per Eq.2 and Eq.3.

$$Accuracy = \frac{N(A \cap B)}{N(B)} \tag{2}$$

$$Redundancy = \frac{N(A - B)}{N(A)} \tag{3}$$

After that we try to compare the relationship edges that we get by applying our algorithm on the concepts that we have found in the feature extraction step with the relationship edges obtained by applying our algorithm on the concepts that we get from the labeled dataset D_1 and D_2. The accuracy and redundancy for the generated relationship edges are calculated as per Eq.2 and Eq.3 where A represents the set of edges generated by our algorithm using our algorithm generated concepts and B represents the set of edges generated by our algorithm using concepts as per D_1 or D_2.

4.2   Accuracy and Redundancy of Concepts

The accuracy corresponding to each of the documents are sorted in increasing order and both the corresponding accuracy and the redundancy with respect to D_1 are plotted against the respective documents in Fig.9(a). The same thing is done with respect to D_2 and the corresponding result is plotted in Fig.10(a). In Fig.9(b) and in Fig.10(b) the documents are sorted according to their respective category, where indices from (1-150), (151-300), (301-450) and (451-600) represent document types Reviews, Boards, Microblogs and Videos respectively.

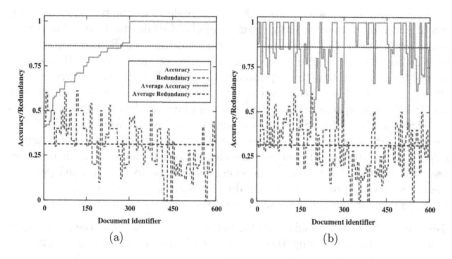

Fig. 9. Accuracy vs. Redundancy for D_1 (a) documents are sorted according to their Accuracy; (b) documents are sorted according to their Types

In Fig.9(a) and Fig.9(b) the *average accuracy* is 0.868 and *average redundancy* is 0.327 (Corresponding to D_1). In Fig.9(a) we see that for high accuracy, the encountered redundancy is pretty much low, which is around 0.28 and for low accuracy the corresponding redundancy is much higher. Now if we look at Fig.9(b) we see that for Reviews the encountered redundancy is higher than mean redundancy, but still we get much better accuracy. In case of Boards, the redundancy is similarly higher than the mean but the accuracy is very poor. For Microblogs and Videos the accuracy is very good and encountered redundancy is much lower than the mean redundancy. Actually in case of Microblogs and Videos, the text portion is short so after removing the stopwords, punctuations and after counting frequency of the resultant concepts we get almost accurate results. But in case of Reviews and Boards, the text portion is much big so we are encountering redundancy more than the *average redundancy* and getting accuracy less than the *average accuracy*.

In Fig.10(a) and Fig.10(b) the *average accuracy* is 0.891 and *average redundancy* is 0.246 (Corresponding to D_2). In this case we see that the redundancy does not vary much and more or less follows the average redundancy value. In Fig.10(b), in case of Boards we see that the accuracy is improved and the redundancy is much lower. So from this, we can infer that if we restrict the user to choose only top 10 concepts, then they are much similar as those obtained by applying our algorithm. This justifies the effectiveness of our approach to find out the concepts or to create WHAT nodes.

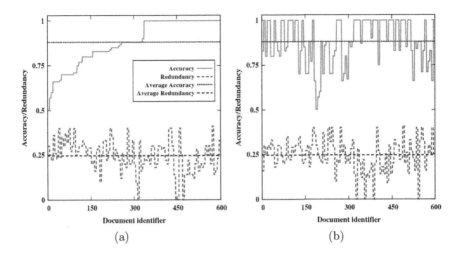

(a)   (b)

Fig. 10. Accuracy vs. Redundancy for D_2 (a) documents are sorted according to their Accuracy; (b) documents are sorted according to their Types

In Fig.11(a) the accuracy and in Fig.11(b) the redundancy corresponding to D_1 and D_2 are plotted against the respective documents. We see that the accuracy and the redundancy corresponding to D_2 is much better than those corresponding to D_1. The reason is, in case of D_1 all possible best concepts are chosen. So sometimes there may be more than 10 concepts in case of D_1. But in case of D_2 there exists at most 10 concepts. In our algorithm, we also calculate the top 10 concepts. So the probability of getting better accuracy and redundancy is higher while compared to D_2 than D_1.

In Fig.11(a) and in Fig.11(b) for some documents the accuracy is lower and the redundancy is higher corresponding to D_2 than those corresponding to D_1. The reason is, for those documents, the concepts that we get using our algorithm are of medium importance. For example:-

Let for a document the set of concepts corresponding to D_1 is {*coke, football, basket, drink, match, ground, India, Pakistan, exciting, television, Messi, Ronaldo*} and corresponding to D_2 it is {*coke, football, basket, drink, match, ground, India, Pakistan, exciting, television*}. If a concept occurs before another then it is more important than the other. So, *coke* is more important than *football*. According to our algorithm we get concepts {*coke, football, basket, drink, match, ground, amazing, India, Ronaldo, Messi*}. So, the accuracy corresponding to D_1 becomes $9/12 = 0.75$ which is greater than the accuracy corresponding to D_2 which is $7/10 = 0.7$. Also the redundancy corresponding to D_1 becomes $1/10 = 0.10$ which is lower compared to that of D_2 which is $3/10 = 0.3$. So we see that $Messi, Ronaldo, amazing$ are not top concepts rather concepts with medium importance. So, we get this type of result.

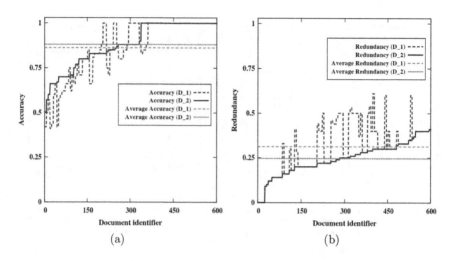

Fig. 11. (a) Accuracy corresponding to D_1 vs. Accuracy corresponding to D_2; (b) Redundancy corresponding to D_1 vs. Redundancy correspondingto D_2

4.3   Accuracy and Redundancy of Relationship Edges

The frequency distribution of edges is shown in Fig.12. For each edge type we have got 3 edge counts. The left most of them corresponds to the edge count when the concepts generated by our algorithm are used (Say C_1). The other two of them corresponds to the edge count when the concepts as per D_1 and D_2 are used respectively (Say C_2 and C_3). We can see that for edge type corresponding to userid (USE), pagerank (PSSE), document reference (DRE) and page reference (PRE) all the 3 counts are same. For edge type CISE corresponding to common interest we see that the difference between C_1 and C_2 as well as C_1 and C_3 are not much. It is because for *Author Description* we get almost same set of concepts either applying our algorithm or using human perception as it consists of limited words. But for edge type DCSE corresponding to document concept similarity we see that the difference between C_1 and C_2 as well as C_1 and C_3 are greater than or equal to 10. It is because for some documents having large text field we lose some concepts and we are not getting relationship edges corresponding to those concepts.

Table 1 shows the types of relationship edges and their corresponding accuracies and redundancies. We see that the accuracy and the redundancy of edges corresponding to PSSE, DRE and PRE are 1 and 0 respectively. The reason is, in those cases we simply establish the syntactical relationship and it does not depend on the concepts a document contains. So there is no possibility of committing error or any type of redundancy in case of those 3 types of relationship edges. We can see that 4 pairs of Authors or WHO nodes are similar in terms of userid. We get this result for both of the two datasets. The accuracy of CISE is very high as that compared to DCSE. The reason is in

Table 1. Accuracy and Redundancy of the edges

Basis of edges	Edge count			Accuracy(A)			Redundancy(R)		
	C_1	C_2	C_3	w.r.t D_1	w.r.t D_2	Average value	w.r.t D_1	w.r.t D_2	Average value
USE	4	4	4	1	1	1	0	0	0
CISE	56	65	62	0.83	0.87	0.85	0.035	0.035	0.035
PSSE	24	24	24	1	1	1	0	0	0
DRE	63	63	63	1	1	1	0	0	0
DCSE	85	113	101	0.707	0.801	0.754	0.058	0.047	0.052
PRE	58	58	58	1	1	1	0	0	0

case of Authors interest we get the concepts more accurately. So, the *average weighted accuracy* and the *average weighted redundancy* of our algorithm for relationship mining are $\frac{4\times1+56\times0.85+24\times1+63\times1+85\times0.801+58\times1}{4+56+24+63+85+58} = 0.9127$ and $\frac{4\times0+56\times0.035+24\times0+63\times0+85\times0.052+58\times0}{4+56+24+63+85+58} = 0.022$. So, we see that for syntactical relationship edges, we get very good result and for semantic relationship edges, we encounter some error as well as redundancy. Yet, the best part is we may not generate all possible relationship edges, but we get negligible false relationship edges because the redundancy value is very low.

4.4 Scalability

Let, n be the size of the RSGR generated. As we are comparing every node with each other of a particular type so the running time is $\mathcal{O}(n^2)$. After that documents come in small chunks relative to the RSGR database. Let, the size of these chunks of documents be m [$m << n$]. But now the nodes stored in the RSGR database are not compared against each other rather the nodes obtained from these small chunks of data of size m are compared against the RSGR database. So now the cost of adding these new chunks of documents is $\mathcal{O}(mn)$. Now due to addition of new documents as the size of the RSGR increases, at some point m becomes much smaller as compared to n. So, now the cost of adding new documents becomes $\mathcal{O}(mn) \approx \mathcal{O}(n)[\because m << n]$.

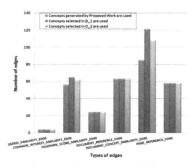

Fig. 12. Frequency of Edges

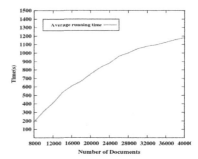

Fig. 13. Average Computation Time

So, in Fig 13 we are adding 2000 documents at a time to the RSGR with an initial size of 8000 documents. Here, we see that it takes significant time to build the RSGR database when the number of documents is low. But, once The RSGR is built the cost of adding different documents to it becomes very less. So, we can claim that RSGR is scalable.

4.5   Applications

We use TitanGraph[18] to store the incoming documents in graph structure as discussed and we maintain the graph database and all its relationships. Whenever new documents arrive we establish new connections among themselves as well as with the existing documents. So the RSGR serves as a source of knowledge. The established relationships can benefit many applications like query regarding highly active Authors, the most popular topics discussed, similar documents, similar Authors, etc. We use SigmaJs[19] to visualize the RSGR generated. Some parts of the RSGR are shown in Fig.14.

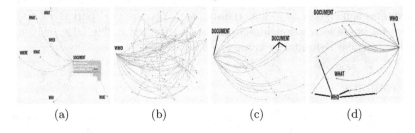

    (a)   (b)   (c)   (d)

Fig. 14. (a)An isolated Document; (b)Most active Author; (c)Document concept similarity edge; (d)Common interest similarity edge

We use different colors representing different types of nodes and relationships. Red, Blue, Green, Yellow, Pink colors represent WHO, DOCUMENT, WHERE, WHEN and WHAT nodes respectively. A single isolated document is shown in Fig.14(a). Fig.14(b) represents the most active Author and different documents created by him. Relationship edges are also colored. Blue represents the document concept similarity edges associated with DOCUMENT and Red represents Common interest relationship edges associated with WHO nodes as shown in Fig.14(c) and Fig.14(d).

5   Conclusion

We have studied mining of different types of relationships from real life online documents to discover syntactical as well as semantic relationships. We propose a two stage framework to mine relationships step by step until we get hierarchies of Authors, Domains, Time and as well as Concepts. We have used Wordnet and

Freebase for this purpose along with different types of api and Quantcast data and to the best of our knowledge it is the first attempt to use them for mining social relationship. We propose a Reduced node Social Graph with Relationships (RSGR) model to establish and store different types of relationships of different social media documents. In this graph we can keep track of relationships of existing documents along with new documents which are constantly added irrespective of their source domain without encountering any redundancy.

Based on the result, we observe for Microblogs and Videos especially for the documents with limited sized text, our approach works well. Interesting problems related to our approach could be how to get a more effective topic mining model irrespective of the size of the document. Another scope is we can use this RSGR to find the credibility or trust of different documents. Clearly much more can be exploited from the knowledge base that we have generated.

Acknowledgment. This research is partially done in IBM's Extreme Blue Lab with Kiran Subbaraman. We thank the anonymous reviewer for labeling the documents.

References

1. Taboada, M., Brooke, J., Tofiloski, M., Voll, K., Stede, M.: Lexicon-based methods for sentiment analysis. Computational linguistics 37(2), 267–307 (2011)
2. Pang, B., Lee, L., Vaithyanathan, S.: Thumbs up?: sentiment classification using machine learning techniques. In: Proceedings of the ACL-02 Conference on Empirical Methods in Natural Language Processing, vol. 10, pp. 79–86. Association for Computational Linguistics (2002)
3. Coppola, B., Moschitti, A., Pighin, D.: Generalized framework for syntax-based relationship mining. In: ICDM, pp. 153–162 (2008)
4. Getoor, L., Taskar, B.: Introduction to Statistical Relational Learning (Adaptive Computation and Machine Learning). The MIT Press (2007)
5. Diehl, C.P., Namata, G., Getoor, L.: Relationship identifcation for social network discovery. In: AAAI 2007, pp. 546–552. AAAI Press (2007)
6. Tang, W., Zhuang, H., Tang, J.: Learning to infer social ties in large networks. In: Gunopulos, D., Hofmann, T., Malerba, D., Vazirgiannis, M. (eds.) ECML PKDD 2011, Part III. LNCS, vol. 6913, pp. 381–397. Springer, Heidelberg (2011)
7. Tang, J., Lou, T.C., Kleinberg, J.: Inferring social ties across heterogeneous networks. In: Proc. the 5th ACM Int. Conference on Web Search and Data Mining (WSDM 2012), Seattle, Washington, February 8-12, pp. 743–752 (2012)
8. Tang, L., Liu, H.: Relational learning via latent social dimensions. In: KDD 2009 Proceedings of the 15th ACM SIGKDD International Conference on Knowledge Discovery and Data Mining, pp. 817–826. ACM, New York (2009)
9. Tang, L., Liu, H., Zhang, J., Nazeri, Z.: Community evolution in dynamic multimode networks. In: KDD 2008: Proceeding of the 14th ACM SIGKDD International Conference on Knowledge Discovery and Data Mining, pp. 677–685 (2008)

10. Wang, C., Han, J., Jia, Y., Tang, J., Zhang, D., Yu, Y., Guo, J.: Mining advisor-advisee relationships from research publication networks. In: KDD 2010, pp. 203–212 (2010)
11. Wu, Z., Palmer, M.: Verb semantics and lexical selection. In: Proceedings of the 32nd Annual Meeting of the Associations for Computational Linguistics, pp. 133–138 (1994)
12. Winkler, W.E.: String Comparator Metrics and Enhanced Decision Rules in the Fellegi-Sunter Model of Record Linkage. In: Proceedings of the Section on Survey Research Methods (American Statistical Association), pp. 354–359 (1990)
13. Jaccard, P.: Distribution de la flore alpine dans le bassin des Dranses et dans quelques régions voisines. Bulletin de la Société Vaudoise des Sciences Naturelles 37, 241–272 (1901)
14. http://opennlp.apache.org/
15. https://www.quantcast.com/top-sites
16. http://wordnet.princeton.edu/
17. http://www.freebase.com/
18. http://thinkaurelius.github.io/titan/
19. http://sigmajs.org/

Clustering Students Based on Student's Performance - A Partial Least Squares Path Modeling (PLS-PM) Study

Yogalakshmi Jayabal and Chandrashekar Ramanathan

International Institute of Information Technology, Bangalore, India
j.yogalakshmi@iiitb.org, rc@iiitb.ac.in

Abstract. Monitoring and evaluation of student's performance is an important task in learning process. The analysis of relation between student performance and other variables in education setting often can be useful in identifying influential factors on performance. This in turn may help to categorize students and provide more focus towards improving their performance. In this paper, the student performance is analyzed based on its relationship with medium of study. More precisely, among 10th grade students (Karnataka SSLC Board) the relationship between performance in their main subjects (Mathematics, Science and Social Science) and their medium of study is analyzed. The analysis is based on the research hypothesis: "The performance in the language of medium of study influences the performance in main subjects". This paper presents a theoretical model that shows how the data related to language of medium of study and the main subjects are modeled and evaluated. The results from the experiments indicate a strong positive plausible causal relationship between medium of study and performance. Clusters of students are identified based on the degree of influence. The contributions in this paper are two fold. 1) Novel clustering algorithm in PLS-PM and 2) Identification of plausible causal indicators for students performance and analysis of the same.

Keywords: students performance evaluation, partial least squares path modeling(PLS-PM), segmentation in PLS-PM.

1 Introduction

Educational data mining is an emerging discipline to study the variety of data coming from educational environments. Different types of studies are being carried out that include analyzing logs of students from educational software, computer adaptive testing, student retention, predicting student performance etc. In all these, identification of the appropriate student model is the key task. In the case of predicting student performance, the different factors related to the performance are identified. These factors in general can be said to relate to performance, but cannot be qualified as causal factors for performance. This is mainly due to the fact that correlation doesn't imply causation.

P. Perner (Ed.): MLDM 2014, LNAI 8556, pp. 393–407, 2014.

The standard approach to answer these inference related questions is to conduct a controlled experiment in which the subjects (here students) are allocated at random and all the subjects are perfectly complaint and all the relevant data are collected and measured without error. With all these, we can then discount "chance" alone as an explanation and any observed effects can be interpreted as causal. In real world, however, such experiments would not be possible practically or economically feasible [1]. In these situations, causal inference must be based instead on observational data. As we move away from ideal experimental setting, more number of aspects of the joint distribution of the variables has to be modeled.

Structural Equation modeling (SEM) is a statistical modeling technique for estimating the plausibility of causal relationships, using a combination of statistical data and qualitative causal assumptions. In SEM, the latent variables (i.e. variables that are not directly measured) are constructed and estimated from several measured variables, each of which taps into one of the latent variables. Among the underlying assumptions of SEM, the important one is the existence of multivariate normality in the data. Often in real world, this property does not always hold true. Alternative to SEM is the soft modeling technique developed by Wold [2]. Soft modeling (PLS-PM) technique refers to the overall situation for which PLS Path Modeling was conceived: working with non-experimental data, working with data matrices having large number of variables or for having causal-predictive modeling. PLS Path modeling technique fits the goal of identifying a causal model from the non-experimental i.e. in here 10th grade data.

Analyzing student performances using PLS-path modeling algorithm, helps to understand if there are plausible causal indicators associated with performance. Additionally, since this algorithm considers correlations that could exist between different variables that are analyzed, this helps in analyzing performance while taking into consideration correlations that are existing with performance and other analyzed variables. Also in traditional clustering, members are mostly grouped based on connectivity, centroid, distributions, density etc. These approaches assume independence among variables. Even if there are correlations existing among variables, this correlation information is not properly made use during clustering. This is due to the reason that the presence of correlated variables leads to collinearity. But PLS-path modeling algorithm in the presence of reflective variables (as explained in section 3) is resistant to collinearity problem [9]. Usually, SEMs/PLS-PM include a number of statistical methodologies that makes it possible to estimate plausible causal relationships as they are based on assumption that variables are correlated. Hence, this information from PLS-PM should be properly used to find maximally coherent clusters. Along these lines, a novel clustering technique "Partitional Segmentation algorithm" is proposed in this paper.

2   Related Work

Predicting/Clustering student's performance is a widely researched area. The search for predicting student's performance in Google results in a number of systems reported, each of which identifies a predictive student model. Most of these systems employ the wide class of data mining algorithms like regression analysis, clustering techniques, neural networks, item response theory based techniques in e-learning platforms etc. For instance, Chen [5] uses regression analysis to find out indicators for performance. Moucary [6] applies neural network approach to predict the student's performance and also uses clustering to cluster the predicted performances to identify the existence of groups, if any. Li [11] finds clusters of students using knowledge components(KDD 2010 student data set). She applies Expectation Maximization(EM) along with k-means to find out clusters. Clustering techniques like EM assume existence of statistical distributions like multivariate normal distribution. Often in real world, assuming multivariate normality does not always hold true.

The presence of different kinds of data in education setting have allowed the following paper authors to apply PLS-path modeling approach. Sellin [3] provides a comprehensive explanation for the suitability of PLS-Path Modeling technique in education related data. Balzano [4] uses PLS-path modeling for evaluating and finding out the factors that greatly influence satisfaction of students. They use feedbacks collected from students to evaluate satisfaction of students. So far to our knowledge, the application of PLS-path modeling approach to model the student performance is not available in the literature. This gap is addressed and is the first contribution in this paper.

Secondly, there are different algorithms available to detect groups in PLS-path modeling algorithm. i.e. using the correlation information. Trujillo in [8] has explained the different group detection algorithms available with their advantages and disadvantages. One among them is REBUS-PLS [10]. Here clustering approach is adapted to find latent groups in data set. These group detection algorithms discussed in [8] are mainly designed for evaluating survey/feedback like data i.e. which have few hundreds/thousands of rows. But in educational data sets, the number of rows are of order of few lakhs. It can be shown that these algorithms do not scale up well. To address this gap, a novel clustering algorithm is proposed. In this paper, to cluster student performances, correlation among variables(under study) is utilized using PLS-path modeling approach along with the novel clustering technique proposed in section 3.1

3   PLS Path Modeling Approach

The PLS-path model is comprised of: structural/inner model and measurement/outer model. Inner model defines relationship between latent variables(lv). Outer model defines relationship between manifest (observed) variables and lv's. The outer and inner models consists of two types of variables namely exogenous and endogenous. The distinction between these two types of variables is whether the variable regresses on another variable or not. In a directed graph of the model, an exogenous variable is recognized as any variable from which arrows only emanate, where the emanating arrows denote which variables that exogenous variable predicts. Any variable that regresses on another variable is defined to be an endogenous variable. In a directed graph, an endogenous variable is recognized as any variable receiving an arrow. Let's assume there is a data set D containing s variables observed in R observations. The s variables can be grouped into M blocks. Every D_m block has 1...n associated manifest variables and associated latent variable(L_m) for that block. The communality is defined as the sum of squared correlation between latent variable and its associated manifest variables. The communality index measures quality of outer model. The redundancy is defined as product of communality and coefficient of determination (R^2) for all endogenous blocks. The redundancy index measures quality of inner model for all endogenous blocks. The PLS-PM approach performs three steps: 1) Estimate latent variables estimates 2) Estimate path co-efficients 3) Calculate the loadings. The loadings are nothing but the correlation between a manifest variable(mv) and its associated latent variable(lv). The important step is computation of lv estimates. The lv estimates are obtained as a linear combination of its own mv's as follows:

$$L_m \, \alpha \, \Sigma_n w_{nm} x_{nm}, \tag{1}$$

where x_{nm} is mv associated to L_mth latent variable and w_{nm} is weight associated to each mv to obtain lv. At start of algorithm, w_{nm} are initialized with arbitrary values and outer estimates for lv are computed using 1. Then for a given outer estimate of each lv obtained in previous step, inner estimate Z_m of each lv is obtained as:

$$Z_m \, \alpha \, \Sigma_{m'} e_{mm'} L_{m'}, \tag{2}$$

where $L_{m'}$ is a lv connected to mth latent variable and $e_{mm'}$ is an inner weight usually obtained using centroid scheme. The symbol α implies lv estimates have to be standardized both in outer and inner estimates. Next step is to update the outer weights w_{nm}. If each mv is considered to be reflecting real world of an underlying concept, that is the lv, then outer weight w_{nm} is the regression coefficient of simple linear regression of each mv on inner estimate of corresponding lv. That is, since inner estimates of lv, being standardized, each outer weight is the covariance between each mv and corresponding lv as follows:

$$w_{nm} = Cov(D_{nm}, z_m). \tag{3}$$

After updating outer weights, they are used to obtain a new outer estimate of lv. These steps are repeated until convergence between inner and outer estimates is reached. The final estimate of lv is then computed using 1. Then the structural relations among endogenous lv's and exogenous lv's are estimated using standard multiple/simple linear regression models. The inner model can be written as

$$L_j = \Sigma_{m \leftrightarrow j} \beta_{mj} L_m + \epsilon_j, \tag{4}$$

where L_m is generic exogenous lv impacting L_j, β_{mj} is ordinary least squares regression co-efficient (path co-efficient) linking mth exogenous lv to jth endogenous lv. Due to page constraint, detailed explanation is not provided. Kindly refer [9] for detailed explanation of PLS-path modeling algorithm.

3.1 Clusters Identification

Cross validation is a process that involves partitioning the data set into complementary subsets, performing analysis on one subset and validating the analysis on other subset(s). To reduce variability, multiple rounds of cross validation are performed using different partitions and validation results are averaged over the rounds. This process helps to limit over fitting of given data in the model being estimated. The global model is estimated using cross validation approach in PLS-path modeling algorithm. The cross validated residuals are computed by finding difference between the observed and the estimated values of variables, at global level, in both outer and inner model. Clusters are formed based on algorithm-PSeg as shown in figure 1. The goal of PSeg is to detect clusters using both inner and outer model. Sharma [7] discusses at length, different model selection criteria available in PLS-PM approach. He shows criteria like Stone-Geisser's Q^2, Akaike Information Criteria(AIC) etc are better compared to using R^2 which is usual approach in PLS-PM approach for model selection. Along these lines, PSeg algorithm makes use of formula based on Stone-Geisser's Q^2 for fitting model in identified clusters. The different steps in PSeg algorithm are as follows:

1. A cluster dendogram is drawn from the cross validated(cv) residuals computed in global model.
2. Initial clusters are formed from initial partition from the above step.

Require: plsmobject
Ensure: Different groups with corresponding path.coef, loadings, GoF indices
  1: resclustbg(plsmobject,...) {Computes cv-residuals,draws dendrogram}
  2: nk =number of groups {number of init grps to be formed from dendrogram}
  3: Gp =Get init gps corresponding to nk
  4: **loop**
  5:     calculate $cvcm(H^2)$ =cv-communality for every group
  6:     calculate $cvr(F^2)$ =cv-redundancy for every group
  7:     **if** any $cvcm, cvr < 0$ **then**
  8:         compute DM_i for every unit to every group
  9:         Find $min(DM_i)$ for a given unit, assign it to that group
 10:     **else**
 11:         break
 12:     **end if**
 13: **end loop**
 14: compute res =path.coef,loadings,GoF indices for every group
 15: **return** res

Fig. 1. Partitional Segmentation(PSeg)

3. The cv-communality, cv-redundancy are calculated for every cluster as in [9]. If calculated values are negative, it implies that lv's are not properly estimated.
4. The members from each cluster are re-distributed until the group's cv-communality and cv-redundancy becomes positive according to step 9 as in PSeg algorithm given in figure 1.

The re-distribution of units is based on a computed distance measure DM_i. The distance measure and various terms associated with distance measure are explained now. The distance measure DM_i is computed as follows:

$$DM_i = \sqrt{\frac{\frac{\sum_{m=1}^{M}\sum_{n=1}^{Mn}e_{inmk}^2}{H^2(D_{nm},L_{mk})}}{\frac{\sum_{r=1}^{R}\sum_{m=1}^{M}\sum_{n=1}^{Mn}e_{inmk}^2}{H^2(D_{nm},L_{mk})}} \times \frac{\frac{\sum_{j=1}^{J}f_{ijk}^2}{F^2(D_{nj},L_{jk})}}{\frac{\sum_{r=1}^{R}\sum_{j=1}^{J}f_{ijk}^2}{F^2(D_{nj},L_{jk})}}}$$ (5)

where, e_{inmk} - cv-outer model residual for ith row in kth group corresponding to nth mv of mth block, $H^2(D_{nm},L_{mk})$ - cv-communality for nth mv of mth block in kth group. This is computed as explained by Tenenhaus in [9], f_{ijk} - cv-inner model residual for ith row in kth group corresponding to jth endogenous block, $F^2(D_{nj},L_{jk})$ - cv-redundancy for nth mv of jth endogenous block in kth group. This is computed as explained by Tenenhaus in [9], m_n, j_n - number of extracted lv, since all lv's are reflective in nature, this will always be equal to 1.

The left term of the product in equation 5 can be considered to measure the predictive relevance in the outer model. The right term of the product can be considered to measure predictive relevance of the inner model. The outer model residual or cv-communality residual, of the ith unit belonging to the kth cluster is obtained as shown in equation 6

$$e_{inmk} = x_{inmk} - \hat{x}_{inmk},$$ (6)

where x_{inmk} is the observed value of the nth mv of mth block, $\hat{x}_{inmk}$ is the estimated value of x_{inmk}. Hence, for ith unit, $\hat{x}_{inmk} = \beta_{nmk(-i)}L_{mk(-i)}$, where $\beta_{nmk(-i)}$ is the cv-loading computed as given in survey paper by Tenenhaus [9], when ith value of nth mv of mth block is missing in kth cluster.L_{imk} is lv estimate of mth block of kth cluster computed, when ith value of variable x_{nm} is missing, as given in equation 7, and as detailed in [9].

$$L_{mk(-i)} = L_{imk} = \Sigma_n w_{nmk(-i)}x_{nmk(-i)},$$ (7)

where $w_{nmk(-i)}$ is the cv-outer weight associated with nth manifest variable of mth block in kth cluster, computed when ith value is missing. The cv-outer weights are obtained from PLS-path modeling algorithm with leave one out cross validation technique. The inner model residual or cv-redundancy residual of the endogenous latent variables on their exogenous variables is obtained as

$$f_{ijk} = x_{inmk} - \hat{x}_{inmk}$$ (8)

where x_{inmk} is the observed value of the nth manifest variable of mth block, $\hat{x}_{inmk}$ is the estimated value of x_{inmk}. Hence, for ith unit, $\hat{x}_{inmk} = \beta_{nmk(-i)}$ $\zeta_{nmk(-i)}L_{m^*k(-i)}$, where, $\beta_{nmk(-i)}$ is the cv-loading, $\zeta_{nmk(-i)}$ is the cv-path coefficient computed when ith value of x_{nm} of kth cluster is missing and L_{im^*k} is the endogenous lv estimate computed when ith value of endogenous x_{nm} of kth cluster is missing. The cv-communality (H^2) as given in eq:9, is computed for every cluster as :

- Sum of Squares of observations for one mv:$SSO_{nm} = \Sigma_i(x_{inm} - \bar{x}_{nm})^2$
- Sum of Squared Prediction error for one mv:$SSE_{nm} = \Sigma_i(x_{inm} - \hat{x}_{inm})^2$
- Sum of Squares observations for for block m: $SSO_m = \Sigma_n SSO_{nm}$
- Sum of Squared prediction errors for block m: $SSE_m = \Sigma_n SSE_{nm}$
- Cv-communality measure for block m:

$$H_m^2 = 1 - \frac{SSE_m}{SSO_m} \tag{9}$$

The mean of the cv-communality indices can be used to measure the global quality of the measurement/outer model if they are all positive for all blocks. The cv-redundancy (F^2) as given in eq:10, is computed for every cluster as:

- Sum of Squared Prediction error for one mv: $SSE'_{nm} = \Sigma_i(x_{inm} - \hat{x}_{inm})^2$
- Sum of Squared Prediction error for one block m: $SSE'_m = \Sigma_n SSE'_{nm}$
- Cv-redundancy measure for the endogenous block m^*:

$$F_m^2 = 1 - \frac{SSE'_{nm}}{SSO_{nm}} \tag{10}$$

The mean of the cv-reducancy indices related to all endogenous blocks can be used to measure the global quality of the structural/inner model if they are all positive. The GoodnessoffFit(GoF) index is computed as

$$GoF = \sqrt{mean(cv - communality_m) \times mean(cv - redundancy_m)} \tag{11}$$

This GoF index takes into account the quality of both outer model and the inner model. Since cross validated residuals are used to draw dendrograms, clusters stand out better and are more visible. The existing set of fit indices in PLS (R^2 and GoF based on R^2) however, have been shown to be woefully inadequate as empirical criteria for model selection. Since both R^2 and the GoF based on R^2 in their current form do not penalize over-parametrization. Along with R^2, the GoF based on R^2 measures are not suitable for model validation as highlighted by Sharma in [7]. Therefore instead of using R^2 based measures for computing distances between members in a cluster, the cv-communality, cv-redundancy and distance measure based on Stone-Geisser's Q^2 are used for forming maximally coherent groups. Unlike R-Squared which has tendency to over fit model, distance measure (DM_i) given in equation 5, used here avoids over fitting the model within every group.

4   Application of PLS-PM to Student Performance Evaluation

The evaluation of 10th grade data using the PLS-path modeling approach is presented here. After preprocessing and cleansing operations, the data set consists of around 1 million observational units. The manifest and their corresponding latent variables are given in table 1 and path diagram in figure 2. The PLS-path modeling algorithm is applied to this dataset followed by the PSeg clustering algorithm. From the results, presence of 5 different clusters can be inferred. Kruskal-Wallis test is performed on bootstrapped path-coefficients to test for the significance of difference among these clusters. The dataset contains categorical variables like School Type and whether school location is urban or rural in Urban-Rural. The PLS-PM and PSeg algorithms are once again applied to check for any moderation effect from these variables along the path of Medium of study → Performance in Main Subjects.

Table 1. Manifest and Latent Variables

Latent Variable	Manifest Variable
Perf. in Medium	L1 MARKS
	L2 MARKS
	L3 MARKS
Perf. in Main Subjects	S1 MARKS
	S2 MARKS
	S3 MARKS

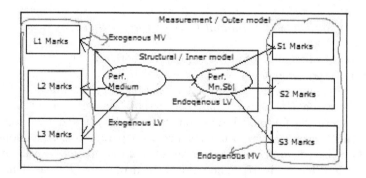

Fig. 2. PLS Path model

5 Results

All the experiments are done using the open statistical software **R**. The package **semPLS** in R is used for generating the global PLS-path model estimates, which are used as input for PSeg clustering algorithm. The results indicate plausible causal relationship of medium on performance in main subjects at global level and it is found as 0.86. Since this value is positive and greater than 0.5, it implies medium of study is heavily influencing performance. The PSeg algorithm given in figure 1 identifies five clusters. The other resultant parameters found at global level and from the groups detected are tabulated in table 2. The GoF indices are based on cv-communality and cv-redundancy, which are in turn based on Stone-Geisser's Q^2. The positive values for Q^2 indicate that model being fit has good predictive accuracy. The GoF indices in table 2, has values > 0 for all the groups, which implies the quality of the model fit is very good. Also seen is, performance in medium of study has varying degree of influence amongst groups. Thus clusters of students are found based on their performance along with finding plausible causal indicators for performance. The values for path coefficients in table 2 for the different clusters identified also show the strong influence of the medium of study on students performances. Additionally, to support this, the GoF indices for different groups are also positive and greater than 0.5, indicating the good quality of the clusters that are identified. The loadings of different manifest variables with their associated latent variable is shown in table 3. These loadings in different clusters also indicate the tight correlation among its members with the cluster.

Table 2. Path Coefficients, CV-Communality, CV-Redundancy, Quality

	Path.coef	Avg.Com-Perf Medium	Avg.Com-Perf Main-Subj	Avg.Red-Perf Main-Subj	Quality	GoF
Global model	0.8649	0.79	0.83	0.62	Rsquared-semPLS	0.75
Cluster 1	0.8406	0.5169	0.5662	0.5612		0.551
Cluster 2	0.8599	0.5338	0.6068	0.6144	sqrt(-cvcom*cvred)	0.5920
Cluster 3	0.8649	0.5424	0.6129	0.6238		0.6003
Cluster 4	0.8755	0.5449	0.6346	0.6449		0.6167
Cluster 5	0.8869	0.5409	0.6218	0.6571		0.6181

Next to test is, whether the difference in terms of path coefficients among the groups is significant. The student's t-test cannot be used here to compare the clusters as the number of clusters are more than two. When there are more than 2 clusters to compare we need to use the Kruskal-Wallis test. More information can be found about kruskal wallis in [12]. Right now, there is only one path

Table 3. Loadings

Loadings	Global model	Cluster1	Cluster2	Cluster3	Cluster4	Cluster5
L1Marks→Perf Medium	0.8819	0.8790	0.8811	0.8910	0.8806	0.8843
L2Marks→Perf Medium	0.8865	0.88455	0.8879	0.8861	0.8993	0.8805
L3Marks→Perf Medium	0.9016	0.8986	0.9013	0.8994	0.9034	0.8870
S1Marks→Perf MainSub	0.9142	0.9135	0.9150	0.92197	0.9302	0.9034
S2Marks→Perf MainSub	0.92667	0.9235	0.9268	0.93326	0.91978	0.92507
S3Marks→Perf MainSub	0.8992	0.8982	0.8991	0.9071	0.9058	0.8854

coefficient available per cluster and comparison cannot be made between single datum. Hence, bootstrapping is carried out. Bootstrapping is a technique that allows assigning measures of accuracy to sample estimates. It falls in the broader class of re-sampling methods. This is implemented by constructing a number of re-samples from observed data, each of which is obtained by random sampling with replacement from the original data. In each cluster, bootstrapping is applied with number of re-samples as 200. The bootstrapped path coefficients are then used to perform kruskal-wallis test. From the results of applying kruskal-wallis test, it is clear that **the difference in the path coefficients among the clusters is significant**. The very small p-value from the kruskal-wallis test, supports this inference which is given below the table 4, along with the first 5 bootstrapped path coefficient values from each cluster.

Table 4. Bootstrapped Path Coefficients

	Cluster 1	Cluster 2	Cluster 3	Cluster 4	Cluster 5
	0.83531	0.86343	0.86499	0.87407	0.90606
	0.83615	0.85732	0.86475	0.86418	0.89669
Bootstrapped Path coefficients	0.86045	0.86015	0.86484	0.87362	0.8879
	0.84104	0.86105	0.86512	0.87534	0.89124
	0.84046	0.85821	0.86463	0.88042	0.88704

kruskal.test(dati)
Kruskal-Wallis rank sum test
data: dati
Kruskal-Wallis chi-squared = 621.3807, df = 4, p-value $< 2.2e - 16$
As mentioned earlier, there are other attributes in the dataset like **School Type** and **Urban-Rural**, which give information on type of the school and whether the school is located in an urban region or rural region. These are categorical in nature. There are 3 and 2 categories available in school type and urban-rural

respectively. The next experiment carried out was to test the moderating effect of these two categorical variables separately on the path perf.in.medium → perf.in.main subjects. The idea is to create dummy variables to get the interaction between the moderating variable and the perf.in medium variable. The process of creating dummy variables is called Dummy Coding. If there are M categories for the moderating categorical variable then, $M - 1$ dummy variables are created along with dummy variables for interaction between the moderating variable and the perf.in medium variable. These are the additional columns added to the dataset. These columns are created according to Dummy Coding as explained in [13]. Here again bootstrapping was applied to test the modelrating effect. From the results shown in table 5 and 6, 0 lies inside most the lower and the upper intervals of the moderating variables. This shows that the path of **Perf.in medium → Perf.in main subj** is not affected by these moderator variables, for all the clusters. Thus from the above experiments, it is evident that perf.in.medium is the highest influential factor, for students perf.in main subjects. Also, the clusters found are shown to be different from each other statistically.

Table 5. Moderation effect of School Type

	Interact.Schty 1 → Perf MainSubj			Interact.Schty 2 → Perf MainSubj			Schtyp 1 → Perf MainSubj		
	Estimate	Lower	Upper	Estimate	Lower	Upper	Estimate	Lower	Upper
Clus1	0.012	−0.05	0.09	−0.03	−0.11	0.06	0.04	−0.02	0.12
Clus2	−0.01	−0.02	0.00	−0.01	−0.03	−0.00	0.02	0.007	0.04
Clus3	−0.01	−0.02	−0.01	−0.03	−0.03	−0.02	0.02	0.02	0.02
Clus4	−0.02	−0.05	0.006	−0.04	−0.08	−0.01	0.04	0.017	0.06
Clus5	−0.02	−0.09	0.04	−0.09	−0.16	−0.02	−0.003	−0.05	0.05

	Schtyp 2 → Perf.in Main Subj			Perf.in Medium → Perf.in Main Subj		
Clus1	0.02	−0.03	0.10	0.86	0.75	0.93
Clus2	0.03	0.01	0.04	0.88	0.86	0.90
Clus3	0.02	0.02	0.02	0.89	0.89	0.90
Clus4	0.01	−0.01	0.04	0.91	0.88	0.95
Clus5	−0.03	−0.09	0.02	0.94	0.85	1.00

5.1 Comparison with Other Clustering Algorithms

Cluster characteristics are studied and compared with other clustering algorithms, by taking equi-sized stratified samples from data for each medium of study(e.g. E−English, H−Hindi, M−Marathi, L−Telugu, T−Tamil, U−Urudu, K−Kannada). Tables 7 and 8 tabulates avg.performances in the medium of study

Table 6. Moderation effect of Urban-Rural

	InteractUrban-Rural1 → Perf MainSubj			Urban-Rural1 → Perf MainSubj			Perf Medium → Perf MainSubj		
	Estimate	Lower	Upper	Estimate	Lower	Upper	Estimate	Lower	Upper
Clus1	0.01	−0.07	0.10	−0.08	−0.14	−0.006	0.83	0.76	0.89
Clus2	0.04	0.02	0.05	−0.07	−0.08	−0.06	0.83	0.82	0.85
Clus3	0.03	0.03	0.03	−0.06	−0.06	−0.06	0.84	0.84	0.84
Clus4	0.01	−0.01	0.04	−0.04	−0.06	−0.01	0.86	0.84	0.89
Clus5	0.06	−0.03	0.16	−0.01	−0.07	0.04	0.83	0.75	0.91

(i.e.medium of study) and main subjects, along with students distribution found within each cluster medium-wise. From table 8 alone, it is difficult to answer research hypothesis mentioned. There is no direct answer available. The only visible information is that English medium students are found to be performing well; for other medium students their performances in medium of study and main subjects are fairly moderate. It is not possible to establish causality between variables or quantify influence of one variable on other in these algorithms. Also, no:of clusters needs to be passed as input. It is well known that, finding right no:of clusters is one of the most difficult tasks in cluster analysis. The no:of clusters passed as input is 5. This is based on cluster dendrogram displayed from **PSeg** algorithm. The figure 3 shows a clear picture of no:of clusters possible in PSeg algorithm as compared to hierarchical clustering algorithm which is directly applied on input data points.

The main advantages of using PSeg are the following:

1. The clusters stand out better in PSeg compared to algorithms like Hybrid Hierarchical clustering, Hierarchical clustering etc.
2. The PSeg algorithm scales well for large data sets. For example: Hcluster, DBScan, CLARA were unable generate clusters in R, when entire dataset was given as input which consists of around1 million rows.
3. The PSeg clustering performs well even in presence of correlated variables, as it takes input from PLS-PM which is resistant to collinearity (in reflective mode).
4. Establishment of possible causal relationship between variables(since clustering is based on PLS-PM approach). For example:
   (a) In cluster 3 from table 7, it can be said that 89% of times, good performance in main subjects is due to good performance in medium of study. As seen from this row, English medium students are dominant in this cluster. Hence, it can be said that for English medium students, good performance in main subjects is attributed to good performance in the medium of study 89% of the times.
   (b) Similarly in cluster 4, Tamil medium students are dominant. It can be observed here that, only 67% of the times performance in main subjects

can be attributed to performance in the medium of study. Hence cluster 4 depicts the characteristic of Tamil medium students and their performance in main subjects is influenced by the medium of study only 67% of the times.

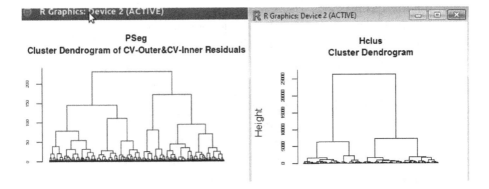

Fig. 3. Comparison PSeg-Hcluster

These kind of statements indicating causality and reasoning for the performance of students are not possible when other clustering algorithms for example: Hierarchical clustering, CLARA are used to find clusters in the given data set.

Table 7. Cluster Characteristics : Partitional Segmentation Algorithm-PSeg

	Average Performance		Students distribution							Path.coef
	Avg.Lng.Perf.	Avg.MSj.Perf.	E	H	K	L	M	T	U	
Cluster 1	34.2078	34.9522	60	85	69	126	87	79	81	0.7572629
Cluster 2	50.6397	49.3163	33	17	38	49	27	25	46	0.8427016
Cluster 3	63.4247	63.9367	119	19	28	23	24	12	12	0.8965214
Cluster 4	45.8541	49.9226	12	39	28	40	31	48	26	0.6721005
Cluster 5	45.5938	51.4004	26	90	87	12	81	86	85	0.763265

6   Conclusion

The paper has demonstrated the applicability of PLS-path modeling algorithm for clustering students based on their performance. The algorithm helps to identify the influential indicators which in turn indicates plausibility of causal relationship of indicators with performance. Also, in identifying clusters, the distance measure based on Stone-Geisser's Q^2 help in identifying the right number

Table 8. Cluster Characteristics : Other Clustering Algorithms

	HCluster Algorithm								
	Average Performance		Students distribution						
	Avg.Lng.Perf.	Avg.MSj.Perf.	E	H	K	L	M	T	U
Cluster 1	23.1048	28.3839	40	63	49	81	72	96	60
Cluster 2	54.623	55.5408	67	81	93	51	63	55	72
Cluster 3	70.5482	64.6411	48	19	20	41	19	17	19
Cluster 4	39.2886	43.3641	33	79	82	73	87	81	95
Cluster 5	83.6453	82.7446	62	8	6	4	9	1	4
Hybrid Hierarchical Algorithm - mutual clustering									
Cluster 1	21.61389	26.09762	35	51	42	65	67	74	45
Cluster 2	77.2898	73.8014	107	24	23	24	23	11	17
Cluster 3	63.7777	58.1154	14	47	76	32	49	34	54
Cluster 4	58.1926	53.1269	52	37	26	37	17	26	23
Cluster 5	38.0545	42.4889	42	91	83	92	94	105	110
Fuzzy C-means									
Cluster 1	77.1537	73.8448	107	23	24	25	23	11	19
Cluster 2	20.1346	24.4807	26	43	36	51	54	66	36
Cluster 3	44.5039	47.7524	29	66	55	44	62	52	69
Cluster 4	33.6246	38.5309	34	56	54	75	63	71	68
Cluster 5	57.5359	57.4763	55	62	81	54	48	50	58
Partioning Around Mediods - PAM									
Cluster 1	19.6429	24.1553	19	42	29	43	45	67	36
Cluster 2	76.7772	75.0866	102	22	25	22	26	9	17
Cluster 3	43.3439	46.9142	12	63	50	48	58	69	27
Cluster 4	57.9809	56.3524	68	67	79	61	51	51	60
Cluster 5	33.0332	38.4253	49	56	67	75	70	54	60
CLARA - Clustering for Large Applications									
Cluster 1	19.1083	24.5884	16	40	27	41	43	71	39
Cluster 2	57.8784	65.1418	24	29	52	4	34	9	29
Cluster 3	53.1054	50.7334	47	85	66	65	53	65	68
Cluster 4	35.1172	40.0324	57	79	88	102	102	90	99
Cluster 5	76.7822	72.5555	106	17	17	37	18	15	15

of groups with maximally coherent members in each group. The clusters found are shown to be statistically different. Thus PSeg clustering technique proposed in the paper has made use of the correlation information amongst variables efficiently which is provided by the PLS-path modeling algorithm.

Acknowledgment. The authors would like to thank Government of Karnataka, India for sharing the students data for analysis.

References

1. Centre for Statistical Methodology, London School of Hygiene and Tropical Medicine, http://csm.lshtm.ac.uk/themes/causal-inference/
2. Wold, H.: Model Building and Scientific Method: A Graphic Introduction. In: Mathematical Model Building in Economics and Industry, pp. 143–158. Charles Griffin and Co. LTD, London (1969)
3. Nobert, S.: Partial Least Squares Modeling in Research on Educational Achievement. Reflections on educational achievement (1995)
4. Simona, B., Laura, T.: Structural Equation Models and Student Evaluation of Teaching: A PLS Path Modeling Study. In: Statistical Methods for the Evaluation of University Systems, Contributions to Statistics, pp. 55–66 (2011)
5. Chen, Y.Y., Mohd Taib, S., Che Nordin, C.S.: Determinants of Student Performance in Advanced Programming Course. In: International Conference for Internet Technology and Secured Transactions, pp. 304–307 (2012)
6. Moucary, C.E., Khair, M., Zakhem, W.: Improving Student's Performance Using Data Clustering and Neural Networks in Foreign-Language Based Higher Education. The International Journal of ACM Jordan II, 27–34 (2011)
7. Sharma, P.N., Kim, K.H.: Model Selection in Information Systems Research Using Partial Least Squares Based Structural Equation Modeling. In: International Conference on Information Systems. Association for Information Systems (2012)
8. Trujillo, G.S.: PATHMOX Approach:Segmentation Trees in Partial Least Squares Path Modeling, PhD thesis (2009)
9. Tenenhaus, M., Vinzi, V.E., Chatelin, Y.M., Lauro, C.: PLS path modeling. Computational Statistics and Data Analysis 48(1), 159–205 (2005)
10. Vinzi, V.E., Trinchera, L., Squillacciotti, S., Tenenhaus, M.: RqEBUS-PLS: A response-based procedure for detecting unit segments in PLS path modelling. Applied Stochastic Models in Business and Industry 24(5) (2008)
11. Li, N., Cohen, W.W., Koedinger, K.R.: Discovering Student Models with a Clustering Algorithm Using Problem Content. In: Proceedings of the 6th International Conference on Educational Data Mining (2013)
12. Kruskal Wallis test, http://en.wikipedia.org/wiki/Kruskal-Wallis_one-way_analysis_of_variance
13. Dummy Coding, http://en.wikiversity.org/wiki/Dummy_variable_(statistics)

Learning of Natural Trading Strategies on Foreign Exchange High-Frequency Market Data Using Dynamic Bayesian Networks

Javier Sandoval[1] and Germán Hernández[2]

[1] Universidad Nacional de Colombia, Universidad Externado de Colombia, Bogotá, Colombia
[2] Universidad Nacional de Colombia, Bogotá, Colombia
{jhsandovala, gjhernandezp}@unal.edu.co
http://www.unal.edu.co

Abstract. This paper constructs a feature vector representing intraday USD/COP transaction prices and order book dynamics using zig-zag patterns. A Hierarchical Hidden Markov Model and its representation as Dynamic Bayesian Network are used to model the market sentiment dynamics choosing from uptrend or downtrend latent regimes based on observed feature vector realizations. The HHMM learned a natural switching buy/uptrend sell/downtrend trading strategy using a Training-Validation framework over one month of market data. The model was tested on the following two months showing promising performance results.

Keywords: Machine Learning, Price Prediction, Hierarchical Hidden Markov Model, Order Book Information.

1 Introduction

Learning profitable trading strategies requires the combination of expert knowledge and information extracted from data.

A novel approach to tackle the financial price prediction problem is to combine expert knowledge with Dynamic Bayesian Networks (DBN) in order to model temporal relationships of complex non-deterministic dynamic systems. This new approach has been successfully applied in other domains and can be also implemented in the financial price prediction problem. Expert knowledge can be used to design a network structure and available data helps to calibrate parameters conditional to the selected model framework.

DBNs represent graphically a stochastic process using Bayesian Networks which include directed edges pointing in the direction of time [10]. The graph is constructed to represent the a priori expert knowledge expressed in a set of junctions which facilitate conditional probability calculations. Following Bengtsson work [1], the entire model can be thought of as a compact and convenient way of representing a joint probability distribution over a finite set of variables.

One of the first special cases of DBNs implemented in the price prediction problem were Hidden Markov Models (HMMs). HMMs assumed that the underlying modeled system exists in one of a finite number of states. The latter states are hidden and are

P. Perner (Ed.): MLDM 2014, LNAI 8556, pp. 408–421, 2014.

responsible for producing a sequence of observable variables. Hassan [6] is one of the first authors who extracted HMMs from speech and image recognition problems and placed them in the stock price prediction domain. In his work [6], a 4-hidden state system was assumed to represent the performance of an airline stock price. Special emphasis was made on today's open, high, low and closing prices and the work aimed at predicting the next day's closing price. Performance was evaluated using MAPE and R^2 between predicted and realized closing values. Results were informally compared to those obtained using an ANN structure. Hassan et al. [7] extended the previous work combining GA and ANN with HMM.

As a natural extension to HMMs, Hierarchical Hidden Markov models (HHMM) were also implemented to represent financial markets and solve price prediction problems. An HHMM is a natural extension because experts identify different levels of time hierarchy when analyzing financial market information. For example, Jangmin et al. [8] presented a 5-state model that described market trends; strong bear, weak bear, random walk, weak bull and strong bull market phases. Each state had different second level abstract states which at the same time, called its own third level variables which were responsible for output emission. Calibration was based on k-means clustering and a semi-supervised training technique. Dataset included daily stock prices of the 20 most active companies of the Korean stock market from 1998 to 2003. Jangmin et al. [8] reported that the HHMM outperformed on average a simple buy-and-hold strategy and a trading strategy following TRIX, a commonly used technical analysis indicator.

As found in Tayal [12] and Wisebourt [15], HHMMs were adapted to high-frequency data. A 2-state model which captured runs and reversals was coupled with a second hidden variable layer which defined the emitted zig-zag observations. Both works implemented asyncronymous time models and recognized regime changes from uptrend to downtrend time periods. The Authors concluded that selected models' performance was considerably affected by market liquidity. Thus, the less liquid an asset was, the worse a knowledge extraction technique performed. The input variable set went from historical prices and volume to elements of the order book.

Applications of DBNs to price prediction is a recently explored study field. Next section will present an HHMM framework and its DBN representation in order to decrease complexity when executing inference. Third and forth sections discuss dataset, methods and experiment. Finally, model's performance and conclusions are provided.

2   The Hierarchical Hidden Markov Model and Its Representation as Dynamic Bayesian Network

As explained in Murphy [11], an HHMM can be represented as a Dynamic Bayesian Network. The state of the whole HHMM is encoded by the vector $Q_t = \{Q_t^1, \cdots, Q_t^D\}$ where D means the number of network levels. Changing to a DBN representation also needs to include a set of indicator variables $F_t^d, d = \{1, \cdots, D\}$ which capture whether the HHMM at level d and time t has just finished. If F_t^d is 1, d will mean the hierarchy level the process is currently at. Additionally, an HHMM has termination states for each level. When transforming to DBN, transition matrices should be expanded to include probabilities of level termination.

Our HHMM will have discrete-valued variables. Therefore, conditional probabilities can be encoded as tabular tables. First, we will have conditional probabilities for $Q_t^d, d = \{1, \cdots, D\}$ variables. Formally,

$$P(Q_t^D = j | Q_{t-1}^D = i, F_{t-1}^D = f, Q_t^{1:D-1} = k) = \begin{cases} \widetilde{A}_k^D(i,j) & f = 0 \\ \pi_k^D(j) & f = 1 \end{cases},$$

$$P(Q_t^d = j | Q_{t-1}^d = i, F_{t-1}^d = f, F_{t-1}^{d+1} = b, Q_t^{1:d-1} = k) = \tag{1}$$

$$\begin{cases} \delta(i,j) & b = 0 \\ \widetilde{A}_k^d(i,j) & b = 1, f = 0 \\ \pi_k^d(j) & b = 1, f = 1, \end{cases}$$

where $\widetilde{A}_k^d$ and $\widetilde{A}_k^D$ represent transition probabilities if the process stays in the same level. π_k^d and π_k^D are initial distributions for level d given that parent variables are in state k. It is important to remember that $\widetilde{A}^d(i,j)(1 - \tau_k^d(i)) = A_k^d(i,j)$ where A represents the automaton transition matrix, $\widetilde{A}$ is the DBN transition matrix and $\tau_k^d(i) := A_k^d(i, end)$, the probability of terminating from state i.

Indicator variables F^d are turned on if Q^d enters final state. F^d conditional probabilities are:

$$P(F_t^D = 1 | Q_t^{1:D-1} = k, Q_t^D = i) = A_k^D(i, end),$$

$$P(F_t^d = 1 | Q_t^{1:d-1} = k, Q_t^d = i, F_t^{d+1} = b) = \begin{cases} 0 & b = 0 \\ A_k^D(i, end) & b = 1, \end{cases} \tag{2}$$

where F_t^d or $F_t^D = 1$ means that the process has signaled an end state in current level and control will be returned to previous level.

The CPD for the first network slice will be $P(Q_1^1 = j) = \pi^1(j)$ for the top level and $P(Q_1^d = j | Q_1^{1:d-1} = k) = \pi_k^d(j), d = 2, \cdots, D$, elsewhere.

Finally, we will have conditional probabilities for the observed node realizations represented as $P(O_t^m | \mathbf{Q}_t) = \alpha_m, 1 \leq m \leq M$ where M means number of elements in the observed variable's alphabet.

Specifically, our DBN has $D=2$ and 2 states in the first level and 4 states in the second level.

$$Q^1(1) = \{\text{Market State 1}\},$$
$$Q^1(2) = \{\text{Market State 2}\},$$
$$Q^2(1), Q^2(4) = \{\text{Negative Feature Producer of Market State 1 and 2}\}, \tag{3}$$
$$Q^2(2), Q^2(3) = \{\text{Positive Feature Producer of Market State 1 and 2}\}.$$

Q^1 has two possible market states that have not been previously marked as uptrend or downtrend. Q^2 will represent feature production related to local positive and negative market conditions. It will be guaranteed that F_T^d be clamped to ensure that all models have finished at the end of the data sequence. Additionally, $F_t^1, 1 \leq t < T$ will be always equal to 0 so the model stays in level 1 and 2 for the whole data sequence.

Figure 1 shows HHMM structure as a 2-level automaton and provides a graphical representation of the selected HHMM expressed as a DBN structure. DBN's observed information will be explained in next subsection.

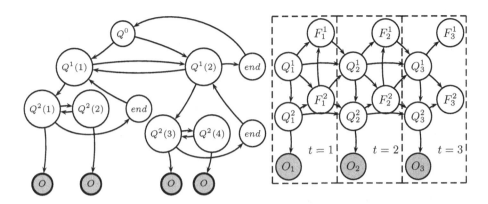

Fig. 1. Left: Proposed 2-level automaton. Right: Proposed HHMM represented as DBN. Q_n^1 is the market regime and Q_n^2 is the feature producer. Elements of the observed variables are explained in section 2.1.

2.1 Observed Feature Vector

We constructed a feature vector that captured information from two valuable sources; transaction and order book dynamics. Transactions are defined as realized market trades. The order book can be defined as the group of orders that has not been executed yet but shows agents' intentions to trade at certain quantities and prices. This information was use to forecast future trend in financial prices because exact future prices are not necessarily needed to create a profitable trading strategy. In the specific case of order book dynamics, several studies had shown that order books have relevant information to improve financial price direction prediction [2,3,4,5,13,14]. Therefore, the feature vector will combine elements from the order book and transaction dynamics.

In order to understand feature vector characterization, we made several definitions. Let be $\{P_y\}, y = \{1, 2, \ldots, Y\}$ a series of market transactions not necessarily homogeneously distributed in time. Transaction durations have not been taken into account and have been left for future research. Let also define $\{E_m, I_m\}, m = \{1, 2, \ldots, M\}$ as a sequence of local extrema from the series P_y at transaction index $I_m = y$, i.e. prices at which the transaction series changes direction. The index set I_m is a strictly increasing series recording positions of extrema transaction prices. The zig-zag process, $\{Z_x\}$, was constructed recording differences between n adjacent local extrema as $Z_x = E_m - E_{m-n}, m = \{2n, 3n, \ldots, M\}$ and $x = \{1, 2, \ldots, X\}$. n controls the number of local extrema price differences accumulated. If $Z_x \geq 0$, the zig-zag is leading upward or is simply called positive and if $Z_x < 0$, the zig-zag is leading downward or is simply called negative. If $n = 1$, we do not accumulate zig-zags. This cumulative zig-zag is useful to reduce model's prediction instability. Figure 2 provides a graphical explanation of the above definitions.

The zig-zag and the extrema price series are complemented with information extracted from the order book dynamics. First, based on limit orders located in the book, we constructed a volume-weighted average limit order book price series as in

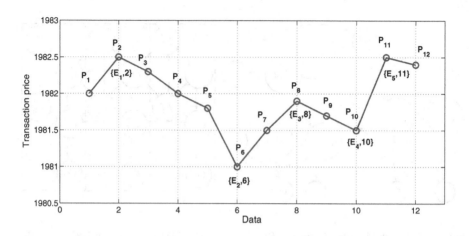

Fig. 2. Transaction price series and the corresponding extrema price and zig-zag processes. Assuming $n = 1$ we can calculate first value of Z process as $Z_1 = E_2 - E_1$.

Kearns [9], $VWAP_y^{Bid}, VWAP_y^{Offer}$. These two series corresponded in time with transaction prices. Trying to capture the most-active part of the order book, the first 10 orders in the bid and offer books were included. $VWAP_y^{Bid}$ and $VWAP_y^{Offer}$ were used to capture books' depth changes between zig-zag realizations. Thus, let define,

$$VWAP_x^{Bid}Spread = \frac{1}{I_{m-1} - I_m} \sum_{y=I_{m-1}+1}^{I_m} [P_y^{Best\ Bid} - VWAP_y^{Bid}],$$

$$VWAP_x^{Offer}Spread = \frac{1}{I_{m-1} - I_m} \sum_{y=I_{m-1}+1}^{I_m} [VWAP_y^{Offer} - P_y^{Best\ Offer}], \quad (4)$$

$$VWAP_xSpread = \frac{1}{I_{m-1} - I_m} \sum_{y=I_{m-1}+1}^{I_m} [VWAP_y^{Bid} - VWAP_y^{Offer}],$$

where $m = x \times n$ with $x = \{2, 3, \ldots, X\}$, I_m is the position of the mth extrema price, $P_y^{Best\ Bid}$ and $P_y^{Best\ Offer}$ are best book order prices when P_y transaction price is observed. This definition differs from the one found in Wisebourt [15] which uses extrema prices instead of best book order prices to calculate $VWAP_x^{Bid}Spread$ and $VWAP_x^{Offer}Spread$.

Next, we defined imbalances in the order book from calculating differences in VWAP Spreads. $\varphi_x = VWAP_x^{Bid}Spread - VWAP_x^{Offer}Spread$. $\varphi_x > 0$ signaled deeper liquidity in the order book offer side. $\varphi_x < 0$ signaled the opposite case. Because our main interest is to capture order book dynamics, we calculated a new variable that took into account local order book dynamics as follows:

$$\tilde{\theta}_x^j = \begin{cases} 1 & \theta_x^j - 1 > \alpha \\ -1 & 1 - \theta_x^j > \alpha \\ 0 & |\theta_x^j - 1| > \alpha, \end{cases} \quad (5)$$

where each $\theta_x^j, j = 1, 2, 3$ will be:

$$\theta_x^1 = \left| \frac{\frac{\varphi_x}{VWAP_x Spread}}{\frac{\varphi_{x-1}}{VWAP_{x-1} Spread}} \right|, \quad \theta_x^2 = \left| \frac{\frac{\varphi_{x-1}}{VWAP_{x-1} Spread}}{\frac{\varphi_{x-2}}{VWAP_{x-2} Spread}} \right|, \quad \theta_x^3 = \left| \frac{\frac{\varphi_x}{VWAP_x Spread}}{\frac{\varphi_{x-2}}{VWAP_{x-2} Spread}} \right|. \quad (6)$$

Using previous definitions we created a discrete feature vector containing three elements that described zig-zag pattern types, transaction and order book dynamics. In particular,[1]

$$O_x = (f_x^1, f_x^2, f_x^3) \quad \text{where,}$$

$$f_x^1 = \begin{cases} 1 & Z_x \geq 0 \text{ local maximum} \\ -1 & Z_x < 0 \text{ local minimum} \end{cases}$$

$$f_x^2 = \begin{cases} 1 & E_{x-4} < E_{x-2} < E_x \text{ and } E_{x-3} < E_{x-1} \text{ local Up-trend} \\ -1 & E_{x-4} > E_{x-2} > E_x \text{ and } E_{x-3} > E_{x-1} \text{ local Down-trend} \\ 0 & otherwise, \end{cases} \quad (7)$$

$$f_x^3 = \begin{cases} 1 & \begin{array}{c} \widetilde{\theta}_x^1 = 1, \widetilde{\theta}_x^2 > -1, \widetilde{\theta}_x^3 < 1, sign(\varphi_x) = -1 \\ \text{and } (sign(\varphi_x) = sign(\varphi_{x-1}) \text{ or } sign(\varphi_x) = sign(\varphi_{x-2})) \end{array} \\ -1 & \begin{array}{c} \widetilde{\theta}_x^1 = -1, \widetilde{\theta}_x^2 < 1, \widetilde{\theta}_x^3 > -1, sign(\varphi_x) = 1 \\ \text{and } (sign(\varphi_x) = sign(\varphi_{x-1}) \text{ or } sign(\varphi_x) = sign(\varphi_{x-2})) \end{array} \\ 0 & otherwise. \end{cases}$$

Using similar notation as in Wisebourt [15], table 1 shows feature vector components. First element in feature vector is the zig-zag type, i.e. maximum or minimum. Second element captures transaction price momentum comparing current maximum or minimum with its recent historical values. Finally, third element captures increasing or decreasing liquidity in the order book. For example, $(1, 1, 1)$ means a local maximum, with a local uptrend and increasing liquidity in the bid order book side.

Table 1. Feature vector observations classified as positive (left) or negative (right)

Symbol	O_x (observation)	Symbol	O_x (observation)
U_1	$(1, 1, 1)$	D_1	$(-1, 1, 1)$
U_2	$(1, -1, 1)$	D_2	$(-1, -1, 1)$
U_3	$(1, -1, 0)$	D_3	$(-1, -1, 0)$
U_4	$(1, 0, 1)$	D_4	$(-1, 0, 1)$
U_5	$(1, 0, 0)$	D_5	$(-1, 0, 0)$
U_6	$(1, 0, -1)$	D_6	$(-1, 0, -1)$
U_7	$(1, -1, 0)$	D_7	$(-1, -1, 0)$
U_8	$(1, 1, -1)$	D_8	$(-1, 1, -1)$
U_9	$(1, -1, -1)$	D_9	$(-1, -1, -1)$

$D_{1:9}$ are observations exclusively produced by $Q^2(1)$ and $Q^2(4)$. $U_{1:9}$ are observations exclusively produced by $Q^2(2)$ and $Q^2(3)$, see figure 1. Thus, the HHMM structure simulates a two-level market in which, first, it enters a market regime and then

---

[1] If φ_x, φ_{x-1} or $\varphi_{x-2} = 0$, it signals equally deep bid-offer order books, so, third feature will be equaled to 0. This will avoid division by zero in theta definition.

within each regime, positive and negative features are produced. It is guaranteed that a positive feature is always followed by a negative featured and viceversa. This structure summarizes expert trader's knowledge of how financial prices evolve.

Using the previous model, we captured foreign exchange dynamics in order to predict short-term future behavior. Next section will present the dataset and its characterization.

3   DataSet

Dataset consisted of three months of tick-by-tick information from the limit order book and transactions of the USD/COP, the Colombian spot exchange rate starting in March, 2012. Data has been extracted from the Set-FX market, the local interbank FX exchange market. Dataset covered 43, 431 transactions. Transactions with similar time stamp have been aggregated into one observation because it is highly probable that they were executed by the same agent. Additionally, data included 658, 059 order book updates of the first 10 orders in each side of the book. Due to liquidity issues, the first and last 10 minutes of the available data were not considered. USD/COP spot interbank exchange market opens at 8 am and closes at 1 pm. It is a semi-blind market, participants only know their counterparts after transitions are executed. USD/COP average daily turnover is 1 billion dollars. Figure 3 shows transaction series during the revised dataset window.

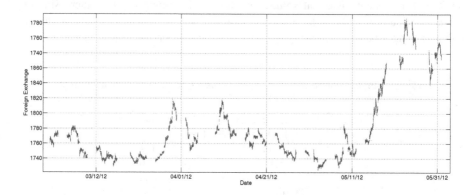

Fig. 3. High-Frequency Colombian foreign exchange rate (USD/COP) from February 2, 2012 to May 31, 2012. Series discontinuities represent weekends and holidays.

4   Methods and Experiment

Different from other studies, see for example Hassan [7], the main objective of this work was to predict market states instead of raw prices or price levels. Therefore, order book and transaction information series were transformed into a feature vector realization series. Though, market states were learned following an unsupervised framework, we expected that during the training phase, the DBN differentiated between two market regimes that were latter marked as uptrend or downtrend.

First, transaction price and order book information were converted into an observed feature series following section 2.1. We used $\alpha = 0.4$ in equation 7 to guarantee that every possible feature vector realization was observed during the training set. We also set $n = 5$ in order to accumulate 3 consecutive negative and positive zig-zags. This cumulative zig-zag covered on average, 10 transactions per observed feature's realization, see table 2.

Table 2. Summary of number of feature vector observations and transactions per feature observation. As expected, when the value of n increases, the number of feature observations and transactions per observation decreases.

Number of Feature Vector Observations/Transactions per Observation			
	Training Set	Validation Set	Testing Set
$n = 1$	$3,244 - 2.9$	$1,571 - 2.99$	$7,812 - 3.06$
$n = 3$	$1,552 - 6.06$	$759 - 6.02$	$3,745 - 6.39$
$n = 5$	$1,020 - 9.22$	$495 - 9.5$	$2,494 - 9.5$
$n = 7$	$771 - 12.20$	$370 - 12.71$	$1883 - 12.71$
$n = 9$	$614 - 15.32$	$298 - 15.79$	$1507 - 15 - 88$

Afterwards, we divided the feature vector observation series in three parts. The first part was used to train the proposed DBN. The second part was the validation set to test the model's generalization ability. Finally, The third part of the data was left to test the model using information that was not provided during the calibration phase. Unfortunately, because data has time structure, we could not shuffle it to repeat the calibration and generalization testing procedure.

Training subset were the first 14 days. Validation set were the next 8 days. 38 days were left for testing the model. First, we used an Expectation Maximization (EM) Algorithm to find the parameters that maximized model's likelihood function over the training set.

When model's parameters were obtained, we moved to implement a Forward-Looking Viterbi inference over the training set. The most probable states of $Q^1_{1:X}$ variables based on observed evidence. Forward-Looking Viterbi inference over observations $1 : X$ is defined as:

$$arg\ max_{Q^1_{1:X}} P(Q^1_{1:X}|O_{1:X}), \qquad (8)$$

where $Q^1_{1:X}$ will be the state realizations of Q^1 variable during the Training data.

Learned states of $Q^1_{1:X}$ variables were marked as uptrend or downtrend based on the average returns obtained in similar states. Formally, we calculated:

$$\bar{R}_{Q^1(j)} = \frac{1}{N_{Q^1(j)}} \sum_{n=1}^{N_{Q^1(j)}} \log P^F_n(Q^1(j)) - \log P^I_n(Q^1(j)), \quad j = 1, 2, \qquad (9)$$

where $P_n^F(Q^1(j))$ and $P_n^I(Q^1(j))$ mean mid-quote best bid-offer at the end (F) and beginning (I) of each consecutive Q^1 realizations in state j. $N_{Q^1(j)}$ is the total number of consecutive Q^1 realizations in state j over the training data set. if $\bar{R}_{Q^1(1)} > \bar{R}_{Q_1(2)}$, state 1 is called uptrend, state 2 is called downtrend and viceversa.

After states $Q^1(j), j = 1, 2$ were marked as uptrend or downtrend in the training set, we solved again the Forward-Looking Viterbi inference problem in the validation set. Then, we recalculated equation 9 over the validation set and simulated a trading strategy that had long positions during an uptrend market and short positions during a downtrend market. Return/risk performance based on return average over standard deviation was also calculated.

The Parameter estimation over the training set, the Forward-Looking Viterbi inference on the training and validation set and the simulated trading strategy in the validation set were repeated several times. We chose the model with the best trading performance in the validation set.

Finally, the selected model performance was measured on the testing set. However, Forward-Looking Viterbi Inference drives to a non-predictable trading strategy because information indexed $x + 1 : X$ will not be available at the moment we decide to trade. Therefore, when testing model's prediction ability, we implemented Viterbi inference without Forward-Looking. Formally, Viterbi inference is defined as,

$$\underset{Q_x^1}{arg\ max}\ P(Q_x^1|O_{1:x}),\quad x = \{1, \cdots, X\}. \tag{10}$$

Afterwards, equation 9 was recalculated over the testing set. Based on Q^1 state's classification obtained from training set, a buy/uptrend and sell/downtrend strategy was evaluated and its trading performance was reported.

Next section will present unconditional and Q^1-state conditional observed feature probability distributions over training and validation sets. We will emphasize the model's ability to differentiate between uptrend and downtrend regimes and the possibility of passing from regime identification to a profitable trading strategy on the testing set. We will also show return distributions on each market regime.

5   Model Performance and Results

We run 50 times the EM algorithm to maximize model's likelihood given observed data over the training set. From this family of optimized models, we selected the one that produced the best trading strategy during the validation set in terms of mean/standard deviation of returns as explained in section 4. Figure 4 shows 10 randomly selected cumulative return series calculated over training and validation sets included the one produced by the selected model. As shown, some runs produced profitable trading strategies during training but they did not show good generalization abilities on the validation set when measured in terms of return/risk performance. The selected model (dotted line) showed the best trading performance on both training and validation sets. Future work will focused on implementing a regularization function directly connected to the maximization steps of the EM algorithm. table 3 summarizes how Q^1 classes were classified.

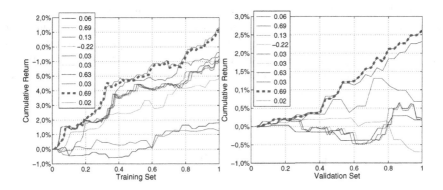

Fig. 4. Observed cumulative returns over training and validation set for 10 randomly selected calibrated models included the chosen one. Class 2 returns were treated with opposite sign as in a short position following findings summarized in table 3. Legends show return/risk performance measure obtained over the validation set.

Unconditional and conditional Q^1 probability distributions of observed features for selected model are shown in figure 5. The first relevant observation is that the most common categories are $U5$ and $D5$. This is an expected result because they capture what we can called noisy realizations, transactions that do not have support on price or order book dynamics. Because $Q^1(1)$ class was marked as an uptrend regime, we can evaluate which features are more common in each regime. Specifically, there is evidence to mark $D2$, $D6$, $D7$, $D9$, $U2$, $U4$, $U7$ and $U9$ as downtrend features and $U1$, $U3$, $U6$, $U8$, $D1$, $D3$, $D4$ and $D8$ as uptrend features. There are some interesting cases. For example, $U2$ is a local maximum with negative momentum in transaction price and increasing liquidity in the bid order book side. However, it favors a downtrend regime. We left for further research if possibly contradicting findings in feature vector realizations are evidence of dominant participants' actions or they are not significant.

The selected model was run on the testing dataset. First, we implemented Forward-Looking Viterbi inference to define a benchmark of maximum performance. Figure 6, upper graph, shows results. Returns obtained from the $Q^1(2)$, the downtrend regime,

Table 3. Summary of return statistics found over training data for selected model. Viterbi Inference implemented equation 8.

	Viterbi Inference	
	Class 1 (Uptrend)	Class 2 (Downtrend)
Observations	37	37
Mean	0.046%	−0.065%
Min	−0.22%	−0.42%
Max	0.38%	0.053%
Std	0.1%	0.098%
Skewness	0.75	−1.62
Kurtosis	5.50	6.41
Type	uptrend	downtrend

class are clearly skewed to the negative side. Returns obtained from the $Q^1(1)$ class, the uptrend regime, are skewed to the positive side. Using Forward- Looking Viterbi inference, the selected model had the ability to find two distinct regimes in the out-sample data. When Viterbi inference was executed and no future information was used, the selected model still preserved the ability to differentiate among two regimes. However, separation seemed less cleared.

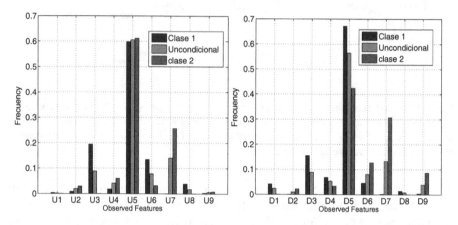

Fig. 5. Observed conditional and unconditional feature probability distributions found over the training set for the selected model. Left: Positive features. Right: Negative features.

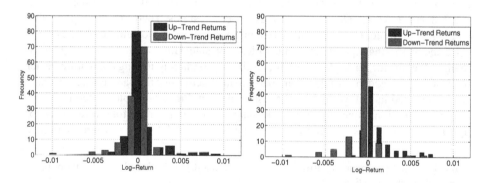

Fig. 6. Out-Sample returns. Left: Equation 8 was used to infer Q^1 states on the whole testing set. Right: Equation 10 was used to infer Q^1 states on the whole testing set. Q^1 states were interpreted as found during training phase.

In order to test if model's regime separation power is significant, we implemented a 2-tailed t-test. The null hypothesis was if uptrend and downtrend return means were equal or if there was evidence of a significant difference. Test results were summarized on table 4. As expected, the return mean difference was stronger when using future information. However, mean equality was still rejected when no future information was

Table 4. Hypothesis testing of return mean comparison conditional to Q^1 state realization

$H_0 : \mu_{Up-T} - \mu_{Down-T} = 0, H_1 : \mu_{Up-T} - \mu_{Down-T} \neq 0.$		
2-tailed t-test $\alpha = 1\%$		
	Forward-Looking Viterbi	Viterbi
p-value	$0,00\%$	$0,50\%$
Action	Reject	Reject
t-statistic	$6,521$	$2,8297$
confident Interval	$[0.103\%; 0.240\%]$	$[0.005\%; 0.120\%]$

Table 5. Summary of return statistics found over out-sample data. Viterbi Inference implemented equation 10. Forward Looking Viterbi Inference implemented equation 8.

	Viterbi Inference		Forward-Looking Viterbi Inference	
	uptrend	downtrend	uptrend	downtrend
Observations	129	128	104	103
Mean	0.04%	-0.020%	0.10%	-0.07%
Min	-0.35%	-1.07%	-0.22%	-1.02%
Max	0.97%	0.45%	0.78%	0.74%
Std	0.19%	0.16%	0.19%	0.19%
Skewness	2.21	-2.70	1.82	-0.95
Kurtosis	9.39	17.73	6.38	11.23

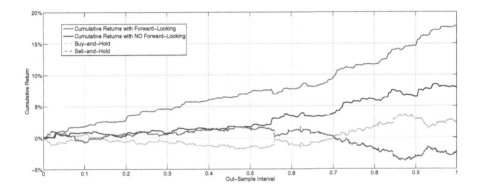

Fig. 7. Cumulative returns over Out-Sample data using Viterbi and Forward-Looking Viterbi. Buy-and-hold and sell-and-hold were provided for comparative reasons. Series are listed based on return performance.

used to classify market regimes. There was evidence that transaction price and order book dynamics helped to predict current market regimes using only available information. Table 5 presents returns statistics calculated over the out-sample dataset.

Figure 7 presents trading results when classification power become a buy/uptrend and sell/downtrend trading strategy. We can observe that future information included in Forward-Looking Viterbi inference gave the best profit results. However, as it was stated before, this is just a non-predictable trading benchmark. When future information was not used, the buy/uptrend and sell/downtrend trading strategy still outperformed passive buy-and-hold and short-and-hold market strategies. We can see that model's performance mimicked the short-and-hold trading result during the first part and the buy-and-hold trading result on the second part of data. This is not surprising because as shown in figure 3, from April 15, 2012 to May 11, 2012, there were a macro downtrend regime followed by a macro uptrend phase until the end of the testing data. The model was able to classify two high-frequency market regimes and this classification became a profitable trading strategy. This is an interesting result for market agents that do not have to worry about market spreads and transaction costs. For example, market makers who provide liquidity. They have the bid-offer spread on their side, and receive a special fee treatment.

6   Conclusions

This work proposed a 2-level HHMM that was converted to a DBN. The proposed model assumed that a specific financial market could be viewed as a complex automaton which entered in two main regimes. Then, each regime cycled through two feature producers, throwing negative and positive observations calculated from transaction and order book data. The former model was tested over the USD/COP foreign exchange rate market using three months of high-frequency data covering transaction prices and tick-by-tick order book information. Data was divided in three groups which represented the training, validation and testing sets. After trained 50 different models over the training set using an EM algorithm that maximized model's likelihood function, the model that showed best generalization abilities over the validation set was selected.

This study used Viterbi inference to classify market regimes between uptrend and downtrend on the testing dataset. Based on previous classification, the selected model produced a profitable buy/uptrend and sell/downtrend trading strategy that outperformed two simple passive strategies, buy-and-hold and sell-and-hold. Market regime classification ability of selected model was statistical significant. No future information was used during the execution of the studied trading strategy. Future work will investigate new order book representations beyond the VWAP concept.

Acknowledgments. We thank Bolsa de Valores de Colombia (BVC) for providing the dataset and the anonymous reviewers for sharing their valuable comments with us. This work was partially supported by Colciencias.

References

1. Bengtsson, H.: Bayesian Networks - A Self- Contained Introduction with Implementation Remarks. Ph.D. thesis, Mathematical Statistics, Lund Institute of Technology (1999)
2. Bouchaud, J.P., Farmer, J.D., Lillo, F.: How Markets Slowly Digest Changes in Supply and Demand. Quantitative Finance Papers 0809.0822, arXiv.org (September 2008),
   http://ideas.repec.org/p/arx/papers/0809.0822.html
3. Dorogovtsev, S., Mendes, J., Oliveira, J.: Frequency of Occurrence of Numbers in the World Wide Web. Physica A: Statistical Mechanics and its Applications 360(2), 548–556 (2006),
   http://www.sciencedirect.com/science/article/
   B6TVG-4GNTCF9-2/2/d9efa60dec0417dacaf9f01a5b4bcc40

4. Eisler, Z., Kertesz, J., Lillo, F.: The Limit Order Book on Different Time Scales. Quantitative Finance Papers 0705.4023, arXiv.org (May 2007),
   `http://ideas.repec.org/p/arx/papers/0705.4023.html`
5. Gu, G.F., Chen, W., Zhou, W.X.: Empirical Shape Function of Limit-Order Books in the Chinese Stock Market. Physica 387(21), 5182–5188 (2008)
6. Hassan, M.: Stock Market Forecasting Using Hidden Markov Model: A New Approach. In: IEEE (ed.) Proceedings of the 2005 5th International Conference on Intelligent Systems Desing and Applications (2005)
7. Hassan, M.R., Nath, B., Kirley, M.: A Fusion Model of HMM, ANN and GA for Stock Market Forecasting. Expert Systems with Applications 33(1), 171–180 (2007),
   `http://www.sciencedirect.com/`
   `science/article/pii/S0957417406001291`
8. Jangmin O, Lee, J.-W., Park, S.-B., Zhang, B.-T.: Stock trading by modelling price trend with dynamic bayesian networks. In: Yang, Z.R., Yin, H., Everson, R.M. (eds.) IDEAL 2004. LNCS, vol. 3177, pp. 794–799. Springer, Heidelberg (2004)
9. Kearns, M., Ortiz, L.: The penn-lehman automated trading project. IEEE Intelligent Systems 18(6), 22–31 (2003), `http://dx.doi.org/10.1109/MIS.2003.1249166`
10. Murphy, K.P.: Dynamic Bayesian Networks: Representation, Inference, and Learning. Ph.D. thesis, University of California, Berkeley (2002)
11. Murphy, K.P., Paskin, M.A.: Linear time inference in hierarchical HMMs. In: Proceedings of Neural Information Processing Systems (2001)
12. Tayal, A.: Regime Switching and Technical Trading with Dynamic Bayesian Networks in High-Frecuency Stock Markets. Master's thesis, University of Waterloo, Canada (2009)
13. Tian, G., Guo, M.: Interday and Intraday Volatility: Additional Evidence from the Shanghai Stock Exchange. Review of Quantitative Finance and Accounting 28(3), 287–306 (2007),
    `http://ideas.repec.org/a/kap/rqfnac/v28y2007i3p287-306.html`
14. Weber, P., Rosenow, B.: Order Book Approach to Price Impact. Quant. Finance 5(4), 357–364 (2005)
15. Wisebourt, S.: Hierarchical Hidden Markov Model of High-Frequency Market Regimes Using Trade Price and Limit Order Book Information. Master's thesis, University of Waterloo, Canada (2011)

Mining Users Playbacks History for Music Recommendations

Alexandr Dzuba and Dmitry Bugaychenko

Odnoklassniki LTD,
Saint-Petersburg State University, Russia
alexandr.dzuba@gmail.com

Abstract. This paper presents a set of methods for the analysis of user activity and data preparation for the music recommender by the example of "Odnoklassniki"[1] social network. The history of actions is being analyzed in multiple dimensions in order to find a number of collaborative and temporal correlations as well as to make the overall rankings. The results of the analysis are being exported in a form of a *taste graph* which is then used to generate on-line music recommendations. The taste graph displays relations between different entities connected with music (users, tracks, artists, etc.) and consists of the following main parts: user preferences, track similarities, artists' similarities, artists' works and demography profiles.

Keywords: music recommendations, taste graph, item similarity.

1 Introduction

In the sphere of music services there are many ways for data extraction which can be useful for generating recommendations, especially in a social networks. Playback history, a user's profile and metadata of the tracks can give an estimate of relations between different entities: users, tracks, artists, etc. Taste graph [1] accumulates these knowledge and helps the recomender to solve different tasks. Above mentioned objects and its relations serve as vertices and edges of the graph. It is illustrated in a following way: a user likes an artist, which is similar to another artist, who has recorded a certain track, and each relation can be weighted by a quantitative metric, based on data analysis. Such edges construct numerous chains of edges between the user and unknown tracks.

The paper [3] is proposing the way to make music recommendations by means of *random walk with restarts* on a stochastic graph. The walk starts with an active user being at the graph vertex and continues to the adjacent vertices. The probability of transition along the edge equals its weight. Thus all weights are normalized. In addition, there is a high probability of return to the initial vertex on each walk step. *Steady state probability* distribution of this walk characterizes the relatedness of initial user and vertices within the connected component [5].

---

[1] www.ok.ru

P. Perner (Ed.): MLDM 2014, LNAI 8556, pp. 422–430, 2014.

However, the application field of this approach is much broader than just its usage for making the track list recommendation. We can use random walks to complement existing music collection with the closest tracks. You can also reorder the given list of songs according to user preferences. Similarity relations between tracks or artists can be used to cluster the tracks.

We present a set of methods, which are used to construct a taste graph in the general purpose social network "OK" (www.ok.ru). The main analyzing object is a history of user activity. It allows to estimate the connection between the user and the item vertices, as well as to identify collaborative correlations between the items (tracks and artists). In addition, we analyze a user profile and extract metadata of music files. Still important notice here is that after the metadata extraction a number of postprocessing phases is always required.

2 Taste Graph Construction

Each part of the taste graph is created using a separate algorithm. The user preferences are constructed by aggregating the history of all his playbacks by means of exponential running average. The artist similarities are calculated in multiple steps: representative tracks are selected from the set of all artist's tracks. Then playbacks for these tracks are being aggregated in order to get a user-artist matrix. In order to reduce the impact of the sparsity, this matrix is being converted to an artist-artist matrix by the usage of simple similarity measure. After that the matrix is being iteratively refined. The track similarity matrix is constructed, being based on the amount of playbacks from the same user in the scope of limited time window, which by turn is being discounted by the overall tracks' popularity. The relation between artists and their compositions is calculated by the analysis of the recent activity around the tracks. In order to avoid the cold-start problem, novel items are being additionally boosted. The demography profiles are constructed by averaging user preferences in the scope of a demography group (defined by age, sex and region) and calculating deviations from the system-wide top. The profiles are used to support new users until the time when enough statistical data is collected to reliably construct their own profiles.

The emphasis on different parts is placed due to practical point of view because it allows different parts of the taste graph to be constructed independently and to be combined afterwards. As will be stated below we assume that after determining the edges weights normalization and balancing will be made [1]. All weights are normalized in order to produce a *stochastic system*. Balancing function β is used to manage impact of different factors on the overall result. This function can be changed at the runtime without re-creation of the graph, what increases the flexibility of the system. We supplement the graph with the balancing vertex θ in order to compensate impact of nodes with small amount of outgoing edges. When the amount of edges for $v \in V$ is below limit, an edge from v to θ is being added to take away weights from the existing edges. Next, we describe the basis on which each part of the graph is made and the methods for edge weighting.

All the computations are implemented in Java using Apache Hadoop and Mahout. The daily audience of the www.ok.ru is more than 40 million users mainly from the Russian Federation, Eastern Europe and Middle Asia, but despite the huge volume of statistics to be analyzed, taste graph is updated on a daily basis. In combination with the on-line storage for the today's user actions it allows to generate relevant recommendations.

2.1   User Preferences

User preferences consist of two parts, the preference for artists and the preference for tracks, which are based on the calculation of playbacks by the user. However, in practice, we often face the situation when once many times listened artists or tracks which are not popular anymore, are still being invariably placed at the top of recommendations because of the big amount of past playbacks. We use exponential moving average to update our preferences:

$$
\begin{aligned}
pref_0(u, i) &= plays_0(u, i) \\
pref_t(u, i) &= \alpha \cdot plays_t(u, i) + (1 - \alpha) \cdot pref_{t-1}(u, i)
\end{aligned}
\tag{1}
$$

where $plays_t(u, i)$ is the number of artist or track i listenings by user u within a t-th month and $\alpha \in (0, 1)$ is a constant smoothing factor. No further action is being done regarding preferences, except the application of the balancing function, which helps to manage the impact of similarities of different types.

2.2   Demography Profiles

The demography profile is used to avoid the cold start problem for new users. When $pref(u) = \{i | pref(u, i) \neq 0\}$ contains not enough elements we use *demographic group* vertex as a start of random walks. Demographic groups $U_1, \ldots U_k$ are disjoint subsets of users U, formed from the profiles with the same values of demographic characteristics (gender, age, region).

$$
d_i^p = \sum_{u \in U_p} pref(u, i)
\tag{2}
$$

In order to extract demographical identity from some group we can compute deviations of top group preferences from the system-wide top swt, where $swt_i = \sum_{u \in U} prefs(u, i)$

$$
prefs(U_p) = \frac{top_n(d^p)}{\|top_n(d^p)\|_1} - \frac{top_n(swt)}{\|top_n(swt)\|_1}
\tag{3}
$$

Above mentioned profiles do not have incoming edges in the graph, therefore as soon as the user gathers enough preferences for independent recommendations, these profiles do not affect the random walks.

2.3 Track Similarities

The standard way of similarities determination is to calculate the collaborative correlations between items based on user ratings. Such methods are established on some similarity measure (usually variations of the Pearson correlation coefficient [2]). Within this approach we need to calculate the metric between hundreds of thousands and millions of tracks, represented by ratings vectors. Algorithms for distributed computation of similar items, such as jobs in the library Apache Mahout, can be time saving, but having a large music catalog and lots of users, demands a powerful machine cluster for timely statistics updating. In order to reduce computational cost we use track *temporal correlations* instead of similarity measure.

Assume $p_{i,j}^u$ is an amount of tracks i and j listenings by the user u in the scope of limited time window. Denote by $p_{i,j}$ the sum of $p_{i,j}^u$ from all users. $p_{i,j}$ reflects how well the tracks are listened together, but this is not enough to conclude that i and j are similar. We need to subtract the popularity of the similar track in order to get pure temporal correlations $t_{i,j}$:

$$t_{i,j} = \frac{p_{i,j}}{\sum\limits_{j=1}^{N} p_{i,j}} - b_j^i, \tag{4}$$

where b_j^i is a *baseline* of the track j adopted to similarity with track i. If $p_j = \sum\limits_{i=1}^{N} p_{i,j}$ and $T^i = \{k | p_{i,k} \neq 0\}$ then

$$b_j^i = \frac{p_j}{\sum\limits_{j=1}^{N} p_j \cdot \mathbf{1}_{T^i}(j)} \tag{5}$$

Thus for calculating similar tracks we normalize rows of $P = \{p_{i,j}\}_{i,j=1}^{N}$ and $B = \{p_j \cdot \mathbf{1}_{T^i}(j)\}_{i,j=1}^{N}$ and compute the *temporal correlation matrix* $T = \{t_{i,j}\}_{i,j=1}^{N} = N - B$.

2.4 Artist Similarities

In case of artists we do not have the problem of big data, what enables the one to use more complex algorithms. Unlike the classical approach [6], where a distance metric between vectors of user ratings is calculated during similar items searching, we start with a vector of artists' common playbacks.

Using the Artist-Artist input instead of the User-Artist one, we can many times increase the vectors' density. This reduces an impact of outliers being revealed in into the input data on the founded collaborative correlations. Also this aggregation eliminates the need to work with the high dimension vectors. The calculations are done in three stages:

1. Pre-filtering of triples *(User, Track, PlayCount)*. We filter the triples with small values of *PlayCount* and users with small amount of statistics. Then we

keep only the reliable tracks. At this point we throw out the tracks which do not pass filtering by density from section 3. Finally we group the playbacks by artist and use the amount of common users as the initial approximation $a_{i,j}^0$ for artist similarity.

2. Iterative calculation of the vector similarity measure between rows of $A^k = \{a_{i,j}^k\}_{i,j=1}^N$. The best results achieved the Euclidean distance and specifically

$$a_{i,j}^k = \frac{1}{1 + \sqrt{\sum_{l=1}^N (a_{i,l}^{k-1} - a_{j,l}^{k-1})^2}} \tag{6}$$

After each iteration elements of the main diagonal are artificially increased for a better convergence:

$$a_{i,i}^k = \alpha \cdot \sum_{i \neq j} a_{i,j}^k, \tag{7}$$

where $\alpha \in (0,1)$ is a constant reducing factor.

3. Similarity lists are filtered by the methods described in section 3.

2.5   Artists' Works

The relations between artists and tracks are selected from the music catalog. Obviously, if the active user vertex is close to the artist vertex, it is reasonable to recommend him the most popular tracks of this artist. But artist's works are not only representing a way to propagate the recommendations of similar artists on the tracks, but are also providing a solution for the cold start problem, so topical for new tracks in a music catalog.

$$w_i = b \cdot \sum_{u \in U} prefs(u, i), \tag{8}$$

where $b > 1$ is a *novelty boost factor*.

Comparing different artists, we can notice large variation in the size of works. In accordance with a stochasticity of the graph, tracks of artists with lots of related songs will be suppressed. We take two steps to avoid this suppression. First, we limit by L the number of adjacent edges with Track-Artist type. Second, for artists with small number of tracks we simulate existence of entry tracks with weights which are close to real. Simulation is gained by adding the edge from the artist j vertex to θ balancing vertex with weight $w_\theta(j)$. Denote by W_j tracks of an artist j:

$$w_\theta(j) = |W_j| \cdot \min_{t \in W_j} a_t \cdot \left(\frac{1}{|W_j| + 1} + \frac{1}{|W_j| + 2} + \dots \frac{1}{L} \right) \tag{9}$$

9 is a model of hyperbolic popularity decay. As shown in fugure 1, the rating of most popular tracks decreases like a hyperbola, so we simulate such decay from the lower track's position to L.

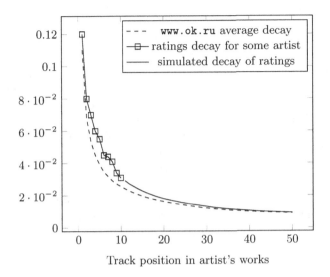

Fig. 1. The weights of Artist-Track edges

2.6 Summary

Along with the graph structure described in this chapter, we have following types of paths from the user to the tracks:

1. $User \to Track \to Track$
2. $User \to Artist \to Track$
3. $User \to Artist \to Artist \to Track$

It is clear that if we have high probability of restarting, the tracks of the second type paths will have much greater impact than tracks reached by the third type paths (for any balancing function). It reflects real relations between entities, but most likely the user is already familiar with works of his favorite artists. In order to increase recommendation novelty we can divide the artist vertex type into two types: an Artist is connected with a Similar Artist who is by turn connected with tracks. So paths in the graph become

1. $User \to Track \to Track$
2. $User \to Artist \to SimilarArtist \to Track$

and the difference between these ways can be controled by balancing function.

3   Similarity Filtration

The methods of outliers filtering for the lists of similar items are presented below. Filtration of similarities with insufficient statistics is close to correlation coefficient shrinkage from [4]. Similarity $s_{i,j}$ between items i and j is discounted like:

$$\hat{s}_{i,j} = \frac{n_{i,j} - 1}{n_{i,j} - 1 + \lambda \cdot min(n_i, n_j)} \cdot s_{i,j}, \tag{10}$$

where $n_{i,j}$ is a number of users who rated both items i and j, n_k is a number of users who rated item k, $\lambda > 0$ is small constant. Note that in different situations, instead of the $min(n_i, n_j)$ arithmetic or geometric mean can be used. Thus the selection function requires additional experiments.

The second filter is used for filtering outliers. To detect outliers we compute sum of similarity values to other elements of items list for each element of this list. In other words, we compute $\tilde{S} = S^2$ where $S = \{s_{i,j}\}_{i,j=1}^{N}$ and filter $s_{i,j}$ with small values of $\tilde{s}_{i,j}$.

The third filter is used to remove similar lists containing some exessively diverse information. As in the case of outliers filtration, we use the similarity to other list items in terms of the subgraph density defined as

$$D = \frac{2|E|}{|V|(|V| - 1)} \tag{11}$$

To check similarity set $s(i) = \{j | s_{i,j} \neq 0\}$ we compute a density D_i of the G subgraph, induced by $s(i)$ vertices. Too low values of D_i tend to indicate non reliable list, which will have a negative impact on the recommendations.

After all filterings, items with short similar lists appear. In order to compensate their increased impact on recommendation we use edge to zero balancing vertex θ. It simulates the presence of missing edges with weights decreasing linearly from a minimum similarity to a 0.

4   Evaluation

We evaluated the graphs using all our users' preferences. Tens of millions of the preferences was splitted into training set and testing set. After that we constructed the graphs from our similarities and baseline similarities. All the elements of the testing set assumed as relevant when calculating recall-precision curves (RPC). Recommendation lists for the users are being compared here. The lists are generated by using random walks with restarts on the graphs [1].

As a baseline for track similarities, we used a measure from [4], including items' baselines and shrunk correlation coefficient. Figure 2 demonstrates good quality of the method, which has been proposed in section 2.3.

We have chosen the data provided by Last.fm API[2] to evaluate similar artists. The graphs are constructed on the base of various Artist-Artist edges and identical Artist-Track edges. Results are shown in figure 3.

---

[2] www.lastfm.ru/api/show/artist.getSimilar

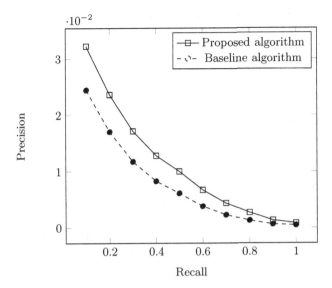

Fig. 2. RPC of the similar tracks

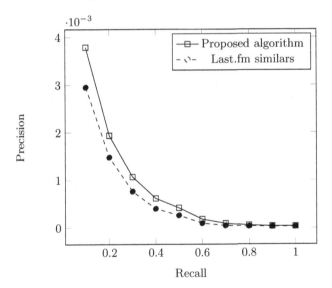

Fig. 3. RPC of the similar artists

Online experiments confirm the decision about quality. After switch from the main graph to the baseline graph, users' activity reduced significantly within the range of 10%. Thereby all the suggested methods demonstrate the ability to recommend relevant content with the quality of state-of-art recommendation methods.

5   Conclusion and Future Work

In this paper we presented the set of methods concerning data analysis, used for evaluation of connections between entities of a music service. The proposed approach allows to calculate these relations independently and to combine them in a taste graph, which provides online recommendations and personalization. At the same time, computational costs for daily graph construction and its updating can be considered to be rather small.

In future we are going to continue working on the taste graph and to pay more attention to the following aspects:

Demographical Adjustments. Large demographic groups can spread collaborative correlations affecting the recommendations of other groups throughout the whole system.

Integration of Other Collaborative Data. Introduction of SVD latent factors to the graph can reduce its size and can make it possible to supplement it with the relations between users without considerable overheads.

References

1. Bugaychenko, D., Dzuba, A.: Musical recommendations and personalization in a social network. In: Proceedings of the 7th ACM Conference on Recommender Systems, pp. 367–370. ACM (2013)
2. Desrosiers, C., Karypis, G.: A comprehensive survey of neighborhood-based recommendation methods. In: Recommender Systems Handbook, pp. 107–144 (2011)
3. Konstas, I., Stathopoulos, V., Jose, J.M.: On social networks and collaborative recommendation. In: Proceedings of the 32nd International ACM SIGIR Conference on Research and Development in Information Retrieval, SIGIR 2009, pp. 195–202. ACM, New York (2009)
4. Koren, Y., Bell, R.: Advances in collaborative filtering. In: Recommender Systems Handbook, pp. 145–186 (2011)
5. Lovász, L.: Random walks on graphs: A survey. Combinatorics, Paul erdos is eighty 2(1), 1–46 (1993)
6. Sarwar, B., Karypis, G., Konstan, J., Riedl, J.: Item-based collaborative filtering recommendation algorithms. In: Proceedings of the 10th International Conference on World Wide Web, pp. 285–295. ACM (2001)

A Method of Crowd-Sourced Information Extraction
From Large Data Files

Indu Mati Anand, Anurag Wakhlu, and Pranav Anand

Sushila Publications, Chelmsford, Masachusetts, 01824 USA
Coloci Inc., Chelmsford, Masachusetts, 01824 USA
UC, Santa Cruz, California USA
indu_anand@uml.edu, indu_m_anand@yahoo.com,
Anurag.Wakhlu@coloci.com, panandt@gmail.com

Abstract. A system and a method is proposed for collecting and aggregating crowd-sourced data from data files based on parameters and measures of relevance of underlying content provided by the intelligent crowd. A user's data may be combined with already existing collective data to generate relevant mark-ups for a document or other consumable data file, such as audio or video. The marked-up version of the document or data fie is then displayed to other users to, *inter alia,* enhance efficiency and comprehension for reading, listening or viewing.

Keywords: Visualization and data mining, Mining Social Media, Crowd-sourcing, knowledge extraction.

1   Introduction

While technology and collaboration have been available for a variety of machine learning and data retrieval applications, the activities related to the consumption and absorption of the information, for example, reading, listening, viewing, remain largely solitary. In this paper, we propose a crowd-sourced method of extracting information from large files of text, audio, image or video data according to user-specified criteria, to assist in efficient consumption of digitally accessible content and to enhance user experience.

Examples of reliance on the wisdom of the crowds abound: from online reviews of goods and services, to measuring the "buzz" for an article on the internet, where the inputs of the previous users in these applications are collectivized into some form of indexing and/or visual depiction. Collaboration among the "crowds" has also been effectively used to put together "wikis" that allow users to collectively edit information.

Unlike other crowd-based utilities that provide information about the content, however, the focus of our methodology is to break down the content by using crowd input to help with a user's actual consumption of the information; essentially, a collaboration between humans and machines, where humans are in the iterative loop from generation –> gathering –> consumption of information, assisting and assisted by, the machines. It keeps humans within the loop, thus defining context more clearly and reliably than purely automatic methods.

P. Perner (Ed.): MLDM 2014, LNAI 8556, pp. 431–436, 2014.

2   A Crowd-Sourced Method of Information Extraction

The central idea of our method (nicknamed "CARP") is to use crowd sourced tools for breaking down content into segments; extracting information from the segments according to the user-specified criteria, e.g., type, depth and manner of extraction; finally, synthesizing this information into a marked-up version of the original.

The last step distinguishes CARP: Crowd-based intelligence, in context, is built from the experience and judgment of prior, similar users. By combining the inputs from the prior users, on demand and in accordance with the criteria specified by a present user (reader/listener/viewer), we let the user guide information extraction, and the extent and manner of the display. For the reading of a document, for instance, the "intelligent crowd" may be the previous readers of the same document who highlighted and/or provided relevant comments. This "intelligent" crowd of prior users, collectively termed here "editors" or "reviewers," may include humans as well as machines, devices, or programs that can provide the information sought in the segments of the data file for user-specified purposes. A user's purpose, for example, may be to scan a document in a given amount of time for the "salient" information, or to elicit a small bit of relevant information hidden in a long essay.

The system allows an editor/reviewer to highlight segments, specify the level at which a segment is judged to be significant by the editor subjectively, specify the purpose for which the segment(s) are significant, and make additional parenthetic comments. Highlights and comments of editors or reviewers are made public unless otherwise instructed. The set of criteria, such as, "relevance" of a highlighted segment for a particular purpose, is defined or augmented by an editor through a flexible administrative process.

In contrast to programs such as, Google Sidewiki and Reframe-It, this scheme of combining the segments highlighted by the editors, allows displaying to the user/reader the most relevant sections on demand. The highlighting and combination maintain the context, without having to rely on keywords, probabilistic weight assignments, regression etc. The parenthetic comments are a secondary but important means to maintain context, and serve to refine or narrow context determination.

The combination of user inputs will typically be effected by using a predefined algorithm. When two editors highlight the same selection, the system refers to a stored matrix to determine how the highlight will appear in display: If the two annotations are identical, the highlight is displayed the common way; if the two disagree in classifying the highlighted section, then the matrix calculates which of the two distinct highlights will be displayed (and how), or calculates and displays a third synthesized version from the two. For this, the system calculates a display value {h} for each highlighted segment, where h may correspond to a color or gray scale value, for example. A simple example is shown in TABLE 1, but a combination algorithm may have different weights assigned to highlights from different editor/reviewers. For instance, weights may be based on the editors' expertise, a higher level of expertise in the area carrying a higher weight. The synthesized collective (it is a "user" in Table 1) generally would be weighted higher than a single editor. The comments and the highlights may be rated and grouped, with awareness of user requirement, for a clearer definition of context. The highlights and comments, however, will generally be collectivized by different methods.

Table 1. Display values computed for "significance" of a segment, based on significance level assigned by two users. Here, 3 = highly significant, 2 = significant, 1 = somewhat significant, and 0 = not significant.

User 1 → / User 2 ↓	3	2	1	0
3	3	2	2	1
2	2	2	1	1
1	2	1	1	0
0	1	1	0	0

This approach is compatible with, and may be used with other common algorithms, such as rating a highlighted segment higher or lower in proportion to the number of occurrences of a particular keyword in the segment. Filtering techniques, e.g., Bayesian or Kalman filtering, may additionally be employed to rank highlights and/or associated comments. The method is also compatible with other statistical and knowledge extraction methodologies, which may be used in conjunction with it.

By keeping humans longer within the loop we can define context at a conceptual level more clearly and reliably than automatic methods, e.g., searching for keyword combinations. This approach may find, for instance, economic analysis buried in a document primarily about a legal concept; the system could categorize the entire document as "Legal" but label the segment under "Economics" if so marked by a previous reader.

3   Basic Command Structure for Text, Audio and Video

Figure 1 shows an illustrative user interface and basic tools for CARP processing of a document, and Figures 2 and 3 describe a CARP implementation. The tools typically include "highlighting" commands to specify: levels or tiers of significance of a highlighted segment; "private" versus "sharable" highlighting; the comments associated with (or without) corresponding segment highlighting; the medium (e.g., text/image/audio/video or other media) of a comment; the nature of a comment (e.g., theme/subject related to the comment, or general evaluative comments etc.).

This method extends to media other than text in self-evident ways whenever the meanings of "segment," "highlight or mark-up" and "significance" etc. can be ascertained or specified and when the set-theoretic notions of union and intersection of two non-identical segments can be discerned or assigned.

A single object within a video may be highlighted, and a segment of a video may be as small as a single pixel. A single object may persist throughout a number of frames of the video. When the highlights are combined, the system may also be instructed to produce a separate video comprised of all the most relevant sections.

For Sound/Audio and for video, the system may identify a segment by the time stamp specifying its beginning and end, and/or specify words/text associated with accompanying nonverbal part of the file. It may identify specific instrument from an ensemble in the audio; or specify a sound, note or syllable.

The system outlined here may be useful in situations, for example, where the analysis of web based data requires a high degree of human sensibility. Problems in social choice or detection of sentiment fall into this category. It can also provide the synthesis of a file across multiple revisions. If there are 5 versions of a document and 14 collaborative editors, a collectivized final copy may be automatically created from their highlighted segment(s).

By showing the highlighted segments within the document, within the underlying content, we allow the user to place relevant sections in the context of the whole, and discern which areas of the file are more significant.

4   Exemplary Schematic Diagrams for Proposed System

FIG. 1. shows a sample user interface screen.

Figures 2 and 3 show sample system implementation. FIG. 2 is an exemplary system that comprises a server and data storage, to which users' computers connect over a communications network, possibly the Internet or a local network of any size. FIG. 3 gives the some details of system-user interactions.

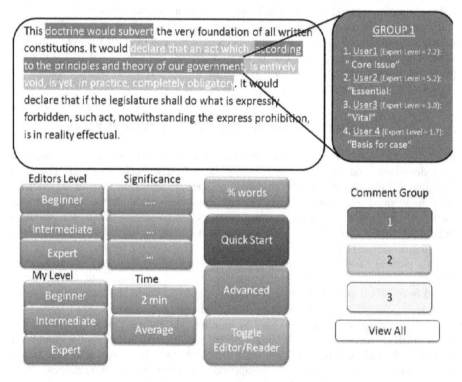

Fig. 1. A sample user interface screen; shading differentiates highlighting by different editors

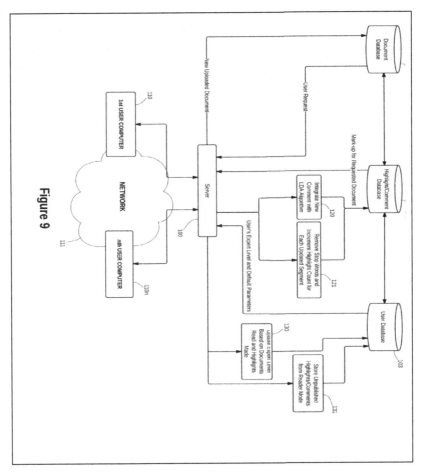

Fig. 2. Schematic of an exemplary system (Reproduced, Figure 9 from Anand, et al., Publication No. WO20121625572A2 (2012))

The system of FIG. 2 comprises a server and data storage, to which users' computers connect over a communications network, possibly the Internet or a local network of any size. A Document Database, Highlight/Comment Databases and a User Database are minimally envisioned. A document for processing may be imported from external storage. When requested for a document, the system also acquires the associated, integrated highlights and comments. The mark-ups may be user- or system- generated, then combined to allow for reading efficiency by time or speed, comprehension, or other user-specified purposes.

FIG. 3 exemplifies user interactions with the system. The user selects a document to view or edit, optionally logging-in for access to stored history and data. The user may toggle between "editor" or reader modes. The editor sends two data sets to the server: the newly added highlights and comments to be integrated with previous version, and editor's updated "user" information. A "Submit" button may be an express command to begin integration process.

In reader mode, the system displays the document with associated highlights and comments. The display, customizable for a reader, may ask, e.g., for mark-ups from specified Expert Levels, or in a particular subject matter, or enable reading at certain speed, or within number of words etc. The reader may ask for non-highlighted text to be obscured, or to view highlights and comments represented visually or aurally, potentially useful for reading on a mobile device.

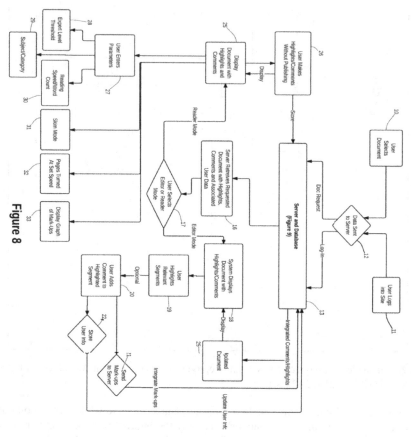

Fig. 3. A sample schematic of system-user transactions (Reproduced, Figure 8 from Anand, et al., Publication No. WO20121625572A2 (2012))

References

Anand, I.M., Wakhlu, A., Anand, P., Anand, I.: A Method And System For Computer-Aided Consumption of Information from Application Data Files. Patent Cooperation Treaty Patent Application, Publication No. WO20121625572A2 (2012)

Modified Bootstrap Approach with State Number Optimization for Hidden Markov Model Estimation in Small-Size Printed Arabic Text Line Recognition

Zhiwei Jiang, Xiaoqing Ding, Liangrui Peng, and Changsong Liu

State Key Laboratory of Intelligent Technology and Systems,
Department of Electronic Engineering, Tsinghua University, Beijing, China
{jiangzw,dxq,plr,lcs}@ocrserv.ee.tsinghua.edu.cn

Abstract. In printed Arabic text line recognition, hidden Markov model brings a facility from no pre-segmentation but leaves a hard work to model estimation. Although bootstrap training can supply good initialization, the bad image quality of small-size samples may make it difficult to find accurate model boundary. This paper introduces a modified bootstrap approach with state number optimization to improve the accuracy of model estimation. Experiments on small-size samples from the APTI dataset show that the modified bootstrap approach in this paper can decrease 13.3% error rate of word recognition and 14% error rate of character recognition than the original one.

Keywords: Hidden Markov model, Optical character recognition, Model estimation, Bootstrap approach, State number optimization.

1 Introduction

Hidden Markov model (HMM) as the statistical tool that can model sequential signals has been used successfully in optical character recognition (OCR) [1, 2]. But in small-size printed Arabic text line recognition, since the quality of text line image decreases heavily (as shown in Fig. 1), it will result in a decrement of recognition rate, as the performance of some systems in the competition of ICDAR 2013[1]. This paper describes a modified bootstrap approach with state number optimization for hidden Markov model estimation. It can improve the performance of HMM-based OCR system through accurate segmentation of samples and proper selection of state number.

Fig. 1. Small-size (left) and large-size (right) samples are zoomed into the same scale

Baum-Welch algorithm, essentially a kind of EM algorithm, is usually used to do model estimation of HMM. However, in practice, quite a lot of training samples are corresponding to more than one unit models, just like one text line containing many

P. Perner (Ed.): MLDM 2014, LNAI 8556, pp. 437–441, 2014.
© Springer International Publishing Switzerland 2014

characters. Except sequential unit model names of the sample, there is no extra segmentation information supplied. So original model estimation algorithm should be replaced with embedded Baum-Welch algorithm [3] and the initialization has to be done with global statistics on all training samples.

However, the sensitivity of initialization from training algorithm often result in model deviation. Bootstrap approach, which can find model boundary in a sample, will be helpful to improve initialization and to relieve model deviation iteratively. Basing on the original bootstrap approach, Yeon-Jun corrects boundary in every iteration, but keeps the state number of model constant [4]. Although state pruning strategy is introduced by Irene, it locates before bootstrap procedure [5].

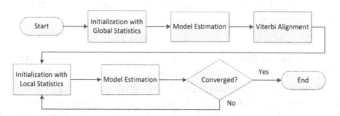

Fig. 2. The flow chart of bootstrap approach used in hidden Markov model estimation

In small-size printed Arabic text line recognition, the image is so blurred that the boundary between two neighbor characters is hard to identify clearly. Furthermore, it is also challenging to choose appropriate state number of HMM. Therefore, boundary correction and state number optimization should be alternatively implemented in this issue. In the rest of this paper, Section 2 shows the theoretical analysis of bootstrap approach and a modified bootstrap approach is introduced in Section 3. Section 4 describes the experiment results and conclusions are made in Section 5.

2   Theoretical Analysis of Bootstrap Approach

Bootstrap approach includes two main steps as shown in Fig. 2, Viterbi alignment to find model boundary and initialization according to the boundary. In this section, theoretical analysis of bootstrap approach in HMM training will be shown.

Since the formula of embedded Baum-Welch algorithm is nearly the same as the one of Baum-Welch algorithm, for the convenience, following discussion could be only focused on the later one. In formula (1), $b_j(O^*)$ represents the estimated probability of state j in HMM when observation is O^* and $\gamma_j(t)$ represents the probability that observation O_t is generated from state j at time t. In addition, the indicator function I is implemented to distinguish whether the current observation O_t is the specific one O^* or not.

$$b_j(O^*) = \frac{\sum_{t=1}^{T} \gamma_j(t) I\{O_t = O^*\}}{\sum_{t=1}^{T} \gamma_j(t)} \tag{1}$$

Obviously, accurate model estimation needs accurate computation of $\gamma_j(t)$. It can be expressed by formula (2). Here, N is the amount of all states and q_t is the actual hidden state that observation O_t should belong to.

$$\gamma_j(t) = \max_{1 \leq i \leq N} \gamma_i(t) \qquad \text{if} \qquad q_t = s_j \qquad (2)$$

Now, there exist two problems. One is how to find the state sequence $\{q_t\}$ corresponding to a sample. The other is how to accomplish the maximum computation.

- **Viterbi Alignment**

When a certain HMM system λ is given, the state sequence $\{q_t\}$ corresponding to a sample is defined as the state path that maximizes the likelihood probability of current sample. The probability can be rewritten in the product form of state transition probability p_{ij} and state emission probability $b_j(O_t)$, as shown in formula (3).

$$\{q_t : t \in T\} = \underset{\{s^{(t)}: t \in T\} \in S^T}{\arg\max} \prod_{t=1}^{T} P(O_t \mid \lambda) = \underset{\{s^{(t)}: t \in T\} \in S^T}{\arg\max} \, b_{s^{(1)}}(O_1) \prod_{t=2}^{T} \left[p_{s^{(t-1)}s^{(t)}} b_{s^{(t)}}(O_t) \right] \quad (3)$$

The path searching problem above is usually solved by Viterbi algorithm. Compared with Viterbi decoding in recognition, here the path to search is identified uniquely by sample label, but not all possible ones. This procedure is called Viterbi alignment and can find the hidden state sequence $\{q_t\}$. However, the reliability   of state boundary is less than the one of model boundary. So in bootstrap, only boundaries between models are reserved for following initialization step.

- **Initialization**

For the purpose of maximizing the probability $\gamma_j(t)$ in formula (2) on all existing states, it is worthy of paying attention to the detailed formula of $\gamma_j(t)$. In formula (4), the two new symbols, $\alpha_j(t)$ and $\beta_j(t)$, represent separately forward probability and backward probability in the period of training HMM.

$$\gamma_j(t) = \frac{\alpha_j(t)\beta_j(t)}{\sum_{i=1}^{N} \alpha_i(t)\beta_i(t)} = \frac{\left[\sum_{i=1}^{N} \alpha_i(t-1) p_{ij} \right] \times b_j(O_t) \times \left[\sum_{k=1}^{N} p_{jk} b_k(O_{t+1}) \beta_k(t+1) \right]}{\sum_{i=1}^{N} \alpha_i(t)\beta_i(t)} \quad (4)$$

In general, state transition probability is initially set to the same value that makes one state has equal opportunity to jump into any probable state. Since p_{ij} and p_{jk} are no longer related to state label j, the unique variable containing j is just $b_j(O_t)$. Therefore, if the initialization of state j can be done only with corresponding samples in the criterion of maximum likelihood, the emission probability $b_j(O_t)$ will reach the maximum value among all states on those samples. Then the maximum of $\gamma_j(t)$ is reached as well. In practice, according to the model boundaries from Viterbi alignment, it is easy to initialize all models one by one through maximum likelihood estimation (MLE).

3   Modified Bootstrap Approach to Training HMM

The flow chart of the modified bootstrap approach with state number optimization is shown in Fig. 3. Compared with the one of original bootstrap approach in Fig. 2, there are two additional steps. Firstly, at the beginning of each initialization step, some HMMs with different state number are prepared separately for the following steps. Then, before each Viterbi alignment step, the trained HMM with the best performance on validation set will be reserved for the next bootstrap iteration.

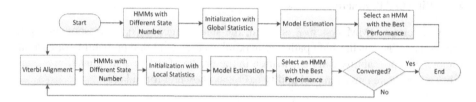

Fig. 3. The flow chart of modified bootstrap approach with state number optimization

In the design of HMM structure, state number is an important parameter to control the complexity of model and to improve system performance. The small-size printed Arabic text line samples are so blurred that accurate model boundary is beneficial to find a good model with proper state number. However, the accurate model boundary aligned come from the performance that a trained HMM can reach. So in this circle problem, bootstrap approach can be treated as a bridge for connecting Viterbi alignment and state number optimization. Eventually, the modified bootstrap steps are executed iteratively until the performance no longer increases.

4   Experiment and Results

The dataset named Arabic printed text image (APTI) is developed for researching on ultra-low-resolution printed Arabic character recognition. There are six sets, the first five available and the last unavailable, in every combination of font, size and style [1].

In the following experiment, font Arabic Transparent and size 6 are chosen to study the performance of modified bootstrap approach in small-size printed Arabic text line recognition. 1000 samples randomly selected from set 4 are kept for validation. Then the rest samples in set 4 and the ones in set 1~3 are collected for model training. Finally, 5000 samples randomly selected from set 5 are treated as test set. Table 1 lists the results of standard training, original bootstrap training and modified bootstrap training. Furthermore, some results of HMM-based systems in ICDAR 2013 are also listed in Table 1. But because the configuration of experiment is different from the one in this paper, the results cannot be compared directly.

From the results in the first two columns, it is known that, after bootstrap training, there is a large improvement and the proper state number changes as well. When model is estimated with the modified bootstrap approach in this paper, the performance is better than before. If the result is translated into error rate, it decreases 13.3% and 14.0% in word and character recognition than original bootstrap training.

Table 1. Experimental results on APTI dataset (above = word recogniton rate; below = character recogniton rate)

State Number	Standard Training	Original Bootstrap	Modified Bootstrap	Reference - ICDAR 2013 [1]		
				THOCR 2	**SID**	**UPV**
8	89.42% 99.00%	95.70% 98.70%	**97.92%** **99.63%**	89.47% 98.28%	94.30% 99.66%	97.25% 99.48%
10	91.06% 99.04%	**97.60%** **99.57%**	with 10-state model at 3rd iteration			
12	**93.28%** **99.15%**	96.98% 99.45%				

5   Conclusions

This paper illustrates a modified bootstrap approach that takes advantage of state optimization to correct Viterbi alignment and to improve the accuracy of model estimation. In spite of the decrement of error rate in recognition is observed, there still exists room to enhance this modified bootstrap approach, like separate state number optimization on each model. Moreover, appropriate image preprocess and robust feature extraction influence HMM-based system on small-size samples as well.

Acknowlegement. This work was supported by the National Natural Science Foundation of China under Grant No. 61032008, 60933010.

References

1. Slimane, F., et al.: ICDAR2013 Competition on Multi-font and Multi-size Digitally Represented Arabic Text. In: 12th International Conference on Document Analysis and Recognition, Washington D.C., pp. 1433–1437 (2013)
2. Philippe, D.: Probabilistic Sequence Models for Image Sequence Processing and Recognition, PhD Thesis, Aachen, Germany (2012)
3. Young, S., et al.: The HTK Book (for HTK Version 3.4), Cambridge University Engineering Department, Cambridge, UK (2009)
4. Yeon-Jun, K., Alistair, C.: Automatic Segmentation Combining an HMM-Based Approach and Spectral Boundary Correction. In: 7th International Conference on Spoken Language Processing, Denver, pp. 145–148 (2012)
5. Irene, A.C., Martin, H., Gerhard, S., Frank, J.: Multiple sequence alignment based bootstrapping for improved incremental word learning. In: 35th IEEE International Conference on Acoustics, Speech, and Signal Processing, Dallas, pp. 5246–5249 (2010)

The Nearest Centroid Based on Vector Norms: A New Classification Algorithm for a New Document Representation Model

Asmaa Mountassir, Houda Benbrahim, and Ilham Berrada

ALBIRONI Research Team, ENSIAS
Mohamed 5 University, Souissi, Rabat, Morocco
asmaa.mountassir@gmail.com, {benbrahim,iberrada}@ensias.ma

Abstract. In this paper, we present a novel model that we propose for document representation. In contrast with the classical Vector Space Model which represents each document by a unique vector in the feature space, our model consists in representing each document by a vector in the space of training documents of each category. We develop, for this novel model, a discriminative classifier which is based on the norms of the generated vectors by our model. We call this algorithm the Nearest Cetroid based on Vector Norms. Our major goal, by the proposition of such new classification framework, is to overcome the problems related to huge dimensionality and vector sparsity which are commonly faced in Text Classification problems. We evaluate the performance of the proposed framework by comparing its effectiveness and efficiency with those of some standard classifiers when used with the classical document representation. The studied classifiers are Naïve Bayes (NB), Support Vector Machines (SVM) and k-Nearest Neighbors (kNN). We conduct our experiments on multi-lingual balanced and unbalanced binary data sets. Our results show that our algorithm typically performs well since it is competitive with the classical methods and, at the same time, dramatically faster especially in comparison with NB and kNN. We also apply our model on the Reuters21578 corpus so as to evaluate its performance in a multi-class environment. We can say that the obtained result (85.4% in terms of micro-F1) is promising and that it can be improved in future works.

Keywords: Machine Learning, Text Classification, Unbalanced Classification, Document Representation, Opinion Mining.

1   Introduction

Nowadays, the amount of available text documents in digital forms (on the web or in data bases) continues to increase exponentially. Consequently, the need to automatically organize and classify these documents becomes more important and, at the same time, more challenging. Text Classification techniques are used in various application domains such as categorization by topic [1] and opinion mining [2].

P. Perner (Ed.): MLDM 2014, LNAI 8556, pp. 442–456, 2014.
© Springer International Publishing Switzerland 2014

Text classification (TC) uses Text Mining, Machine Learning and Natural Language Processing techniques to get meaningful information from documents. TC consists in assigning predefined categories to free-text documents. For this purpose, several and different machine learning methods are applied, including Naïve Bayes, Decision Trees, Support Vector Machines, and k-Nearest Neighbors [3].

In the literature, there are several attempts to compare the most commonly used classification algorithms in TC tasks. Some studies are based on empirical evaluations [1][4][5][6] while others are based on theoretical analyses of the different methods [7][8]. Typically, the studied algorithms include Naïve Bayes (NB), k-Nearest Neighbours (k-NN), Rocchio, Support Vector Machines (SVM), Decision Trees (DT) and Neural Networks (NNet). The results of comparison differ from study to another since the used data collections and the adopted text representation models are not always the same. For example, Aas & Eikvil [4] find that Rocchio, NB, k-NN, DT and SVM seem to be effectively competitive. Yang & Liu [1] conclude that SVM and k-NN outperform NB and NNet when the number of positive training instances per category is small, and that all the methods are competitive when the categories are sufficiently common. Mountassir et al. [6] find that NB and SVM are effectively competitive, but the effectiveness of k-NN depends on the used data set. According to their interpretation, the more the data set is homogeneous, the more k-NN is effective.

Most of the performed comparative studies focus on the effectiveness rather than the efficiency of the studied algorithms [1][4][5][6]. In fact, the major issue that we seek to address is to enhance classification results; which means that we have effective classifiers. Nevertheless, the requirement of a classifier is also an important issue to tackle. For example, if we have two classifiers with the same effectiveness, we would choose the faster and the less complex one.

Since most of the standard algorithms used in TC tasks handle vectors generated by the Vector Space Model [9], we think that these algorithms remain computationally expensive. Indeed, TC suffers from the common problem of huge dimensionality and, consequently, that of vector sparsity. Even there are several attempts to overcome these problems by the application of some standard and sophisticated Reduction Dimensionality techniques [10][11][12], the number of features remains strong especially when dealing with large data collections. Moreover, and even these techniques may decrease dramatically the dimensionality, we note that the application of such techniques is also expensive and we have to take into account their complexity.

In this paper, we try to overcome these two problems (i.e. huge dimensionality and vector sparsity) by the proposition of a quite different classification framework that consists of a novel document representation model and a suitable discriminative classification algorithm. The proposed representation model generates for each document its representing vector with respect to training documents of each category (i.e. the number of generated vectors for each document corresponds to the number of categories). Afterward, the norms of these vectors are used by our proposed classifier to classify unseen documents. We call our algorithm the Nearest Centroid based on Vector Norms. We evaluate the performance of our framework by comparing its effectiveness as well as its efficiency with those of some standard classification algorithms when used with the classical document representation approach. As classical classifiers, we study Naïve Bayes, Support Vector Machines and k-Nearest Neighbors [1]. We focus on binary classification by considering a problem of Opinion Mining [2] which consists in classifying opinionated documents as either positive or negative. We use as data collections six bilingual (Arabic and English) data sets for Opinion Mining. Three of these data sets are balanced while the others are unbalanced with different rates of imbalance. On each data set and for each classifier, we record the obtained classification result, the time (in seconds) taken by the classifier for representation, learning and testing, and the number of features. Our major goal is to build a faster and a less complex classifier with better (or even similar) classification results with regard to the state-of-the-art classifiers. We extend our classifier to a multi-class problem by the use of the one-versus-all strategy. We use for this aim the benchmark Reuters21578 corpus.

The remainder of this paper is organized as follows. In the second section, we give an overview of the classical document pre-processing and representation. The third section describes the proposed classification framework (i.e. the novel document representation model and the new classification algorithm). In the fourth section we present the data collections that we use. The fifth section gives more details about our experiments and discusses the obtained results. The last section concludes the paper and presents some future works.

2   Classical Document Pre-processing and Document Representation

Before we can use Machine Learning techniques to classify a set of text documents, it is necessary to first pre-process these documents and then map them onto a vector space. The first step is called document pre-processing. The second step is known as document representation. We give below some details about these two steps.

2.1   Classical Document Pre-processing

Document pre-processing consists in cleaning and normalizing text documents so as to prepare them to document representation step. We present below some common tasks of pre-processing phase.

• Tokenization: This task used to transform a text document to a sequence of tokens separated by white spaces. We can tokenize a given text into either single words or phrases depending on the adopted model. The most common model is the bag-of-words which consists in splitting a given text into single words. We can also find the n-gram word model [13].

• Stemming: Word stemming is a pseudo-linguistic process which removes suffices to reduce words to their word stem [14]. For example, the words "writing", "writer" and "writes" would be stemmed to the same word "writ".

- Stop Word Removal: Typically, stop words refer to function words such as articles, prepositions, conjunctions, and pronouns, which provide structure in language rather than content. Such words do not have an impact on category discrimination.
- Term Frequency Thresholding: This process consists in eliminating words whose frequencies are either above a pre-specified upper threshold or below a pre-specified lower threshold. This process helps to enhance classification performance since terms that rarely appear in a document collection will have little discriminative power and can be eliminated. Likewise, high frequency terms are assumed to be common and thus not to have discriminative power either.

2.2   Classical Document Representation

This step consists in mapping each document onto a vector representation. These vectors are generated by the use of the Vector Space Model [9]. Terms (single words or phrases) retained after the pre-processing stage are called features. For a given document, its corresponding vector is obtained by computing the weight of each term with respect to that document. We can find various weighting schema where the most used ones are presence, frequency-based and TFIDF weightings [2][6].

Let $\{f_i\}$ be the features set, and $n_i(d)$ be the weight of feature f_i regarding document d. Each document d is mapped onto the vector $d := \{n_i(d)\}$. Basically, the dimensionality of the native feature space is so strong; it can be tens or hundreds of thousands of terms for even a moderate-sized data set [10]. This is considered as a serious problem in classification tasks since it makes some classifiers intractable because of their complexity [11] on one hand. On the other hand, it affects the classification effectiveness since most of native features are noisy or do not carry information for classification [16]. This is why it is recommended to eliminate useless features by the use of Dimensionality Reduction techniques.

Once we represent text documents by their vectors in the feature space, these vectors can serve as input for classification algorithms so as to learn classifiers from training document vectors and, as a test step, to classify unseen documents.

3   The Proposed Classification Framework

The classification framework that we propose consists of a novel document representation model with a suitable classification algorithm.

3.1   Novel Document Representation Model

As mentioned in the introduction, the main idea behind the proposed document representation that we propose is to represent each text document by a vector in the space of training documents for each category. Consequently, each document has as many vectors as the number of categories. In the following, we give further details on the proposed model. For simplification, we consider a problem of binary-class text classification. The described procedure can be generalized for a multi-class problem.

Our model consists in first splitting training set D into two sub-sets D^+ and D^- which include respectively positive and negative training documents. Given the feature set that we extract from training documents, we aim to build for each feature its two vectors v^+ and v^- that might represent it in D^+ and D^- respectively. The dimensionality of v^+ corresponds to the number of positive training documents (i.e. cardinality of D^+). It is the same for v^-. For a feature f, its corresponding v^+ is obtained as follows: the ith component of v^+ corresponds to the weight of f with regard to the ith document of D^+. This weight has as value 1 if f appears in that document; and 0 otherwise. Vector v^- is obtained likewise. After building for all features their v^+ and v^-, we proceed to building for each document its corresponding V^+ and V^-. For a document d, we obtain its V^+ by summing vectors v^+ that correspond to all features which appear in d. In other words, we consider just features that occur in d and we sum their associated v^+ vectors. The resulting vector is called V^+ and is associated to document d. We proceed likewise to associate vector V^- to document d. Note that this representation process is applied on both training and test documents.

This novel representation, which consists in generating for each document its corresponding V^+ and V^-, helps to overcome the problem of huge dimensionality since the number of training (positive and negative) documents is usually much lower than the number of extracted features from training documents.

As we can notice, for a document d (either training or test), its corresponding V^+ used to represent d in the positive training document space. V^+ shows how many features are shared between document d and each positive training document. Likewise, its corresponding V^- used to represent d in the negative training document space. V^- shows how many features are shared between document d and each negative training document. So it is clear that, among weighting schema, sole the presence-based weighting is meaningful. Even if we have data sets represented with other weighting schema, it is necessary to convert them into a binary weighting so that we can apply the present model. Algorithm 1 illustrates the steps of our model.

It is of worth to note that vectors V^+ and V^- can be obtained with a much simpler method by considering matrices. Let M^+ and M^- be matrices that represent respectively positive and negative training documents in the feature space (i.e. the ith line in M^+ represents the vector of the ith positive training document in the feature space). Let D be the matrix that represents test documents in the feature space. We note by W^+ the matrix that results from the product of D and the transpose of M^+. It is the same for W^- which results from the product of D and the transpose of M^-. We can see that W^+ lines (respectively W^- lines) are exactly the V^+ (respectively V^-) of test documents. Likewise, V^+ (respectively V^-) of positive training documents (for instance) can be obtained by the multiplication of M^+ by the transpose of M^+ (respectively by the transpose of M^-). We can do similarly to compute V^+ and V^- for negative training documents.

Algorithm 1. Novel Document Representation Model

Input: Number of positive training documents n^+
Number of negative training documents n^-
Number of test documents M
Number of features N in training documents
Positive training set $D^+ = \{d^+_i\}$ (i=1.. n^+)
Negative training set $D^- = \{d^-_i\}$ (i=1.. n^-)
Test set $T = \{t_i\}$ (i=1.. M)
Feature set $F = \{f_i\}$ (i=1.. N)
//construct for each feature its corresponding vectors
for each feature $f \in F$
   $v^+(f) = (w_f(d^+_1),\ldots, w_f(d^+_{n+}))$ //$w_f(d^+_i)=1$ if f appears in d^+_i, 0 otherwise
   $v^-(f) = (w_f(d^-_1),\ldots, w_f(d^-_{n-}))$ //$w_f(d^-_i)=1$ if f appears in d^-_i, 0 otherwise
end for
//construct for each document its corresponding vectors
for each document $d \in D$

$$V^+(d) = \sum_{i=1}^{N} \delta_i(d) * v^+(f_i)$$

$$V^-(d) = \sum_{i=1}^{N} \delta_i(d) * v^-(f_i) \quad //\delta_i(d)=1 \text{ if } f_i \text{ appears in } d, 0 \text{ otherwise}$$

end for
for each document $d \in T$

$$V^+(d) = \sum_{i=1}^{N} \delta_i(d) * v^+(f_i)$$

$$V^-(d) = \sum_{i=1}^{N} \delta_i(d) * v^-(f_i) \quad //\delta_i(d)=1 \text{ if } f_i \text{ appears in } d, 0 \text{ otherwise}$$

end for

3.2   The Nearest Centroid Based on Vector Norms Algorithm

The classification algorithm that we propose is a binary classifier and belongs to the family of discriminative classifiers. Our classifier is different from the standard classifiers as it handles the L_p norms of vectors (V^+ and V^-) associated to documents instead of handling the vectors themselves. This is why we call the proposed algorithm the Nearest Centroid based on Vector Norms (NCVN). It is built especially for the proposed document representation model described above.

It is of worth to note that, for a given document, the norm of its associated V^+ (respectively V^-) can be viewed as a way to quantify the degree of similarity between this document and the whole positive (respectively negative) training documents. The use of vector norms helps to reduce dramatically the complexity of computation.

The principle of our algorithm is as follows. To each training document, our classifier assigns as score the quotient of the norm of its V^+ and the norm of its V^-.

Afterward, the classifier determines the centroid of each class by averaging the scores of its training documents. We denote by $score^+$ (respectively $score^-$) the average of positive (respectively negative) training document scores.

For a given test document, our classifier assigns as score the quotient of the norm of its V^+ and the norm of its V^-. If the computed score is closer to $score^+$ than $score^-$, the test document is deemed positive. Otherwise, it is deemed negative. This classification principle is quite similar to that of the classical centroid classifier. The main difference is that the centroid classifier handles vectors while our classifier handles points (i.e. one dimension).

Algorithm 2. Novel Classification Algorithm

Input: Number of positive training documents n^+
　　　　Number of negative training documents n^-
　　　　Number of test documents M
　　　　Number of features N
　　　　Training set $D=D^+ U D^-=\{d_i\}$ (i=1.. n^++n^-)
　　　　Positive vectors for training documents $\Delta^+_D=\{V^+(d_i)\}$ (i=1.. n^++n^-)
　　　　Negative vectors for training documents $\Delta^-_D=\{V^-(d_i)\}$ (i=1.. n^++n^-)
　　　　Positive vectors for test documents $\Delta^+_T=\{V^+(d_i)\}$ (i=1.. M)
　　　　Negative vectors for test documents $\Delta^-_T=\{V^-(d_i)\}$ (i=1.. M)
for each document $d \in D$
　score(d) = $\|V^+(d)\| / (\|V^-(d)\|+\varepsilon)$
　//ε is a smoothing value that we add to avoid the problem of division by 0. It is set to 10^{-7}.
end for

$$score^+ = \sum_{i=1}^{n^+} score(d_i^+) / n^+ \text{ //score of positive centroid}$$

$$score^- = \sum_{i=1}^{n^-} score(d_i^-) / n^- \text{ //score of negative centroid}$$

for each document $d \in T$
　score(d) = $\|V^+(d)\| / (\|V^-(d)\|+\varepsilon)$
　//ε is a smoothing value that we add to avoid the problem of division by 0. It is set to 10^{-7}.
　if | score(d) - $score^+$ | < | score(d) – $score^-$ |
　　class(d)=positive
　else
　　class(d)=negative
　end if
end for
Output: class(d) for $d \in T$

Our algorithm has one parameter p which corresponds to the degree of the computed L_p norms. The value of this parameter is to be set empirically.

4   Data Collection

To evaluate our classification framework in a binary-class context, we use Arabic and English data sets for Opinion Mining. For the multi-class problem, we use Reuters21578. We present below each of the employed data sets.

We use three Arabic data sets built, from Aljazeera's website forums (www.aljazeera.net), by Mountassir et al. in [6][17]. We call these data sets[1] respectively DS1, DS2 and DS3. DS1 [17] consists of 468 comments about movie reviews toward a famous historical series. DS2 [6] is a collection of 1003 comments about 18 sport issues. DS3 [17] is a collection of 611 comments about a political issue (all the documents of this data set are about the same issue). These three data sets are labeled manually by one annotator.

We use also the Opinion Corpus for Arabic (OCA[2]) built by Rushdi-Saleh et al. [18] which consists of 500 movie-reviews collected from several Arabic blog sites and web pages. This data set is labeled automatically on the basis of rating systems.

We use as English data set the polarity dataset v2.0[3] built by Pang et al. [19]. It consists of 2000 movie-reviews collected from IMDb website (www.imdb.com). This data set is also labeled automatically on the basis of rating systems. In the following, we call this data set IMDB. We also use SINAI corpus[4] built by Rushdi-Saleh et al. [20] and that consists of 1846 product reviews.

We classify these six data sets as either balanced or unbalanced. In Table 1, we summarize the structure of each data set. We show for each data set its language, the distribution of each class (i.e. number of documents and percentage over the whole data set) as well as the total number of documents.

Table 1. Structure of the binary data sets

	Data Set	Language	Positive	Negative	Total
Balanced	DS2	Arabic	486 (48.6%)	517 (51.4%)	1003
Data sets	OCA	Arabic	250(50%)	250 (50%)	500
	IMDB	English	1000 (50%)	1000 (50%)	2000
Unbalanced	DS1	Arabic	184 (39.4%)	284 (60.6%)	468
Data sets	DS3	Arabic	149 (24.4%)	462 (75.6%)	611
	SINAI	English	1701 (92.2%)	145 (7.8%)	1846

Concerning the multi-class data set, we use the restriction of Reuters21578 ModApté to the largest 10 categories[5], i.e. Earn, Acquisition, Money-fx, Grain, Crude, Trade, Interest, Ship, Wheat and Corn. This data set consists of 12902 news wires published in 1987.

---

[1] These data sets can be obtained by contacting the first author
[2] http://sinai.ujaen.es/wiki/index.php/RecursosEn
[3] http://www.cs.cornell.edu/people/pabo/movie-review-data/
[4] http://sinai.ujaen.es/wiki/index.php/RecursosEn
[5] http://www.cs.waikato.ac.nz/ml/weka/datasets.html

5   Experiments

We aim by our experiments to compare between the combination of the new document representation with the new classification algorithm and the combination of the classical document representation with three classical classifiers (i.e. NB, SVM and k-NN). These experiments are described in the present section. First, we detail the experimental design, then we present and discuss the different results.

5.1   Experimental Design

As sampling strategy, we resample all the binary data sets by randomly splitting them into two sets where 75% represent the training part and 25% represent the test part. We repeat this split 25 times so as to generate 25 samples for each data set. This sampling is known as the hold-out strategy. For the Reuters data set, we use the ModApté split which consists of 9603 of training documents and 3299 test documents.

As pre-processing, we apply the stemmer of Khoja and Garside [15] on the Arabic data sets, and the Porter stemmer[6] [21] on the English data sets (including Reuters). We use, as weighting scheme, binary weights which are based on term presence. As feature selection, we remove terms that occur once in each data set. Since our documents are written by mere internet users, this thresholding may help us to clean documents from, among others, typing errors made by these internet users [17]. This thresholding is applied also on Reuters though it is not written by internet users. Finally, we specify that we adopt the standard bag-of-words model and that we remove stop words.

For our pre-processing and classification tasks, we use the data mining package Weka[7] [22]. As aforementioned, we compare the performance of our classifier (NCVN) with that of three standard classifiers, namely Naïve Bayes (NB), Support Vector Machines (SVM) and k-Nearest Neighbours (k-NN). For the NCVN classifier, we use as norm degree the value of p that gives the best results. The p values we test range from 1 to 5. As we adopt a presence-based weighting (i.e. binary weighting), we use the Multivariate Bernoulli model for NB [23]. For SVM classifier, and as our binary data sets are of small sizes, we use the Linear Sequential Minimum Optimization (SMO) [24]. Concerning k-NN classifier, we use a linear search and a cosine-based distance [25]. Among the tested odd values of k (which range from 1 to 31), we choose those giving the best results.

As evaluation measures, we use the micro-averaged F1 and the macro-averaged F1 [26]. The difference between macro-averaging and micro-averaging is that the micro-average performance scores give equal weight to every document, and are therefore considered a per-document average, while the macro-average performance scores give equal weight to every category, regardless of its frequency, and is therefore a per-category average.

---

[6] http://snowball.tartarus.org/algorithms/porter/stemmer.html
[7] http://www.cs.waikato.ac.nz/ml/weka/

In addition to these two measures, we adopt for unbalanced data sets the g-mean metric which is also known as g-performance [27]. According to Kubat & Matwin [27], g-performance is suitable for unbalanced classification since it maximizes the accuracy of the two classes in order to balance both classes at the same time.

Finally, we specify that, to evaluate our model on a multi-class problem, we adopt the strategy one-versus-all for the Reuters corpus since our classifier is a binary algorithm. Other strategies will be investigated in future works. We do not apply standard algorithms on Reuters since one can refer to reported results in the literature to make comparisons.

5.2   Results

As aforementioned, we are interested in evaluating the performance of our classification framework in terms of both effectiveness and efficiency. So, while conducting our experiments of classification, we record the obtained results of classification (in terms of micro-F1, macro-F1 and g-mean), as well as the time (in seconds) taken by the classifier for learning and testing. We also record the dimensionality of the used feature vectors by the classical classifiers.

In Table 2, we show the classification results that we obtain on each data set by the application of each classifier. The recorded results are in terms of micro-F1 and macro-F1 for all the data sets. The results obtained on unbalanced binary data sets are recorded also in terms of g-mean. In front of each result, we indicate the corresponding standard deviation. We also show for each data set the number of features. We compute for each data set the time in seconds taken by each classifier for both learning and test. We specify that for NCVN, we include time required also for the new representation model. Note that the reported time in Table 2 may give an insight on how fast is our algorithm in comparison with the standard classifiers. We specify that we use, for our experiments, a machine with dual core 1.8GHz CPUs and 4GB of memory. Finally, we note that we mark in bold the best result recorded on each data set with regard to each evaluation measure.

As a first remark, we can see that recorded results in terms of both micro-F1 and macro-F1 are even the same for balanced data sets. However, for unbalanced data sets, there are clear differences between results recorded regarding the three evaluation metrics. As we advanced, the micro-F1 is per-document score while the macro-F1 is per-category one. So, it is evident that on unbalanced data sets, the macro-F1 will be worse than the micro-F1 and can be considered as a more reliable measure for skewed data sets. Yet, we can see that results in terms of g-mean are even worse than the macro-F1. Recall that g-mean balances the two classes at the same time. Hence we think this metric is the most suitable for skewed data sets and it gives information about the sensitivity of a given algorithm toward imbalance [17].

Table 2. Classification results for all classifiers on all data sets

		Algo	Micro-F1	Macro-F1	G-mean	Time (s)	#Features
Binary Balanced Classification	DS2	NB	64.5 ± 3.6	64 ± 3.7	-	37	1769
		SVM	66.3 ± 1.9	66.2 ± 2	-	23	
		kNN	**67.3 ± 2.3**	**67.3 ± 2.3**	-	1213	
		NCVN	67 ± 3.7	67 ± 3.6	-	6.2	
	OCA	NB	79 ± 6.1	79 ± 6.2	-	48	3175
		SVM	88.1 ± 2.9	88 ± 2.9	-	6.1	
		kNN	85.8 ± 3.1	84.7 ± 3.2	-	899	
		NCVN	**88.6 ± 3**	**88.8 ± 2.9**	-	4.1	
	IMDB	NB	79.5 ± 1.9	79.3 ± 2	-	1652	14540
		SVM	**83.2 ± 1.4**	**83.2 ± 1.4**	-	178.1	
		kNN	78.1 ± 1.7	78 ± 1.7	-	>100000	
		NCVN	81.5 ± 1.9	81.5 ± 1.9	-	85	
Binary Unbalanced Classification	DS1	NB	70 ± 5.3	68.6 ± 5.1	**68.4 ± 5**	16	1133
		SVM	65.7 ± 3.7	64 ± 4	63.5 ± 4.4	3	
		kNN	**70.4 ± 4.9**	68.3 ± 5	67.5 ± 4.7	259	
		NCVN	70.2 ± 3.7	**71.6 ± 3.7**	68.3 ± 5.3	1	
	DS3	NB	69 ± 5.4	58 ± 4.6	53.7 ± 4.7	47	2233
		SVM	70.8 ± 3.3	59.7 ± 4.9	55 ± 7	6.4	
		kNN	**72.5 ± 3**	58.8 ± 4.1	53.7 ± 6	936	
		NCVN	72.2 ± 3.2	**61 ± 4.8**	**55.3 ± 6.8**	2.2	
	SINAI	NB	**95.2 ± 1.2**	**83 ± 3.5**	79.4 ± 4	348	3939
		SVM	94.8 ± 0.7	80 ± 3.4	73.7 ± 5	20	
		kNN	92.5 ± 0.8	68.1 ± 3.5	56.1 ± 6	18896	
		NCVN	92.8 ± 1	80.2 ± 2.3	**85 ± 3.2**	22.2	
Multi-Class Classification	Reuters	NCVN	85.4	74.3	-	-	9259

To make a comparison of the reported results in Table 2, we use the paired t-test [28]. We report in Table 3 the decision of comparison after computing the statistic t. Recall that we have 25 samples for each data set. So, we consider that two compared classifiers do not have the same accuracy when |t|>2.064. The reported decision can be '>' (i.e. the first classifier is better than the second classifier), '<' (i.e. the first classifier is worse than the second classifier) or '~' (i.e. the first classifier is the same as the second classifier). We perform t-test with regard to each evaluation measure.

Table 3. Comparison of classifiers for binary data sets using t-test

		Algorithms	Micro-F1	Macro-F1	G-mean
Binary Balanced Classification	DS2	NB vs NCVN	<	<	-
		SVM vs NCVN	~	~	-
		kNN vs NCVN	~	~	-
	OCA	NB vs NCVN	<	<	-
		SVM vs NCVN	~	~	-
		kNN vs NCVN	<	<	-
	IMDB	NB vs NCVN	<	<	-
		SVM vs NCVN	>	>	-
		kNN vs NCVN	<	<	-
Binary Unbalanced Classification	DS1	NB vs NCVN	~	<	~
		SVM vs NCVN	<	<	<
		kNN vs NCVN	~	<	~
	DS3	NB vs NCVN	<	<	~
		SVM vs NCVN	~	~	~
		kNN vs NCVN	~	~	~
	SINAI	NB vs NCVN	>	>	<
		SVM vs NCVN	>	~	<
		kNN vs NCVN	~	<	<

As Table 3 shows, our classifier is typically competitive with the classical algorithms. On the balanced data sets, NCVN is either better or similar to the other classifiers except on IMDB where it is outperformed by SVM. This difference is up to 5.2% in terms of macro-F1. On the unbalanced data sets, the comparison depends on the considered metric. If we consider the micro-F1, NCVN is worse than both SVM (difference of 2%) and NB (difference of 2.4%) just on SINAI. For the macro-F1, NCVN is outperformed just by NB (2.8%) on SINAI. But if we focus on g-mean, NCVN is either better or similar to the other classifiers. We can see that NCVN outperforms NB with a difference of 5.6%. The difference between NCVN and SVM can achieve 11.3%. Finally, the difference between NCVN and kNN can be up to 28.9%. All these differences are observed on SINAI in terms of g-mean. So, as a general conclusion about the effectiveness of the proposed algorithm, we can say that it performs well and that it can beat the most popular classifiers especially on unbalanced data sets.

Note that we cannot say whether NCVN is sensitive to imbalance since it gives good results on DS1 and SINAI but a poor result on DS3 in terms of g-mean. This remains to be studied separately in next works. However, it is more likely that this poor result is due to the special characteristic of DS3. Though this data set is collected from a discussion about one issue, internet users seem tackle other, even far, issues beyond the discussed topic. So, we can say that this data set remains difficult to be classified by machine learning techniques.

Concerning the time taken by the classifiers, it is clear from Table 2 that NCVN is dramatically faster than NB and kNN. It is in most of cases faster (even slightly) than

SVM. The sole exception is observed on SINAI where SVM is faster than NCVN by a difference of 2 seconds. Recall that, on this data set, SVM is worse than NCVN with a difference of 11.3%. So, a loss of 2 seconds with a gain of 11.3% in performance for NCVN would not influence our finding related to the performance of our classifier in terms of both effectiveness and efficiency.

Concerning the multi-class environment, we get as best result 85.4% in terms of micro-F1 and 74.3% in terms of macro-F1 for Reuters. Recall that this difference is due to the skewed distribution among the ten classes of this collection. We consider the obtained result as promising as a preliminary evaluation of the new binary algorithm in a multi-class environment. We have as major future direction to further study this aspect to see whether other strategies (rather than one-versus-all) could help to enhance the obtained results in the present study. Moreover, we will use other benchmark multi-class corpora so as to have solid conclusions.

6   Conclusion and Future Works

The present paper introduces a novel classification framework that consists of a new document representation model and a discriminative classification algorithm. Our major goal by developing this framework is to overcome the problems related to huge dimensionality and vector sparsity which are considered as the bottleneck of text classification problems. The proposed document representation model is based on representing each text document by a vector in the space of training documents of each category instead of representing it by a unique vector in the feature space. To classify unseen documents, our classifier handles the norms of these vectors rather than handling the vectors themselves. This is why the proposed classifier is less complex than the standard classification algorithms. We evaluate the performance of our framework by comparing its effectiveness as well as its efficiency with those of some standard classification algorithms used together with the classical document representation. The studied algorithms are Naïve Bayes (NB), Support Vector Machines (SVM) and k-Nearest Neighbours (k-NN). We conduct our experiments on bi-lingual (Arabic and English) balanced and unbalanced data sets for Opinion Mining. The results show that our classifier is typically competitive with the standard algorithms and substantially faster especially in comparison with NB and k-NN. We also extend our algorithm to a multi-class problem by the use of the one-versus-all strategy. We get a promising result of up to 85.4% in terms of micro-F1.

As future works, we look forward to studying the sensitivity of our classifier toward imbalance since the obtained results in the present study do not show this aspect clearly. We also aim to further study the capacity of our algorithm to perform well in multi-class problems by the use of other strategies rather than one-versus-all and by the use of other benchmark corpora. Finally, we will extend our framework to very-large scale classification problems.

References

1. Yang, Y., Liu, X.: A re-examination of text categorization methods. In: Proceeding of ACM SIGIR Conference on Research and Development in Information Retrieval, pp. 42–49 (1999)
2. Pang, B., Lee, L., Vaithyanathain, S.: Thumbs up? Sentiment classification using machine learning techniques. In: Proceedings of the Conference on Empirical Methods in Natural Language Processing, pp. 79–86 (2002)
3. Sebastiani, F.: Machine learning in automated text categorization. ACM Comput. Surv. 34(1) (2002)
4. Aas, K., Eikvil, L.: Text Categorization: A Survey. Report No. 941, pp. 82–539 (June 1999) ISBN 82-539-0425-8
5. Khan, A., Baharudin, B., Lee, L.H., Khan, K.: A Review of Machine Learning Algorithms for Text-Documents Classification. Journal of Advances in Information Technology 1(1), 4–20 (2010)
6. Mountassir, A., Benbrahim, H., Berrada, I.: A cross-study of Sentiment Classification on Arabic corpora. In: Bramer, M., Petridis, M. (eds.) Research and Development in Intelligent Systems XXIX, pp. 259–272. Springer, Heidelberg (2012a)
7. Bhavsar, H., Ganatra, A.: A comparative Study of Training Algorithms for Supervised Machine Learning. International Journal of Soft Computing and Engineering 2(4), 74–81 (2012) ISSN: 2231-2307
8. Harish, B.S., Guru, D.S., Manjunath, S.: Representation and Classification of Text Documents: A Brief Review. IJCA Special Issue on "Recent Trends in Image Processing and Pattern Recognition" RTIPPR, 110–119 (2010)
9. Salton, G.: Automatic Text Processing: The Transformation, Analysis, and Retrieval of Information by Computer. Addison-Wesley, Reading (1989)
10. Yang, Y., Pedersen, J.O.: A comparative study on feature selection in text categorization. In: Proceedings of the Fourteenth International Conference, ICML 1997 (1997)
11. Dash, M., Liu, H.: Feature selection for classification. Intelligent Data Analysis 1(3) (1997)
12. Hoi, S.C.H., Wang, J., Zhao, P., Jin, R.: Online feature selection for mining big data. In: Proceedings of the 1st International Workshop on Big Data, Streams and Heterogeneous Source Mining: Algorithms, Systems, Programming Models and Applications, BigMine 2012, New York, NY, USA, pp. 93–100 (2012)
13. Shannon, C.: A mathematical theory of communication. Bell System Technical Journal 27 (1948)
14. Smeaton, A.F.: Information retrieval: Still butting heads with natural language processing. In: Pazienza, M.T. (ed.) SCIE 1997. LNCS, vol. 1299, pp. 115–139. Springer, Heidelberg (1997)
15. Khoja, S., Garside, R.: Stemming Arabic text. Computer Science Department, Lancaster University, Lancaster (1999)
16. Lewis, D.D.: Representation and Learning in Information Retrieval. Ph.D. thesis, Department of Computer and Information Science, University of Massachusetts, USA (1992)
17. Mountassir, A., Benbrahim, H., Berrada, I.: An empirical study to address the problem of Unbalanced Data Sets in Sentiment Classification. In: Proc. of IEEE International Conference on Systems, Man and Cybernetics (SMC 2012), Seoul, Korea, pp. 3280–3285 (2012b)

18. Rushdi-Saleh, M., Martin-Valdivia, M.T., Urena-Lopez, L.A., Perea-Ortega, J.M.: Bilingual Experiments with an Arabic-English Corpus for Opinion Mining. In: Proc. of Recent Advances in Natural Language Processing 2011, Hissar, Bulgaria (2011a)
19. Pang, B., Lee, L.: A sentimental education: Sentiment analysis using subjectivity summarization based on minimum cuts. In: Proceedings of the 42nd Annual Meeting of the ACL, ACL 2004, Barcelona, Spain (July 2004)
20. Rushdi-Saleh, M., Martin-Valdivia, M.T., Urena-Lopez, L.A., Perea-Ortega, J.M.: Experiments with SVM to classify opinions in different domains. Expert Systems with Applications 38, 14799–14804 (2011b)
21. Porter, M.F.: An algorithm for suffix stripping. Program 14(3), 130–137 (1980)
22. Witten, I.H., Frank, E.: Data Mining: Practical machine learning tools and techniques, 2nd edn. Morgan Kaufmann, San Francisco (2005)
23. McCallum, A., Nigam, K., Employing, E.M.: pool-based active learning for text classification. In: Machine Learning: Proceedings of the Fifteenth International Conference (ICML 1998), pp. 359–367 (1998)
24. Platt, J.: Fast training on SVMs using sequential minimal optimization. In: Scholkopf, B., Burges, C., Smola, A. (eds.) Advances in Kernel Methods: Support Vector Learning, MIT Press, Cambridge (1999)
25. Salton, G., McGill, M.: Modern Information Retrieval. McGraw-Hill, New York (1983)
26. Yang, Y.: An evaluation of statistical approaches to text categorization, Inform. Retr 1, 1–2 (1999)
27. Kubat, M., Matwin, S.: Addressing the Curse of Imbalanced Data Sets: One-Sided Sampling. In: Proceedings of the Fourteenth International Conference on Machine Learning, pp. 179–186 (1997)
28. Dieterich, T.G.: Approximate statistical tests for comparing supervised classification learning algorithms. Neural Computation 10(7), 1895–1923 (1998)

SimiLay: A Developing Web Page Layout
Based Visual Similarity Search Engine

Ahmet Selman Bozkır and Ebru Akçapınar Sezer

Hacettepe University, Computer Science and Engineering Department, Ankara, Turkey
selman@cs.hacettepe.edu.tr, ebruakcapinarsezer@gmail.com

Abstract. Web page visual similarity has been a trend topic in last decade. Furthermore, effective methods and approaches are crucial for phishing detection and related issues. In this study, we aim to develop a search engine for web page visual similarity and propose a novel method for capturing and calculating layout similarity of web pages. To achieve this, web page elements are classified and mapped with a novel technique. Furthermore, an extension of well known bag of features approach named spatial pyramid match has been employed via histogram intersection schema for capturing and measuring the partial and whole page layout similarity. Promising results demonstrate that spatial pyramid matching kernel can be used for this field.

Keywords: Web page visual similarity, spatial pyramid match kernel, bag of words.

1   Introduction

To date, document similarity has been a hot topic and there exist vast amount of studies about it. However with the advent of the Internet, World Wide Web (WWW) has become of the most import information source and repository today [1, 2] and it began to change the way of understanding on document concept. Moreover, in last two decades, web pages have significantly replaced the well known document entities with digital HTML and various interactive contents. Therefore document similarity and related visual document similarity problems have just transformed to another platform. As stated in [3], the physical layout features of a document include significant amount of data that can be used for different purposes such as categorizing by type or genre in a corpus (e.g. newspapers, scientific papers) and document clustering [4]. On the other hand, analysis of the web pages' visual and structural features can serve various benefits such as useful information extraction [5], phishing detection [6], smart web archiving [7], web page clustering [8] and search engine optimization [10].

One of the hot topics of web page analysis is finding out similarities between web pages which constitute the basis of *web page similarity*. Currently, web page similarity studies can be classified into two groups: (1) web content similarity and (2) web page visual similarity. While content similarity studies are focusing on retrieving web pages similar to queried page in topic, content or structural perspective; the visual

P. Perner (Ed.): MLDM 2014, LNAI 8556, pp. 457–470, 2014.

similarity researches aim to challenge with different problems such as anti-phishing. Phishing is defined as a form of online fraudulent activity in which an attacker aims to steal a victim's sensitive information, such as an online banking password or a credit card number [9]. As reported in [11], in a typical phishing attack, large number of spoofed e-mails are sent to potential victim users who are aimed to convinced that the e-mail is coming from a legitimate company or organization. In many cases these e-mails request the users to change, update or validate their account information. In order not to draw attention a valid explanation is provided by the attackers. Moreover the redirected web page always mimics the legitimate organization web site for the success of attack. If the potential victim is not aware of these kinds of fraudulent events or careless at that moment, there exists a substantial probability of being per-suaded ("phished"). As reported in [11], according to statistics of anti-phishing work group, phishing problem has grown significantly in last years. Gartner Inc.'s 2006 report [12] states that $2.8 billion dollar was lost to phishing attacks in respective year. It is also reported that more than 20.000 phishing pages have been created each month in 2008 [6]. Therefore the importance of anti-phishing methods and solutions is growing more and more year by year.

On the other hand, with the advent of electronic commerce, many new online busi-ness constitutions have been established. Currently, many good and service are being purchased via online services. At this point, various concerns such as trust, security, usability and comfort come into the prominence. In [13], Gehrke and Turban state that web page user interface quality is the one of the major success factor which determines buyers' willingness to buy on the Net. As the functionality and user expe-rience play important role in web site success, companies have found themselves in a race to serve the most eye-catchy designs, easiest and most functional pages. Moreo-ver web page design is being considered as one of the main items of their trademark. This has led the creation of novel and original web page design which had encouraged the birth of various design companies and online web page design template dealers (e.g. templatemonster.com). Nonetheless, copying or stealing a web page design or layout is a trivial task for a web page designer/implementer and it is possible to see one organization's page design in another web page. Due to time limitations, lack of inspiration, laziness or malicious intentions, it has become a rare but serious habit among the web page designers and implementers. In essence, the main problem is here is the lack of a search engine which the designers can check their designs out for originality or cheating. In many cases it is very hard to detect a very similar page designs on all over the Internet. Yet, it could be very easy in case of a smart web page search engine existence which captures the visual features and allows the users to retrieve visually similar pages to a given query page.

To date, visual web page similarity has been a trend topic mainly due to anti-phishing researches. Anti-phishing studies also employ other methodologies (e.g. blacklist, DOM analysis) but visual page similarity is significantly considered as a key for detection. In section 2, the methods and studies related to this subject are comprehensively introduced. However, as reported in [10], the recognition of the visual structural information in web pages is a field which is still unexplored by the scientific community. Furthermore, according to our best knowledge, in the literature there exists no search engine which retrieves visual similar pages to a query page.

As the phishing pages must mimic the legitimate web page, they have to have the same visual appearance and colors as in original page. Nonetheless, as we are seeking a solution to detect visual similarity in more comprehensive aspect, we have focused on analysis of visual layout instead of the color, style and texture information in web pages. One of the main distinguishing aspects of our study then a typical anti-phishing solution is that, it aims to discover not only complete visual similarity, but partial similarity too. It primarily focuses on revealing of partial and complete layout similarity for search engine purposes and aims to be used as phishing detection as auxiliary solution.

In this study, we introduce a novel web page layout based visual similarity search engine which named as "SimiLay". Briefly, SimiLay decomposes the *rendered* visual elements into five categories and evaluates them as visual words. At the following stage, it employs a computationally effective extension of an bag of features" (BOF) image representation method named *spatial pyramid* [14] for measuring the layout similarity. BOF methods are based on orderless collections of quantized local image descriptors which discard spatial information resulting computationally and conceptually simpler way than many alternative approaches [15]. As stated in [15], in principle, a BOF method represents images as orderless collection of local features and they have various application fields such as image classification and video search. However they have lack of ability to capture the *spatial relations*. Therefore in this study, Lazebnik's [14] spatial pyramid matching schema which is an extension to well-known BOF methodology has been utilized. The main benefit of spatial pyramid based matching is partitioning the feature space into increasingly fine sub-regions and computing histograms of local features found inside each sub-region in 2D space [14]. Thus, it provides the ability of matching visual items in *holistic* fashion. A more detailed presentation of spatial pyramid matching can be found in [14]. As a result, SimiLay presents visual element based visually similar web page search engine that is invariant to color and underlying HTML code changes.

In the rest of this paper, related works about web page visual similarity and anti-phishing literature are reviewed in section 2. Section 3 describes the followed methodology. The detail of the application that is subjected to this paper is presented in section 4 while section 5 investigates the results. Finally, section 6 concludes the paper.

2   Related Work

The past decade has seen the growing number of web page and visual web page similarity studies. For various purposes (e.g. information extraction, content similarity search, anti-phishing and clustering) different approaches have been invented. A typical web page is composed of HTML (hyper text markup language) code. HTML is a XML tree like markup language lacking of semantic information and is used to locate page elements (*rendering*) on browser which results a DOM (Document Object Model) tree. This process can also be thought of transforming one dimensional code to 2D geometric surface. After completion of rendering stage, texts, images, form elements, and non HTML contents (e.g. Flash) are all drawn on browser page with accurate positioning. However, it is well-known fact that same visual appearance can be obtained with different HTML element arrangements and tricky manipulations. Thus, web page visual similarity studies are generally branched into two main categories: (1) DOM analysis and (2) pure image analysis based. While DOM based methods employ different methods for better visual structure segmentation, image based methods rely on pure image analysis, segmentation and identification. In this chapter, some of these studies in the literature are introduced.

In Cai et al. [16], a vision based page segmentation algorithm (VIPS) which partitions coherent, semantically related and visually consistent blocks at DOM tree, has been proposed. VIPS fundamentally segments the page by revealing visual separators by making full use extracting suitable blocks from HTML DOM tree with top-down fashion. Guo et al. [17] introduced a web page partitioning method by clustering the visual items based on *spatial locality* concept independent from DOM tree. To achieve this, Guo et al. follow the way of transforming DOM tree to block tree and then partition tree at next stage by finding visual separators which results semantically related blocks. In [2,10], Alpuante and Romero suggest a method which compresses DOM tree to obtain a normalized tree that effectively represents visual structure of the web page. As stated in [10], the transformation stage is based on previously classified html-tags and the tree edit distance was used to measure the distance between to web pages' normalized DOM trees. In [9], Medvet et al. propose a pair-wise web page comparer for anti-phishing purpose. They first extract a pair of feature set from page elements for each web page: (1) style and coordinate information of textual areas and (2) each image's color histogram with 2D Haar wavelet transformation. Thus, two signatures of page are generated and compared with various distance functions such as Euclidean and Levenshtein distance metrics. As a result, if the calculated similarity exceeds the previously determined threshold level then the phishing alert is given. In another study, Tombros and Ali [18] investigate the factors affecting web page similarity in three different aspects: (1) textual HTML tags, (2) structural layout and (3) query terms contained within pages. As a result they conclude that the combination of these features yields better results than singular way. On the other hand, an automatic way of analyzing and discovering the semantic structures in web pages followed by detection of visual similarities of contents is aimed in [1]. However they first start with the assumption of same content categories share same kinds of subtitles and records on web page. Furthermore their work is highly dependent to HTML tags.

In [6], Hara et al. propose a visual similarity based phishing detection without victim site information system which captures the images on suspicious web page and send them to check whether or not they exist on Internet via open source ImgSeek [19] service. If the suspicious page shares a number of same images which exceeds the defined threshold and does not have the same URL of legitimate web page then a phishing alert is raised. In [20], a top-down web page visual segmentation method based on pure image segmentation is developed. In their paper, Pnueli et al. point out that while traditional DOM tree based methods work well in HTML pages, they fail in dynamic HTML pages [20]. In their study, Kudelka et al. [21] investigates and compares two methodological points of view on visual similarity of web pages. While they introduce the visual segmentation and graph matching in the first part, the functional object model concept is presented at the second chapter. With this study they compare and state the pros and cons of non semantic and semantic segmentation on web pages. Besides, they categorize all visual entities in three categories: (1) text, (2) image and (3) mixture which we had inspired in this study. On the other hand, there exist some other very unique studies such as [22] one considers web pages as one indivisible object with the motivation of super signal concept of Gestalt theory and compares them with algorithmic complexity theory. To achieve this, they compress the web pages and measure the normalized compression distance for phishing detection.

3   Methodology

When literature about visual similarity of web pages is reviewed, it can be seen that the proposed approaches are employing some well known methods and techniques. Briefly they can be listed as follows: (1) DOM tree generation and manipulation, (2) graph matching between detected potential regions/segments and (3) image based color or texture similarity search via popular image descriptors (e.g. SIFT [7]). Apart from image analysis, we argue some facts about existing studies and methods done so far. First of all, HTML is a developing and evolving standard for WWW. As of 2014, HTML 5 is the ultimate HTML version which the various browsers are still being developed for satisfying its requirements. According to our proposal, the dependency of HTML tags should be kept as low as possible for better generalization and futuristic solutions. Second, many studies consider and employ the whole page area for similarity analysis. Yet, as noted in [7] the most significant information is certainly not uniformly distributed over whole page and it is reported that most important information is usually presented in the visible part of web pages without scrolling. On the other hand, several studies focus on segmenting the web page considering either HTML content or screenshot image. Nonetheless, as Lazebnik et al. point out in [14], there should be a way to develop an image representation that utilize low-level features to directly infer high-level semantic information without going through the intermediate step of segmentation.

According to these observations, we propose a novel approach to detect and measure the layout similarity between pages. The motivation of our methodology has the proposals and assumptions listed below:

- As pointed out in [14], people can recognize scenes by considering them in "holistic" manner, while overlooking most of the details of the constituent objects. The proposed approach should provide ability of complete and partial layout similarity capturing.
- Currently, many of the web pages contain dynamic loaded parts with the advent of web technologies. Therefore, it should be flexible and robust to dynamic and static HTML contents.

- It should consider the web page as different functional parts. It also should have the ability of capturing visual cues such as separators. For this reason, we divide and map the visual page elements with five different categories: (1) textual contents, (2) static image regions, (3) form elements, (4) animation or motion regions and (5) blank spaces. Furthermore, the importance of these categories should be set.
- As we intent to design and build a visual similarity search engine for web pages, it must extract, capture and recognize the features of page layout instead of color and textures.
- It should generate a comparable signature that is representing the visual layout of web page.
- It should enable us to consider the upper part of web page instead of whole page.

In this study, we first generate DOM tree upon retrieving the whole web page. At the second stage, all page elements (HTML tags) are decomposed and categorized into five categories (text, images, form, animation, blank) with the motion regions (animations) detection via page screenshot captures (five shots with four seconds interval). Then the page boundaries (left, right and bottom) are detected. According to our observations, due to embedded javascript based images sliders or complex tag arrangements; it is not a trivial task to detect the right side boundary of web pages. Therefore we applied K-means clustering algorithm to cluster right most 75 elements with $k = 3$ parameter which results three clusters. The cluster which has the lowest standard deviation is accepted as the right side boundary of web page. At the next stage, we detect the coordinates of all leaf HTML nodes on DOM tree and mapping them to 2D and 1024 x 1024 dimensional matrix by five type categorization stated above. While textual tags (e.g. TR, TD, DIV, LI) are all classified and mapped as one type *text*, the elements such button, checkbox or radio button are all considered as form elements and mapped as *form* type. With the advent of web, animations are being used more than past and there exist several ways of creating and implementing motion based elements on web pages (e.g. Flash, Silverlight or Javascript/Jquery based image animations). Thus, we followed the way of capturing 5 screenshots of web page with 4 seconds interval. Then we compare them with the initial page image. With this technique, the motional regions over page are detected and mapped as *anim* type. On the other hand, tags such as IMG, CANVAS, MAP and SVG are all mapped as *img* type. Besides, the other unmapped regions are all evaluated as blank spaces and mapped as *map* type.

Our principle and novel approach is based on transforming the one dimensional HTML code to *visual words* space by mapping them with the methodology stated above (See Fig. 1 for demonstrative expression) and then utilizing a bag of words (visual features) methodology which satisfies the requirements listed above. As we assume that the most significant information is presented on top of web pages we restricted our region of interest (ROI) with top 1024 pixels. For the simplicity of calculations, we also accepted 1024 pixels as the width of ROI.

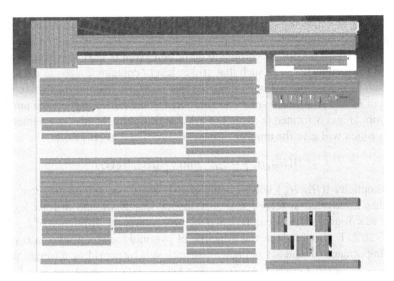

Fig. 1. Visual mapping of html tags according to their types

As stated in [14, 23] locally orderless images play significant role in visual perception. Moreover, Lazebnik et al. in [14] note that in computer vision literature, orderless methods based on bags of features have recently demonstrated impressive levels of performance for image classification. Likewise these methods are robust to clutter, viewpoint changes, occlusion and non-rigid deformations. In essence, as reported in [14] the underlying and unifying theory of subdivision and disordering techniques in the image analysis literature is proposed by the concept of *locally orderless images* of Koendrik and Van Doorn [23]. However, as we stated before, orderless bags of features based methods have lack of capturing spatial relationships. Nonetheless, Lazebnik et al. proposed an extension of orderless bags of features methodology called *spatial pyramid matching* (SPM) which partitions the image or any kind features into increasingly fine sub-regions and concatenating multi resolution histograms of features found inside each sub-region [14]. This partitioning is depicted in Fig 2.

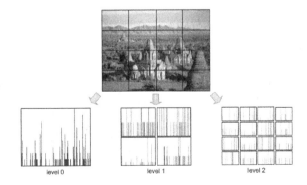

Fig. 2. A schematic illustration of SPM [24]

Combining the spatial pyramid schema with the works of Grauman and Darrel on pyramid match kernel [25], *spatial pyramid match kernel* was built via histogram intersection kernel concept. According to definition given in [25], if we build an array of grid at resolutions 0,...., L, such that grid at level l contains 2^l cells in each dimension which products a total $T = 2^{dl}$ cells. At this point, let H_M^l and H_N^l denote the histograms of M and N at this resolution. Thus, $H_M^l(i)$ and $H_N^l(i)$ express the number of points from M and N located in ith cell of grid. At this point the histogram intersection function which will give the number of matches is defined as [14] Eq.1.

$$I(H_M^l, H_N^l) = \sum_{i=1}^{T} \min \left(H_M^l(i), H_N^l(i) \right) \tag{1}$$

For simplicity $I(H_M^l, H_N^l)$ will be denoted as I^l. Here, one should note that number of matches discovered at level l also contains the matches found in finer level $l+1$. Yet, the new found matches can be calculated by getting the difference of $I^l - I^{l+1}$ for $l=0, 1, ..., L-1$. One fundamental principle of pyramid matching schema is penalizing the importance of matches in higher coarser levels by providing a weight factor to each intersection level l. Thus, for each level l the weight is calculated as $\frac{1}{2^{L-l}}$ that is inversely proportional to cell width at that level [14]. Combining all of these, we obtain the *pyramid matching kernel* as follows:

$$K^L(M, N) = I^L + \sum_{l=0}^{L-1} \frac{1}{2^{L-l}} (I^l - I^{l+1}) \tag{2}$$

In this study we employ orthogonal approach as used in Lazebnik et al.[14]'s study which means performing pyramid matching kernel in 2D space. In details, quantization stage (labeling all HTML leaf nodes in five categories) produces D discrete channels with the assumption that the items of one channel can be matched only with items belonging to same channel. Each channel c outputs two sets of vectors X_M and Y_M which contains coordinates of type c found in feature space. Finally, we obtain the kernel by summing up all individual channel kernels as can be seen below in Eq.3 [14]. As stated in [14], this kernel has the property of maintaining the continuity to *visual vocabulary* paradigm. A more detailed explanation of pyramid matching kernel can be found in [25]. Besides, comprehensive information about spatial pyramid kernel is given in [14].

$$K^L(M, N) = \sum_{c=1}^{C} \kappa^L(M, N) \tag{3}$$

In our study we employ 3 levels pyramidal histogram intersection which results $1+4+16+64 = 85$ cells which overlay a grid of $1024*1024$ pixels. The regions below the top most 1024 pixels are discarded. By using 5 different channels it results a 425 dimensional vector. This vector actually represents the visual layout signature of web page. Finally the comparison between any two pages S_x and S_y is done by histogram intersection of these two pages. More similar pages yield higher histogram intersection values. This value constitutes the main similarity metric of visual similarity.

4   Application

SimiLay contains three main parts: (1) a wrapper, (2) web based search engine inter-face and (3) a distributer. All parts are implemented on C#.NET. Wrapper is an appli-cation which loads the web page, extracts the DOM tree, classifies the tags and maps HTML nodes on grid. It also takes the snapshots of pages with four seconds interval. Finally, the signatures of the web pages are generated and written to central corpus database. Mozilla GeckoFx [26] has been employed for HTML rendering and DOM tree extraction.

A leaf tag or group of leaf tags may be child of the same ancestor type. For this reason we developed a *tree pruning* function which scans all leaf nodes and their ancestors and prunes the child leaf node(s) to capture bigger region in the name of better mapping. In the lack of this function, it is indispensible sometimes possible to incorrectly validate a textual region as blank region. On an Intel Core i7 1,73Ghz 4 Core, 4 GB memory computer, it is determined that creating the signature of a page takes 25 seconds in average including motion capturing (page loading is excluded due to variable network conditions and page size). Similay search engine (see Fig. 3) is implemented on ASP.NET [27] platform. With one's query page input (Url of the page), it retrieves the similar page results and rank them according to similarity score.

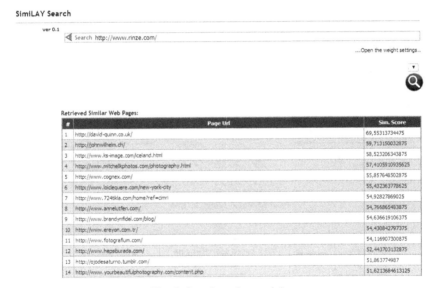

Fig. 3. Search engine module

For efficient computation and utilizing the cores of CPU, we implemented a wrap-per distributor. The main function of distributor is reading the page Urls and distribut-ing them to newly run wrappers. Degree of parallelization can be set on distributor user interface. The system schema and work flow of SimiLay is depicted on Fig. 4.

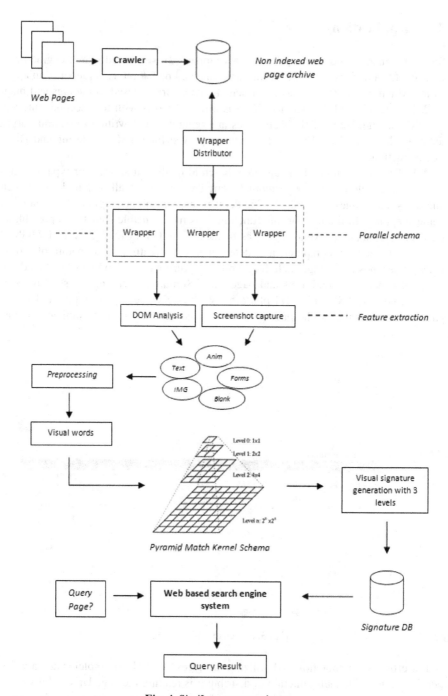

Fig. 4. SimiLay system schema

5   Evaluation and Discussion

According to our best knowledge, there exist no standard well-known dataset for web page visual similarity measurement and as can be seen in the literature of anti-phishing, several studies utilize the web site *phishtank.com* for gathering daily phish cases. In *phishtank,* the legitimate and fake web pages are listed. However, for measuring our approach's success, *phishtank's* lists are not adequate. As phishing pages share same color and page layout in almost perfect condition and we are primarily searching the layout similarity among different web pages, we concluded to build our test dataset. For this reason, we randomly picked up 100 pages. During the process of selection, apart from the other pages, we chose four different layout groups totally containing 26 pages. The in-group pages are validated and marked as visually similar by human users. These 26 pages build our ground truth dataset. The remaining pages are neither so different nor so similar to our ground truth web pages.

To evaluate the success of retrieval results, we first queried each page against to ground-truth dataset and whole corpus at the second stage. Our expectation was always to retrieve the in-group sets on the top of the retrieval. For the performance measure, we applied Average Normalized Rank (ANR) [15]. As the expected pages are listed on top of retrieval result, ANR value tends to 0. For instance, if ANR of a query result is 0.8, this can be interpreted as relevant pages of this query is in the top of 8%. A perfect ANR result is always zero. A detailed explanation about ANR can be found in [15].

The retrieval results are listed in Table 1. First column lists the web page URLs. While second column presents the retrieval order of the in-group pages by considering only ground truth data, third column gives the calculated ANR for this page query. Similarly, fourth and fifth columns are providing same information by considering whole corpus (100 pages).

According to results, 46% of the queries retrieved the pages with the success of 0.25 ANR score in corpus showing that relevant web pages are retrieved in top of 25% of the whole corpus. This ratio is rising up to 61% in ground truth only set.

As can be seen on Table 1, while group 3 shows the best performance among the other groups, group 2 presents the worst one. If we investigate the reasons of this, we reveal one truth. To date, HTML has evolved and reached to its fifth release. During the years it has been enriched with new standards such as CSS [28]. With the advent of CSS, web page developers have started to locate image elements with different techniques. As the IMG tag is still being used by majority, many developer have chose the way of embedding images as background images via CSS rules. This method may have advantages for better and modern design. However this technique makes current SimiLay wrapper impossible to detect and map image regions. Due to implicit declaration, it is a very exhaustive task to detect them correctly. In essence, current SimiLay wrapper uses some smart tricks to detect the boundary rectangle of items. A similar problem sometimes occurs in some web pages which employs java-script based animation/motion regions. Therefore, we can conclude that new style HTML implementations can cause indispensable problems. For the future work, we are planning to apply a smarter way of tag coordinate calculation.

Table 1. Experiment result with ground truth and whole corpus data

Web Page Url	Retrieval order in GT dataset	ANR in GT dataset	Retrieval order in whole corpus dataset	ANR in whole corpus
Group 1				
http://www.rinze.com/	1,2,5,9,16,19	0,192	1,5,10,35,58,78	0,276
http://www.brandynfidel.com/blog/	1,2,3,4,7,8	0,025	1,2,3,4,8,10	0,011
http://blog.artversion.com/web-design-company/why-is-web-design-the-most-important-part-of-marketing/	1,2,6,7,8,9	0,076	1,2,10,15,17,20	0,073
http://johnwilhelm.ch/	1,2,3,5,6,7	0,019	1,2,3,5,6,8	0,006
http://ojodesaturno.tumblr.com/	1,2,8,15,16,18	0,250	1,2,21,60,61,67	0,318
http://david-quinn.co.uk/	1,3,6,13,20,23	0,288	1,16,33,53,81,88	0,418
http://www.alejandropg.com/	1,3,5,6,7,14	0,096	1,3,11,19,21,54	0,146
Group 2				
http://www.loiclequere.com/new-york-city	1,2,8,18	0,121	1,2,37,67	0,242
http://www.ks-image.com/iceland.html	1,2,11,22	0,166	1,3,57,88	0,347
https://marion-luttenberger.squarespace.com/	13,15,18,19	0,352	65,78,84,85	0,755
http://www.mitchellkphotos.com/photography.html	1,2,14,19	0,166	2,8,68,85	0,382
http://www.yourbeautiful photography.com/content.php	15,16,23,24	0,435	60,63,87,94	0,735
Group 3				
http://www.dolusepetim.com/	1,2,3,4,5,7,8	0,006	4,5,12,15,16,21	0,086
http://www.hepsiburada.com/	1,2,3,4,7,12	0,051	1,2,3,4,16,59	0,106
http://www.fotografium.com/	1,2,3,4,7,12	0,051	1,2,3,4,17,57	0,105
http://www.annelutfen.com/	1,2,3,4,7,11	0,044	2,3,4,5,12,51	0,093
http://www.ereyon.com.tr/	1,2,3,4,6,12	0,044	1,2,3,4,14,53	0,093
http://www.724tikla.com/	1,2,3,4,7,12	0,051	1,2,3,4,24,63	0,126
http://www.sanalpazar.com/	1,2,3,5,6,12	0,330	1,4,5,7,9,56	0,101
Group 4				
http://www.magiclick.com	5,9,10,11,12,14	0,256	8,20,25,26,32,53	0,238
http://www.cognex.com/	4,9,10,12,21,24	0,378	8,42,45,60,87,96	0,528
http://www.noritsu.com/	2,3,6,10,18,21	0,250	4,6,13,31,83,93	0,348
http://www.interneer.com/	1,2,3,12,15,17	0,185	6,10,13,67,84,89	0,413
http://www.decathlon.com.tr/	1,8,10,12,20,24	0,346	62,85,89,93,94,100	0,836
http://www.god.de/god/	7,8,9,11,12,13	0,250	16,17,22,39,43,45	0,268
http://www.ikea.com.tr/	1,2,5,14,18,20	0,250	2,8,21,54,84,90	0,396

As we stated before, one of the objectives of our study and approach is to provide a flexible way to weighting the importance of each visual word. In essence, histogram vectors of spatial pyramid match kernel provide this flexibility. Currently we are setting the weights of categories as follows: (1) *blank:* 1.55, (2) *text:* 0.75, (3) *image:* 1.1, (4): *form elements:* 1 and (5) *animation:* 1. As the textual regions may contain different font size and types we decided to decrease the importance of textual regions which means providing textual property invariance. In contrast to text areas, we increased the weight of blank spaces as several studies suggest the importance of horizontal and vertical separators. With this setting, we believe that we are boosting the importance of white spaces on pages and make the algorithm to work more sensible against to visual separators. However we have not optimized these parameters. As a future work, we are also planning to make an optimization on visual words weight settings.

6   Conclusion

In this study we designed and implemented a web page similarity search engine with a novel method combined with a bag of features approach. According to our best knowledge, there exists no study which considers and tries to solve this problem with the followed approaches. Viewing the web page as a collection of different functional object categories and mapping them to a multi resolution pyramid schema provides us several benefits and flexibilities. Yet, it yields also good and promising results. By this study, we can conclude that spatial pyramid matching kernel schema is a suitable solution with the combination of our visual word decomposition approach.

References

1. Yang, Y., Zhang, H.J.: HTML Page Analysis Based on Visual Cues. In: Proceedings of Sixth International Conference on Document Analysis and Recognition (2001)
2. Alpuante, M., Romero, D.: A Visual Technique for Web Pages Comparison. Electronic Notes in Theoretical Computer Science 235, 3–18 (2009)
3. Eglin, V., Bres, S.: Document Page Similarity based on Layout visual saliency: Application to query by example and document classification. In: Proceedings of the Seventh International Conference on Document Analysis and Recognition (2003)
4. Wan, X.: A Novel Documents Similarity Measure based on Earth Mover's Distance. Information Sciences 177, 3718–3730 (2007)
5. Kang, J., Choi, J.: Recognising Informative Web Page Blocks Using Visual Segmentation for Efficient Information Extraction. Journal of Universal Computer Science 14(11), 1893–1910 (2008)
6. Hara, M., Yamada, A., Miyake, Y.: Visual Similarity-based Phishing Detection without Victim Site Information. In: Proceedings of Computational Intelligence in Cyber Security 2009, pp. 30–36 (2009)
7. Law, M.T., Gutierrez, C.S., Thome, N., Gançarski, S., Cord, M.: Structural and Visual Similarity Learning for Web Page Archiving. In: Proceeding of CBMI, pp. 1–6 (2012)

8. Bohunsky, P., Gatterbauer, W.: Visual Structure-based Web Page Clustering and Retrieval. In: Procedings of the 19th International Conference on World Wide Web, pp. 1067–1068 (2010)
9. Medvet, E., Kirda, E., Kruegel, C.: Visual-Similarity-Based Phishing Detection. In: Proceedings of SecureComm 2008 (2008)
10. Alpuente, M., Romero, D.: A Tool for Computing the Visual Similarity of Web Pages. In: Proceedings of Applications and the Internet, SAINT (2010)
11. Rosiello, A.P., Kirda, E., Kruegel, C., Ferrandi, F.: A Layout-Similarity-Based Approach for Detecting Phishing Pages. In: Proceeding of Security and Privacy in Communications Networks and the Workshops, pages, pp. 457–463 (2007)
12. Gartner Press Release. Gartner Says Number of Phishing E-mails Sent to U.S. Adults early Doubles in Just Two Years (2006),
    http://www.gartner.com/it/page.jsp?id=498245
13. Gehrke, D., Turban, E.: Determinant of successful website design: Relative importance and recommendations for effectiveness. In: Proceedings of the 32th Hawaii International Conference on System Sciences (1999)
14. Lazebnik, S., Schmid, C., Ponce, J.: Beyond Bags of Features: Spatial Pyramid Matching Recognizing Natural Scene Categories. In: Proceedings of IEEE Computer Society Conference on Computer Vision and Pattern Recognition (2006)
15. O'Hara, S., Draper, B.A.: Introduction to The Bag of Features Paradigm for Image Classification and Retrieval, CoRR abs/1101.3354 (2011)
16. Cai, D., Yu, S., Wen, J.R., Ma, W.Y.: VIPS: a Vision-based Page Segmentation Algorithm, Technical Report MSR-TR-2003-79, Microsoft Research (2003)
17. Guo, H., Mahmud, J., Borodin, Y., Stent, A., Ramakrishnan, I.V.: A General Approach for Partioning Web Page Content Based on Geometric and Style Information. In: Proceedings of Document Analysis and Recognition – ICDAR 2007 (2007)
18. Tombros, A., Ali, Z.: Factors Affecting Web Page Similarity. In: Proceedings of ICIR 2005, pages, pp. 487–501 (2005)
19. ImgSeek (January 28, 2014), http://www.imageseek.net
20. Pnueli, A., Bergman, R., Schein, S., Barkol, O.: Web Page Layout Via Visual Segmentation, Technical Report HPL-2009-160 (2009)
21. Kudelka, M., Takama, T., Snasel, V., Klos, K.: Visual Similarity of Web Pages, Advance. Intelligent and Soft Computing 67, 135-146 (2010)
22. Chen, T.C., Dick, S., Miller, J.: Detecting Visually Similar Web Pages: Application to Phishing Detection. ACM Transactions on Internet Technology 10(2) (2010)
23. Koenderink, J., Doorn, A.V.: The structure of locally orderless images. IJVC 31(2/3), 159–168 (1999)
24. Lazebnik, S., Schmid, C., Ponce, J.: Spatial Pyramid Matching,
    http://www.cs.unc.edu/~lazebnik/publications/
    pyramid_chapter.pdf
25. Grauman, K., Darrell, T.: Pyramid match kernels: Discriminative classification with sets of image features. In: Proceedings of ICCV (2005)
26. Mozilla GeckoFx.Net (January 28, 2014),
    https://code.google.com/p/geckofx/
27. ASP.NET (January 30, 2014), http://www.asp.net/
28. What is CSS? (February 1, 2014), http://www.w3.org/Style/CSS/

Record Linkage Using Graph Consistency

Marijn Schraagen and Walter Kosters

Leiden Institute of Advanced Computer Science
Leiden University, The Netherlands
{M.P.Schraagen,W.A.Kosters}@liacs.leidenuniv.nl

Abstract. This paper provides a method for automated record linkage in the historical domain based on collective entity resolution. Multiple records are considered for linkage simultaneously, using plausible record sequences as a substitute for pair-wise record similarity measures such as string edit distance. The method is applied to the problem of family reconstruction from historical archives. A benchmark evaluation shows that the approach provides a computationally efficient way to produce family reconstructions which are useful in practise. Further improvements in linkage accuracy are expected by addressing data issues and linkage assumption violations.

1 Introduction

Databases with personal information, such as hospital files or historical archives, generally contain multiple records involving the same individual person or combination of individuals. The process of identifying sets of records involving the same people (or entities in general) is known as *record linkage*. Traditionally, record linkage is performed using pairwise comparison of records. Alternatively, database records can be represented as nodes in a graph, and the occurrence of a person in two records can be represented as an edge between the corresponding nodes. The consistency of this graph with respect to domain constraints can be used as a measure of record linkage quality. Additionally, graph topology and consistency can suggest previously undiscovered record links. Graph consistency checking uses information from multiple records simultaneously, providing additional evidence for linking which is not available from pairwise record comparison.

This paper presents a method for automatic family reconstruction using domain-based graph consistency constraints. Records are linked based on a valid sequence of events, which limits the influence of string similarity computation and the difficulties associated to similarity measures. In the record linkage process small subgraphs of candidate record links are constructed. Family reconstruction is performed by merging subgraphs according to provided constraints.

The paper is structured as follows: Section 2 contains references to related work. In Section 3 the data used in the current experiments is described. To perform the family reconstruction, an initial matching procedure is applied to create a first approximation or *seed* of the family structure in a dataset of historical records. The seeded partial families are used for the actual family reconstruction.

P. Perner (Ed.): MLDM 2014, LNAI 8556, pp. 471–483, 2014.
© Springer International Publishing Switzerland 2014

Section 4 provides details of this procedure. In Section 5 domain constraints for linkage are discussed. Section 6 provides an evaluation of the method on a benchmark data set, and Section 7 concludes.

2   Related Work

Record linkage using graph features is known in the literature as collective entity resolution or multiple instance learning (MIL). In historical record linkage MIL has been applied to families in census records [1], using matching family members as compensation for non-matching individuals in the same families. A method using pedigree relations as classification features for historical record linkage is provided in [2]. In author disambiguation [3] common unambiguous co-author names are used to cluster ambiguous authors. A topology-oriented approach can be found in [4], using constraints on nodes and edges for approximate matching of subgraph patterns. Domain-specific constraints are applied in, e.g., [5], for mapping of database or ontology definitions based on the structure and the semantics of the fields in these definitions.

Pairwise record linkage has been subject of research for several decades, both in social sciences and in computer science. Early examples can be found in [6] and the very influential paper [7], respectively. Recent overviews of the area are presented in, e.g., [8], [9] and [10]. Most current approaches apply some kind of string similarity metric as a decision criterium for matching, which is generally syntax-based (Levenshtein edit distance and derivatives), set-based (e.g., Jaccard index), or phonetically based (Soundex, Double Metaphone). These metrics are not always sufficient to capture variations and errors in strings (see, e.g., [11]). The current linkage approach uses graph consistency as a partial replacement for string similarity, which consequently reduces the problems associated to the use of string similarity metrics.

3   Data

The linkage method presented in this paper is applied to civil certificates from a database called *Genlias*[1]. This database contains birth, marriage and death certificates issued in The Netherlands. Genlias spans a time period from the start of the Dutch civil registration around 1800 until the early 20st century (the availability of recent records differs per certificate type due to privacy laws). The digitization process is not yet completed, which causes missing data problems for data mining approaches. However, the current contents of the database is sufficient for development purposes.

All certificates contain the date and place of the event. Birth certificates contain the first name and family name of the child, mother and father. Marriage certificates contain the names of the bride and bridegroom and their parents. Death certificates contain the names of the deceased person and the parents.

---

[1] Currently available on www.wiewaswie.nl (in Dutch).

Figure 1 contains examples of the different certificate types. For the current research an approximate total of 4 million birth certificates, 3 million marriage certificates and 7.5 million death certificates is used. The different amounts for birth and death certificates are a clear example of missing data (in this case for birth certificates). However, certificates are added to the database in groups according to municipality and time period. To reconstruct a family a limited spatial and temporal context is generally sufficient, therefore the influence of missing data is limited for the current problem. For evaluation a benchmark database of the Dutch city of Coevorden is used. This database contains around 6,200 families consisting of around 22,000 people in total. However, the benchmark database has a larger time span than the Genlias database, which means that only part of the benchmark overlaps with the civil certificates. In the evaluation 1,076 marriages and 5,183 births from the benchmark are used.

Death certificate
date: 01-07-1831
location: Schokland
latitude: 52.63345
longitude: 5.77769
deceased: Antje Bruins
gender: female
age: 27
father: Albert Bruins
mother: Dirkje Sijmens
partner: Paulus Gillot
remarks: wife of the constable

Marriage certificate
date: 17-01-1846
location: Hodenpijl
latitude: 51.98333
longitude: 4.31667
bridegroom: Jan Vermeulen
age: 27
birth place: Abtsregt
parents: Arent Vermeulen, Jannetje Eijk
bride: Aaltje Rijn
age: 32
birth place: Stompwijk
parents: Jacob Rijn, Anna Santen
remarks: bride widow of Melis Haasteren

Birth certificate
date: 03-11-1846
location: Hodenpijl
latitude: 51.98333
longitude: 4.31667
child: Melis Vermeulen
father: Jan Vermeulen
mother: Aaltje Rijn

Fig. 1. Example civil certificates

4   Method

The linkage approach consists of two main parts. First, potential families are identified using a seeding algorithm. In the second step event consistency is used to create the final reconstruction.

To seed the family reconstruction, all birth certificates are included into sets of certificates with exactly equal names for the parents (see Algorithm 1). This process is illustrated in Figure 2. A sample of input data from the Genlias database is shown in Figure 2a, consisting of four birth certificates and a marriage certificate. A birth certificate contains the names of the child and the parents. In Figure 2a, the names of the parents are used to postulate a hypothetical marriage certificate (dotted boxes). As a domain constraint, the hypothetical marriage event must take place before the birth event. The hypothetical certificate functions as a slot where an actual marriage certificate can be inserted in subsequent steps of the algorithm. In the example, the seeding step results in two partial families which are shown in Figure 2b. These partial families have been obtained as follows: first, the 1846 birth certificate is considered. The parent names *Jan Vermeulen, Aaltje Rijn* have not been encountered before (Algorithm 1, line 2), therefore a new family is added. For the 1848 certificate a new family is added with the parent names *Jan Vermeulen, Alida Rijn*. For the 1849 certificate, the parent names are previously encountered (in the 1846 certificate). If certificates are found with the same parent names but different timeframes, for example more than 20 years between births, a family is hypothesized for each timeframe (Algorithm 1, line 10, see Figure 3 for an example). The 1849 certificate is however within the same timeframe as the 1846 certificate and therefore the new certificate can be added to the existing set (line 9). The 1848 and 1850 certificates are combined into a partial family in the same way.

The family reconstruction process is outlined in Algorithm 2. The algorithm combines two or more partial families and a parent couple (represented by a marriage certificate) into a full family. It is assumed that a partial family can be linked to a marriage certificate in case the names of the family parents exactly match the names of the marriage couple. This assumption can be used as a basic linkage method for birth and marriage certificates (Algorithm 2, line 4). Basic

---

Algorithm 1. Family seeding. Variables: $num[f, m]$ contains the number of different families with parent names $\langle f, m \rangle$, $C^i_{f,m}$ is the set of children for the ith family with parent names $\langle f, m \rangle$.

```
1: for all a = ⟨child c, father f, mother m, . . .⟩ ∈ births do
2:     if ⟨f, m⟩ ∉ parents then
3:         parents ← parents ∪ {⟨f, m⟩}
4:         num[f, m] ← 1
5:         C¹_{f,m} ← {a}
6:         families ← families ∪ {C¹_{f,m}}
7:     else
8:         if ∃k, a' : (a' ∈ Cᵏ_{f,m}, δ_{year}(a, a') ≤ θ) then
9:             Cᵏ_{f,m} ← Cᵏ_{f,m} ∪ {a}
10:        else
11:            num[f, m]++
12:            C^{num[f,m]}_{f,m} ← {a}
```

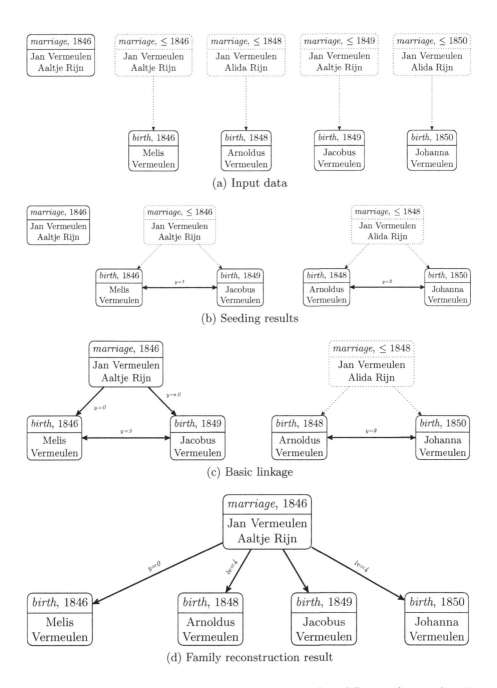

Fig. 2. Family reconstruction example. Relevant event date difference (in years) and name edit distance is indicated by y and lv, respectively.

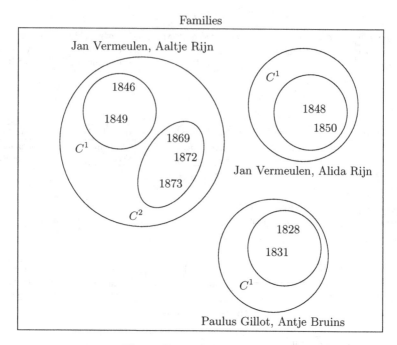

Fig. 3. Example family seeds

linkage is illustrated in Figure 2c, where one of the partial families is linked to the 1846 marriage certificate which is an exact match on the names *Jan Vermeulen, Aaltje Rijn*. A more interesting case is the use of the assumption for non-exact matching. Candidate links between a marriage certificate q and a partial family $C_{f,m}^k$ are selected by exact matching on one of the partners (line 2). Using the basic linkage assumption, q is associated to zero or more partial families $C_{a,b}^i$ (line 10) where the name of the bride b is different from the name of the mother m in the couple $\langle f, m \rangle$. In the example, the partial family with $f = Jan\ Vermeulen$ and $m = Alida\ Rijn$, consisting of the 1848 and 1850 birth certificates, potentially matches the marriage certificate with $a = Jan\ Vermeulen$ and $b = Aaltje\ Rijn$ based on an exact match between a and f (line 2). The algorithm performs a consistency check between $C_{f,m}^k$ and $C_{a,b}^i$ before combining these partial families. A combined family C' (line 11) is considered consistent if there is at least one year (line 12) and at most θ' years (line 15) between consecutive births. In the example, the partial family $C_{a,b}^i$ associated to the marriage couple $\langle a, b \rangle$ consists of the 1846 and 1849 birth certificates. The four births in the combined family all differ by at least one year and the maximum difference between two consecutive certificates is two years, which is below θ' for any reasonable value of this threshold. Therefore, the two partial families are linked into a full family consisting of five certificates (Figure 2d). This approach avoids having to decide on the similarity between the strings *Aaltje* and *Alida*, as the similarity requirement is replaced by an event sequence consistency check.

In case the marriage certificate is not associated to any partial family a similarity measure can be used (line 18). If the partner name differs to a large extent (line 6), the consistency check might not be sufficient. In the next section these cases are considered further. The functions DIFF and DIFF′ are arbitrary similarity functions intended to incorporate domain constraints as well as general-purpose similarity measures.

Algorithm 2. Family reconstruction. Selection of candidates is based on the father (line 2). A symmetrical algorithm is used for $b = m$.

1: **for all** $C^k_{f,m} \in$ families **do**
2:     **for all** $q = \langle$bridegroom a, bride $b, \ldots\rangle \in$ marriages $: a = f$ **do**
3:       **if** $b = m$ **then**
4:          links $\leftarrow$ links $\cup \{(C^k_{f,m}, q)\}$
5:       **else**
6:          **if** DIFF$(b, m) > \theta$ **then**
7:            *reject*
8:          **else**
9:            **if** $\langle a, b\rangle \in$ parents **then**
10:              **for all** $i : C^i_{a,b} \in$ families **do**
11:                 $C' \leftarrow C^k_{f,m} \cup C^i_{a,b}$
12:                 **if** $\exists c, c' \in C' : c \neq c'$ **and** $\delta_{year}(c, c') = 0$ **then**
13:                    *reject*
14:                 **else**
15:                   **if** $\forall c \in C' \exists c' : c \neq c'$ **and** $\delta_{year}(c, c') \leq \theta'$ **then**
16:                     links $\leftarrow$ links $\cup \{(C^k_{f,m}, q)\}$
17:            **else**
18:               **if** DIFF′$(C^k_{f,m}, q) \leq \theta''$ **then**
19:                 links $\leftarrow$ links $\cup \{(C^k_{f,m}, q)\}$
20:               **else**
21:                 *reject*

---

5   Additional Domain-Based Linkage

The linkage method described above uses a domain-based collective entity resolution strategy to reduce the dependence on string similarity measures in the linking process. The assumption behind the method is that a large string distance for certain record fields between a set of candidate matches can be compensated if a plausible link graph for the set of candidate matches can be constructed. Applied to the current problem this means that a large distance for the name of one parent is compensated by a reasonable birth sequence based on the name of the other parent. However, birth sequence consistency is not always sufficient to replace string similarity, especially in case both the first name and the family

name of the second parent are different between records. The string similarity
check in Algorithm 2 (line 6) is intended to filter candidate matches where the
names of the second parent are completely different. However, even if the second parent is a different person, the first parent might still be the same. This
situation occurs when one of the parents has died and the other parent is re-
married to another person, which was very common in the 19th century. This
additional domain knowledge can be used to improve the linkage algorithm. In
case two different people are assumed, a death certificate may be present for the
individual that was mentioned first. This death certificate provides support for
the original match, and the birth sequence check can proceed as before. Figure 4
provides an example of this procedure. A candidate match is found between an
1833 birth certificate and an 1830 marriage certificate. However, the mother on
the birth certificate, which is *Riemda Kofman*, is completely different from the
bride on the marriage certificate, which is *Antje Bruins* (birth parents are again
shown as a hypothetical marriage above the birth certificate). The procedure
locates a death certificate for *Antje Bruins* from 1831. This provides evidence
that indeed the father *Paulus Gillot* is the same person as the bridegroom. The
birth sequence, consisting of an 1831 certificate and an 1833 certificate, confirms
this match. Note that the actual 1833 marriage certificate for the new marriage
can be found using the standard algorithm described above.

Another example of domain knowledge for record linkage is the use of event
location. In the 19th century mobility was limited, therefore a large distance
between two events can be used as evidence that these events should not be
linked. To incorporate this knowledge, a geographical coordinate (consisting of
latitude and longitude) has been added to each municipality in the Genlias data
set. Using the standard haversine formula (see, e.g., [12]) a straight line distance
between two municipalities can be computed. This distance is used as a threshold
in the reconstruction procedure (Algorithm 2, as part of the difference function
in lines 6 and 18).

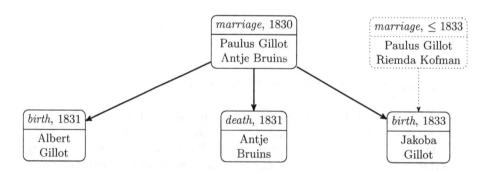

Fig. 4. Linkage using death certificate

6   Benchmark Results

The application of Algorithms 1 and 2 on the Genlias data set is compared to a benchmark consisting of a manual family reconstruction for the small Dutch city of Coevorden. The benchmark is extracted from the website *Coevorder Stambomen*[2] (English: Coevorden family trees), which presents family trees as natural language text. The text on the website has been generated from a database using a fixed sentence structure, therefore parsing is relatively straightforward. In some cases the parser is not entirely accurate, see Subsection 6.2 for further analysis. Out of range records have been removed from the benchmark. This means that the municipality where the event is filed is not present in the Genlias database (either not yet entered or foreign), or that the event is outside of the Genlias time span. However, in some cases it can be difficult to automatically discover that a record is out of range, which complicates evaluation of algorithm results (see Subsection 6.2).

Civil certificates   Coevorden benchmark

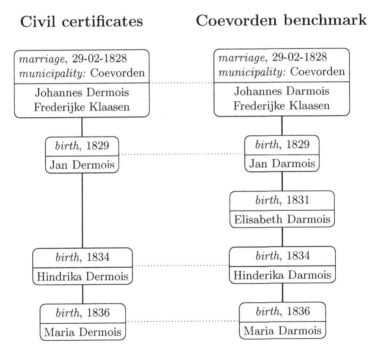

Fig. 5. Example mapping between algorithm results and benchmark records. The family size is 4, of which 3 children are matched. Mapping is based on date and municipality, which allows for various spelling differences between mapped certificates.

The benchmark procedure consists of two steps. First, marriage certificates from Genlias are mapped to benchmark marriages by exact matching of date

---

[2] http://coevern.nl/stambomen/

and municipality. For multiple matches edit distance is used to select the best match. Second, the families associated to mapped marriages are compared using the year of birth. The union of birth years of both families is considered to be a complete family. The intersection of birth years from the Genlias family and the benchmark family represents the match between the two families. An example is presented in Figure 5.

Table 1. Similarity by set size. Sets denoted by A contain the children of a family reconstructed by the algorithm, sets denoted by B contain the children of the corresponding family in the benchmark. Column $A = \emptyset$ denotes benchmark families that the algorithm failed to capture.

$\|A \cup B\|$	$A = \emptyset$	$0, A \neq \emptyset$	1	2	3	4	5	6	7	8	9	10	11	12	13
						$\|A \cap B\|$									
0	0	10													
1	12	0	104												
2	13	1	23	109											
3	8	0	5	16	99										
4	18	1	8	10	12	84									
5	13	0	8	3	7	17	74								
6	11	1	8	2	7	8	22	83							
7	6	0	6	1	4	3	7	13	55						
8	7	0	3	4	4	3	1	3	9	35					
9	9	0	1	1	2	0	0	0	3	3	22				
10	4	1	3	2	1	2	0	1	0	2	6	14			
11	3	0	3	1	1	0	0	1	0	0	0	1	4		
12	0	0	0	1	1	1	1	0	2	1	2	0	2	4	
≥ 13	1	0	2	0	1	0	1	0	1	1	0	0	0	1	1

The result of this comparison is summarized in Table 1. The rows represent the size of the complete family, the columns represent the size of the match. Consider for example the families with two children (146 families in total). For this family size 13 cases are not discovered by the algorithm. One family has no children in common between the reconstruction and the benchmark. In this case the reconstruction contains one child for this family and the benchmark also contains one child, however these children are not the same (indicated by different year of birth). The union of the children therefore has size two while the intersection has size zero. In 23 cases only one child is found in common, generally because the benchmark contains two children and the reconstruction has discovered only one of them. Finally, 109 families are matched on both children. The remainder of Table 1 can be interpreted in the same way: the diagonal represents a perfect match between the reconstruction and the benchmark, the cases close to the diagonal are near-perfect reconstructions. The example in Figure 5 is part of the 12 families for which $|A \cup B| = 4$ and $|A \cap B| = 3$, i.e., four unique children in total of which three are matched. Note that birth sequence consistency could not be used for the families of size 0 or 1 (first two rows), the additional linkage methods from Algorithm 2 have been applied to these cases. The families represent a total of 5,183 birth certificates, of which 74.9% is matched by the algorithm. If the families without any match (column $A = \emptyset$ in Table 1) are not considered 83.5% of the birth certificates is correctly matched. The lack of matches for these families can be caused by data issues (see Subsection 6.2), which means that these birth certificates are not available for linkage and therefore these cases should be discarded in the evaluation of algorithm accuracy. In other cases the lack of matches is indeed caused by the design of the linkage algorithm, however both causes also apply to partially matched families. Therefore some adjustment of the percentage of correct matches is justified, however the amount of adjustment is difficult to estimate.

The event consistency algorithm is intended to replace string similarity computation, both from a conceptual and a computational point of view. This is most useful if the generated matches indeed differ in string representation. For the benchmark evaluation 41.3% of the families contain at least one marriage-birth event pair with different string representations (i.e., a non-exact match). Considering the set of all matches 28.0% of the cases is a non-exact match.

6.1   Implementation Analysis

Algorithms 1 and 2 require a data structure to store a total of nearly 15 million birth, marriage and death events and the families reconstructed from those events, as well as a look-up procedure for set elements. This has been implemented in C++, using the standard library `map` and `multimap` containers. These containers are generally implemented as a binary search tree, which allows look-up in logarithmic time. Operations on the event and family sets account for a large part of the memory load and processing time of the algorithm. The C++ implementation is tested on a 3.16 GHz processor running 64-bit Linux with 6GB memory. The seeding algorithm runs in 22 minutes on the full civil certificate dataset, the reconstruction itself runs in 24 minutes. This kind of performance is exceptional for record linkage approaches, which generally take several hours or even days to process datasets with millions of records.

6.2   Analysis of Matching Errors

In Table 1 the column marked $A = \emptyset$ contains families that the algorithm failed to link to a marriage certificate. An overview of different causes for a sample of 30 errors is given in Table 2. Note that some errors have multiple causes, which accounts for a total of 32 causes.

Error 1 is due to the indexing method that requires an exact name match of one of the parent names on a birth certificate to one of the partners on a marriage certificate (Algorithm 2, line 2). This is a computation issue which is not necessarily a constraint on the reconstruction procedure. Error 4 represents cases in which both the first name and the family name of a birth parent are different, and the edit distance to the corresponding names on the marriage certificate is above some threshold (note that edit distance is not involved in case only one

Table 2. Causes of reconstruction errors

#	error description	frequency
1	name variation for both parents	13
2	out of range	6
3	parser error	6
4	large difference for one parent	5
5	certificate incomplete	1
6	year difference $> \theta'$	1
total		32

name differs). Error 6 represents cases where the model of birth sequences fails. These three types of errors are actual shortcomings of the algorithm. The other types are examples of missing or incorrect data and cannot be attributed to the design of the algorithm.

7   Discussion and Future Work

The algorithm described in this paper provides an approach to record linkage based on the relations between multiple records. This allows for a strongly reduced dependence on string similarity metrics. Results on the benchmark set show that the amount of errors in a practical setting remains limited. Further analysis indicates that a significant part of the errors is caused by missing or incorrect data, although it is difficult to differentiate between data issues and method design flaws. However, this method could be considered as complementary to other record linkage approaches to increase matching coverage.

The design of the algorithm is influenced by the domain of Dutch 19th century civil certificates. However, the event sequences are incorporated into the algorithm in a general way, which is also applicable to other domains. Consider for example an online travel agent interested in identifying repeated customers in his sales records. As domain constraints a customer is unlikely to book overlapping trips, or to have booked two trips only with several years in between. This can be combined with plug-in similarity measures on person names or, e.g., destination characteristics. An implementation of this record linkage task is relatively straightforward using the presented method.

In future work, the candidate selection process can be refined to increase the coverage of the method. The consistency model can be extended to improve accuracy of the matching, for example by imposing a constraint that a child may not have the same name as the other (living) children in the family. Additionally, the influence of various threshold values could provide further insight into the problem of historical family reconstruction as well as the group linkage problem in general.

References

1. Fu, Z., Zhou, J., Christen, P., Boot, M.: Multiple instance learning for group record linkage. In: Tan, P.-N., Chawla, S., Ho, C.K., Bailey, J. (eds.) PAKDD 2012, Part I. LNCS, vol. 7301, pp. 171–182. Springer, Heidelberg (2012)
2. Ivie, S., Pixton, B., Giraud-Carrier, C.: Metric-based data mining model for genealogical record linkage. In: Proceedings of the IEEE International Conference on Information Reuse and Integration, pp. 538–543. IEEE (2007)
3. Bhattacharya, I., Getoor, L.: Collective entity resolution in relational data. ACM Transactions on Knowledge Discovery from Data 1(1), Article 5 (2007)
4. Zampelli, S., Deville, Y., Dupont, P.: Declarative approximate graph matching using a constraint approach. In: Proceedings of the Second International Workshop on Constraint Propagation and Implementation, pp. 109–124 (2005)
5. Madhavan, J., Bernstein, P., Doan, A., Halevy, A.: Corpus-based schema matching. In: Proceedings of the 21st IEEE International Conference on Data Engineering, pp. 57–68. IEEE (2005)
6. Winchester, I.: The linkage of historical records by man and computer: Techniques and problems. The Journal of Interdisciplinary History 1(1), 107–124 (1970)
7. Fellegi, I., Sunter, A.: A theory for record linkage. Journal of the American Statistical Association 64(328), 1183–1210 (1969)
8. Goiser, K., Christen, P.: Towards automated record linkage. In: Proceedings of the Fifth Australasian Conference on Data Mining and Analytics, pp. 23–31. Australian Computer Society, Inc. (2006)
9. Winkler, W.: Overview of record linkage and current research directions. Technical report, U.S. Census Bureau (2006)
10. Christen, P.: Data Matching: Concepts and Techniques for Record Linkage, Entity Resolution, and Duplicate Detection. Springer (2012)
11. Bilenko, M., Mooney, R., Cohen, W., Ravikumar, P., Fienberg, S.: Adaptive name matching in information integration. Intelligent Systems 18(5), 16–23 (2003)
12. Robusto, C.: The cosine-haversine formula. The American Mathematical Monthly 64(1), 38–40 (1957)

Investigating Long Short-Term Memory Networks
for Various Pattern Recognition Problems

Sebastian Otte[1], Marcus Liwicki[2], and Dirk Krechel[1]

[1] University of Applied Sciences Wiesbaden, Germany
{fotte,krechelg}@ecmlab.de
[2] German Research Center for AI, Kaiserslautern, Germany
marcus.liwicki@dfki.de

Abstract. The purpose of this paper is to further investigate how and why long short-term memory networks (LSTM) perform so well on several pattern recognition problems. Our contribution is three-fold. First, we describe the main highlights of the LSTM architecture, especially when compared to standard recurrent neural networks (SRN). Second, we give an overview of previous studies to analyze the behavior of LSTMs on toy problems and some realistic data in the speech recognition domain. Third, the behavior of LSTMs is analyzed on novel problems which are relevant for pattern recognition research. Thereby, we analyze the ability of LSTMs to classify long sequences containing specific patterns at an arbitrary position on iteratively increasing the complexity of the problem under constant training conditions. We also compare the behavior of LSTMs to SRNs for text vs. non-text sequence classification on a real-world problem with significant non-local time-dependencies where the features are computed only locally. Finally, we discuss why LSTMs with standard training methods are not suited for the task of signature verification.

1 Introduction

Introduced in the 1990s [10] the long short-term memory network (LSTM) has become quite popular for many domains in pattern recognition, including speech [8] and handwriting [5] recognition, protein homology detection [9], or even music improvisation [1]. Often LSTMs outperform standard recurrent networks (SRNs) and they won several pattern recognition competitions including several IAPR handwriting recognition competitions[1].

However, while it can be easily seen on the experimental results that LSTMs outperform other RNNs architectures as well as other approaches like hidden Markov models, dynamic time warping, or hybrid approaches, the question of *How* and *Why* remains. We need to understand the *pros* and *cons* of LSTM in

---

[1] For complete list of the won competitions see
http://www.idsia.ch/~juergen /vision.html

P. Perner (Ed.): MLDM 2014, LNAI 8556, pp. 484–497, 2014.

order to wisely apply it on new pattern recognition tasks. This is more favorable than just applying LSTMs on any problem and looking for the performance. Recently, for example, [19] applied LSTM on the task of signature verification. Extensions of this work have later been published as a book [20]. While their experiments show that the training and testing error of LSTMs decreases faster than that of SRNs, the application of RNNs on this problem in general is unsuitable. We will discuss this issue further in Section 4.3.

Note that this is not the first study to analyze the behavior of LSTMs. A comparison of different recurrent neural network architectures has been performed several times (see the latest work of Alex Graves [6]). Most of the analysis papers focus on theoretical problems or the application to speech or handwriting recognition. While these works give good insights into the behavior of LSTM, more analysis is needed to better assess the potential especially on other practical pattern recognition tasks. Therefore this paper will focus on a new problem relevant to several pattern recognition tasks and two application domains, i.e., mode detection and signature verification.

The rest of this paper is organized as follows. First, Section 2 shortly describes the LSTM model and its main differences to SRNs. Second, Section 3 gives an overview of theoretical and analytical studies on LSTMs performed in the past. Third, Section 4 presents novel experiments performed with LSTMs to further study the behavior on realistic pattern recognition tasks. Finally, Section 5 discusses the main findings and gives an outlook to the future work.

2   Long Short-Term Memory

An improved RNN architecture are *long short-term memory* nctworks (LSTMs) [10]. LSTMs overcome the main problem of traditional RNNs where the error vanishes dramatically over time while training. This results in a very restricted event learning horizon for latter approaches, and makes it them very difficult or even impossible to link and associate chronologically widely spaced input events. By simulating kind of differentiable memory cells, LSTMs are able to "trap" the error inside, thus they can learn long time-lag problems (see next section).

The fundamental idea of the model is that a self-referential linear cell, the *Constant Error Carousel* (CEC), carries an inner state. The CEC or also called *memory cell* is protected from outer influences through surrounding gates, which learn to handle the inner state.

We use LSTMs in their extended version with forget gates [2] and peephole connections [4]. Fig. 1 shows an LSTM block with one inner memory cell. Note that LSTM blocks can also consist of more than one memory cell. In our experiments the gates are activated with the standard *sigmoid function* $(1 + e^{-x})^{-1}$ and cell inputs and outputs are squashed using *tangens hyperbolicus*.

In earlier versions LSTMs have been trained using an approximate gradient (the *truncated gradient*), which is calculated with a combination of Backpropagation Trough Time (BPTT) [21] and Real Time Recurrent Learning (RTRL) [15]. Thereby, the gradient is truncated after each time step such that the error

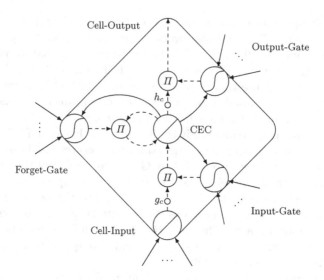

Fig. 1. Visualization of an LSTM block[2]. A self-referential linear cell (CEC) carrying an inner state is protected from external influences by surrounding gates, which learn when to forget overwrite or output the conserved state. The dashed lines indicate non-weighted connections where all others are weighted.

flows back in time only through the CEC. But it has been shown, that the exact gradient can also be computed with BPTT [8], which is more common today.

When implementing LSTMs most parts are straight forward to SRNs. In fact, LSTM can be seen as a specific RNN structure. However, the multiplicative cells bring in significant changes, especially during training of the network. In standard cells with sum-integration the incoming error is only scaled with the cell-local activation derivation and is from then on the same for all predecessors cells (barring the multiplication with the connection weights). With product-integration this is different. The incoming error is scaled with the entire input $\prod_i x_i^t$, where x_i^t is the output of the i-th predecessor at time t. The error that flows back to each predecessor i must then be divided by its output x_i^t. A complete explanation of the LSTM equations is given in [8].

3   Previous Studies

In [10] the theoretical foundations of LSTM are introduced and many theoretical studies have been performed. For the purpose of completeness the main highlights will be summarized in the following. The training complexity of LSTM is similar SRNs, i.e., $O(W)$, where W is the number of weights thus making LSTM very efficient. On several theoretical problems (mentioned in parenthesis in the following) the importance of the gates (Reber Grammar) and the memory cell

---

[2] This arrangement of the cells is inspired by the visualization in [7].

(very long time lag problem). It has been shown that LSTM can store information over more than 1000 time steps, they can add or multiply widely separated continuous values in long sequences with high precision, and they are able to distinguish sequences based on the order of two widely separated symbols. Most of these problems are impossible to solve with SRN because of the vanishing gradient problem or the imprecise storage of values over longer time lags.

In the discussion followed in [10] some remaining disadvantages and some advantages of LSTM have been described. Relevant disadvantages are that the number of parameters is increased, however, only by a factor and not polynomial (16 in the worst case). Furthermore some theoretic problems still cannot be solved by LSTM, e.g., the strongly delayed XOR problem. Advantages on the other hand are that LSTM can solve long time lag problems with noise and continuous values and that they can generalize well. Furthermore, on theoretical problems the need for parameter fine tuning is not so urgent. However, as stated in [10] this has to be validated on realistic problems as well.

Later studies have tested the ability to learn Simple Context Free Grammars (CRF) [3] and non-regular grammars [16]. LSTM are able to generalize detecting sequences of the form $a^n b^n$, $a^n b^m B^m a^n$, and even $a^n b^n c^n$, i.e., only short sequences with $n <= N$ were presented during training and the LSTM was tested on sequences with $n > N$. As has been shown, SRNs fail already if n is only slightly larger than N during testing, e.g., $n = 14, N = 10$. LSTM, on the other hand could generalize to $n = 500$ if N was as small as 40, only. [3] show interesting graphics visualizing the ability of LSTM to count. "Counters of potentially unlimited size are automatically and naturally implemented by linear units, the CECs of standard LSTM, originally designed to overcome error decay problems plaguing previous RNNs." [16]. A CEC can be used to count up and later down and the other gates protect the CEC from irrelevant data.

Regarding the real-world pattern recognition problems, [8] analyzed several LSTM and other RNN structures for the task of speech recognition. The ability of LSTM to bridge over longer time lags has been found to be very important for phoneme classification (details in [8]).

In summary, many theoretical questions have already been answered in the past. Furthermore, the behavior of LSTM has been analyzed on speech data.

Conclusively for this section, Table 1 gives an overview to some important sequential pattern recognition problems including which has been learning successfully with LSTMs. These include problems which have already been mentioned as well as those which are contributed in the remainder of this paper. In addition to the usual input modalities (sequence length, size of the feature vector, etc.), it is shown which specific difficulties have to be overcome to learn each of the problems.

In the next section we will perform the new experiments, which give novel insights into the abilities of LSTM on pattern recognition problems and the differences to SRNs.

Table 1. Overview on some sequence pattern recognition problems. Problems which analysis has been contributed in this paper are filled lighter gray. The darker gray indicates the leading (novel) aspect for each particular analysis.

Problem	Task	Mapping	Seq. Len.	Feat.	Classes	Special Difficulties
Adding	Regr.	Seq. to Val.	10 - 500	2	–	Precise value memorizing over long time periods
Timelag	Classif.	Seq. to Lab.	1 - 50	3	3	Various event positioning with noise, long time-lags
Grammars	Classif.	Seq. to Lab.	10 - 1 000	1	2	Discrete production rules
Mode Detection	Classif.	Seq. to Lab.	10 - 1 000+	5	2	Very high within-class variability
Signature verification	One-class classif. (regressive evaluation)	Seq. to Val.	100 - 300	9	2	Within-class variability vs inter-class variability
HWR/ASR	Classif.	Seq. to Seq.	10 000+	27	26 - 80	Input sequence length differs from output sequence length
Mackey-Glass Time-Series Pred.	Pred.	Seq. to Val.	3 000	1	–	Function learning based on noisy samples

4   Experiments and Analysis

In the following we present three pattern recognition problems by which we reveal new aspects involved in learning with LSTMs and their behavior in direct comparison to SRNs. To make a fair comparison we use LSTMs and SRNs with the same amount of parameters (similar number of weights and equivalent ground architecture as well as the same training method). Furthermore, we use the same training parameters, because often LSTMs can converge faster and achieve higher recognition rates when the training parameters are changed. More details on that follow in 4.2

Our SRN architecture is a two-layered RNN with a fully connected recurrent hidden layer. The hidden layer consists of k units using the *hyperbolic tangent* (tanh) for activation. In the remainder of this paper, we refer such a network as RNN(k). The LSTM architecture is the same as for SRNs, however, the k hidden units are replaced by k LSTM blocks. Analogously we denote the used architecture as LSTM(k). Both architectures are depicted in Figure 2.

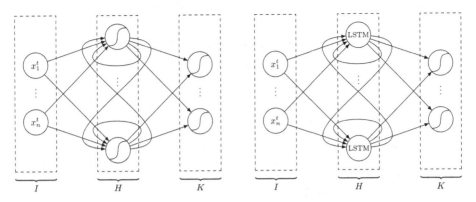

Fig. 2. Our standard recurrent architecture (left) and our LSTM architecture (right). I, H and K denote the input, hidden and output layer. In the LSTM architecture all cells of the hidden layer are replaced by LSTM blocks. Each connection targeting a LSTM block refers actually four connections (one for each gate and the cell-input.

4.1 LSTM for Sequence Classification

Many problems in pattern recognition try to label sequences depending on the information contained therein. In realistic data, however, the position of the information is unknown. Consider, for example, the problem of finding out if a specific event happened during a long time period. We have constructed the following problem to model such sequences and analyze the behavior of SRNs and LSTMs on incrementally increasing the input sequence length under constant training conditions i.e. learning rate, momentum rate, number of training samples. This allows us to gain some isolated insights on the effect of raising the problem's complexity, thereby addressing the upper mentioned positioning aspect.

The task of our toy problem is to classify three different types of sequences over $\mathbb{R}^3$. These sequences are generated as described by the following points:

- Each sequence contains some uniform distributed ground noise (in our experiments of the interval $[-0.05, 0.05]$). Additionally, a special segment is placed at a random position.

- This special segment consists of three vectors $(x_i, y_i, z_i)^T$, where z_i is a constant value (1.0) distinguishing the sequence segment from ground noise (it is set to (0.0) outside of these segments) and the $(x_i, y_i)^T$ vectors are forming a unique sequence corresponding to one of the three classes.

- Characteristic segments we use in our experiment are:

$$\begin{pmatrix} 0.0 \\ 0.0 \end{pmatrix} \begin{pmatrix} 0.5 \\ 0.5 \end{pmatrix} \begin{pmatrix} 1.0 \\ 1.0 \end{pmatrix} \mapsto \text{class 1}$$

$$\begin{pmatrix} 1.0 \\ 1.0 \end{pmatrix} \begin{pmatrix} 0.5 \\ 0.5 \end{pmatrix} \begin{pmatrix} 0.0 \\ 0.0 \end{pmatrix} \mapsto \text{class 2}$$

$$\begin{pmatrix} 1.0 \\ 0.5 \end{pmatrix} \begin{pmatrix} 0.5 \\ 0.5 \end{pmatrix} \begin{pmatrix} 0.0 \\ 0.5 \end{pmatrix} \mapsto \text{class 3}$$

The chosen values provide a compromise of difficulty and learnability. Although these sequences can be separated uniquely, but therefor the vector order (class 1 vs. class 2) and also small differences (class 3 vs. class 1 and 2) must be considered by the classifier. Note that, the ground noise also affects the characteristic segments including the z_i component.

For our experiments we are using an RNN(8) and an LSTM(4) with 104 respectively 132 weights. The time-lag is increased from $T = 0$ to 50 additional to the sequence length. The training is performed using BPTT with momentum term for both architectures. Therefor we constantly use the a learning rate of 0.01, a momentum factor of 0.9 and training set with 300 samples. The error is measured on a separately generated testing set after 10 (50, 100) epochs of training. An additional training of 100 epochs using an approximate exponential function [18] is performed.

Note that with a constant number of training samples not only the time a state must be kept increases, also the number of similar samples corresponding to the same class decreases. The experiment has been repeated about 10 times to compensate random-initialization dependent performance fluctuations, so that the following discussion refers therefore always on average values.

The Figures 3 and 4 show the experimental training results. It is obvious to see that the LSTM generally outperforms the SRN in almost all cases. Noteworthy, the graphs do not depict the maximum performance of the LSTM. With a larger training set, a LSTM can easily handle the problem for $T > 100$ where producing suitable results, while SRNs are unable to handle even short sequences with $T < 10$. The restricted limit of $T < 10$ for SRNs that has been frequently mentioned in the literature also appears in the graphs of the experiment. Although the result that LSTMs dominate SRNs is not unexpected, other details are more interesting in the context of this paper.

The first aspect is the speed of convergence. It is noticeable that after just 10 iterations the LSTM produces a significantly smaller error than the SRN. This starts already with a time-lag of $T < 5$. But the advantage grows with the number of iterations.

The performance gain in the LSTM results confirms the well known fact that LSTMs are able to learn long-term dependencies and SRNs are not. But keeping in mind the upper hint (decreasing number of similar training samples), it becomes clearly that LSTMs generalize dramatically better than SRNs.

Further we can observe that LSTMs are quite more robust regarding the training parameter. It is not unusual when training neural networks with gradient

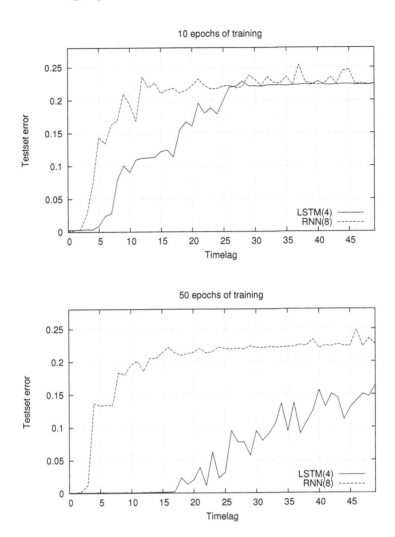

Fig. 3. Results of time-lag experiment comparing classic RNNs and LSTMs after 10 iterations (top) and 50 iterations (below). LSTMs provides a significant higher training convergence speed and generalisation ability than RNNs.

descent a detailed manual parameter optimization is required to yield sufficient results. But even with a time-lag of $T > 10$ the LSTM performs well in spite of a non-optimal setup.

When comparing both graphs in Figure 4 we can observe another interesting aspect of the behavior of LSTMs. While the SRN is affected only marginal on using an approximation for the exponential function, the LSTM behaves much more sensitive and this all the more so the greater the time lag is. This is even

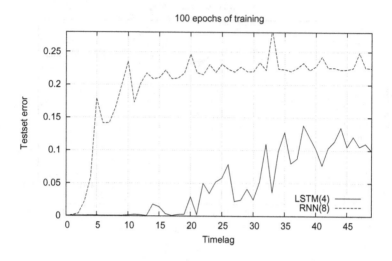

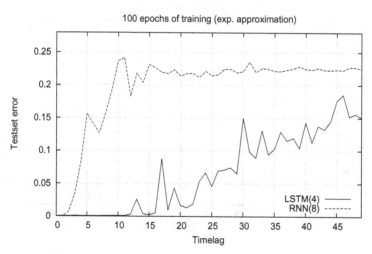

Fig. 4. Results of time-lag experiment comparing classic RNNs and LSTMs using the native exponential function (top) and an approximate exponential function (below). While the training parameter keeps constant, the time-lag is increased iteratively.

more interesting when we recall that in contrast LSTMs are highly insensitive to noised input data.

The upper observation leads to the conclusion that a high numerical precision of the provided mathematical functions is essential for LSTMs to unfold their learning power.

4.2   LSTM for Mode Detection

Recently, we have applied LSTM on the real-world task of online mode detection [14], i.e., the task of distinguishing online handwritten text from setched shapes. Samples are given as sequences over $\mathbb{R}^2$. From such a stream of points features are computed only locally. Locally here means that only three successively points are used to compute one feature vector of 5 features. Linking or combining long-term dependencies between input events is left to the network architectures. As shortly described in Table 1, the difficulty of this problem is the very high within-class variability.

The experiment is based on the benchmark of the IAMonDo database [11]. On all data including partial subsets LSTMs significantly outperform SRNs. Furthermore, when investigating the hardest subset, i.e., diagrams containing text and shapes, the best SRN achieves a mean accuracy of 87.23 % while the best LSTM achieves 93.65 %. In the following, we further analyze the behavior during the learning of the problem.

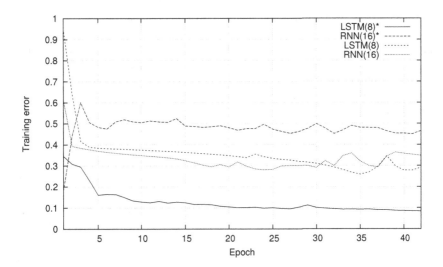

Fig. 5. Results of diagram experiment. The graph shows the training error during gradient descent for different network architectures. Plots annotated with * refers to a learning rate of 10^{-3} instead of 10^{-5}.

Fig. 5 shows the training error over 42 epochs on four different networks, all trained with momentum of 0.9. Networks annotated with * are trained with learning rate of 10^{-3} while the others are trained with 10^{-5}. As we can see with low learning rate the SRN and the LSTM possess nearly the same convergence behavior. However, the recognition rate of the LSTM is on average about 0.3 % higher [14], which again points out that LSTMs generalize better than SRNs.

When increasing the learning rate, the training error of the SRN begins to oscillate (although the first gradient step minimizes the error well), while the LSTM not only keeps stable, it also benefits from the higher learning rate. Here the recognition rate is about 6.3 % higher than for any other SRN applied on this problem.

On one hand, this experiment shows us that findings of the behavior of LSTMs regarding aspects like generalization, parameter robustness and convergence speed obtained on artificial toy problems can be proven in real-world scenarios. On the other hand, the results suggest, that LSTMs can handle problems containing high within-class variability.

4.3 LSTM for Signature Verification

As mentioned above, [19] performed experiments using LSTM for the signature verification, i.e., in that case the task of identifying if a signature is from the genuine author or if an unskilled forger tried to imitate the reference signatures. Instead of reporting the results in a state-of-the-art manner, i.e., using the accuracy, the equal error rate (EER), or the calibrated log likelihood-ratios ($\hat{C}_{llr}^{min}$) [12], they just reported the error of the LSTM-outputs, which is of no practical value.

In order to investigate the practical value of LSTM for signature verification, we have reproduced these networks. However, a better normalization and feature extraction procedure was applied. The system takes a complete text line as input, which is first normalized with respect to skew, writing width and baseline location. Normalization of the baseline location means that the body of the text line (the part which is located between the upper and lower baselines), the ascender part (located above the upper baseline), and the descender part (below the lower baseline) will be vertically scaled to a predefined size each. Writing width normalization is performed by a horizontal scaling operation, and its purpose is to scale the characters so that they have a predefined average width. To extract the feature vectors from the normalized images, a sliding window approach is used. The width of the window is one pixel and nine geometrical features are computed at each window position. Thus an input text line is converted into a sequence of feature vectors in a 9-dimensional feature space. The nine features correspond to the following geometric quantities. The first features represent the average gray value of the pixels in the window and their center of gravity. Furthermore, the second order moment in vertical direction is taken. In addition to these global features, the location of the uppermost and lowermost black pixel are used. Their positions and their gradients, determined by using the neighboring windows, are taken. Feature number eight is the number of black-white transitions between the uppermost and lowermost pixel in an image column. Finally, the proportion of black pixels to the number of pixels between these two points is used. For a detailed description of the features see [13].

We performed experiments on the latest Signature Verification dataset SigComp2011 [12]. As in [19] we could observe that the training converges faster using LSTM compared to SRNs and that the test error is significantly lower.

However, when performing tests using the measures mentioned above (EER and $\widehat{C}_{llr}^{min}$), the results become much worse.

While the LSTM was performing perfect if forgeries of the questioned signatures are also used for training (as also done in [19]), in more realistic scenarios they usually fail. In real case no sample forgeries will be available. Thus the network has to learn the signature based on other information. We have tried to (i) train a general network to distinguish forgeries from genuine signatures on a large training set, (ii) adopt a network which works perfectly on the training set to a specific writer, or (iii) train the network on genuine signatures only, but all attempts failed. In the best case we received an EER of 23.75 % which is clearly below the state-of-the-art performance of 3 %. Thus we can conclude that standard LSTM trained with standard mechanisms are not appropriate for the task of signature verification.

5   Conclusion

In this paper we have investigated LSTM especially on sequential learning problems relevant to pattern recognition. We have confirmed findings of researches in the past and made several novel insights.

First, we have investigated the behavior of LSTM on a novel difficult sequence classification problem and found out that LSTM are much better in generalizing under bad training conditions than SRNs. Thereby, we also figured out that LSTMs are significantly more robust regarding training parameters. It is interesting that the performance on sequences of length much larger than 10 is very good, keeping in mind that the training configuration is fixed and position of the interesting information in the sequences are random. Although it was confirmed that LSTMs are insensitive to noisy input data, but in contrast we could show that LSTMs seemingly losses a lot of their power when using an approximate exponential function, while SRN are thereby lesser affected.

An application of LSTMs to real-world problems requiring the mentioned abilities is a promising future task. Also, the behavior regarding varying numerical precision shall be analyzed more in detail thereby in the context of realistic problems. This seems to be an important research task e.g. consider the situation when implementing LSTMs on embedded devices where high precision floating point types are not natively supported.

Second, we have analyzed the convergence of LSTM for difficult mode detection scenarios. LSTMs converge much faster than SRNs and perform significantly better on the test set. We also confirmed again for this real-world problem that the need for parameter fine tuning is not so urgent for LSTM as for SRNs even on practical problems if enough training data is available.

Third, we have investigated the applicability of LSTM on the task of signature verification. In contrast to previous work we conclude that standard LSTM are not suited for this task. While we confirm the low testing error, the use of state-of-the-art metrics in signature verification analysis reveal that the systems perform much worse than state-of-the-art approaches. In future, the use of auto-encoding LSTM might be an interesting research topic to overcome this problem. Therefore, the recently introduced Evolino training [17] seems to be promising.

References

1. Eck, D., Schmidhuber, J.: Finding temporal structure in music: Blues improvisation with lstm recurrent networks. In: Proceedings of the 2002 IEEE Workshop on Neural Networks for Signal Processing XII, pp. 747–756. IEEE (2002)
2. Gers, F.A., Schmidhuber, J., Cummins, F.: Learning to forget: Continual prediction with LSTM. Neural Computation 12, 2451–2471 (1999)
3. Gers, F.A., Schmidhuber, J.: LSTM recurrent networks learn simple context free and context sensitive languages. IEEE Transactions on Neural Networks 12, 1333–1340 (2001)
4. Gers, F.A., Schraudolph, N.N., Schmidhuber, J.: Learning precise timing with LSTM recurrent networks. Journal of Machine Learning Research 3, 115–143 (2002)
5. Graves, A., Liwicki, M., Fernandez, S., Bertolami, R., Bunke, H., Schmidhuber, J.: A novel connectionist system for unconstrained handwriting recognition. IEEE Transactions on Pattern Analysis and Machine Intelligence 31(5), 855–868 (2009)
6. Graves, A.: A comparison of network architectures. In: Graves, A. (ed.) Supervised Sequence Labelling with Recurrent Neural Networks. SCI, vol. 385, pp. 49–58. Springer, Heidelberg (2012)
7. Graves, A., Bruegge, B., Schmidhuber, J., Kramer, S.: Supervised Sequence Labelling with Recurrent Neural Networks. Ph.D. thesis, Technische Universitaet Muenchen (2008)
8. Graves, A., Schmidhuber, J.: Framewise phoneme classification with bidirectional LSTM and other neural network architectures. Neural Networks: The Official Journal of the International Neural Network Society 18(5-6), 602–610 (2005)
9. Hochreiter, S., Heusel, M., Obermayer, K.: Fast model-based protein homology detection without alignment. Bioinformatics 23(14), 1728–1736 (2007)
10. Hochreiter, S., Schmidhuber, J.: Long Short-Term Memory. Neural Comput 9(8), 1735–1780 (1997)
11. Indermuehle, E., Liwicki, M., Bunke, H.: IAMonDo-database: an online handwritten document database with non-uniform contents. In: 9th Int. Workshop on Document Analysis Systems, pp. 97–104 (2010)
12. Liwicki, M., Malik, M.I., Heuvel, C., Chen, X., Berger, C., Stoel, R., Blumenstein, M., Found, B.: Signature Verification Competition for Online and Offline Skilled Forgeries (SigComp 2011). In: 2011 International Conference on Document Analysis and Recognition (ICDAR), pp. 1480–1484. IEEE (2011)
13. Marti, U.V., Bunke, H.: Using a statistical language model to improve the performance of an HMM-based cursive handwriting recognition system. Int. Journal of Pattern Recognition and Artificial Intelligence 15, 65–90 (2001)
14. Otte, S., Liwicki, M., Krechel, D., Dengel, A.: Local feature based online mode detection with recurrent neural networks. In: 2012 International Conference on Frontiers in Handwriting Recognition, ICFHR (2012)
15. Robinson, A.J., Fallside, F.: The utility driven dynamic error propagation network. Tech. Rep. CUED/F-INFENG/TR.1, Cambridge University Engineering Department, Cambridge (1987)
16. Schmidhuber, J., Gers, F., Eck, D.: Learning nonregular languages: A comparison of simple recurrent networks and lstm. Neural Computation 14 (2002)

17. Schmidhuber, J., Wierstra, D., Gagliolo, M., Gomez, F.: Training recurrent networks by evolino. Neural Comput. 19(3), 757–779 (2007)
18. Schraudolph, N.N.: A fast, compact approximation of the exponential function. Neural Computation 11, 4–11 (1998)
19. Tiin, C., Omlin, C.: LSTM recurrent neural networks for signature veri cation. In: Proc. Southern African Telecommunication Networks & Applications Conference (2003)
20. Tiin, C.: LSTM Recurrent Neural Networks for Signature Veri cation: A Novel Approach. LAP LAMBERT Academic Publishing (2012)
21. Williams, R.J., Zipser, D.: Gradient-Based learning algorithms for recurrent networks and their computational complexity (1995)

Resolution of Geographical String Name through Spatio-Temporal Information

Luca Mazzola[1,*], Pedro Chahuara[1], Aris Tsois[1], and Mauro Pedone[2]

[1] European Commission, Joint Research Center *(JRC)*
Via Enrico Fermi 2749, Ispra, Varese, Italy – I-21027
{luca.mazzola,pedro.chahuara,aris.tsois}@jrc.ec.europa.eu
[2] SERCO
pedone.mauro@gmail.com

Abstract. Entity resolution of geographical references is essential for the analysis of spatial temporal information when data come from heterogeneous sources. In the ConTraffic project we work on the analysis of trajectory of moving objects (e.g. commercial containers and vessels) and, as part of data processing step, the right location for textual references must be determined. In this paper we present an application of Bayesian networks that leverage the information of already resolved references in order to estimate the right entity corresponding to a geographical location. Contextual information of objects that have followed similar trajectories is used as well. Our approach is suitable to perform entity resolution efficiently even when the database contains millions of movements. The results we obtained prove that our method is useful in cases where string similarity methods are unable to provide a solution.

1   Introduction

In the domain of mobility data management it is quite common to find entities that are described by several different textual descriptions giving place to ambiguities when trying to identify them. This ambiguity when describing entities can be originated because data comes from external systems, having different semantics and different data models. The need for data interpretation and integration of entities names stored in spatio-temporal data warehouses is essential to produce unified and valuable results for the purpose of a given task [1]. The *ConTraffic* pilot project running at the Joint Research Centre of the European Commission is one of such systems that needs to transform and integrate data from many sources. ConTraffic deals with information on the status and movement of cargo containers. The aim of this pilot project is to develop new techniques and processes to help authorities in their effort to control the flow of cargo containers. It is estimated that today more than 20 million cargo containers are used worldwide to transport about 90% of the world's cargo. The key concept in ConTraffic is the usage of available data sources, currently not

---

* Corresponding author.

P. Perner (Ed.): MLDM 2014, LNAI 8556, pp. 498–512, 2014.

exploited by authorities, and apply state of the art analysis techniques on the collected mobility data [2].

The basic information unit in ConTraffic is the *Container Status Message* (CSM). Each single CSM record follows a de–facto structure [3], and describes a logistic event of a particular cargo container like the loading of a container on a vessel or the delivery of a full container to its final destination. Such events can be represented by an ontology [4] and can be analyzed using a semantic approach to extract valuable relations [5]. Each CSM contains a unique identifier of the container following the ISO standard [6], a date (and sometimes precise time) and a textual description of the location where the event took place, besides other fields. In order to be able to use the CSMs in various analysis processes, and in order to be able to integrate the data from the different sources, the textual location description in each CSM must be mapped to a real-world location. In ConTraffic we use the *UN/LOCODE* [7] taxonomy, that is a set of location for trades and transport purposes collected by the United Nations, as a table of real-world locations.

In this paper we address the issue of text strings resolution for strings that represent locations of containers. The ConTraffic system aims at the segmentation of temporal sequence of container events in order to obtain a timed list of locations reached by a container in a certain interval of time. This analysis is necessary for risk analysis of the goods transported in the container. Each obtained segment is known as a *trajectory* . The trajectory estimation excludes the messages whose declared location cannot be accurately resolved, since the storage of trajectories with uncertain location points could negatively impact the risk analysis procedure. This fact explains the importance of string resolution into location, in order to provide a more complete and realistic trajectory estimation.

A string similarity based entity resolution is not feasible in several cases because the description of the location does not match a *UN/LOCODE* location, for instance when the declared location of a container shows a description of a terminal port instead of a real location. Another case is when a declared location description could be related with many geographical entities in the world. We believe that relational information within the sequence of events of the container whose location we want to resolve is essential in order to precisely estimate the right location. The method we propose relies on contextual information about the container and the assumption that most containers in a location share a small set of possible previous and next locations. In ConTraffic, based on the hundreds of millions of CSM records available, one can reconstruct millions of trajectories that represent the movement of containers. This is possible by using the location description and the date/time of each CSM records whose location description are not ambiguous. In this work, we present the application of statistical machine learning in order to learn a model that allows us to estimate the location of a container by means of an analysis of a set of containers belonging to trajectories having similar features.

1.1 Related Work

Named entity resolution is a important task in order to clearly identify the identity of a reference when the information comes from heterogeneous sources. It has several applications in domains where data is extracted from unstructured or semi-structured textual descriptions such as publication references; but also when the data is organized by means of structured languages.

The most naive approach to resolve entities names is comparing references according to a measure of string similarity. Several N-grams based methods [15] can score the similarity of textual reference. Despite its efficiency and simplicity for implementation, using the textual description of the reference as the only attribute can lead to a poor efficacy. In most real applications, several different entities can have very similar names or even share the same name. That is why most of the works on entity resolution agree on the fact that other attributes of the entity besides its name can largely improve the results of the resolution. Furthermore, the relational information of the target among entities in the database is also relevant.

The methods that have been applied for entity resolution can be organized in two main trends: the unsupervised approach that generated clusters of references where a cluster is expected to group references of the same entity, and those that rely on a learning corpus and treat the resolution as a classification problem. The first case have the advantage of avoiding the step of learning a model and annotating a corpus. Gooi et al. [16] have proposed the use of agglomerative clustering for cross-document co-reference resolution using Kullback-Leibler divergence measure to form the clusters. Li et al. [17] have developed the framework SeReMatching that creates clusters from references of tuples (a vector of attributes of the reference), while assigning a weight to every attribute. An important development on this direction has been the work of Bhattacharya et al. [18] that includes the relational information of the entity. For example, when resolving publication authors names, if two authors references are quite similar and they are associated to the same list of publications is highly probable that these reference correspond to the same author. Their approach allows to collectively resolve several type of entities as well. An approach that exploits relation information by Markov Logic is presented in [19] where the model is composed of weighted logical rules which can be used to make probabilistic inference about the group of references corresponding to the same entity. Based on the facts that the weights are not learn and the structure of the formulas is given a priori, the approach remains unsupervised. The main disadvantage of these unsupervised methods is the need to make the similarity evaluation of many pairs of references in order to create the clusters. It means, if E is the set of references, the similarity is measured for every $(e_i, e_j) \in E \times E$. In applications dealing with millions of references, clustering methods can be inappropriate when a short time response is required.

Supervised methods, on the other hand, offer better inference performance in terms of accuracy and efficiency. In this case the classification problem can be seen in two ways: first, given a reference to resolve and an entity the classification result is a binary value match/not match; or second, given a reference to resolve the possible classes are candidate entities. The best strategy is usually domain dependent: the number of entities, the degree of certainty of the information, and the available learning corpus must be taken into account. Some classical methods of machine learning in the domain are Bayesian networks and SVM [20,21]. Nguyen et al. [22] have resorted to Conditional Random Fields for named entity resolution in the task of bacteria biotopes extraction. When resorting to supervised approaches, one of the limitations is the need of a learning corpus that in some cases is not available or is effort demanding to annotate.

In the present work we focus on the special case of spatial-temporal entity resolution, which has not been largely treated in the literature. Supervised and unsupervised methods can be applied to this case provided that the resolution method allows to model spatial and temporal relation among entities. For instance, Zhang et al. [23] show the application of entity relation graphs for person identification in videos using the location and the activities performed by subjects. Martins et al. [24] have presented a system to recognize geographical and temporal entities from textual documents. In our work we focus in spatio–temporal entity resolution in big datasets. In a previous work of Con-Traffic, Mazzola et al. [25] have already proposed the use of trajectory for entity resolution in spatio–temporal data.

2   Problem Presentation

The issue we try to address is when a text string referencing a location within a CSM, the location of a moving object, cannot be matched with confidence using string similarity to any particular location in the locations list (the UN/LOCODE in the ConTraffic case). There are three cases in which an entity cannot be resolved:

- **Single Match:** There is only one location candidate for the reference to be resolved, however the confidence is not high, so a manual verification is required.
- **Multiple Match:** There are several location candidates and one of them must be manually chosen. It is possible that the right location is not among the candidates.
- **Fail:** No candidates were found for the given reference.

The main objective of our work is to propose a method that can resolve location references in cases where string similarity cannot be applied because of the described problems. We require a method that provide always a set of candidates containing the right location. Every candidate should be given with a numerical score and the right location is expected to have a high score.

We count on a database storing, for every container k , an ordered set of CSMs, $C_k = \{c_1, ..., c_n\}$, each c_i contains the information of a specific container event in tuple with the structure $c_i = <x, d, t, s>$ representing the followings informations of the event: the location(x), the description(d), the date and time(t), and the status(s) (full or empty) of the container. In most cases x is a character string referencing a geographic entity in the world, but sometimes it encodes something else, such as the name of a transport facility, the name of a consignor or even an acronym. This descriptions can be interpreted by people/companies involved in the transportation chain, but they are not suitable to rightly estimate the trajectory of a container.

We store also a list of locations(with UN/LOCODE) in the database $L = \{l_1, ..., l_m\}$ and the aim is, for every c_i, to find the right location l_i that is referenced in the CSM c_i by $attribute_x(c_i)$. The domain of $attribute_x(c_i)$ is any combination of alpha numeric characters of any size. Ideally the aim would be to find a function $f(c_i) = l_j$, where $l_j \in L$. However, wrong matches of locations can have a strong negative impact in final applications exploiting the database. That is why we opted to obtain a set of candidates with a probability of being the right location and to automatically choose the best candidate if its score is higher than a certain threshold or to allow a human expert to decide among the candidates.

Besides, we consider important to leverage the temporal spatial relations among the CSM of the same container. Indeed, the elements in the ordered set C_k describe consecutive points in the trajectory of a given container k. In order to disambiguate the location of a CSM we include in the analysis the surrounding (previous and next ones) valid geographical locations. We reference each consecutive couple of location a transportation leg, representing the elementary information brick for our computation in ConTraffic. Consequently, our function becomes: $f(c_{i-1}, c_i, c_{i+1}) = \{(l_1, w_1), ..., (l_s, w_s)\}, l_i \in L, w_i \in [0, 1]$ and includes not only the location reference of past and previous events, but also the information of the entire event. In this way the contextual information of the container event is taken on account, such as the time of these events and the kind of event they belong to. The number of candidates, s, must be set to a small value and every candidate must be obtained with a confidence value.

3   Methods

The approach we follow is based on supervised machine learning. A classifier is a suitable alternative to implement the function , since it can take into account the contextual information of a CSM with respect to past and future information. As a requirement to our system, we need to implement a method whose output gives a set of candidate locations together with a probability for each of them. We have chosen to use Bayesian networks since this method allows to represent relation dependencies among the entities of the domain, and to model the degree of certainty through the use of probabilities that are obtained from learning data.

The two main differences of our approach with classical applications of supervised models are: first, that our learning data is taken from CSMs that are already related with a location in the database, then there is no need of an additional learning corpus; and second that the classification model must be created dynamically for each CSM that has an unreferenced location because given the possible combination of triples of locations it is not feasible to create a model a priori for each possible case. This imply that before performing the classification, it is necessary to select a sample of learning records that is representative of the whole population and whose size is enough to avoid under-fitting, and at the same time allows fast learning and inference.

3.1 Bayesian Networks

A Bayesian Network(BN) is a powerful technique to make statistical inference with uncertain data. A BN is a directed acyclic graph where the nodes are variables and the arcs represent the conditional dependency among the variables. The conditional relations are given by a table containing the conditional probability $P(X_i \mid Pa(X_i))$ for each variable X_i, where $Pa(X_i)$ is the parent set of X_i. Under a conditional independence assumption, the joint probability distribution of $X = (X_1, ..., X_n)$ is represented by $P(X = x) = \prod_{i=1}^{n} P(X_i \mid Pa(X_i))$

3.2 Attributes

The attributes we need in order to estimate the right location described in a CSM must express the spatio–temporal relations of the current CSM with the previous and next CSM after the global temporal ordering of the set. The idea is to exploit the information of other containers that followed the same trajectory.

Table 1. Attributes for Location classification

Attribute	Description	Domain
Prev_loc	Previous location (transportation leg start)	Numeric UN/LOCODE
Next_loc	Next location (transportation leg end)	Numeric UN/LOCODE
Diff_prev	Difference (in weeks) from the previous geolocalized CSM	$[1, 12] \in N$
Diff_next	Difference (in weeks) from the following geolocalized CSM	$[1, 12] \in N$
Event	Description of the event (e.g. container charged in a vessel, discharged, transported inland)	Code of the event $[1, 90] \in N$
Load_status	Weather the container is empty or full with goods	E/F

Table 1 shows the attributes used in the classification process; they start with the spatial relationship with the previous (*Prev_loc*) and following (*Next_loc*) record. Then the difference of time between the desired location and the origin (*Diff_prev*) and the destination (*Diff_next*) of the container is considered: it is important since, given that sample records and analysed CSM share the same origin and destination, the most similar the interval time in the sample records is to the analysed CSM the higher the probability that these two entities correspond to the same trajectory. The granularity for temporal distance has proved to be more efficient for fast inference when rounded to weeks, instead of being expressed in days. Also the message reported (*Event*) and the status of loading of the container (*Load_status*) are important for the similarity discrimination. From our experience, these dimensions amongst all the others are enough to create a basic "signature" of records.

3.3 Dataset Description

The dataset is composed of more than a billion records of CMS, in their original form (raw) and in an elaborated one that is the result of the *extract, transform, load* (ETL) process. The ETL section is a complex pipeline that is in charge of transforming each collected record into the most suitable form to make sense of the data and to extract higher level information. The major steps in this pipeline are the *event description cleaning and reconciliation*, the *load_status association*, the *location cleaning*, the *vessel resolution*, the *obsolete record removal*, the *future event marking*, and eventually the *duplicate removal*.

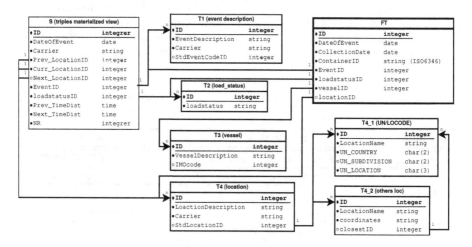

Fig. 1. Diagram of database structure on which our experiments relied

The ETL process relies on a data set organized as a star, with the fact table (*FT*) as the center. All the other tables (*T_x*) are lookup tables of different depth for the dimensions that specify the context of the event registered in the fact table. The *FT* contains, in addition to the identifications to link with the leafs, the date of the event, the collection date, and an unique identification of the container. The most important lookup tables are represented by the events descriptions (*T_1*), the load_status (*T_2*, that indicates which status the container is reported with: *full, empty,* or *unknown*), the names of the vessel (*T_3*), and the locations ones(*T_4*).

The latter dimension is represented as a logical view composed by two different tables: one that stores the location mapped into UN/LOCODE and one that allows to represent locations that are currently outside that taxonomy (this allows us not to force the aggregation of locations into the better known one, that is part of the taxonomy). As a side effect, this approach to locations mapping grants the possibility to make sense also of very close events, like the one happening in the port area and the one immediately following it, in the receiving storage facility.

In particular the current work is devoted to improve the **location cleaning** part of the ETL, in fact the result of our cleaning procedure will eventually modify the link of the record with the locations lookup table (T_4).

The additional view we created for the current task stores triples of consecutive events in different locations and all the additional dimensions used for its interpretation, as described in the next section. This is justified by reasons of manageability and performances. For this task a database object, known as "materialized view" was adopted: it works as a normal table and can be queried on needs having the triple physically stored in the data space, but on the same time it is updatable on needs with a specific SQL query like a view. In this way, the update can be operated by a scheduled job with the requested frequency, without affecting the response time and the database server load for each query that uses the stored data.

3.4 Learning Corpus

In the database of ConTraffic we have created a table S that stores the number of occurrences for every triple of locations that appears consecutively at least once in a container trajectory, see Fig. 1. The purpose of this table is to store the representativity of every triple in the database. Given a triple c_{i-1}, c_i, c_{i+1} where the location of c_i is unknown, the procedure to create the classifier to estimate this location starts by taking a random sample of triples from the database where the first and the last locations are respectively $c_{i-1}(Location)$ and $c_{i+1}(Location)$, where the probability of a record to be chosen is proportional to its representativity in the table S. We have limited empirically the size of the sample, for manageability reasons, to some thousands records randomly extracted from the full set. The variable dimension of the set allowed guarantees the possibility to extract a representative set also based on the absolute number of distinct samples present for the specific examined location. This is useful for considering less frequent but really simile examples, despite the risk, indeed limited for big numbers as our case, of affecting the requested statistical representativity.

3.5 Entity Resolution Process

The procedure to resolve a location reference starts by obtaining a random sample from the database of triples of events that serve as a learning corpus. Given a triple $(c_{prev}, c_{curr}, c_{next})$ where the reference to be resolved is

$attribute_x(c_{curr})$, we take from the database a random set of triples, T, having the following property: $\forall t^i = (t^i_{prev}, t^i_{curr}, t^i_{next}) \in T, attribute_x(t^i_{prev}) = attribute_x(c_{prev}), attribute_x(t^i_{next}) = attribute_x(c_{next})$. Additionally, all the reference in a triple t_i must be already related with a location in the database. It is important to take only a sample since the number of movements in the database is so big that learning a model with all of them would be too expensive. We rely on examples of containers that have followed the same trajectory and, by including contextual information, we ensure that those examples have been obtained under the same circumstances of the event for which we want to resolve the location. It is clear that the final set of candidates will be a subset of the locations referenced by every $attribute_x(t^i_{curr})$. The process is resumed in the Figure 2.

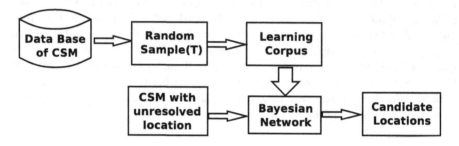

Fig. 2. Diagram of the overall method for location resolution

Using the set T we created the learning corpus with the structure described in 3.2. Then a Bayesian network allows us to give a probability for the locations that are referenced in the learning set of triples. After the inference is performed, the locations are ranked according to their probability and we take the first 5 locations as our candidate set. We have tested empirically that the sum of probabilities for the first five locations is usually close or bigger than 95%. This is why we considered five an appropriate size for the candidates set.

In the cases where one of the candidates location has a probability of being the unreferenced string is higher than a certain threshold, the resolution is carried out automatically. However, if the probability of every candidate is below the threshold, a human expert must select one location among the candidates.

4   Experiments

For testing the performance of the proposed method we chose some representative cases in our data source for different carriers. The representativity is given by the number of container movements that share the same unresolved location. We then tried to find manually the corresponding locations in the UN/LOCODE table for these cases and to asses the accuracy of the solution proposed by our new method. Manual verification is a task that involves comparing the unresolved case with similar container trajectories and determining the most likely location by analyzing the trajectory in a map. The manual verification effort required for our test dataset was equivalent to about half a man–month work.

4.1 Evaluation

The evaluation was based on the presence and position among the candidate locations of the right location resulted from the manual analysis. On the hundred cases analyzed, half was locations already cleaned with the previous methodology in use in our group. For this we simulated the complete cleaning process through their conversion to unclean in a parallel development environment. Two tests have been performed. In the first one, 100 already resolved locations have been modified to appear in the database as unresolved. While in a second test, we have taken references that cannot be resolved using only string similarity and require a method to provide a set of alternatives (see section 2).

4.2 Test1: Already Manually Resolved Locations

The references used in this test have previously been resolved by string similarity and expert knowledge. They correspond to cases having one of the following two characteristics: (1) they belong to the "*SINGLE*" case (see section 2) and then the reference was confirmed by the human expert, or (2) they belong to the "*MULTIPLE*" case and one of the alternative was rightly chosen by a human expert. The purpose of this test is to demonstrate that the performance of our proposed method is at least as good as the comparison of string similarity in cases where the judgment of the human expert is needed. We have performed the test with 100 references. Only in one reference the right location appears as the second best candidate in the set of alternatives given by our method. With the other 99 references, the right location was ranked first. Despite the limited number of location, it is important to remember that these entries are taken in descending order of representativeness in our CSM dataset, thus the cleaning of this top 100 cases resulted in a relevant coverage in term of cleaned movements.

4.3 Test2: Resolved Locations

In this test we used references that belong to one of the following case: (1) they are classified as "*MULTIPLE*" but the set of alternatives given through string similarity does not include the right location, or (2) they are classified as "*FAIL*" so the human expert has no alternative to choose from. With this test we intend to evaluate what would be the potential improvement of using the proposed method in the cases where string similarity fails at proposing the right location. The results for this test are given in the Table 2.

In 12 of the references the location cannot be estimated even manually since the textual description is too ambiguous. It can be, for instance, the description

Table 2. Results for random sample of unresolved references

Right location	Case		
	FAIL	MULTIPLE	Total
Is the first candidate	20	36	56
Among the candidates	6	10	16
Is not a candidate	6	3	9
Cannot be resolved	10	2	12
Not enough data to estimate	4	3	7
Total	46	54	100

of a generic place, like *Port terminal 1*. As described above the method create a classification model using similar trajectories under the same context, however in some cases those trajectories are two few (less than 100) in order to create a classification model, there are 7 references like this. Fore more of half of the references the right location was ranked first among the candidates, and only in 9 cases the method provided false candidates. It suggests that the proposed method is an excellent complement for the string similarity tool.

4.4   Results

The performances reported in the table below showed a quite good success rate, with about 80% of the cases where the first location proposed is the best candidate even at the manual inspection (with the limits of 90% in the group of simulated and around 70% for the other). Extending the analysis the second class (the one that consider the first 5 results proposed) the success rate increases till 93,5%. Even if the additional 13% requires more attention, it is based on an already existing category for the user of the system, identified as "MULTIPLE". To be worth to notice is the fact that the preceding cleaning approach adopted –using a string matching approach on the location description– classified, in the same dataset, a percentage close to 20% as "FAIL", not having any proposal for the cleaning of that locations.

The differences in performances registered amongst the cases real and simulated one can be explained by the fact that the cleaned one (simulated) represents by definition the easiest and most clear cases to be solved. The proposed approach, stressing the contextual information of the single event, the time distance between events, and the temporal sequence of events (limited to three steps: one preceding and one following the event under analysis), is able to create a model to classify a freshly entered location.

In respect of the previously adopted one, that relied just on the information contained in the record analyzed, the improved method can discriminate better the similarity of the cases already present in our dataset with the analyzed one. It was also able in some cases to propose a candidate location for the cleaning that even at the manual inspection was not emerged; furthermore this candidate proposal was then recognized by the human being as a possible locations, being supported by further inspections.

A Practical Case. Sometimes, the locations suggested by our application does not include the one that the user would have naturally selected. Fig. 3 presents a case where the location that was associsated manually to the descriptor *"BUENA PARK, BUENA PARK, ORANGE, CALIFORNIA, UNITED STATES"* was the UN/LOCODE US.CA.BUN (UN_country, UN_Subdivision, and UN_Location), corresponding to the string *"Buena Park"*.

Fig. 3. An Example: the two locations suggested by the method *"Long Beach"* and *"Los Angeles"*, represented on a Google Map© together with the candidate manually identified *"Buena Park"*. They are in the same region and can be considered equivalent on a long scale trajectory identification and validation.

Despite the fact this an easy task for an human being, it is a challenging one for an automatic procedures, due to some factors:

- the objective location is a small one in the middle of two others deeply used for good transportation, namely US.CA.LGB (*"Long Beach"*) and US.CA.LAX (*"Los Angeles"*),
- the unknown location does not represent a port or a final destination, but an intermediate step of an inland transport segment for the sending or delivering logistic chain,
- the frequency and total number of trip passing for the location under exam are not comparable, being some order of magnitude inferior, in respect with the two close bigger ones. This makes its representativeness in the S table of triples very low.

Anyway, it is worth to observe that the two locations suggested with higher confidence as possible candidates in the specific case to the attention of the human being in charge of the cleaning procedure ("Long Beach" and "Los Angeles") are close enough to the "correct" one. In fact they can be adopted as good substitutes when trying to use this data to have an overview and make inference over a container trip, that takes place all around the globe, being really relevant and interesting when and where involving long or intercontinental segments. For this reasons, we think that our solution could be beneficial for the task in object as an additional informative source, particularly considering the fact that the strings encoding the names of the locations have no practical needs to show an high similarity index, and this is in fact tha main dimension normally used by human being working on this task.

5   Conclusions and Future Work

We presented an application of supervised machine learning that helps to find the location referenced by text description in container status messages. The main idea is the use of relational information contained within the sequence of messages that represent a trajectory. We proved that a learning corpus containing trajectories under the same circumstances is useful to approximate the location that is referenced in a message. A supervised approach results appropriate since the use of unsupervised methods would lead to the generation of too many clusters, and consequentely a low efficiency, given the number of entities and references we deal with. Furthermore, the problem of annotating a learning corpus is not present in the process we follow for entity resolution as the classification model is implemented dinamically using a subset of the whole dataset.

In our project, most of the references can be resolved through the use of string similarity techniques, which is a simpler and faster solution. Thus, we do not intend to replace this basic approach with our proposed method, but to apply supervised machine learning only when the right location cannot be estimated with enough certaintly based on the textual definition due to a low similarity between the description of the reference and the name of the location.

Our experimental results show that it is possible to find the referenced location when the reference is, for instance, a generic name of a transportation facility. The method relies on the fact that containers trips having the same origin and destination, are prone to share similar intermediate locations. Using similar trips to resolve intermediate locations has the logical consequence of restricting the set of candidates to locations that are geographically in between the two locations referenced in the previous and the next messages, which is impossible relying only on the textual description. Moreover the temporal aspect also plays an important role since intermediate locations having trajectories where the time of transportation from previous location and to next location are similar to the unresolved case would have a higher probability.

As mentioned above, a container status message is composed of several fields besides location that contain also text references, such as the description of the event. The name of the vessel where the container travels is also declared in the message, although we did not included it in our definition to maintain simplicity. These elements could also be subject to entity resolution using supervised machine learning. It would lead to an integrated solution of entity resolution with multiple types of entities.

References

1. Mazzola, L., Eynard, D., Mazza, R.: GVIS: a framework for graphical mashups of heterogeneous sources to support data interpretation. In: 3rd IEEE Conference on Human System Interactions, HSI 2010, pp. 578–584 (2010)
2. Varfis, A., Kotsakis, E., Tsois, A., Donati, A.V., Sjachyn, M., Camossi, E., Villa, P., Dimitrova, T., Pellissier, M.: ConTraffic: Maritime container traffic anomaly detection. In: MAD 2011 Workshop Proceedings, p. 113 (2011)
3. US Government Printing Office,
   `http://www.gpo.gov/fdsys/granule/CFR-2012-title19-vol1/`
   `CFR-2012-title19-vol1-sec4-7d/content-detail.html`
   (last accessed on April 10, 2013)
4. Villa, P., Camossi, E.: A description logic approach to discover suspicious itineraries from maritime container trajectories. In: Claramunt, C., Levashkin, S., Bertolotto, M. (eds.) GeoS 2011. LNCS, vol. 6631, pp. 182–199. Springer, Heidelberg (2011)
5. Camossi, E., Villa, P., Mazzola, L.: Semantic-based Anomalous Pattern Discovery in Moving Object Trajectories. arXiv preprint arXiv:1305.1946 (2013)
6. International Organization for Standardization. ISO 6346:1995. Freight Containers–Coding, identification and marking,
   `http://www.iso.org/iso/catalogue_detail?csnumber=20453`
7. United Nations Code for Trade and Transport Locations,
   `http://www.unece.org/cefact/locode/welcome.html`
   (last accessed on April 12, 2013)
8. Güting, R.H., Böhlen, M.H., Erwig, M., Jensen, C.S., Lorentzos, N.A., Schneider, M., Vazirgiannis, M.: A foundation for representing and querying moving objects. ACM Transactions on Database Systems (TODS) 25(1), 1–42 (2000)
9. Sehgal, V., Getoor, L., Viechnicki, P.D.: Entity resolution in geospatial data integration. In: Proceedings of the 14th Annual ACM International Symposium on Advances in Geographic Information Systems, pp. 83–90. ACM (2006)

10. Kang, H., Sehgal, V., Getoor, L.: GeoDDupe: a novel interface for interactive entity resolution in geospatial data. In: 11th International Conference on Information Visualization, IV 2007, pp. 489–496. IEEE (2007)

11. Martins, B., Anastácio, I., Calado, P.: A machine learning approach for resolving place references in text. In: Geospatial Thinking, pp. 221–236. Springer, Heidelberg (2010)

12. Krumm, J., Horvitz, E.: Predestination: Inferring destinations from partial trajectories. In: Dourish, P., Friday, A. (eds.) UbiComp 2006. LNCS, vol. 4206, pp. 243–260. Springer, Heidelberg (2006)

13. Abdessalem, T., Moreira, J., Ribeiro, C.: Movement query operations for spatio-temporal databases. Proc. 17emes Journées Bases de Données Avancées (2001)

14. Winkler, W.E.: Overview of Record Linkage and Current Research Directions. Research Report Series, RRS (2006)

15. Kondrak, G.: N-gram similarity and distance. In: Consens, M.P., Navarro, G. (eds.) SPIRE 2005. LNCS, vol. 3772, pp. 115–126. Springer, Heidelberg (2005)

16. Gooi, C.H., Allan, J.: Cross-document coreference on a large scale corpus. In: Susan Dumais, D.M., Roukos, S. (eds.) HLT-NAACL 2004: Main Proceedings, Association for Computational Linguistics, Boston, Massachusetts, USA, pp. 9–16 (2004)

17. Li, Y., Wang, H., Gao, H.: Efficient entity resolution based on sequence rules. In: Shen, G., Huang, X. (eds.) CSIE 2011, Part I. CCIS, vol. 152, pp. 381–388. Springer, Heidelberg (2011)

18. Bhattacharya, I., Getoor, L.: Collective entity resolution in relational data. ACM Transactions on Knowledge Discovery from Data 1(1), 1–36 (2007)

19. Singla, P., Domingos, P.: Entity resolution with markov logic. In: ICDM, pp. 572–582. IEEE Computer Society Press (2006)

20. Nadeau, D., Sekine, S.: A survey of named entity recognition and classification. Lingvisticae Investigationes 30(1), 3–26 (2007)

21. Rao, D., McNamee, P., Dredze, M.: Entity linking: Finding extracted entities in a knowledge base. In: Poibeau, T., Saggion, H., Piskorski, J., Yangarber, R. (eds.) Multi-source, Multilingual Information Extraction and Summarization. Theory and Applications of Natural Language Processing, pp. 93–115. Springer, Heidelberg (2013)

22. Nguyen, N.T.H., Tsuruoka, Y.: Extracting bacteria biotopes with semi-supervised named entity recognition and coreference resolution. In: Proceedings of the BioNLP Shared Task 2011 Workshop, BioNLP Shared Task 2011, pp. 94–101. Association for Computational Linguistics, Stroudsburg (2011)

23. Zhang, L., Vaisenberg, R., Mehrotra, S., Kalashnikov, D.V.: Video entity resolution: Applying er techniques for smart video surveillance. In: PerCom Workshops, pp. 26–31 (2011)

24. Martins, B., Manguinhas, H., Borbinha, J.: Extracting and exploring the geo-temporal semantics of textual resources. In: 2008 IEEE International Conference on Semantic Computing, pp. 1–9 (2008)

25. Mazzola, L., Tsois, A., Dimitrova, T., Camossi, E.: Contextualisation of Geographical Scraped Data to Support Human Judgment and Classification. In: European Intelligence and Security Informatics Conference (EISIC) 2013, August 12-14, pp. 151–154 (2013) doi: 10.1109/EISIC.2013.33

26. Cao, H., Wolfson, O.: Nonmaterialized motion information in transport networks. In: Eiter, T., Libkin, L. (eds.) ICDT 2005. LNCS, vol. 3363, pp. 173–188. Springer, Heidelberg (2005)

Detecting Programming Language from Source Code Using Bayesian Learning Techniques

Jyotiska Nath Khasnabish, Mitali Sodhi, Jayati Deshmukh,
and G. Srinivasaraghavan

International Institute of Information Technology, Bangalore
Bangalore - 560100, India
{jyotiskanath.khasnabish,mitali.sodhi,jayati.deshmukh}@iiitb.org,
gsr@iiitb.ac.in

Abstract. With dozens of popular programming languages used world-wide, the number of source code files of programs available online for public use is massive. However most blogs, forums or online Q&A websites have poor searchability for specific programming language source code. Näive thumb rules based on the file extension if any are invariably used for syntax highlighting, indentation and other ways to improve readability of the code by programming language editors. A more systematic way to identify the language in which a given source file was written would be of immense value. We believe that simple Bayesiam models would be adequate for this given the intrinsic syntactic structure of any programming language. In this paper, we present Bayesian learning models for correctly identifying the programming language in which a given piece of source code was written, with high probability. We have used 20000 source code files across 10 programming languages to train and test the model using the following Bayesian classifier models – Naive Bayes, Bayesian Network and Multinomial Naive Bayes. Lastly, we show a performance comparison among the three models in terms of classification accuracy on the test data.

Keywords: bayesian learning, naive bayes, bayesian network, multinomial naive bayes, programming language identifier, source code detection.

1 Introduction

As web-based hosting services such as Github [1], Bitbucket [2] for sharing software development projects with revision control are gaining popularity among developers, the number of source code files (we use the term 'sources' in the rest of this paper to mean 'source code files') available online is growing exponentially. There is an abundance of source code snippents shared by users in popular online forums dedicated to troubleshooting and solving programming related questions such as Stack Exchange [3]. However most search engines seem to have difficulty finding sources in specific programming languages essentially due to lack of automatic source language detection. Often most websites and forums use Näive thumb rules to guess the language by the extension used and/or

P. Perner (Ed.): MLDM 2014, LNAI 8556, pp. 513–522, 2014.

tag based methods to identify the keywords. But with thousands of online forums available, the source language for most of the sources remain unidentified
and they appear simply as plain text files. It would be immensely helpful if systematic tools were used to correctly identify the programming language in which
a source or even a code snippet was written. This would naturally lead to more
widespread code reuse and better search engine visibility. Source code search
engines like SearchCode [4] or Codase [5] can greatly benefit from this model
by being able to index sources by detecting the language. In this paper we propose a Bayesian learning based model for detecting the underlying programming
language from a source or code snippet.

In our work, we are proposing a supervised learning model based on Bayesian
learning techniques for detecting the programming language in which a source
was written. We have implemented the model using three different learning techniques — Näive Bayes Classification, Bayesian Network and Multinomial Näive
Bayes Classification. We have used over 20,000 sources written in ten different
programming languages – C, C++, C#, Objective-C, Java, Python, Ruby, Perl,
HTML, JavaScript for training our classifier models. After the classification is
complete, we run each of the classification models on the test set of sources and
measure accuracy of each of the classifiers.

The rest of the paper is organized as follows. In Section 2, we discuss popular
online source code highlighting tools available and related work that has been
done in this area. Then in Section 3, we explain each of the classification models
in detail and discuss why the model is suited for the programming language
identification problem. In Section 4, we describe our training data and how
we train our classifiers using the training data available. Finally we draw the
comparison among the three different classification techniques used in terms of
classification accuracy and measure the performance of our models in Section 5.

2   Related Work

There are existing tools available such as Google Code Prettify [6] that allow
syntax highlighting in source code snippets using heuristics. Grammars for each
of the programming languages supported are already pre-specified in the application. But typically these are unable to detect the programming language of the
code snippet. Also, there are tools like SyntaxHighlighter [7] or Highlight [8] that
syntax-highlight code snippets in blog posts using the set of pre-defined keywords
available. Tools like SourceClassifier [9] use Näive Bayes classifiers for detecting
the programming language from a source. As corroborated by our results in
this paper, these turn out to be rather inadequate, since the accuracy of these
methods is well below what would be considered acceptable in practice. Klein
et al. [10] propose a statistical technique for programming language detection
based on detection of comment blocks or strings from the source and performing
statistical analysis on special characters such as brackets, first word in a line,
last character, operators, punctuations etc. This method beats the performance
of SourceClassifier but is still rather unacceptable in practice with an accuracy

that falls well below 50%. Source code search engines such as SearchCode [4] and Codase [5] have millions of source codes indexed and provide searchability for specific keywords. However, they mainly use fast indexing techniques on the keywords for producing the results. They do not necessarily identify the programming language. These techniques may work well for retrieving sources in a particular programming language but would not work for applications such as automatic syntax highlighting.

3   Classification Models

We used three well known classification models for the problem using a feature vector consisting of token occurrences in the sources. By tokens we mean the standard 'token's as in a typical programming language parser, preceding a syntax analysis. Each source is characterized by its feature vector.

3.1   Using Näive Bayes Classifier

Näive-Bayes Classification is a well known technique for classification of datasets into discrete categories. Näive-Bayes expects a Binomial feature vector for every data instance, assumes independence among the features and arrives at Maximum Likelihood estimates of probability that a new data point belongs to a certain category. Given a set of features $F_1, F_2, \ldots, F_n$ for a source the standard Näive-Bayes way of computing the maximum likelihood probability that the source is written in the language L is as given below:

$$p(L|F_1, F_2, \ldots, F_n) = \frac{p(L)p(F_1|L) * p(F_2|L) * \ldots * p(F_n|L)}{p(F_1, F_2, \ldots, F_n)} \qquad (1)$$

where $p(L)$ is simply the ratio of the number of sources of the given language L to the total size of the dataset, $p(F_i|L)$ is the ratio of the number of sources of language L with feature F_i to the number of the sources in language L and $p(F_1, F_2, \ldots, F_n)$ is the ratio of the number of sources with all the features $F_1, \ldots, F_n$ to the total size of the dataset. The independence assumption may be appropriate for languages that do not share too many tokens in which case occurrence or non-occurrence of tokens can in fact be treated as more-or-less independent events. This observation is indeed borne out by our experimental results that we present later in Section 4. As observed in some earlier studies on Näive-Bayes Classifiers by I. Rish [11] using Monte-Carlo simulations, Näive-Bayes works well for low-entropy feature distributions and for certain nearly-functional feature dependencies. We do have reasons to believe that programming language sources if fact have low-entropy feature distributions, with token occurrences as features. Also as noted by Rennie J. D., Shih L., Teevan, J. & Karger D. R. [12], heuristic improvements on the performance of a Näive-Bayes is in fact possible and such improvements result in performance as impressive as Support Vector Machines. Considering the ease of implementation and less demand on the size of the dataset of a Näive-Bayes we suspected that Näive-Bayes is very likely to work well for our scenario.

3.2   Using Bayesian Networks

The next natural choice was to use a Bayesian Network. Co-occurrences and a predictable syntactic order of tokens is typical of any programming language. A Bayesian belief network would be a natural choice to model these dependencies. Bayesian Network is a directed graph that naturally models conditional dependencies between attributes of the datapoints in a dataset. The survey by David Heckerman [13] [14] discusses graphical models in great detail. A Bayesian treatment of the kind that Bayesian Belief networks allow us to do, does provide robustness against missing keywords, enables incorporating priors for conditional dependence among keywords for specific languages, and offers a ready-made protection against overfitting.

In our problem, the keywords from each programming language are the nodes of the network making a total of 1056 nodes. The nodes are dependent on other nodes considering the fact that certain keywords in a language are succeeded by particular keywords. For example, if the keyword 'if' is taken for C, C++, Java language, then it is very likely that it will be succeeded by the 'else' keyword, while for languages like Python, 'if' keyword is often followed by an 'elif' or an 'else' keyword. So in our test data, whenever we get an 'if' keyword followed by the 'elif' keyword, Python will get a much higher likelihood score than C or Java.

3.3   Using Multinomial Naive Bayes Classifier

Multinomial Näive Bayes is a Bayesian Classifier where the number of occurrences of each token is also accounted for. We would expect this to work better than the Näive Bayes model that builds on a binary 'Bernoulli' value for each attribute (token) indicating occurrence or non-occurrence of the token. Multinomial Näive Bayes has traditionally been used for document classification. the programming language identification problem is very similar in spirit, with each language having a characteristic distribution of token occurrences across sources. McCallum et al. [15] shows that multinomial model outperforms the Bernoulli model when the size of the text document is large and incorporates a large vocabulary.

Kibriya et al. [16] discusses how given a document, Multinomial Näive Bayes (MNB) computes the probabilities for different classes along with several improvement over MNB for better classification accuracy. The set of classes is defined by C and the size of the vocabulary is defined by N. Using Bayes' rule, the maximum likelihood of assigning a test document t_i to class c is given by,

$$Pr(c|t_i) = \frac{Pr(c)Pr(t_i|c)}{Pr(t_i)}, c \in C \qquad (2)$$

Finally, given f_{ni} the count of word n in test document t_i and $Pr(w_n|c)$ the probability of word n for class c, the probability for obtaining a document like t_i in class c is calculated as,

$$Pr(t_i|c) = \alpha \prod_n Pr(w_n|c)^{f_{ni}} \tag{3}$$

for a constant α. We discuss the performance of MNB and compare it with other classification algorithms in Sections 4 and 5.

4 Data Analysis and Results

For building our model we collected over 20,000 sources from publicly available repositories at Github. Using the Github search and a custom built crawler sources were extracted from over 100 repositories in Github [1] in 10 different programming languages – C, C++, C#, Java, Objective-C, Python, Ruby, Perl, HTML and JavaScript. A manual sampling was done on the sources to check the correctness of the crawler and the mapping of the downloaded sources to the corresponding programming languages. Table 1 shows the number of sources of each programming language in our dataset.

Table 1. Numbers of Sources used for Training

Programming Lanuage	Number of Source Codes
C	1916
C++	2185
C#	1560
Objective-C	1527
Java	1693
Python	2028
Ruby	2008
Perl	1176
HTML	1820
JavaScript	3734
Total	19647

Next a list of key tokens for each of the programming language was collected. Lists of tokens are easily available for most programming languages, typically at documentation sites such as Oracle Java documentation [17] for Java. The number of tokens for languages like C, Java or C++, is usually much more than for languages like HTML. Function and class names found in Standard libraries that typically come with many of the languages was kept out of the list of tokens. For example Java has hundreds of default and user-created string functions. Many of these are not likely to be characteristic of the programming language and are often non-standard. There were a total of 1056 keywords for all the 10 programming languages.

After the keyword list has been generated, we then built the training data for Weka [18] to use. Two of the popular formats that Weka uses are – 'ARFF' (Attribute-Relation File Format) and 'CSV' (Comma-separated values). Our Python script takes the sources from their respective directories and produces a CSV file that is the input for Weka. The CSV file generated is a large matrix — one row for each source and one column for each of the 1056 keywords across all the programming languages along with one last column for the source programming language. The procedure to generate CSV file from the source codes is shown in Algorithm 1.

```
1  foreach keyword file do
2  |   foreach keyword do
3  |   |   add keyword to keyword_list
4  |   end
5  end
6  write keyword_list to CSV file, append 'language' as last column
7  foreach source file do
8  |   tokenize content using NLP Word Tokenizer
9  |   foreach token do
10 |   |   if token in keyword_list then
11 |   |   |   increment token frequency by 1
12 |   |   |   append token to list
13 |   |   end
14 |   |   else
15 |   |   |   set token frequency as 0
16 |   |   |   append token to list
17 |   |   end
18 |   end
19 |   append language of source file to the list
20 |   write list to CSV file
21 end
```

Algorithm 1. Procedure for Generating the CSV file

We show the results of our experiments, with the three classifiers — *NaiveBayes*, *BayesNet* and *NaiveBayesMultinomial* below.

Table 2 shows the confusion matrix for the Näive Bayes Classifier. The confusion matrix shows how many sources were correctly classified for each language and how many sources were misclassified. Out of 2392 source files which we used as test data, 1973 files were correctly classified giving an accuracy of 82.48%.

In Table 3, shows the confusion matrix for the Bayesian Network Classifier. The Bayesian Network Classifier is significantly more accurate over the Näive Bayes Model. Out of 2392 files used for test data, 2143 source files have been correctly identified and 249 files have been incorrectly identified giving an accuracy of 89.59%.

Table 2. Confusion Matrix for Näive Bayes Classifier

c	java	python	cpp	ruby	html	javascript	objective_c	cs	perl	classified as
168	0	0	11	0	0	1	2	9	2	c
0	290	0	5	0	0	0	1	1	0	java
0	0	231	3	52	0	1	6	6	1	python
0	0	1	175	5	0	11	0	34	4	cpp
0	0	3	1	374	0	9	0	0	1	ruby
0	0	0	0	0	127	0	0	167	0	html
0	2	0	1	3	0	151	0	44	4	javascript
0	0	1	1	0	0	1	217	1	0	objective_c
0	0	0	10	0	0	1	0	140	0	cs
0	0	0	2	0	0	7	0	4	100	perl

Table 3. Confusion Matrix for Bayesian Network

c	java	python	cpp	ruby	html	javascript	objective_c	cs	perl	classified as
145	0	0	20	0	0	2	5	11	10	c
1	284	0	2	0	0	1	1	8	0	java
0	0	254	0	26	0	3	1	3	13	python
17	0	1	174	1	0	0	0	31	6	cpp
0	0	4	1	382	0	1	0	0	0	ruby
0	0	0	0	0	294	0	0	0	0	html
0	0	0	1	1	0	138	0	16	49	javascript
1	0	0	0	0	0	2	212	3	3	objective_c
1	0	0	2	0	0	0	0	148	0	cs
0	0	0	1	0	0	0	0	0	112	perl

Table 4. Confusion Matrix for Multinomial Naive Bayes

c	java	python	cpp	ruby	html	javascript	objective_c	cs	perl	classified as
158	8	3	16	0	0	2	4	1	1	c
0	297	0	0	0	0	0	0	0	0	java
0	2	293	2	3	0	0	0	0	0	python
11	7	2	202	0	0	0	0	7	1	cpp
1	1	1	1	378	3	0	0	0	3	ruby
0	0	0	0	0	294	0	0	0	0	html
0	1	5	2	5	3	175	12	0	2	javascript
1	6	4	0	0	0	0	210	0	0	objective_c
0	5	0	19	0	0	7	0	120	0	cs
0	0	0	0	2	2	0	0	0	109	perl

Finally, in Table 4, we show the confusion matrix after using Multinomial Näive Bayes Classifier. The results show that Multinomial Näive Bayes Classifier has performed better than Bayesian Network Classifier model. This in turn establishes the superiority of Multinomial Näive Bayes model over Näive Bayes model. Out of 2392 files which were used as test data, MNB Classifier is able to correctly classify 2236 source files, where only 156 files have been incorrectly classified. This gives us the accuracy of 93.48%, which is the best result among all the Bayesian classifier models used in our experiment.

In Figures 1, 2 we show the results we have obtained from our experiments. Figure 1 shows the percentage accuracy of each classification model based on the test data we have provided. Figure 2 shows the performance of each learning model for every programming language in terms of accuracy.

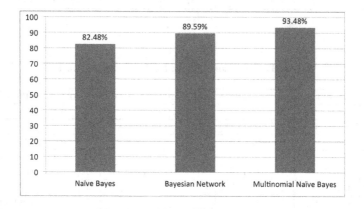

Fig. 1. Accuracy of Each Classification Model

In Figure 1, we show the percentage accuracy of each classification model. We use a separate test data in form of a CSV file which has been generated using the procedure mentioned in Algorithm 1. We have used the CSV file as a test dataset and measured the accuracy of each model in Weka. From the results, we see that Naive Bayes has the least accuracy among 3 models with 82.48% accurate classification. Bayesian Network model performs better than Naive Bayes with 89.59% accuracy but falls short of Multinomial Naive Bayes model which has 93.48% accuracy. This shows that MNB Classifier has performed best among the rest of the classifiers.

In Figure 2, we show how each classifier performs for each of the programming language. Based on the Confusion Matrix we have obtained in Table 2, 3 and 4 we measure the accuracy of each classifier model for every programming language. After we have calculated the correctness percentage, we have plotted the data in 2. From the figure, we see that Multinomial Naive Bayes classifier outperforms the other two classification models in most of the cases. But in some of the cases like for C or Objective-C, Naive Bayes Classifier performs better and for Ruby, Perl and C# Bayesian Network performs better than the rest.

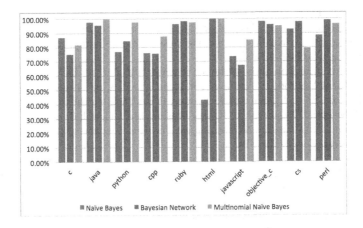

Fig. 2. Accuracy of Each Model for every Programming Language

5   Conclusion

In this paper we presented three Bayesian Learning models for programming language detection from source code. We have discussed the existing tools that are currently available and their drawbacks. The existing tools are mainly used for highlighting syntax in a source where the language is determined by the extension of the source file. But we have taken a machine learning approach to train our model with a dataset of over 20,000 sources across ten programming languages. We report our results after using three classification techniques — Naïve Bayes Classifier, Bayesian Network and Multinomial Naïve Bayes Classifier. The results are extremely encouraging — we have been able to achieve very high levels of accuracy in programming language detection. This also opens up possibilities of using appropriate techniques for domain specific languages.

We foresee widespread appplication of our findings — for detecting the huge number of sources which are found in blogs, forum posts and online Q&A websites and in indexing programming language specific sources better. Also such detection models are essential for increasing the performance of online source code search engines and increasing the searchability of sources by popular search engines.

References

1. Github, http://www.github.com
2. Atlassian Bitbucket, http://bitbucket.org
3. Stack Exchange, http://stackexchange.com
4. SearchCode, http://searchcode.com
5. Codase, http://codase.com
6. Google Code Prettify, http://code.google.com/p/google-code-prettify/
7. SyntaxHighlighter, http://alexgorbatchev.com/SyntaxHighlighter/

8. Highlight.js, `http://highlightjs.org`
9. SourceClassifier, `https://github.com/chrislo/sourceclassifier`
10. Klein, D., Murray, K., Weber, S.: Algorithmic programming language identification. CoRR abs/1106.4064 (2011)
11. Rish, I.: An empirical study of the naive bayes classifier. In: IJCAI 2001 Workshop on Empirical Methods in Artificial Intelligence, vol. 3, pp. 41–46 (2001)
12. Rennie, J.D., Shih, L., Teevan, J., Karger, D.R.: Tackling the poor assumptions of naive bayes text classifiers. In: ICML, Washington DC, vol. 3, pp. 616–623 (2003)
13. Heckerman, D.: A tutorial on learning with bayesian networks. Tech. rep., Learning in Graphical Models (1996)
14. Heckerman, D., Geiger, D., Chickering, D.M.: Learning bayesian networks: The combination of knowledge and statistical data. Tech. Rep. MSR-TR-94-09, Microsoft Research, Redmond, WA
15. McCallum, A., Nigam, K.: A comparison of event models for naive bayes text classification. In: AAAI 1998 Workshop on Learning for Text Categorization, pp. 41–48. AAAI Press (1998)
16. Kibriya, A., Frank, E., Pfahringer, B., Holmes, G.: Multinomial naive bayes for text categorization revisited. In: Webb, G.I., Yu, X. (eds.) AI 2004. LNCS (LNAI), vol. 3339, pp. 488–499. Springer, Heidelberg (2004)
17. Oracle Java Documentation, `http://docs.oracle.com/javase/tutorial/`
18. Weka, `http://www.cs.waikato.ac.nz/ml/weka/`

Creation of Bi-lingual Social Network Dataset
Using Classifiers

Iqra Javed and Hammad Afzal

Department of Computer Software Engineering
National University of Sciences and Technology, Islamabad, Pakistan
iqra.mscs17@students.mcs.edu.pk, hammad.afzal@mcs.edu.pk

Abstract. This paper presents an approach towards creation of topic focused short text (social data) dataset using classification. With emerging use of internet, social networks have turned as the most advanced tool for information sharing among communities. Different communities from different backgrounds use globally renowned social networks often using and promoting their own cultures and languages. Hence, such information exchange turns social networks into multi-lingual information hubs. There are a number of behavioral and demographic oriented analytical studies reported that use data from social networks, but most of the studies are performed using English. In this study, we have focused on development of topic oriented bi-lingual dataset that can be used as corpus to perform further analytical studies. The languages focused are English and Roman-Urdu (which is spoken by about 8 million active users of social network). The main contribution is bi-lingual classifier which is used to create English and Roman-Urdu classified tweets dataset.

Keywords: Bi-Lingual Classification, Twitter Dataset, Language Resources.

1 Introduction

Recently, there is an increasing trend of internet usage throughout the globe. The facilities provided by internet such as e-mail, online surfing and most recent and effective of them, the social networks, have become a vital part of everyone life. Twitter is a micro-blogging social networking web site and is rapidly growing with over 500 million active users till 2012 [1] that generate approximately 50 million posts a day [2]. The statistics show that such social media can provide a powerful tool to analyze public behavior as people tend to share their experiences, feelings about certain events, institutions and products etc. Jensen and Zhang analyzed over 1.5 million twitter posts, and conclude that approximately 19% of the posts discuss different brand names, while 20% include some expression of brand analysis, and 50% of these tweets discussed positive and 33% have critical views of any company or product [2].

The popularity of social media has also extended to developing countries such as Pakistan. During a survey[1] it is analyzed that there are about 30 million internet users

---

[1] http://ansr.io/blog/pakistan-market-trends-2013-online-mobile-social/

P. Perner (Ed.): MLDM 2014, LNAI 8556, pp. 523–533, 2014.

in Pakistan; whereas twitter is ranked as 9th most used website in Pakistan[2]. Social networks, particularly Twitter, hold the potential to serve as opinion reviewing and result predicting tool. By using specific time and geographical boundaries, issues and problems (regarding some event, accident or natural disasters) of different communities can be addressed easily. A lot of research has been carried out in this domain for; however, a vast majority of Twitter users are not well versed in English and often use their local languages to express their opinions. In order to mine the views of such users and evaluate information in their tweets, it is foremost important to categorize these messages into their communication languages and create the (cross lingual) lexical resources that can contribute towards understanding of such tweets. Based on translation dictionaries, sentiment classification of cross-lingual text has drawn major attention of research community [3]. Using such lexical resources, the research carried out for English can be well utilized for other languages as well.

This research aims to address the problem of bi-lingual classification of topic focused tweets dataset. We have created a corpus of approx 82000 tweets, collected from particular geographical and temporal boundaris, focusing on a particular topic, i.e. political association of people of Pakistan. We have performed languagee classification to separate the English Tweets from Roman-Urdu tweets (Urdu language messages written using English alphabets) and performed evaluation of our classification results. Our classifier showed 96% accuracy and 97% F-measure. The percentage of Roman-Urdu tweets is approx 26% which is fairly a large amount of tweets and therefore, corpus thus created can be utilized in other studies that involve analysis of data from the communities that do not use English language as medium of expression.

2   Background

Recently, many classification models have been introduced for topic classification using social networks. [4,5] proposed classification models showing significant interest in text mining tasks such as information retrieval and text analysis. [6] proposed a technique based on Wikipedia for extraction of bi-lingual lexicon. Wikipedia is used in order to establish links between information gathered from multiple domains in different languages. By using these links, translation dictionaries are created. The proposed technique is tested for two language pairs (English-French and English-Roman) on datasets from four different domains (cancer, Corporate Finance, Wind Energy and Mobile Technology). The best results comprise F-measure of 86% in domain of wind energy using French-English language pair.

[7] proposed a generative model based approach, Bilingual Latent Dirichlet Allocation (Bi-LDA) that addresses the problem of knowledge transfer across comparable and multilingual corpus. Gibbs sampling is used for training of the proposed model and SVM is selected as classification model. For proposed model's training and evaluation, two datasets from Wikipedia are used. First dataset is based on three interlanguage linked articles from Spanish, Italian and French. Other classification dataset comprises of topics from domains of books, videogames, sports, programing languages and films whereas language of these documents can be English, French, Italian and Spanish. Best classification accuracy of 86.5% observed for Bi-LDA on topic of programing languages using English-French language pair.

---

[2] http://www.dawn.com/news/642962/pakistan-needs-to-tweet-more

[8] proposed a topic classification technique using grouping of tweets on their content basis. Tweets are classified as hot topics according to the frequent population of tweets on relative topics and geo-location information associated with tweet text. However, due to semantic fluctuations, the proposed classification technique does not work particularly good enough as tweets can use multiple words to refer same event.

From above it is observed that the approaches for language classification rely on large training data written in formal language (that is without typographical mistakes). However, web based text messages are not formally written and comprises of lots of cross lingual words based on culture and ethical background of the message senders.

3   Methodology

3.1   Collection of Tweets Dataset

Twitter search API[3] is used which allow tweets retrieval using multiple parameters that include certain keywords, hashtags, locations, languages and time boundary. Generally hashtags are the popular words used in tweets that refer the trendy topics followed by "#". These hashtags are used with tweets to specify specific topic and hence often help in easy categorization of popular tweet topics. Similarly keyword based search retrieves tweets containing keywords and also those tweets having relative hashtags (keyword followed by "#"). The subject of "political segmentation/association of public in Pakistan" is selected to as topic of our dataset. Tweets related to four major political parties; Pakistan Tehreek-e-Insaaf (PTI), Pakistan Muslim League Noon (PML-N), Pakistan Peoples Party (PPP) and Muttahida Qaumi Movement (MQM) from within the radius of 20 miles of five major cities of Pakistan (Islamabad, Lahore, Karachi, Peshawar and Quetta) are collected. Overall time span for dataset collection is from Dec 2012 to March 2013.

The collected dataset comes with lots of writing style variations including case variation, digits as well as special character insertions. These tweets are transformed to their lowercase representation and are then saved in a text file along with their basic search parameters. The dataset comprises huge amount of noisy data, especially due to non-political tweets attached to political hashtags. 89,000 tweets are collected (along with noisy data). The details of statistics regarding the number of tweets collected from various cities and about different parties are presented in Table 1.

Table 1. Tweets Dataset Statistics

	PTI	PML	PPP	MQM	
Islamabad	7768	1690	3323	3441	16222
Lahore	8991	7111	4969	6783	10612
Karachi	9477	3991	10746	10076	34290
Peshawar	6161	1121	1801	1529	27854
Queta	37	13	33	10	93

---

[3] https://api.twitter.com/1.1/search/tweets.json

Party	City	Language	Twitter Text
PML	Lahore	Roman Urdu	:#pml-n k workers ny aik din pehley hi jalsey ki tyariyan mukamal kr lien ~aajtak news
PTI	Islamabad	Roman Urdu	:#isf k shahinoo k liya khuskhabri, #isf 14 march ko #lahore, 16 march ko #karachi aur 18 march ko #islamabad main seminar kar rahe hain #pti
PTI	Lahore	English	:rt @usaman_musharaf should be brought back to pakistan: sc l http:Vvt.coVfj6auoxtiz
MQM	Karachi	English	:rt @peshavar: #mqm next jalsa will be held in #karachi on mar18! @ceraa_h
PPP	Islamabad	Roman Urdu	:islamabad: \nsabiq governor state bank dr. ishrat hussain ko nigran wazir e azam banai janne ka imkaan zarai.\n#ppp

Fig. 1. Snapshot of tweets datasets from different cities, belonging to different parites and expressed in English and Roman-Urdu

Tweet contents depend upon its writer's background, particularly his/her location and surrounding conditions. Similarly, the language of tweet message depends upon user's background (in particular, education) and hence as a result the collected dataset came up with contents comprises of two major languages; English and Urdu written with English script (termed as Roman-Urdu). A snapshot of tweets belonging to different parties and written in different languages is presented in Figure 1.

3.2   Classification of Tweets

Tweet dataset thus collected comprises of tweets in English and Romain Urdu and contains noisy data as well. On manual observation, it was found that noise is mainly due to the real-state advertisement tweets attached to hashtags of political parties. We performed two-fold classification over dataset (see Fig.2. classification system workflow). First, classification of tweet either belonging to political or non-political class is carried out. This classification results in removal of noisy data (non-political tweets in our case) from dataset. Second, classification of tweet's language is performed in order to detect the language of tweet either as English or Roman-Urdu. After first classification, number of topic focused (political tweets) extracted are 82,224 (non-political tweet messages removed as noisy data were 6,847) (see details in sec 4.1). In language classification 60,462 tweets are classified as English and 21,762 as Roman-Urdu tweet messages (see details in sec 4.2). Both classifications are initially carried out on keyword basis. In case of language classification, lexicons are also involved.

3.2.1   Classification of Tweets Based on Topic of Interest
The tweet dataset retrieved contains a lot of non-political data such as advertisement messages attached with the hashtags of different political parties and cities. Such tweets have no contribution towards semantic analysis of political data and thus, turn into noise for this research work. In order to remove this noise, based upon the manual observation; some prominent non-political keywords have been selected to differentiate between political and non-political tweets. These keywords are Bahria Town,

DHA, Villas, Estate, Kanal, Marla, sale, plot, buy and purchase. By using these key-words in political tweets extraction process, tweets have been classified as belonging to either political class or non-political class.

3.2.2   Classification of Tweets Based on Language

Dataset comprising political tweets comprises of words from two different languages English and Roman-Urdu both written in English script (a very small amount of tweets written in Urdu script is also observed but is not topic of interest for this re-search). Language extraction process is introduced in order to classify tweet either belonging to English class or Roman-Urdu class.

It is observed from the collected dataset that there are no fixed patterns for writing Roman-Urdu. These words are written in a way they sound and seem to their writers.

We used common English words such as adjectives, adverbs, conjunctions, prepo-sitions, pronouns, verbs as English language keywords for language classification (see details of keywords in Appendix-1[4]). Keyword matching technique is applied for classification that resulted in approx. 81 percent tweets classified correctly for lan-guage. However, 81% correctly classified data is not suitable to be used as corpora to perform further analysis tasks (such as opinion and sentiment mining) and therefore, we further improved the classification result as described below.

In order to improve the classification, we used some additional parameters using natural language processing (NLP) techniques. This classification process is divided into following three steps.

In first step, tweets are tokenized. Next, in addition to traditional stop word remov-al; tokens containing special characters, digits and hyperlinks are also removed. In addition to these tokens, most occurring proper nouns and common occurring words (in English and Roman-Urdu) are removed as such tokens are often used in Roman-Urdu script as well and thus do not contribute in language classification of tweets and if not properly filtered then cause ambiguity in tweets language classification. In fol-lowing step, each token from tweet is searched in manually created existing lexicon (consist of 2,000 most common English words) and if this token does not exists in this lexicon then it is searched in WordNet. On finding the existence of token either in lexicon or WordNet, the token is considered as English token. In third step, for each tweet its number of English tokens counted and then divided by total number of tweet's token in order to get tweet's weight with respect to English content in it. Weight of a tweet can be computed as follows:

$$Weight_{tweet} = \frac{W_{eng}}{W_{Total}} * 100 \tag{1}$$

where,

W_{eng}= number of English tokens (words) in tweet.

W_{total} = total number of tokens (words) in tweet.

A test dataset of 300 random tweets is selected. Various thresholds are tested on bi-lingual classifier empirically using test dataset in order to get optimum threshold weight (that was computed at 35%). So, tweets with weight above 35% are classified as English whereas those tweets having weight below 35% are classified as belonging to Roman-Urdu class (see details in sec. *Language Classifier Performance*).

---

[4] http://www.englishclub.com/vocabulary/common-words.htm

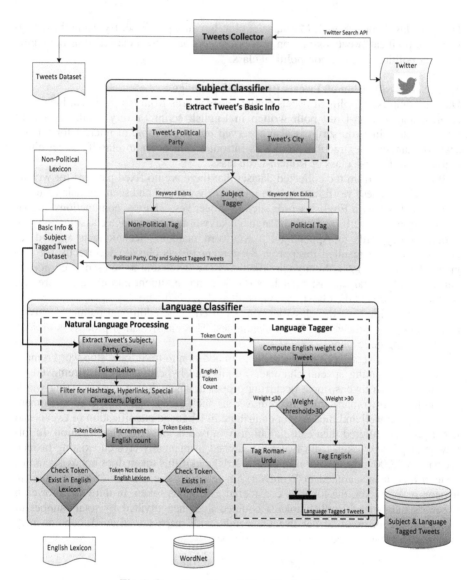

Fig. 2. Overall working of Classification System

4   Evaluation

4.1   Subject Classifier Performance

We select a test dataset comprising 300 tweets (manually tagging them as Political and Non-Political) and used them to evaluate performance of subject classifier. It is observed that our simple keyword based classifier left very small amount of noise of 2% in dataset, i.e. 98% of tweets belonged to our topic (Pakistani politics).

The classifier is then applied on whole dataset, and noise due to non-political tweets has been removed that comprises about 6,847 tweets messages while tweets classified as political were about 82,224 in total. The distribution of political tweets among four political parties is represented in Table 2.

Table 2. Distribution of Political Tweets among political parties

City	Number of Political Tweets			
	PTI	PML	PPP	MQM
Islamabad	7296	836	2937	2915
Lahore	8025	5952	4674	6695
Karachi	9316	3929	10041	9387
Peshawar	6090	1121	1579	1338
Queta	37	13	33	10

4.2   Language Classifier Performance

A test dataset of random 300 tweets is selected. Bi-lingual classifier is tested using multiple weight thresholds in order to attained best performance for language classification of tweet based dataset. [9] Suggests some evaluation measures that are used for analyzing the performance of our bi-lingual classifier. The performance of language classifier at different thresholds is measured using evaluation measures of specificity, sensitivity (recall), precision, accuracy, error rate and F-measure. A complete comparison of these weight thresholds tested on classifier are shown in Table.3.

Table 3. Comparison of bi-lingual classifier based on various Evaluation metrics using multiple thresholds

Threshold	Sensitivity/ Recall (%)	Specificity (%)	Precision (%)	F-Score (%)	Accuracy (%)	Error Rate (%)
30	98	81	89	93	92	8
35	99	91	95	97	96	4
40	81	92	94	87	86	14
50	67	97	97	79	79	21
60	33	99	98	50	59	41

Table 3 shows basic performance of evaluation parameters as maximum sensitivity (99%) attained when classifier weight threshold was set at 35% while it is observed that sensitivity is reduced on increasing threshold. Maximum specificity (99%) and precision (98%) were observed when threshold was set at 60 while specificity reduced as threshold is decreased. An accuracy of 96% is observed at threshold 35. Similarly, F-score's maximum value (97%) was also observed at threshold fixed at 35. Error rate also reduced to percentage of 4% at threshold.

From Table 3, it is observed that by change in classifier's threshold, some evaluation parameter measures decrease whereas other parameter measures increases at the same time. By fixing classifier's threshold to 35, an optimum combination of these parameters is attained with minimum error rate of 4% and maximum value of recall and F-measure.

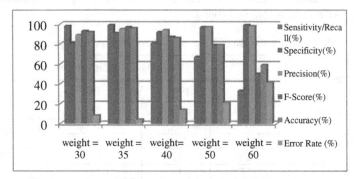

Fig. 3. Bi-lingual classifier Performance on multiple weights

5   Results

On the basis of bi-lingual classification, it is observed that tweets in English contribute a major share of 60,426 tweets in dataset where 21,762 tweets are classified as Roman-Urdu political tweets (see details in Table.4). However, this bi-lingual classification exists throughout the entire tweet dataset irrespective of political party and city as shown in Table.4. Therefore, political tweet dataset is divided into two datasets based on the tweets language. These datasets can be individually processed for

Table 4. Most optimized Language Classification results using bi-lingual classifier

City	Number of English Tweets			
	PTI	PML	PPP	MQM
Islamabad	5586	544	2026	1969
Lahore	6319	4509	3513	5122
Karachi	7149	2236	7190	7067
Peshawar	4415	459	1219	1091
Queta	19	5	21	3

City	Number of Roman-Urdu Tweets			
	PTI	PML	PPP	MQM
Islamabad	1710	292	911	946
Lahore	1706	1443	1161	1573
Karachi	2167	1693	2851	2320
Peshawar	1675	662	360	247
Queta	18	8	12	7

sentiment/ opinion mining or other behavioral analysis studies (e.g. to find dominance of political parties in different cities).

After processing through two iterations of classification (subject classification and language classification), tweets dataset is classified under four main classes as shown in Table.5. Almost 7.6% noise is removed during first iteration of tweet's subject classification. During second iteration of language classification, 26.4% political tweets categorized as roman-Urdu while a contribution of 73.6% English tweets is observed in political dataset (see details in Table.5).

Table 5. Classification of tweet dataset

Tweets Collection span	Tweets Dataset	Political Tweets	Non-Political Tweets	English Political Tweets	Roman-Urdu Political Tweets
5th/Dec/2012- 1th/May2013	89071	82224	6847	60462	21762

6   Conclusion and Future Work

In this paper we proposed a keyword based approach for bi-lingual classification of topic focused short text messages from twitter. We proposed a lexicon based binary classifier that predicts the language of tweet as English and non-Englis (we used Roman-Urdu). Our classification achieved 96% accuracy in predicting the language. The dataset thus prepared is comprises of 21,762 tweets in Roman-Urdu script. This huge amount of tweets can well be used in opinion-analysis and other behavioral analysis studies which, otherwise, used to be ignored as focus had been only English tweets, and therefore, the local communities used to be ignored. One such study is reported in [10] where we used this dataset to mine the public opinion before elections about different political parties.

While there is still some space for improvement in bi-lingual classifier, for instance, there is little amount of tweets in dataset, which shares both English as well as Roman-Udu contents in them. In our future work, we aim to improve the classifier so that it can classify the portion of contributed language in every single tweet. So, it will be easy to perform sentiment analysis on precisely classified tweets dataset.

References

1. Measuring tweets, http://blog.twitter.com/2010/02/measuring-tweets.html
2. Jensen, B.J., Zhang, M., Sobel, K., Chowdury, A.: Twitter power: Tweets as electronic word of mouth. Journal of the American Society for Information Science and Technology 60(11), 2169–2188 (2009)
3. Mihalcea, R., Banea, C., Wiebe, J.: Learning multilingual subjective language via cross-lingual projections. In: ACL 2007 (2007)
4. Hofmann, T.: Probabilistic latent semantic analysis. In: Proceedings of Uncertainty in Artificial Intelligence, UAI, Stockholm (1999)

5. Blei, D.M., Ng, A.Y., Jordan, M.I.: Latent dirichlet allocation. Journal of Machine Learning Research 3, 993–1022 (2003)
6. Bouamor, D., Popescu, A., Semmar, N., Zweigenbaum, P.: Building Specialized Bilingual Lexicons Using Large-Scale Background Knowledge. In: Proceedings of the 2013 Conference on Empirical Methods in Natural Language Processing, Seattle, Washington, USA, October 18-21, pp. 479–489 (2013)
7. Smet, W.D., Tang, J., Moens, M.-F.: Knowledge Transfer across Multilingual Corpora viaLatent Topics. In: Proceedinds of the 15th Pacific Asia Conference on Advances in Knowledge Discovery and Data Mining, Part1, Shenzhen, China, pp. 549–560 (May 2011)
8. Ishikawa, S., Arakawa, Y., Tagashira, S., Fukuda, A.: Hot Topic Detection in Local Areas Using Twitter and Wikipedia
9. Powers, D.M.W.: Evaluation: From Precision, Recall and F-Measure Toroc, Informedness, Markedness& Correlation. Journal of Machine Learning Technologies 2(1), 37–63 (2011)
10. Javed, I., Afzal, H.: Opinion analysis of bi-lingual event data from social networks. In: Proceedings of International Workshop on Emotion and Sentiment in Social and Expressive Media (ESSEM), Italy. Ceur Workshop Proceedings, vol. 1096, pp. 164–172

Appendix

English Keywords Used For Language Classification.

ADJECTIVES	ADVERBS	CONJUNCTIONS
Good	Up	And
New	So	That
First	Out	But
Last	Just	Or
Long	Now	As
Great	How	If
Little	Then	When
Own	More	Than
Other	Also	Because
Old	Here	While
Right	Well	Where
big	Only	After
high	Very	So
different	Even	Though
small	Back	Since
large	There	Until
next	Down	Whether
early	Still	Before
young	In	Although
Important	As	Nor
Few	Too	Like
Public	When	Once
Bad	Never	Unless
Same	Really	Now

PREPOSITIONS	PRONOUNS	VERBS
Of	It	Be
In	I	Have
To	You	Do
For	He	Say
With	They	Get
On	We	Make
At	She	Go
From	Who	Know
By	Them	Take
About	Me	See
As	Him	Come
Into	One	Think
Like	Her	Look
Through	Us	Want
After	Something	Give
Over	Nothing	Use
Between	Anything	Find
Out	Himself	Tell
Against	Everything	Ask
During	Someone	Work
Without	themselves	Seem
Before	Everyone	Feel
Under	Itself	Try
Around	Anyone	Leave

Author Index